Ascent to the Einstein Equations

Michael Ruhrländer

Ascent to the Einstein Equations

Spacetime, Gravitational Waves, Black Holes and more

 Springer

Michael Ruhrländer
Wiesbaden, Germany

ISBN 978-3-662-72671-6   ISBN 978-3-662-72672-3   (eBook)
https://doi.org/10.1007/978-3-662-72672-3

This book is a translation of the original German edition "Aufstieg zu den Einsteingleichungen," 2nd edition, by Michael Ruhrländer, published by Springer-Verlag GmbH, DE in 2021. The translation was done with the help of an artificial intelligence machine translation tool. A subsequent human revision was done primarily in terms of content, so that the book will read stylistically differently from a conventional translation. Springer Nature works continuously to further the development of tools for the production of books and on the related technologies to support the authors.

Translation from the German language edition: "Aufstieg zu den Einsteingleichungen" by Michael Ruhrländer, © Der/die Herausgeber bzw. der/die Autor(en), exklusiv lizenziert durch Springer-Verlag GmbH, DE, ein Teil von Springer Nature 2021. Published by Springer Berlin Heidelberg. All Rights Reserved.

This Springer imprint is published by the registered company Springer-Verlag GmbH, DE, part of Springer Nature.
The registered company address is: Heidelberger Platz 3, 14197 Berlin, Germany

If disposing of this product, please recycle the paper.

Preface

Content Updates

The present edition is an improved, slightly expanded and translated version of the second German edition. Compared to the first German edition the content has been expanded by three chapters. In Chap. 10, as a conclusion of the part about tensor calculation in Euclidean space, the general inertia tensor in three dimensions is discussed.

In Chaps. 25 and 26 further applications of the General Theory of Relativity to objects outside our solar system are presented. In Chap. 25 we deal with the topic of "gravitational waves", which has gained a lot of attention in recent years due to the first successful measurement of gravitational waves on Earth in 2015 (announced in February 2016, honored with the Physics Nobel Prize 2017). In Chap. 26 we venture into the interior of a star, examine the balance between pressure and gravity, and derive the so-called "interior" Schwarzschild solution. We have summarized these two chapters together with Chap. 27 about static spherically symmetric black holes in a new part V.

Acknowledgments

I would like to express my sincere thanks to all readers of the first edition who contacted me and pointed out improvements and inconsistencies. Their contributions and suggestions have made the present second edition more detailed, more understandable and more error-free. My thanks also go to Springer-Verlag, especially to Caroline Strunz and Bettina Saglio for the excellent collaboration.

Although I have made great efforts to make the book error-free, there will probably be some (hopefully few) erroneous places in the text. The possible error locations and the corresponding corrections can be viewed in an errata list on my homepage `http://www.michael-ruhrlaender.de`. There I will also publish any additional materials for the book.

About this Book

Albert Einstein first published the **General Theory of Relativity (GR)**, his theory of gravity, in 1915. In his time and even today, the GR is considered difficult to understand, especially because of the mathematically demanding representations. Although it is no longer the case, as it is anecdotally attributed to the British astronomer Sir Arthur Stanley Eddington, who in the 1920s, when asked by a journalist whether it was true that there were only three people in the world who understood the General Theory of Relativity, is said to have replied with the counter-question: "Who is the third?" Today, the GR is a standard lecture in physics studies, but it is usually only offered in the graduate or master's degree program. This means that those who first deal with the GR, are usually advanced physics students who have already studied for five to six semesters and have learned the basic areas of physics (mechanics, electrodynamics, special theory of relativity, quantum mechanics and statistical physics) and have built up the necessary mathematical knowledge. And this is also the reason why most textbooks on the General Theory of Relativity are only understandable to readers who have already acquired this prior knowledge in physics and mathematics.

Here a completely different approach is chosen. The minimum physical and mathematical prerequisites are only what is taught in the upper grades of high schools or in technical colleges. Everything else that is necessary for understanding the General Theory of Relativity is introduced carefully and in detail. With enough willingness to learn and perseverance, even those interested in natural sciences can understand the contents of the book, who have not taken an advanced course in physics or mathematics or whose school days are further in the past. Also physics students, who are dealing with the GR for the first time, will find here an easy and detailed introduction to the topic.

The book offers the simplest possible introduction to the *quantitative* GR, i.e., it describes the theory in its mathematical formulation, similar to the form in which Einstein published it. Therefore, the reader has to deal with a whole series of physical phenomena and mathematical techniques, to get to the point where he understands the contents of Einstein's theory of gravity also in terms of formulas.

Origin

The author is not a recognized expert in GR, i.e., not one who - like most authors of books on relativity theory - can boast that he has conveyed the contents of the theory in a series of years in lectures at universities. Although I am a trained mathematician, after my studies and doctorate I decided for a professional path outside of universities and then worked for almost 30 years "in the industry", in areas of work that had nothing to do with physics or higher mathematics. What has accompanied me throughout my professional life, is the interest in scientific, especially physical questions.

And so this work was created because I wanted to deal intensively with Einstein's General Theory of Relativity once again in my advanced age. I wanted to understand this theory in a depth that goes beyond a popular scientific framework. So I started looking for suitable literature for me, but quickly realized that there were two completely different approaches to conveying the contents of the theory. On the one hand, there is a multitude of books and articles that attempt to explain the basic ideas and concepts as well as possible consequences of the theory in everyday language. These explanatory approaches usually do without formulas, so they remain qualitatively descriptive, but are sometimes very good at conveying the physical backgrounds and possible areas of application. The other category consists of those textbooks and technical articles that an advanced physics student takes and works through in parallel with the lectures. These textbooks are often formulated in a "modern", very abstract formula language and were mostly incomprehensible to me at first. In other words: there was no suitable book for me that I could use to penetrate the General Theory of Relativity quantitatively in self-study. I had to rely on selecting and working through the appropriate passages from a variety of different readings, which consisted of popular science presentations, mostly English-language physics textbooks, and lecture scripts and teaching materials available on the Internet. In addition, I attended lectures on the theory of relativity as a guest listener and thus further expanded my acquired knowledge.

This book is intended to significantly shorten the path I took to acquire the basic content of the General Theory of Relativity and to close the existing gap between the popular science and the "high science" presentations.

Since I taught mathematics for prospective engineers at a university of applied sciences for several years, I know quite well what physical and mathematical prior knowledge someone brings who starts such a study. And therefore it was my endeavor to choose the level so that people with similar knowledge can dock where the book begins. Nevertheless, it should be pointed out that reading/working through will be very strenuous for most readers, even if the initial requirements are rather low. Because the book quickly picks up speed and quickly penetrates areas that are usually no longer covered in school.

Subject

This book deals with gravity, also called **gravitational force**. Gravity is one of the four so-called **physical fundamental forces** (in addition to gravity, these are the **electromagnetic force** and the **weak** and **strong nuclear force**) and is characterized by the fact that it is omnipresent, i.e., you cannot switch off gravity. It ensures that galaxy clusters form, that the stars in our Milky Way do not fly apart, that the planets orbit the Sun and the Moon orbits the Earth, that the apple falls from the tree, etc.

What does gravity have to do with the theory of relativity? You may already know that there are two fundamentally different theories of relativity, both developed by Einstein. The so-called **Special Theory of Relativity (SR)**, which Einstein published in 1905, does away with our intuitive understanding of space and time that follows common sense. It is - as we will show in part III - understandable in its mathematical formulation in its basic concepts at intermediate level. Nevertheless, it requires a complete rethinking of time and space that is contrary to normal intuition and is probably for this reason still incomprehensible to many people today.

The General Theory of Relativity is based on two foundations. On the one hand, there is the so-called **Newtonian Theory of Gravitation**, which was developed by Isaac Newton in the 17th century and in the following centuries formed the basis for all celestial mechanical calculations (e.g. orbits of the planets, discovery of new planets) and for Earth-related phenomena (e.g., solar and lunar eclipses, the formation of tides, the change of seasons). On the other hand, GR is based on the Special Theory of Relativity, but is formulated in a mathematical language that has so far prevented it from being made accessible to a larger audience.

The book follows the approach of a textbook in its diction, as it aims to make quantitative statements about the General Theory of Relativity. Therefore, many mathematical expressions can be found in the individual chapters, which may initially deter some readers. But, and this should be emphasized, in this book the easiest and often longer path is always chosen, which leads slowly and leisurely into the nevertheless not to be underestimated heights. Steeper passages and other possible shortcuts, which usually require a more advanced climbing technique (i.e., higher mathematics), are always avoided where there is an easier detour.

Basically, in this book we let ourselves be guided by the physical phenomena, so we always try to understand the physical content first, and then in the next step work out the corresponding mathematical representations. In deriving the formulas, we strive to ensure that *every* step is understandable, i.e., it is shown in detail, how conclusions and transformations result. Finally, we interpret the information contained in the derived formulas again physically, so that a further deepened understanding of the relationship between physical content and mathematical representations can be built up.

Audience

For which readership is this book written? Well, these are people who have a basic interest in natural sciences, especially physical questions, and who can "bite into a matter", who therefore have a great willingness to perform and a considerable perseverance. So, for example, pupils, who intend to take up a scientific or technical study, or students with other disciplines than physics. But especially also physics students, who are dealing with the General Theory of Relativity for the first time and are looking for an easy entry. Or, like myself, people who in their youth perhaps studied engineering, chemistry, mathematics or similar and who consider dealing with physics as their hobby.

Prerequisites

Let's now come to the question of what one must bring along in order to follow our approach, which of course cannot start "from scratch". You don't need to know anything about the theory of relativity, not even the special one. Everything necessary for understanding the General Theory of Relativity from the Special Theory is extensively presented in this book. Since the General Theory of Relativity is an extension of Newton's theory about gravity, it is very helpful if basic knowledge from classical Newtonian mechanics (e.g., energy, law of gravitation, planetary orbits) is present or can be quickly reactivated. If you have no knowledge of classical mechanics, you do not necessarily have to refer to other books. Because in the initial chapters of this book, the basic terms and concepts from the Newtonian theory, as far as they are necessary for further understanding, are presented. In other words: Basic physical knowledge is helpful, but not absolutely necessary.

The situation is somewhat different with prior knowledge in mathematics. A level of knowledge equivalent to the middle upper level of a high school or technical college is necessary. It is assumed that you have knowledge of fractions, powers, roots, logarithms, solving equations as well as simple geometry and trigonometric functions (sine, cosine, tangent, etc.). In addition, it would be very helpful if you know or can quickly learn how to calculate limits of functions, differentiate and integrate functions in one variable, and what laws of differential and integral calculus exist. From geometry or linear algebra knowledge should be present, such as how to calculate with vectors in space, what scalar and vector products, matrices and determinants are and how to calculate with them. But the same applies to the mathematical prerequisites as stated above: All mathematical basics that go beyond what has just been described as necessary are presented in this book, albeit briefly, so that you basically do not need to have another mathematics book at hand. In addition, you will find in the appendix (chap 29) those mathematical formulas and physical laws that are used in this book but are not explained or derived in detail, i.e., in this appendix are the things that you should "actually" already bring with you. Of course, it is up to you to delve deeper into both physics and mathematics. For this purpose, there are dedicated literature recommendations at the end of each part of the book on the topics addressed, all of which can also be found in the **bibliography** at the end of the book.

Title

The title of the book "Ascent to the Einstein Equations" is deliberately chosen. We have a strenuous and long journey ahead of us, divided into stages that must be reached in order to follow the respective next section. So you cannot read the book punctually or section by section (unless you already have good knowledge of our subject). So it's like climbing a mountain: First, the base camp (Newtonian mechanics) must be reached. In the base camp, new climbing techniques (vector and tensor calculation) are learned and practiced, which are then needed for the ascent to the intermediate camp (Special Theory of Relativity). In the intermediate camp, the techniques are perfected (expansion of vector and tensor calculation), and then the summit (Einstein equations) is targeted and climbed. The summit of our venture is thus the Einstein equations, in which the General Theory of Relativity is highly compressed. To take a brief look at it in advance, they are

$$R_{\mu\nu} - \frac{1}{2} g_{\mu\nu} R = 8\pi\, G\, T_{\mu\nu},$$

consisting of only a few expressions, and yet one can write a whole book about it to convey what they ultimately mean. Why actually Einstein equations, where there is only one equation? This is due to the compressed notation: The (lower) indices μ (Greek letter My) and ν (Greek letter Ny) can *each* take the four values $\mu, \nu = t, x, y, z$ (what exactly this means will become clear later). That is, the expression $\mu\nu$ stands for all possible combinations of t, x, y, z, i.e., $\mu\nu = tt$ or $\mu\nu = tx$ or $\mu\nu = ty$ or also $\mu\nu = zz$ etc. Written out in more detail, the equations are thus

$$R_{tt} - \frac{1}{2}g_{tt}\,R \;=\; 8\pi\,G\,T_{tt}$$

$$R_{tx} - \frac{1}{2}g_{tx}\,R \;=\; 8\pi\,G\,T_{tx}$$

$$R_{ty} - \frac{1}{2}g_{ty}\,R \;=\; 8\pi\,G\,T_{ty}$$

$$\vdots \qquad \vdots \qquad \vdots$$

$$R_{zz} - \frac{1}{2}g_{zz}\,R \;=\; 8\pi\,G\,T_{zz}.$$

So there are initially $4 \cdot 4 = 16$ individual equations, but some of them say the same thing, ultimately leaving *ten* independent equations.

When we have reached the summit, there is still some time to look in particular at the quantitative consequences of the General Theory of Relativity, i.e., where it differs so much from its predecessor (the Newtonian Theory of Gravitation) that it can explain observed phenomena for which the Newtonian theory has no answers. This is a very wide field, think for example of the various cosmological models with which one can describe our entire universe, or the depiction of black holes or the detection of gravitational waves. All topics that one could write entire books about. In this book, we primarily focus on the "classical" topics, i.e., on physical phenomena in our solar system. We discuss the earliest historical solution to the Einstein equations, the so-called (**outer**) **Schwarzschild solution**, and use it to describe, for example, the perihelion shift of Mercury, the light deflection in the gravitational field of the Sun, and as a modern application, GPS navigation. Afterwards, we examine some phenomena related to gravitational waves, which have received more attention in recent years, after the results of the first measurement of gravitational waves on Earth were published in 2016. Subsequently, we look at the mechanisms by which stars collapse and how to use the (**interior**) **Schwarzschild solution** to describe the interior of a star. In the final sections we examine some properties of static black holes and thus move on to the most mysterious objects in the universe.

Structure

This book consists of six parts.

- In part I the basic elements of Newtonian mechanics (including laws of falling, momentum, force, work and energy, rotational movements) are introduced. This is followed by a detailed presentation of Newton's law of gravitation, including the derivation of the possible orbits of planets and comets in the solar system. This lays the foundation for a deeper understanding of the most important physical phenomena associated with Newtonian gravity.

- In part II the **vector and tensor calculus** necessary for understanding the theory of relativity is introduced, but for the simplest case of the two-dimensional flat plane. This part is predominantly mathematical.

- In part III three topics are covered. First, the phenomena of Special Relativity are presented with as simple mathematical means as possible, then a further deepening of tensor calculus follows, and finally the newly learned, extended tensor calculus is applied to the already achieved results of SR, leading to a new formulation of the physical laws in SR.

- Part IV begins with a description of physical phenomena under the influence of a gravitational effect, which makes the need for an extension of both Newton's theory of gravitation and the Special Theory of Relativity apparent. Afterwards, tensor calculus is further developed and generalized so that the new laws of General Relativity, i.e., in particular the Einstein equations, can be formulated. Finally, some consequences of Einstein's theory in our solar system are discussed.

- In part V some selected phenomena of cosmological objects are discussed. We investigate how gravitational waves are generated and how they can be measured on Earth. Afterwards, we describe the collapse of a star and calculate its interior equilibrium. Finally, we deal with static black holes and show how cleverly chosen coordinates can make statements about the interior of a black hole.

- Part VI consists of an appendix in which physical laws, mathematical formulas, and tables with physical units and size specifications of cosmological objects are listed for reference.

An important aspect in acquiring the quantitative description of General Relativity is learning the necessary (mathematical) techniques and modes of representation with which the physical laws of GR are formulated. As just explained, tensor calculus plays a decisive role in this. However, there are also (usually preliminary) additional topics from mathematics that are already necessary for the formulation of Newtonian mechanics or the Special Theory of Relativity. We distinguish between two categories of mathematical tools and skills:

1. Tools and skills that the readers should (actually) bring with them are presented relatively briefly and each as numbered **Remark** with the heading „**Mathematical Foundations** (short **MF**:)" introduced and ended by the symbol □.

2. Tools and skills that are new to learn and practice, are integral parts of the book and are also as numbered Remark with the heading „**Mathematical Tools** (short **MT**:)" marked. They are derived in detail and described and practiced extensively. These sections are also ended by the symbol □.

If formulas are used in the text without derivation, it is assumed that these are known to the reader. However, he can look them up again in the appendix (Chap. 29). Also, the physical constants used in this book (e.g., the speed of light c) or real physical quantities (e.g., the mass of the Sun in kg) can be found in an appendix (Chap. 30).

All mathematical tools are always defined where they are used for the first time from a physical point of view. This has the advantage that the mathematics stays close to the physics and it becomes clear for which physical reasons this or that mathematics is needed. The disadvantage of this approach is that the mathematics is not conveyed compactly, not in one piece, but served in small bites.

Pedagogical Notes

In this book, I have refrained from moving detailed calculations to appendices or separate "boxes" as I did not want to disrupt the flow of reading. The detailed calculations are thus integrated into the text and can be followed immediately, if the reader wishes. Of course, there is always the option to skip the details and proceed with the next steps, even though, in the author's opinion, such an approach does not generate a deep understanding of the thought processes and emerging structures. The detailed presentations of each individual calculation should give the reader, if he can follow them, sufficient confidence that he has delved deeper into the material. It is also advisable to try to derive some of the formulas and laws independently and then compare them with those in the book. This book therefore does not contain any exercises with which the reader can check whether he has really understood the parts he has worked through.

When reading physical literature, I have always found it annoying to have to flip back when the text refers to a much earlier result that I no longer had fully at hand. Therefore, in this book, references to long past results are given, but the corresponding formulas are mostly also repeated at the reference points, so that flipping back can be avoided at least in cases where a recognition of the former is present. If a fact is referred to that was derived in the same chapter, then usually no repetition takes place, as I assume that the memory of it is still fresh.

To return to the analogy of mountain climbing: The pace also plays a role in a mountain climb. A famous mountaineer once answered a journalist's question about how he climbs mountains, simply with "slowly". And you should heed that. Expect to read or work through about two pages per day on average, i.e., you will need almost a year to successfully complete the expedition.

Physical Laws

When we talk about laws in the following, as is usual in the natural sciences, a model statement is always meant. A **physical law** is a hypothesis about how observed or not yet observable phenomena can be explained within the framework of a physical theory. Physical laws can therefore *never* claim to be true. The correctness of physical models cannot be proven, the models can only be falsified, which usually happens by contradictory experimental findings. Scientific progress then consists in finding an extension of the existing laws and then again validating them by experiments or correct predictions.

Notes on Notation

Textual Emphasis

In the text, newly introduced terms are highlighted with **bold** letters. If a passage is to be particularly emphasized, we write the relevant words in *italicized* letters. Important physical statements are framed with a box.

Coordinates and Indexing

The terms and definitions used in this section will all be discussed and explained in detail in the course of the book, i.e., this section serves more as a reference than as a definition.

- In two-dimensional Euclidean space, we use Cartesian coordinates x, y or alternatively polar coordinates r, φ to uniquely identify a point in the plane.

- In three dimensions, we use Cartesian coordinates x, y, z, spherical coordinates r, ϑ, φ or cylindrical coordinates r, φ, z.

- In four-dimensional spacetime, time is added as a coordinate and we use Cartesian coordinates t, x, y, z or spherical coordinates t, r, ϑ, φ or also cylindrical coordinates t, r, φ, z

All coordinates are also used for indexing physical quantities. For example, we write the three components of a three-dimensional vector $\vec{v}$ as v^x, v^y, v^z, when we are in the Cartesian coordinate system. We use the abbreviated notation v^i, when we mean a (any) component of the vector. The superscript letters x, y, z and i are also called indices, which can also be subscripted. The convention in physics for indices is to use small Latin letters, e.g. i, j, k, for the indices x, y, z or r, ϑ, φ in two or three dimensions and small Greek letters, e.g. μ, ν, ρ for the indices t, x, y, z or t, r, ϑ, φ in spacetime. An arbitrary Cartesian component $A_{\mu\nu}$ of an indexed four-dimensional object can therefore be any element from the following selection:

$$\{A_{tt}, A_{tx}, A_{ty}, A_{tz}, A_{xt}, A_{xx}, A_{xy}, A_{xz}, A_{yt}, A_{yx}, A_{yy}, A_{yz}, A_{zt}, A_{zx}, A_{zy}, A_{zz}\}$$

We also abbreviate the coordinates themselves with i, j, k in Euclidean space or with μ, ν, ρ. For example, if a point particle moves through space, i.e., its position changes over time, we write briefly $i(t)$, when we want to name an arbitrary coordinate of the particle at time t. The same applies to the derivative of an arbitrary coordinate with respect to time, which we briefly note as di/dt.

Contents

List of Figures

Part I.

The Worldview of Gravitation before Einstein

In this part, we go back and look at what Einstein's predecessors found out about gravitation. We do this not only out of historical interest, but also because, on the one hand, the old approaches already contain a lot of the thought content on which Einstein built his General Theory of Relativity. On the other hand, the formulaic description of the so-called Newtonian mechanics and the early laws of gravitation provide a good opportunity to introduce some of the fundamental mathematical concepts, which we will also need again and again in the later chapters, to introduce and anchor in physical laws.

According to our approach, Part I contains almost all **mathematical foundations**, but also some new **mathematical tools**, which are usually not covered in school education anymore.

As far as gravitation is concerned, we do not start from scratch, but skip the Earth-centric world views of, for example, Aristotle and Ptolemy and start in the early 17th century. In this pre-Newtonian time, there were two separate approaches dealing with the phenomenon of gravity, without the researchers of that time being aware that these were different aspects of a single physical law. On the one hand, the movements of celestial bodies were investigated with astronomical observations, which became possible in a completely new quality due to the invention of the telescope, and on the other hand, the laws of falling on Earth, essentially motivated by ballistic experiments with cannonballs.

1. Kepler's Laws

After the invention of the telescope, astronomers were able to determine the locations of the Sun, the Moon, the planets, and the stars much more accurately. Galileo Galilei (1564-1642) found among other things four "planetis" orbiting Jupiter, and interpreted them as moons of Jupiter, thus as a small planetary system. With the available observational data, Johannes Kepler (1571-1630) was finally able to find laws with which he could predict the position of the planets. He found that the planets do not move on circular orbits, as postulated in the Copernican world view, but on elliptical orbits around the Sun.

Remark 1.1. **MF: Ellipse I**

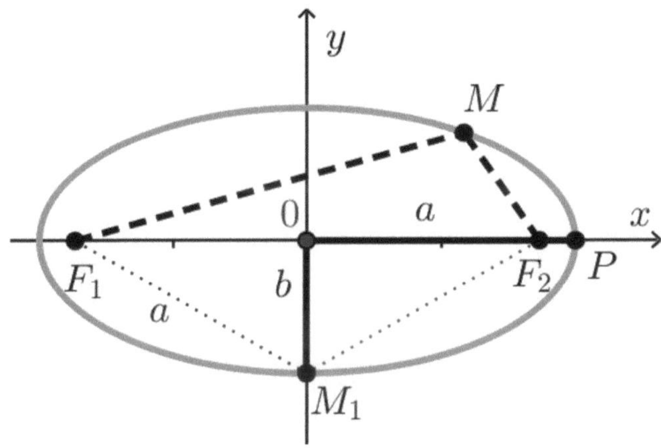

Figure 1.1.: Ellipse with semi-axes a and b

An **ellipse** is the set of (light gray) points in Fig. 1.1, for which the sum of their distances to the two **foci** F_1 and F_2 is constant. That is, for any arbitrary point M on the ellipse, the sum of the dashed lines $\overline{F_1M} + \overline{F_2M}$ is constant. The distance a from 0 to P is called **major semi-axis**, the distance b from 0 to M_1 **minor semi-axis**. We calculate the sum of the distances from point P to the two foci and get

$$\overline{PF_2} + \overline{PF_1} = \left(a - \overline{0F_2}\right) + \left(a + \overline{0F_1}\right) = 2a,$$

since $\overline{0F_2} = \overline{0F_1}$. This means that the sum of the distances of each point on

© The Author(s), under exclusive license to Springer-Verlag GmbH,
DE, part of Springer Nature 2026
M. Ruhrländer, *Ascent to the Einstein Equations*,
https://doi.org/10.1007/978-3-662-72672-3_1

the ellipse to the two foci is always $2a$. The point M_1 has the same distance to the two foci, i.e., it holds

$$\overline{M_1 F_1} = \overline{M_1 F_2} = a,$$

as the dotted lines in Fig. 1.1 show. If a and b are the same size, then the two foci coincide and the ellipse becomes a circle. $\square$

Remark 1.2. **MF: (x, y)-Coordinate System**
In Fig. 1.1 also appears for the first time a **coordinate system**, which is characterized by two perpendicular lines, called axes.

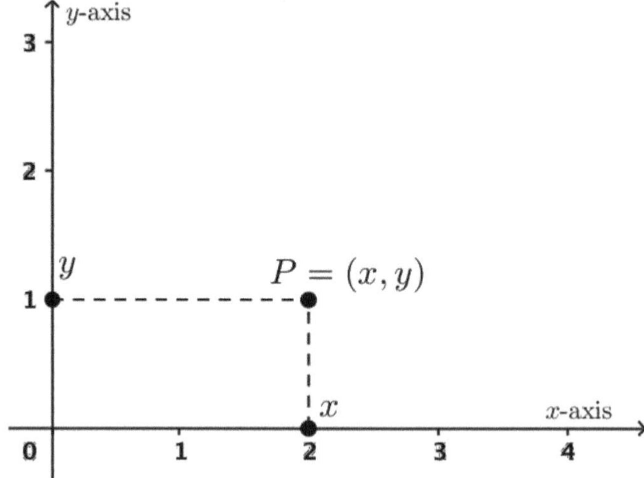

Figure 1.2.: Point in (x, y)-plane

The x-axis lies horizontally and points to the right (as indicated by an arrow), the y-axis lies vertically and points upwards. Coordinate systems play a major role in our further explanations. Only they allow numerical calculations, e.g. on the position of a point. To do this, imagine that the x and y axes are divided into units (e.g. km), so that after defining the origin (where the two axes intersect) any other point $P = (x, y)$ of the (x, y)-plane (in our example the orbital plane of the planet) can be uniquely identified by its coordinates, which are the **projections** (dashed lines in Fig. 1.2) onto the x and y axes. If a point is to the right of the origin, its x-coordinate is positive, to the left accordingly negative. If it is above the origin, its y-coordinate is positive, below accordingly negative. The position of a point is expressed by the coordinate symbol (x, y), i.e., it is customary to also denote any point on the x-axis with

x, the same applies to y. The origin therefore has the coordinates $(0,0)$, but is usually briefly marked as "0" in the figures. $\square$

1.1. First Kepler's Law

> All planets move in elliptical orbits around the Sun. The Sun is in one of the two foci of the respective elliptical orbits.

Kepler's first law is based on the heliocentric worldview of Copernicus, but does not place the Sun exactly in the center, but slightly outside of it. Let's take a closer look at the movements of the planets.

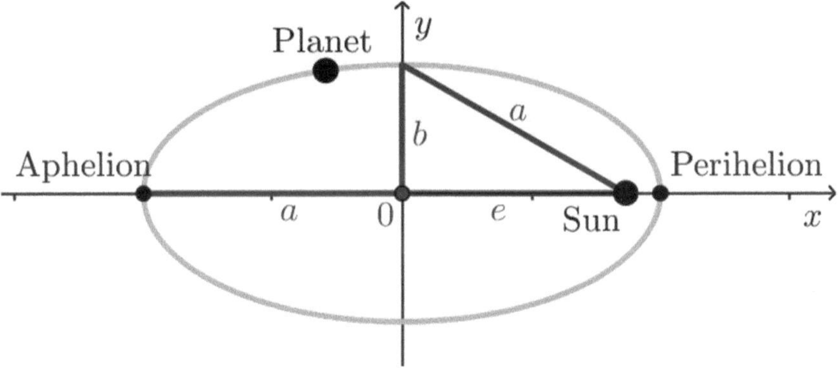

Figure 1.3.: Orbit of a planet

Fig. 1.3 shows the elliptical orbit of a planet. The **perihelion** is the point where the planet comes closest to the Sun. The point where the distance to the Sun is greatest is called **aphelion**. The distance e is the distance of the foci (the Sun is in one focus) to the center of the ellipse.

To calculate the distance of the two foci of the ellipse to the origin, we apply the **Pythagorean theorem** (see Chap. 29) to the right-angled triangle formed by the sides (a, b, e). It holds

$$e^2 + b^2 = a^2,$$

from which

$$e = \sqrt{a^2 - b^2}$$

follows. This clearly defines the position of the Sun, which is always at the point $(e, 0)$. The second focus, not marked in the graphic, has the coordinates

$(-e, 0)$. The **numerical eccentricity**

$$\varepsilon = \frac{e}{a} = \frac{\sqrt{a^2 - b^2}}{a} = \sqrt{\frac{a^2 - b^2}{a^2}} = \sqrt{1 - \frac{b^2}{a^2}} \tag{1.1}$$

indicates how much an ellipse deviates from the circular orbit. For the circular orbit, the relationship $\varepsilon = 0$ follows because $a = b$. The orbit of the Earth has a small eccentricity of 0.017, so it deviates only slightly from a circular orbit, while Mercury's orbit has the highest eccentricity among the planets in the solar system at 0.205. The ellipses in the illustrations have an eccentricity of about 0.85, so they are intended to be exaggerated examples rather than representations of real planetary orbits.

Note: When we divide by a quantity in the calculations in this book, we *always* assume that this quantity is *different from zero*. In the present case, this means that the semi-axes a, b are strictly greater than zero, so we are dealing with "real" ellipses and not with so-called **degenerate cases**.

Remark 1.3. **MF: Ellipse II**
We now want to determine all possible positions (x, y) of a planet moving on an elliptical orbit around the Sun.

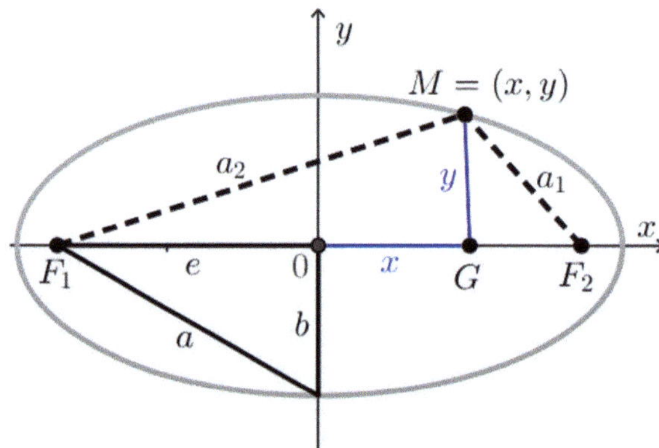

Figure 1.4.: Calculation of the coordinates of the planet

In Fig. 1.4 a point M with coordinates x and y, i.e., $M = (x, y)$ is marked, which is at a distance

$$a_1 + a_2 = 2a$$

from the two focal points F_1 and F_2. If we apply the Pythagorean theorem to the right-angled triangles (F_1, G, M) and (G, F_2, M) and note that

$$\overline{F_1 G} = e + x$$

and

$$\overline{GF_2} = e - x,$$

we get on the one hand

$$y^2 = (a_1)^2 - \left(\overline{GF_2}\right)^2 = (a_1)^2 - (e - x)^2$$

and on the other hand

$$y^2 = (a_2)^2 - \left(\overline{F_1 G}\right)^2 = (a_2)^2 - (e + x)^2 . \tag{1.2}$$

Setting the last two equations equal to each other gives

$$(a_2)^2 - (e + x)^2 = (a_1)^2 - (e - x)^2 .$$

Expanding on both sides results in

$$(a_2)^2 - \left(e^2 + 2ex + x^2\right) = (a_1)^2 - \left(e^2 - 2ex + x^2\right) .$$

Simplifying leads to

$$(a_2)^2 - 2ex = (a_1)^2 + 2ex$$

and further to

$$(a_2)^2 - (a_1)^2 = 4ex.$$

We use the third binomial formula for the left side (see Chap. 29):

$$a^2 - b^2 = (a + b)\,(a - b)$$

and get

$$(a_2 + a_1)\,(a_2 - a_1) = 4ex.$$

It was $a_1 + a_2 = 2a$, so it follows

$$2a \cdot (a_2 - a_1) = 4ex.$$

And from this

$$a_2 - a_1 = \frac{2ex}{a}.$$

Since $a_1 = 2a - a_2$, it follows

$$a_2 - (2a - a_2) = \frac{2ex}{a},$$

so finally

$$a_2 = \frac{ex}{a} + a. \tag{1.3}$$

This inserted into (1.2) results in

$$
\begin{aligned}
y^2 &= (a_2)^2 - (e+x)^2 \\
&\underset{1.3}{=} \left(\frac{ex}{a} + a\right)^2 - (e+x)^2 \\
&= \left(\frac{e^2 x^2}{a^2} + 2\frac{ex}{a}a + a^2\right) - (e^2 + 2ex + x^2) \\
&= \frac{e^2 x^2}{a^2} + 2ex + a^2 - e^2 - 2ex - x^2 \\
&= \frac{e^2 x^2}{a^2} - x^2 + a^2 - e^2.
\end{aligned}
$$

Now we use that $a^2 - e^2 = b^2$, and obtain

$$
\begin{aligned}
y^2 &= \frac{e^2 x^2}{a^2} - x^2 + b^2 \\
&= \frac{e^2 x^2 - a^2 x^2}{a^2} + b^2 \\
&= \frac{\left(e^2 - a^2\right) x^2}{a^2} + b^2 \\
&\underset{e^2 - a^2 = -b^2}{=} \frac{-b^2 x^2}{a^2} + b^2.
\end{aligned}
$$

So after division by b^2 we get

$$
\frac{y^2}{b^2} = -\frac{x^2}{a^2} + 1
$$

and finally the general **equation of an ellipse**

$$
\frac{x^2}{a^2} + \frac{y^2}{b^2} = 1. \tag{1.4}
$$

Thus, we have found the equation for all possible positions (x, y) of the planet in its orbit. That is, all points (x, y) that satisfy the above equation form the elliptical orbit, with the origin of the coordinate system at the center of the ellipse. $\square$

The 1. Kepler's law does not make any statement about the speed at which the planet traverses its elliptical orbit. Kepler observed that the speed of the planets increases as they approach the Sun and decreases as they move away from the Sun. The speed is highest at perihelion and lowest at aphelion. He also found a quantitative relationship.

1.2. Second Kepler's Law (Area Law)

> The straight line connecting the Sun and a planet sweeps out equal areas in equal times.

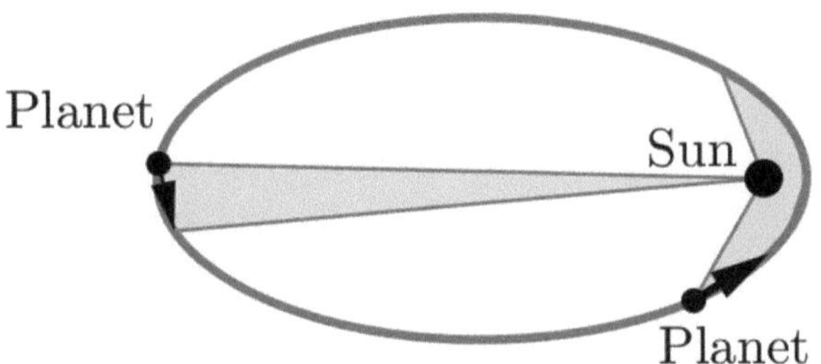

Figure 1.5.: 2nd Kepler's Law, Area Law

The areas marked in gray in Fig. 1.5 are therefore equal if the planet takes the same time on its orbit to generate these areas.

Finally, Kepler established a third law that connects the dimensions of different planetary orbits with their respective orbital periods.

1.3. Third Kepler's Law

> The squares of the orbital periods of two planets are proportional to the cubes of their major semi-axes.

So if T_1, T_2 are the orbital periods and A_1, A_2 are the major semi-axes of the elliptical orbits of two planets, then:

$$\frac{(T_1)^2}{(T_2)^2} = \frac{(A_1)^3}{(A_2)^3} \tag{1.5}$$

or

$$\frac{(A_1)^3}{(T_1)^2} = \frac{(A_2)^3}{(T_2)^2} = C = constant.$$

The constant C is called the **Kepler constant**. However, it is not universally valid, but depends on the respective system consisting of the central star and planets. For our solar system, it is

$$C_S = 3.362 \cdot 10^{18}\,\mathrm{m}^3/\mathrm{s}^2. \tag{1.6}$$

In the Earth-Moon system with the Earth as the central body, it is

$$C_E = 1.010 \cdot 10^{13} \, \text{m}^3/\text{s}^2.$$

1.4. Physical Quantities and Units

With the Kepler constant, a real **physical quantity** has appeared for the first time. In general, the laws of physics describe relationships between physical quantities such as length, time, force, energy, or temperature. Such quantities must be clearly defined and can be measured exactly. Measuring means comparing the quantity with a corresponding, precisely defined **unit**. For example, when we measure the distance between two points, we compare it with the unit of length, e.g., the meter. If the distance is 20 meters, then it is 20 times as long as the unit meter. *Every* physical quantity Q can be defined as the product of a number with a unit, and we write abstractly

$$Q = \{Q\} \, [Q] \, ,$$

where $\{Q\}$ denotes the number and $[Q]$ denotes the unit. We always write physical quantities with *italic* letters.

SI and Natural Units

In the first part of the book, we use the **International System of Units (SI)** (see Chap. 30) for the representation of the units of physical quantities. In the SI system seven base quantities are defined, including the length with the unit meter (m), the time with the unit second (s), and the mass with the unit kilogram (kg) (for the others, see Tab. 30.1 on page 581). All physical quantities used in this book are represented by the seven base quantities. For example, the force F, which - as we will see later - is the product of mass and acceleration, is measured in the unit N (Newton), and it holds:

$$[F] = \text{N} = \text{kg} \cdot \text{m}/\text{s}^2,$$

since the acceleration has the unit m/s^2. For further explanations, see also Tab. 30.2

When describing the theory of relativity, it makes sense to use another unit system, the so-called **natural unit system**, because it makes the expressions simpler and more uniform. This unit system is introduced in Part III (see also Sect. 30.2 on page 582). In other words: we use the SI unit system in this text up to and including Chap. 12, from then on we will mainly use natural units.

Dimension

Each physical quantity is assigned a **dimension**. In the International System of Units, the dimensions of the base quantities are named like the base quantities, so for example, the dimension of the base quantity length is also called length. The symbol of a dimension is an upright, sans-serif capital letter, for the length the letter L, for the mass M and for the time T. The dimension of any physical quantity Q can be represented as a product of powers of the dimensions of the base quantities, e.g.

$$\dim Q = \mathsf{T}^{\alpha} \cdot \mathsf{L}^{\beta} \cdot \mathsf{M}^{\gamma},$$

where the Greek exponents α, β, γ can be positive or negative integers (including zero). For example, the dimension of force is equal to mass times length times (time to the power of -2), in formula language

$$\dim F = \mathsf{M} \cdot \mathsf{L} \cdot \mathsf{T}^{-2}.$$

If all exponents are equal to zero, then

$$\dim Q = 1,$$

and Q is then called **dimensionless**. The dimension of a physical quantity is *independent* of the chosen unit system. For example, the physical quantity speed v has the dimension

$$\dim v = \mathsf{L} \cdot \mathsf{T}^{-1},$$

which is independent of the choice of units m/s or km/h (kilometers per hour). Further details can be found in section 30.1 on page 581.

In physical equations, different quantities are usually linked by formulas. The dimensions of the physical quantities on both sides of the equation *must* match. This means that to check the correctness of physical equations, it is often advisable to first compare the dimensions of the sides. As an example, we consider the equation for the length s:

$$s = vt + at, \tag{1.7}$$

where t denotes time, v speed, and a acceleration. The dimension of the left side is L, on the right side both summands must also have the dimension L. Now

$$\dim vt = \mathsf{L} \cdot \mathsf{T}^{-1} \cdot \mathsf{T} = \mathsf{L},$$

but

$$\dim at = \mathsf{L} \cdot \mathsf{T}^{-2} \cdot \mathsf{T} = \mathsf{L} \cdot \mathsf{T}^{-1},$$

so Eq. (1.7) is *not* a correct physical equation.

When applying complicated functions (such as exponential functions, logarithmic functions, or trigonometric functions) to a physical quantity Q, we *always* assume that the independent variables and the values of the function are *dimensionless*, i.e., real numbers. In physical laws, the first condition is often met, as the dimensionlessness of the arguments is guaranteed by *composite* physical quantities. As an example, we consider the description of an oscillation by

$$y = \sin(\omega t).$$

Here, ω denotes the **angular frequency** (explained later in Eq. (2.13)) and t the time. The dimensions are:

$$\dim \omega = \mathsf{T}^{-1}, \dim t = \mathsf{T} \Rightarrow \dim(\omega t) = 1,$$

and thus the quantity $\sin(\omega t)$ is also dimensionless.

In the following calculations, we usually proceed in such a way that we first derive the formulas without units (as in (1.5)) and only add the corresponding units at the end or after inserting specific values for the variables, as in (1.6).

2. Laws of Falling

In this chapter, we deal with movements that can occur at constant or even changing speed. We will deal with the description of movements without questioning their cause. This happens in Chapter 3, when we show *why* things fall as they do.

We introduce the concepts of instantaneous speed and instantaneous acceleration and derive how this allows movements in space and time to be described. Our considerations lead to the laws of falling on Earth, which were predominantly established by Galilei in his investigations of the trajectories of cannonballs.

To derive these laws, we need a bit more mathematics: We introduce the **derivative of a function** with respect to time and deal with its inverse, the so-called **integral calculus**. In addition, we expand the consideration of coordinate systems introduced in the last chapter and take initial steps towards **vector calculus**.

2.1. Motion in One Dimension

To simplify our consideration of motions, we imagine that the position of an object, whose movements we want to study, can be characterized by specifying the coordinates of *one* point in space at any time. We call such an idealized object a **particle** (or **point mass**). It therefore has no spatial extension, its mass is concentrated in one point. In many cases, this idealization does not represent a significant limitation. For example, for some purposes, it is useful to imagine the Earth as a particle moving around the Sun. In such cases, one is only interested in the movement of the Earth's center of mass, the size and rotation of the Earth are disregarded.

We initially limit ourselves to motion in one dimension, i.e., motion along a straight line. A simple example of one-dimensional motion is a car driving on a flat, straight, and narrow road. For such a motion, there are only two possible directions of movement, a positive (forward) and a negative (backward).

For graphical illustration we use for the first time the **spacetime diagrams** frequently used in (Special) Relativity Theory. These are coordinate systems whose horizontal axis represents "space" (in most cases, for simplification reasons, only the x-coordinate) and whose vertical axis represents time. The

M. Ruhrländer, *Ascent to the Einstein Equations*,
https://doi.org/10.1007/978-3-662-72672-3_2

depicted motion of a particle is called the **worldline** of the particle. Fig. 2.1 shows the worldline of a particle that was at point $x = 1$ at time $t = 0$ and remains there for all eternity.

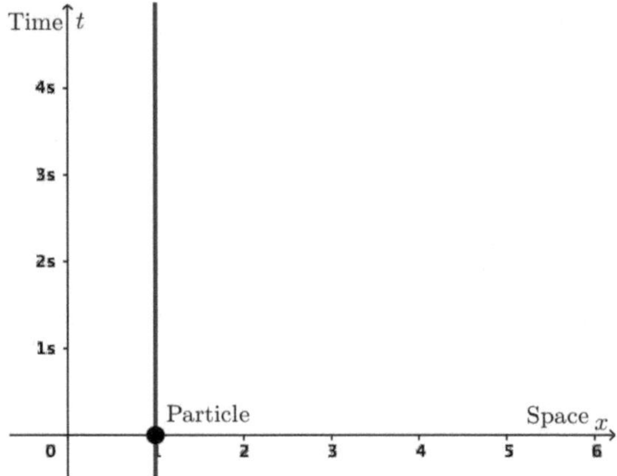

Figure 2.1.: Worldline of a particle that does not move

Velocity and speed

In describing motions, velocity is of high importance. We are familiar with the concept of **average speed** in everyday life. We form the ratio of distance traveled to the time spent:

$$\text{average speed} = \frac{\text{total distance}}{\text{total time}}.$$

Fig. 2.2 shows a particle (e.g., a hiker) moving at a constant speed within 3 seconds from the starting point $x = 1\,\text{m}$ to the target $x = 4\,\text{m}$. If you calculate the average speed AS, you form the difference between the target and starting point and divide by the difference between the end time and start time:

$$AS = \frac{\text{target point} - \text{starting point}}{\text{end time} - \text{start time}} = \frac{4\text{m} - 1\text{m}}{3\text{s} - 0\text{s}} = \frac{3\text{m}}{3\text{s}} = 1\,\frac{\text{m}}{\text{s}}$$

So, the hiker is moving at a constant speed of 1 meter per second or 3.6 km/h, which in this case is also his average speed.

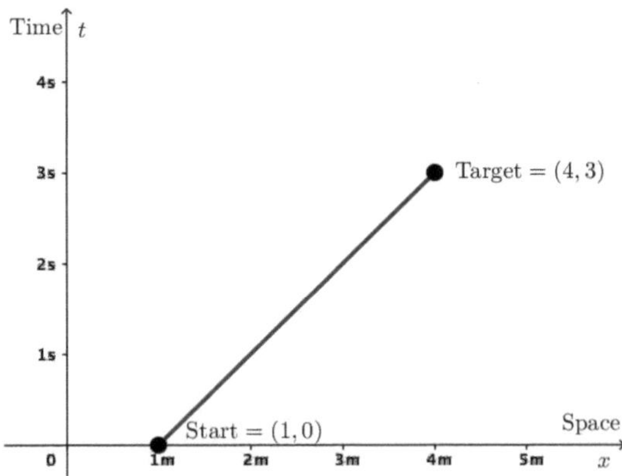

Figure 2.2.: Worldline of a particle with constant speed

Of course, the hiker does not have to walk at a constant speed, he can speed up or slow down, take a break or sprint. For the calculation of the average speed, all this does not matter, since you only need the initial and final positions as well as the initial and final times. Fig. 2.3 shows such a movement with non-constant speed. The hiker thus walks to point A at $1,5\,\mathrm{m}$ per second, takes a break there for one second (so the speed is zero) and then walks again at $1,5\,\mathrm{m}$ per second until the target. His average speed is, as in the last diagram, 1 meter per second.

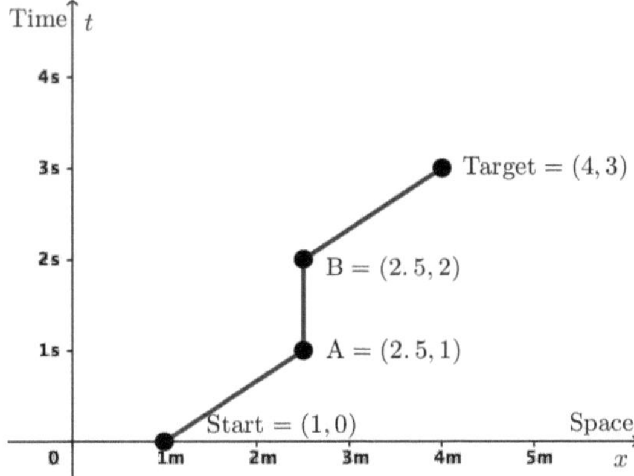

Figure 2.3.: Worldline of a particle with non-constant speed

We can conclude:

> If a particle moves at a constant speed, its worldline is a straight line. If a particle moves at varying speeds, its worldline is uneven, which we want to call **curved**. The worldline of a particle is flatter the faster the particle is, and steeper the slower it is. If it does not move at all, its worldline is vertical.

We now want to consider how to determine the instantaneous speed, simply called **current speed**. In practice when driving a car, this is very simple, you look at the speedometer. But the question is, how does the car calculate the current speed? Well, roughly speaking, the car determines an average speed at any time, by measuring two very close waypoints, forming their difference and dividing by the difference of the corresponding two very short successive time points. A bit more precise and mathematically correct, we want to assume that the car at time t_1 is in position $x_1(t_1)$. The place x, where the car is at a certain time t, is thus a **function**, dependent on time t, and we write for it $x = x(t)$.

Remark 2.1. **MF: Real-valued functions**
A **real-valued function** f is understood to be a rule that assigns *exactly one* real number y to each real number t, and for this we write in short form $y = f(t)$. For example, the rule that assigns the square to each number t is a function, which is written in short form

$$y = f(t) = t^2. \tag{2.1}$$

On the other hand, the rule that assigns to each (positive) number t that number y which, when multiplied by itself, gives t, is *not* a function, as the example $t = 4$ shows, since y can be both 2 and -2 and is therefore not uniquely determined. The quantity t is called the **variable** (or the **argument**) of the function. Functions that take real numbers as values are also called **numerical** or **scalar**. The letter x is often chosen as the variable of the function f, i.e. $y = f(x)$, if another quantity (e.g. location) is to serve as an argument instead of time t.

In Section 1.4 we discussed that the two sides of physical equations must be dimensionally equal. Therefore, we must pay attention to this condition when using functions in physical contexts. For example, if in Eq. (2.1) the size y represents a length and the variable t represents time, then the dimensions of the sides do not match. Eq. (2.1) is then mathematically correct, but not physically. $\square$

After a short time span Δt, the car is at position $x_2 = x_1 + \Delta x$ at time $t_2 = t_1 + \Delta t$. The Greek capital letter Δ ("Delta") always symbolizes a *difference* between two quantities in this text. If we form the average speed AS between these two points, we get

$$AS \ from \ x_1 \ to \ x_2 = \frac{x_2 - x_1}{t_2 - t_1} = \frac{x_1 + \Delta x - x_1}{t_1 + \Delta t - t_1} = \frac{\Delta x}{\Delta t}.$$

The **current speed** of a particle at time t_1 is now defined in such a way that the time difference Δt is made increasingly smaller (which of course also makes Δx smaller), a so-called **limit process** is carried out and $\Delta t \to 0$ ("Δt tends / converges to zero") is written for this. In this limit process the average speed is observed, and if this also has a limit, this limit is defined as the current speed:

$$\text{current speed at time}\ t_1 = \text{limit of}\ \frac{\Delta x}{\Delta t}\ \text{for}\ \Delta t \to 0$$

If the usual abbreviation for the limit, namely lim from the Latin Limes ($=$ border), is used, i.e.

$$\text{limit of}\ \frac{\Delta x}{\Delta t}\ \text{for}\ \Delta t \to 0 = \lim_{\Delta t \to 0} \frac{\Delta x}{\Delta t},$$

and the designation of $v(t_1)$ for the instantaneous speed at time t_1 is used, then one gets

$$v\left(t_1\right) = \lim_{\Delta t \to 0} \frac{\Delta x}{\Delta t}.$$

Remark 2.2. **MF: The derivative of a function**
A function $f\left(x\right)$ is called **differentiable** (or **derivable**) at a point x_0, if the so-called **difference quotient**

$$\frac{f\left(x\right) - f\left(x_0\right)}{x - x_0}$$

has a limit for $x \to x_0$. This limit is called the **(first) derivative** of f with respect to x at the point x_0 and is written as $f'\left(x_0\right)$ or $df\left(x_0\right)/dx$, i.e.

$$f'(x_0) = \frac{df\left(x_0\right)}{dx} = \lim_{x \to x_0} \frac{f\left(x\right) - f\left(x_0\right)}{x - x_0}.$$

The symbol $df\left(x_0\right)/dx$ is intended to remind of "infinitesimally small" changes in numerator and denominator. Is it clear, which point x_0 is meant, one also writes briefly df/dx or dy/dx for $f'\left(x_0\right)$. If you set $\Delta x = x - x_0$, you can alternatively write

$$f'(x_0) = \lim_{\Delta x \to 0} \frac{f\left(x_0 + \Delta x\right) - f\left(x_0\right)}{\Delta x}.$$

If f is a function that depends on time t, in physics instead of $f'(t_0)$ the notation $\dot{f}(t_0)$ is usually chosen.

If the function $f(x)$ is differentiable everywhere, then one can also consider the first derivative as a function $f' = f'(x)$. If this function is differentiable, i.e., if the limit

$$\lim_{x \to x_0} \frac{f'(x) - f'(x_0)}{x - x_0},$$

exists, this limit is called the **second derivative** of f at the point x_0 and is written

$$f''(x_0) = \frac{d^2 f(x_0)}{dx^2} = \lim_{x \to x_0} \frac{f'(x) - f'(x_0)}{x - x_0}$$

or

$$\ddot{f}(t_0) = \frac{d^2 f(t_0)}{dt^2} = \lim_{t \to t_0} \frac{\dot{f}(t) - \dot{f}(t_0)}{t - t_0},$$

if the function depends on time. In a similar form, the third, fourth, etc. derivatives are defined. $\square$

The current speed is thus the first derivative of the function $x = x(t)$ with respect to t at the time t_1

$$v(t_1) = \lim_{\Delta t \to 0} \frac{\Delta x}{\Delta t} = \frac{dx(t_1)}{dt} = \dot{x}(t_1).$$

Before we now perform some simple calculations of current speeds, let's first look at a geometric interpretation of the instantaneous velocity.

Fig. 2.4 which contains a distance-time diagram and *not* a spacetime diagram (note the axis labels), shows a section of the particle curve $x(t)$. The distances $\Delta x = x_2 - x_1$ and $\Delta t = t_2 - t_1$ as well as the connecting line of the two points (t_1, x_1) and (t_2, x_2) lying on $x(t)$ are drawn in.

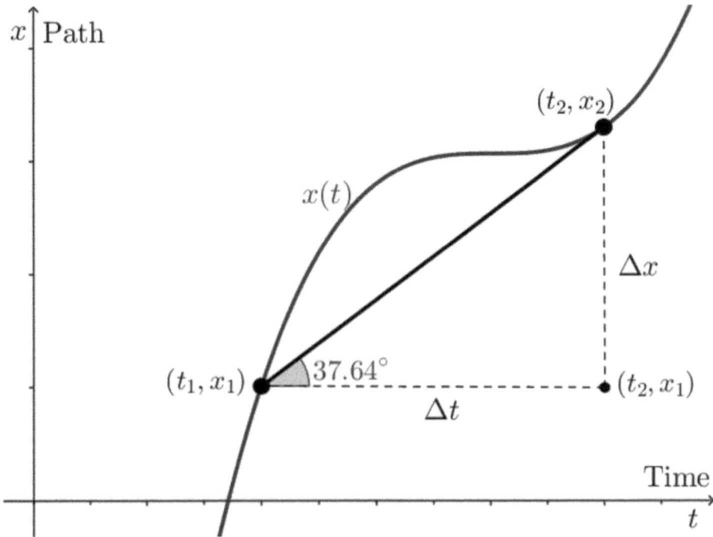

Figure 2.4.: Slope of the secant

The ratio of Δx to Δt, i.e., the average speed between t_1 and t_2, represents the slope of the straight connecting line between (t_1, x_1) and (t_2, x_2), which is also called **secant**. Because if we look at the right-angled triangle formed by the points (t_1, x_1), (t_2, x_2) and (t_2, x_1), and remember that the slope of the connecting line is the tangent of the **slope angle** $37, 64°$, then in this triangle

$$\frac{\Delta x}{\Delta t} = \frac{\text{opposite}}{\text{anjacent}} = \tan\left(37, 64°\right).$$

As we approach t_2 closer and closer to t_1, the slope of the secant approaches more and more the slope of the **tangent** to the curve $x\left(t\right)$ at the point (t_1, x_1). In Fig. 2.5 another point (t_3, x_3) is shown between (t_1, x_1) and (t_2, x_2). Its connecting line to (t_1, x_1) has the slope angle $61, 03°$. One can imagine that the slope angles of connecting lines, which come even closer to (t_1, x_1), increasingly approach the slope angle $71, 16°$ of the tangent at the point (t_1, x_1).

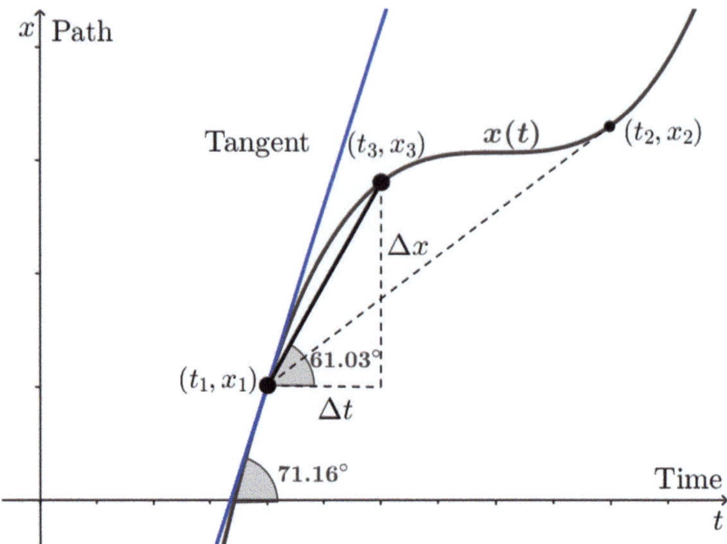

Figure 2.5.: Slope of the tangent

> The current speed for a specific point in time is the slope of the tangent to
> the curve $x(t)$ at this point in time. This is determined by the 1st derivative
> of the position $x(t)$ with respect to time at this point in time.

We want to calculate some examples of currrent speed. Let's start with the
simplest one.

Example 2.1.

A particle moves at a constant speed v_0. Of course, we expect that the current
speed v_0 is the same at all points in time. Let t_1 be the point in time at which
we want to calculate the current speed. Assuming that the particle started at
$x = 0$ at time $t = 0$, then it has reached the point $x(t_1) = v_0 \cdot t_1$ at time t_1. If
we form

$$x(t_2) = x(t_1 + \Delta t) = v_0 \cdot (t_1 + \Delta t),$$

it follows that

$$\Delta x = x_2 - x_1 = v_0(t_1 + \Delta t) - v_0 t_1 = v_0 \Delta t$$

and thus for the average speed

$$\frac{\Delta x}{\Delta t} = \frac{v_0 \Delta t}{\Delta t} = v_0,$$

i.e., the average speed does not depend on Δt anymore, which in turn means
that

$$v(t_1) = \lim_{\Delta t \to 0} \frac{\Delta x}{\Delta t} = \lim_{\Delta t \to 0} v_0 = v_0$$

applies, i.e., the instantaneous speed is also equal to v_0. $\square$

Example 2.2.
In the next example, we assume that the curve of the particle is given by $x(t) = t^2$. We want to determine the current speed after $10\,\mathrm{s}$, assuming again that the particle was at the origin at $t = 0$. If we consider Δx in general, we obtain for any point in time t_1 as above:

$$
\begin{aligned}
\Delta x &= x(t_1 + \Delta t) - x(t_1) = (t_1 + \Delta t)^2 - (t_1)^2 \\
&= (t_1)^2 + 2t_1 \Delta t + (\Delta t)^2 - (t_1)^2 = 2t_1 \Delta t + (\Delta t)^2
\end{aligned}
$$

So, for the average speed, we get

$$
\frac{\Delta x}{\Delta t} = \frac{2t_1 \Delta t + (\Delta t)^2}{\Delta t} = 2t_1 + \Delta t,
$$

i.e., the first term does not depend on Δt anymore and the second term converges to zero as Δt tends to zero. So, we have

$$
v(t_1) = \lim_{\Delta t \to 0} \frac{\Delta x}{\Delta t} = \lim_{\Delta t \to 0} (2t_1 + \Delta t) = 2t_1.
$$

So, for the derivative of $x(t) = t^2$ with respect to the time

$$
\dot{x}(t) = \frac{dx(t)}{dt} = v(t) = 2t,
$$

i.e., the particle becomes faster as time increases. If we substitute $t = 10\,\mathrm{s}$, the speed after 10 seconds is

$$
v(10) = 2 \cdot 10\,\frac{\mathrm{m}}{\mathrm{s}} = 20\,\frac{\mathrm{m}}{\mathrm{s}} = 72\,\frac{\mathrm{km}}{\mathrm{h}}. \square
$$

Acceleration

If the current speed of a particle changes over time, it experiences an **acceleration**. The **average acceleration** $\langle a \rangle$ in a certain time interval $\Delta t = t_2 - t_1$ is defined by the ratio

$$
\langle a \rangle = \frac{\Delta v}{\Delta t},
$$

where $\Delta v = v_2 - v_1$ is the change in instantaneous speed in this time interval. The SI unit for acceleration is meters per second per second

$$
[\langle a \rangle] = \frac{\mathrm{m}}{\mathrm{s}^2}.
$$

The **current acceleration** $a(t)$ is defined as the limit of the average acceleration for ever smaller time intervals, i.e.

$$a(t) = \frac{dv}{dt} = \lim_{\Delta t \to 0} \frac{\Delta v}{\Delta t} = \dot{v}(t).$$

Acceleration is thus the 1st derivative of speed with respect to time. Since speed itself is the 1st derivative of the position $x(t)$ with respect to time, the instantaneous acceleration is the 2nd derivative of the position with respect to time and we write

$$a(t) = \dot{v}(t) = \frac{dv}{dt} = \frac{d^2 x}{dt^2} = \ddot{x}(t).$$

If a particle moves at a constant speed v_0, i.e., if $v(t) = v_0$ for all times, then $\Delta v = 0$ and thus $a(t) = 0$.

Example 2.3.
For the above example with $x(t) = t^2$ we get with

$$\Delta v = v(t + \Delta t) - v(t) = 2 \cdot (t + \Delta t) - 2t$$

for the average acceleration

$$\langle a \rangle = \frac{\Delta v}{\Delta t} = \frac{2 \cdot (t + \Delta t) - 2t}{\Delta t} = \frac{2 \cdot \Delta t}{\Delta t} = 2,$$

i.e., the average acceleration no longer depends on Δt. Thus, for the instantaneous acceleration we get

$$a(t) = \lim_{\Delta t \to 0} \frac{\Delta v}{\Delta t} = 2,$$

so the particle moves with the constant acceleration $2\,\mathrm{m/s^2}$. $\square$

Often one is confronted with the reverse problem, i.e., the acceleration or the speed is given and one wants to find the motion line $x(t)$ of the particle. For this we have to apply a method called **integration**, which is the reverse of derivation.

Remark 2.3. **MF: Primitive functions and integrals**
Let $f(x)$ be a real function, then a real function $F(x)$, whose derivative is equal to $f(x)$, is called a **primitive function** (also called **antiderivative**) of $f(x)$. Since the derivative of a constant function is zero, one can add a constant to any primitive function and get another primitive function. Primitive functions are therefore not unique, they differ by an additive constant. For example,

$$F(x) = \frac{x^2}{2} + 5$$

is an antiderivative of $f(x) = x$ just like

$$F(x) = \frac{x^2}{2}.$$

For the antiderivative $F(x)$ we also use the notation

$$F(x) = \int f(x)\, dx + C$$

and call $F(x)$ the **integral** of $f(x)$ and the constant C the **constant of integration.** $\square$

So, if the acceleration $a(t)$ is known as a function of time, it is first necessary to find a function $v(t)$ whose first derivative corresponds to the acceleration. If, for example, the acceleration is constant, i.e.,

$$a(t) = \frac{dv}{dt} = a_0 = constant,$$

then the velocity is a function of time, whose derivative is just the constant a_0, i.e., $v(t)$ is an antiderivative of $a(t)$. If we call the constant of integration $C = v_0$, we get as a general solution

$$v(t) = \int a_0\, dt + v_0 = a_0\, t + v_0, \tag{2.2}$$

since the derivative of the function $f(t) = a_0\, t$ is equal to $\dot{f}(t) = a_0$. The constant v_0 has a physical meaning, it is the initial velocity at time $t = 0$ since

$$v(0) = a_0 \cdot 0 + v_0 = v_0.$$

The function $x(t)$ for the position of the particle is accordingly that function, whose derivative gives the velocity

$$\frac{dx}{dt} = v(t) = a_0\, t + v_0.$$

To find $x(t)$ we can consider each term separately. The function, whose derivative leads to the constant v_0, is $v_0\, t$ plus an arbitrary constant of integration x_0. If we differentiate the function

$$x(t) = \frac{1}{2} a_0\, t^2$$

with respect to time, we get - similar to above for $x(t) = t^2$ - the result $\dot{x}(t) = a_0\, t$. Summarized for the position of the particle follows

$$x(t) = \frac{1}{2} a_0\, t^2 + v_0 t + x_0. \tag{2.3}$$

In the solution, there are thus two constants of integration: the initial velocity v_0 and the initial position $x(0) = x_0$. They are therefore called **initial conditions**. The general solution thus depends not only on the acceleration a_0, but also on the initial conditions at time $t = 0$. Since - as we will see - the acceleration of a particle is determined by the forces acting on the particle, we can in principle calculate its position for all times if we know the forces and its initial position and initial velocity. This is then referred to as the **solution of the equation of motion**. Eq. (2.3) thus represents the solution of the equation of motion in the case of constant acceleration.

Free Fall

If you throw an object vertically upwards or downwards, you can, if you neglect air resistance and other influences, find that the object is accelerated downwards at a certain constant rate. This rate is called **gravitational acceleration** (or **Earth acceleration**) and is denoted by g. The Earth acceleration is approximately equal to

$$a = -g = -9.81\ \mathrm{m/s^2}$$

near the surface of the Earth. The negative sign results from the fact that we have chosen a coordinate system in which the y-axis (i.e., the positive direction) points upwards. The acceleration due to gravity points downwards towards the center of the Earth and is therefore negative.

It was probably Galileo who first calculated the acceleration due to gravity and also found out that the rate at which bodies fall does not depend on their weight. It is said that he dropped two iron balls, one much heavier than the other, from the Leaning Tower of Pisa. Most people in his time (and probably many today) would have expected the heavier ball to fall faster than the lighter one. But no: both balls reached the ground at the same time. This discovery by Galileo, as much as it contradicts "common sense" (doesn't it take more effort to lift a heavy stone than a light one? Why shouldn't the heavy one also fall faster?), has been confirmed in countless experiments since then. We will later see that it can be derived from the **Newton's laws** together with the so-called **principle of equivalence**.

Example 2.4.
We want to calculate an application example for free fall. A ball is thrown vertically upwards with an initial velocity of $v_0 = 30\,\text{m/s}$. How much time passes until it reaches its highest point? What distance does it cover upwards to this point? We choose the origin of our coordinate system at the starting point of the ball and again the upward direction as positive. As the ball flies upwards, its speed decreases until it becomes zero. At this moment, the ball is at its highest point. According to (2.2), the unknown ascent time T_1 is given by the known quantities a, v_0 and v:

$$v\left(T_1\right) = aT_1 + v_0 = -gT_1 + v_0,$$

from which, because $v\left(T_1\right) = 0$

$$0 = -gT_1 + v_0, \quad d.h. \quad T_1 = \frac{v_0}{g} \tag{2.4}$$

follows, so in our example

$$T_1 = \frac{v_0}{g} = \frac{30\,\text{m/s}}{9.81\,\text{m/s}^2} = 3.06\,\text{s}.$$

The distance the ball covers in this time, we find with Eq. (2.3):

$$x\left(T_1\right) = \frac{1}{2}a\,T_1^2 + v_0\,T_1 + x_0.$$

Noting that we have chosen the coordinate system so that $x_0 = 0$ applies, it follows with $a = -g = -9.81$

$$x\left(3,06\right) = \frac{1}{2}\left(-9.81\right)\cdot\left(3.06\right)^2 + 30\cdot 3.06 = -4.905\cdot 9.3636 + 91.8 = 45.87\,\text{m}.\;\square$$

2.2. Motion in Two Dimensions

We now want to drop the restriction that the particle can only move along one axis. Displacement, velocity, and acceleration are now considered quantities that can have both a magnitude and a direction in space. Such quantities are called **vectors**.

Remark 2.4. **MF: Vector calculus**
So we are dealing with the so-called **vector calculus** and for simplicity's sake we initially focus on the two-dimensional case, the (x, y)-plane, which is again defined by an (x, y)-coordinate system. Vectors can be represented by two

points P_1 and P_2 in space. The point P_1 is called the starting point and the point P_2 is the endpoint of the vector. We graphically represent vectors as arrows and denote them either with small letters with an arrow above them $(\vec{a})$ or by the starting and ending point with an arrow above them $\left(\overrightarrow{P_1 P_2}\right)$, as shown in Fig. 2.6.

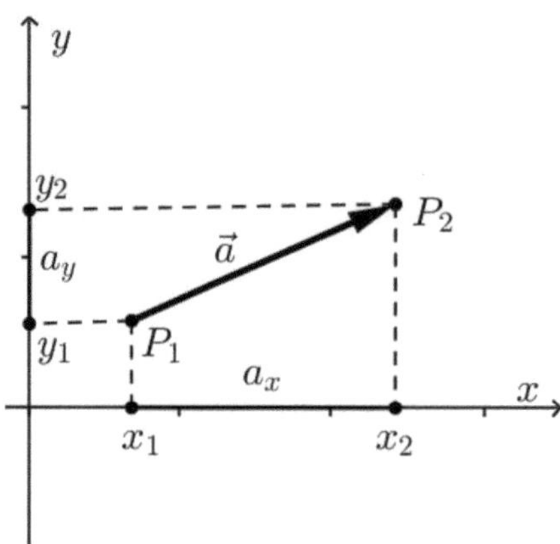

Figure 2.6.: Vector as a connection between two points P_1 and P_2

Now, if the coordinates of the points P_1 and P_2 are given by $P_1 = (x_1, y_1)$ and $P_2 = (x_2, y_2)$, the vector $\vec{a}$ can also be represented by its **components**

$$\vec{a} = \begin{pmatrix} a_x \\ a_y \end{pmatrix} = \begin{pmatrix} x_2 - x_1 \\ y_2 - y_1 \end{pmatrix}.$$

While we write the coordinates of the points in the plane *in sequence*, e.g. $P = (x, y)$, the components of vectors are written *one below the other*

$$\vec{a} = \begin{pmatrix} a_x \\ a_y \end{pmatrix}.$$

If $P_3 = (x_3, y_3)$ and $P_4 = (x_4, y_4)$ are two further points with the property

$$\begin{pmatrix} x_4 - x_3 \\ y_4 - y_3 \end{pmatrix} = \begin{pmatrix} x_2 - x_1 \\ y_2 - y_1 \end{pmatrix},$$

the vector $\vec{a}$ can also be written as

$$\vec{a} = \begin{pmatrix} a_x \\ a_y \end{pmatrix} = \begin{pmatrix} x_4 - x_3 \\ y_4 - y_3 \end{pmatrix}.$$

A vector $\vec{a}$ is uniquely defined by its components a_x, a_y, which can be any real numbers, after choosing a coordinate system. In the graphical representation, two *arbitrary* points P_1, P_2 can be chosen as starting and ending points, provided they fulfill the condition:

$$\begin{pmatrix} a_x \\ a_y \end{pmatrix} = \begin{pmatrix} x_2 - x_1 \\ y_2 - y_1 \end{pmatrix}$$

In other words:

> All arrows in the plane that have the same length and the same direction represent the same vector.

The **zero vector** is defined by

$$\vec{0} = \begin{pmatrix} 0 \\ 0 \end{pmatrix}.$$

The **length of a vector** $\vec{a}$ is denoted by a, it results from Fig. 2.6 with the Pythagorean theorem to

$$a = \sqrt{a_x^2 + a_y^2} = \sqrt{(x_2 - x_1)^2 + (y_2 - y_1)^2}. \tag{2.5}$$

If $P_1 = (0,0)$ is the origin of coordinates and $P = (x, y)$ is an arbitrary point in space, then the vector

$$\vec{r}(P) = \overrightarrow{0P} = \begin{pmatrix} x \\ y \end{pmatrix}$$

is called the **position vector** to P and is written, if it is clear which point P is meant, instead of $\vec{r}(P)$ briefly $\vec{r}$. So one can uniquely represent the set of points in the plane by position vectors. The **unit vector**s of the plane are the (position) vectors

$$\vec{e}_x = \begin{pmatrix} 1 \\ 0 \end{pmatrix}, \vec{e}_y = \begin{pmatrix} 0 \\ 1 \end{pmatrix},$$

they lie on the x or y axis, have the length 1 and are also called **Cartesian basis vectors**.

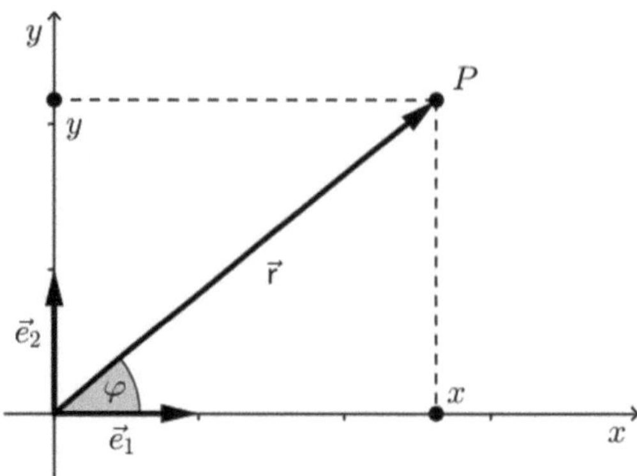

Figure 2.7.: The position vector and Cartesian unit vectors

If φ denotes the angle between the vector $\vec{r}$ and the positive x-axis, one can read from Fig. 2.7 that

$$\cos\varphi = \frac{\text{Adjacent}}{\text{Hypotenuse}} = \frac{x}{r}$$

and

$$\sin\varphi = \frac{\text{Opposite}}{\text{Hypotenuse}} = \frac{y}{r},$$

so

$$x = r\cos\varphi \ \text{ und } \ y = r\sin\varphi$$

apply. This notation of a vector

$$\vec{r} = \begin{pmatrix} x \\ y \end{pmatrix} = \begin{pmatrix} r\cos\varphi \\ r\sin\varphi \end{pmatrix} \tag{2.6}$$

is called representation in **polar coordinates**. For the magnitude of $\vec{r}$ follows with (2.5)

$$r = \sqrt{r^2\cos^2\varphi + r^2\sin^2\varphi} = r\sqrt{\cos^2\varphi + \sin^2\varphi}$$

and from this the important relationship

$$\cos^2\varphi + \sin^2\varphi = 1. \tag{2.7}$$

Two vectors

$$\vec{a} = \begin{pmatrix} a_x \\ a_y \end{pmatrix}, \ \vec{b} = \begin{pmatrix} b_x \\ b_y \end{pmatrix}$$

can be added

$$\vec{a} + \vec{b} = \begin{pmatrix} a_x + b_x \\ a_y + b_y \end{pmatrix}$$

as well as multiplied by a number c (also called: by a **scalar** c):

$$c \cdot \vec{a} = \begin{pmatrix} c \cdot a_x \\ c \cdot a_y \end{pmatrix}$$

These two vector operations can also be visualized graphically. In the addition, one attaches a copy of the vector $\vec{b}$ to the tip of the vector $\vec{a}$ and obtains the sum as the vector of the diagonal of the parallelogram spanned by $\vec{a}$ and $\vec{b}$, see Fig. 2.8. The multiplication of a vector by a number c means a stretching $(c > 1)$ or a compression of the vector $(c < 1)$. If $c < 0$, the arrow changes its direction, the tip and the arrow start exchange their places.

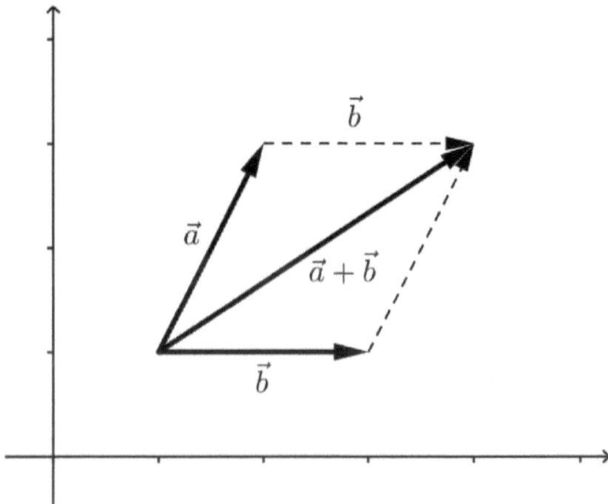

Figure 2.8.: Addition of two vectors

Every vector $\vec{a}$ can be written as a **linear combination** of the basis vectors:

$$\vec{a} = \begin{pmatrix} a_x \\ a_y \end{pmatrix} = \begin{pmatrix} a_x \\ 0 \end{pmatrix} + \begin{pmatrix} 0 \\ a_y \end{pmatrix} = a_x \begin{pmatrix} 1 \\ 0 \end{pmatrix} + a_y \begin{pmatrix} 0 \\ 1 \end{pmatrix} = a_x \, \vec{e}_x + a_y \, \vec{e}_y \ \square \tag{2.8}$$

Trajectory, Velocity and Acceleration Vector

Remark 2.5. **MF: Vector-valued Functions**
A function that uniquely maps a number x to a (two-dimensional) vector is called **vector-valued**. The notation for a **vector-valued function** is $\vec{f}(x)$. The arrow over f is intended to indicate that the values of the function are vectors. In component notation, using the so-called **component functions** f_1 and f_2, we get:

$$\vec{f}(x) = \begin{pmatrix} f_1(x) \\ f_2(x) \end{pmatrix}$$

The component functions are *real-valued* functions and also depend on the variable x. A vector-valued function can also be written as a linear combination of the basis vectors

$$\vec{f}(x) = f_1(x)\,\vec{e}_x + f_2(x)\,\vec{e}_y.$$

This representation shows that only the components depend on the variable x, the Cartesian basis vectors remain unchanged. If one wants to examine the location of a particle at any arbitrary time t in a physical application, the notation

$$\vec{r}(t) = \left(\begin{array}{c} x(t) \\ y(t) \end{array} \right)$$

is often chosen instead of

$$\vec{f}(t) = \left(\begin{array}{c} f_1(t) \\ f_2(t) \end{array} \right). \square$$

We now want to examine the motion of such a particle. The particle has a position in the plane at each fixed time t, which can be represented by a position vector:

$$\vec{r}(t) = x(t)\,\vec{e}_x + y(t)\,\vec{e}_y = \left(\begin{array}{c} x(t) \\ y(t) \end{array} \right)$$

Since the position of the particle can change over time, the components of the particle's position vector are no longer fixed numbers, but depend on the time variable t. In Fig. 2.9 the bold curve represents the actual path that the particle traverses in the plane. This path is also called a **trajectory**. Each point of the trajectory, i.e., each location of the particle, is specified by giving an x- and y-coordinate.

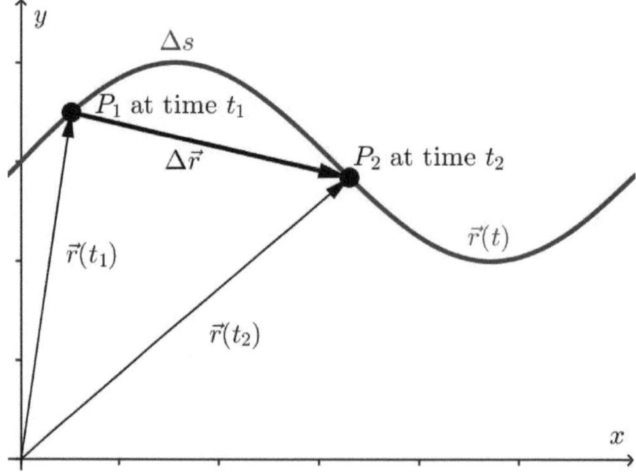

Figure 2.9.: Particle path, trajector in the (x, y)-plane

The particle is at point $P_1 = (x(t_1), y(t_1))$ at time t_1. At a later point in time t_2, it is located at the point $P_2 = (x(t_2), y(t_2))$. The vector of displacement from P_1 to P_2 indicates the spatial change of the position vector:

$$\Delta \vec{r} = \vec{r}(t_2) - \vec{r}(t_1)$$

The ratio of displacement vector to time interval $\Delta t = t_2 - t_1$ represents the *vector of average velocity*:

$$\langle \vec{v} \rangle = \frac{\Delta \vec{r}}{\Delta t}$$

In Fig. 2.9 we see that the magnitude of the displacement vector is not equal to the actual path Δs of the particle. The displacement vector is smaller than this distance, but as the time interval becomes smaller it increasingly approaches the actual distance that the particle travels along the curve. The direction of $\Delta \vec{r}$ (and thus also of $\Delta \vec{r}/\Delta t$) approaches the direction of the tangent to the curve at point P_1. We define the vector of current velocity at time t_1 as the limit of the vector of average velocity for Δt approaching zero:

$$\vec{v}(t_1) = \lim_{\Delta t \to 0} \frac{\Delta \vec{r}}{\Delta t} = \frac{d\vec{r}}{dt} = \dot{\vec{r}}(t_1)$$

The vector of current velocity is thus the derivative of the position vector with respect to time. *Its direction points along the tangent* to the curve that is traversed by the particle in space. It therefore always points in the direction of the particle's motion. To calculate the components of the instantaneous velocity, we must decompose the position vector into its components:

$$\Delta \vec{r} = \vec{r}(t_2) - \vec{r}(t_1) = \left(\begin{array}{c} x(t_2) - x(t_1) \\ y(t_2) - y(t_1) \end{array} \right) = \left(\begin{array}{c} \Delta x \\ \Delta y \end{array} \right)$$

Now we divide the vector $\Delta \vec{r}$ by Δt and obtain for the components

$$\frac{\Delta \vec{r}}{\Delta t} = \frac{1}{\Delta t} \left(\begin{array}{c} \Delta x \\ \Delta y \end{array} \right) = \left(\begin{array}{c} \Delta x / \Delta t \\ \Delta y / \Delta t \end{array} \right),$$

from which

$$\vec{v}(t_1) = \lim_{\Delta t \to 0} \left(\begin{array}{c} \Delta x / \Delta t \\ \Delta y / \Delta t \end{array} \right) = \left(\begin{array}{c} dx/dt \\ dy/dt \end{array} \right) = \left(\begin{array}{c} \dot{x}(t_1) \\ \dot{y}(t_1) \end{array} \right) = \left(\begin{array}{c} v_x(t_1) \\ v_y(t_1) \end{array} \right)$$

follows. For the trajectory of a particle, we have thus calculated the first derivative, the same applies for the general case.

Remark 2.6. **MB: The derivative and the integral of a vector-valued function**

If

$$\vec{f}(x) = \begin{pmatrix} f_1(x) \\ f_2(x) \end{pmatrix}$$

is a vector-valued function, then the derivative of $\vec{f}$ with respect to x is defined by

$$\vec{f}'(x) = \begin{pmatrix} f_1'(x) \\ f_2'(x) \end{pmatrix},$$

i.e., the derivative of a vector-valued function is again a vector-valued function with the component functions f_1' and f_2'. It is obtained by taking the derivatives of the component functions f_1 and f_2 and consider them as components of a vector. If the variable is time t, we also write $\dot{\vec{f}}(t)$ instead of $\vec{f}'(t)$. Higher derivatives of a vector-valued function are defined analogously. If you want to find the integral of a vector-valued function, you proceed similarly. You integrate the component functions individually and consider these integrals as components of a vector

$$\int \vec{f}(x)\, dx = \begin{pmatrix} \int f_1(x)\, dx \\ \int f_2(x)\, dx \end{pmatrix}. \quad \square$$

The vector **average acceleration** is defined analogously as the ratio of the change in current velocity $\Delta\vec{v}$ to the time interval Δt:

$$\langle \vec{a} \rangle = \frac{\Delta\vec{v}}{\Delta t}$$

The vector of **current acceleration** is the derivative of the velocity vector with respect to time:

$$\vec{a} = \lim_{\Delta t \to 0} \frac{\Delta\vec{v}}{\Delta t} = \frac{d\vec{v}}{dt} = \dot{\vec{v}} = \ddot{\vec{r}}$$

To determine the components of the current acceleration, we proceed exactly as in the derivation of the current velocity

$$\vec{v}(t) = \begin{pmatrix} v_x(t) \\ v_y(t) \end{pmatrix} = \begin{pmatrix} dx/dt \\ dy/dt \end{pmatrix}.$$

This results in

$$\vec{a}(t) = \lim_{\Delta t \to 0} \frac{\Delta\vec{v}}{\Delta t} = \begin{pmatrix} d^2x/dt^2 \\ d^2y/dt^2 \end{pmatrix} = \begin{pmatrix} \ddot{x}(t) \\ \ddot{y}(t) \end{pmatrix} = \begin{pmatrix} a_x(t) \\ a_y(t) \end{pmatrix}.$$

The velocity vector can change its magnitude, its direction, or both. We speak of acceleration when the velocity vector varies in *any* way. A particle can move with a velocity of constant magnitude and still be accelerated. This is the case, for example, when the particle moves at the same speed on a circular path. Here, the direction changes (constantly), but not the magnitude of the velocity!

Projectile Motion

An important application of motion in two dimensions is that of a body that is thrown or shot into the air and can then move freely. We want to assume again the ideal situation, that air resistance, rotation of the Earth, etc. have no influence on the motion of the projectile. The motion should be determined exclusively by the assumed constant gravitational acceleration g and the initial conditions.

Galileo was probably the first to discover in his fall experiments, that in projectile motions the horizontal and vertical components of the motion are independent of each other, always provided that no other horizontally or vertically acting forces influence the motion. Sometimes this independence of the motions is also called the earliest version of the **principle of relativity** . One can illustrate this principle as follows. Let's imagine we are at a train station and a train passes us at a constant speed. On the train is a man who throws a ball vertically into the air. From the man's point of view there is a situation as we have described above. For him, there is only the vertical throwing motion, and he can determine the flight height and the fall time with the above means. For us, however, the horizontal position of the man (and thus of the ball) changes due to the speed of the train, i.e., we observe that the movement of the ball also has a horizontal component. Nevertheless, the ball falls at the same speed for us and the man, i.e., the two directions of movement are independent.

Now let's imagine that the ball is thrown with an initial velocity $\vec{v}_0$, which has both a vertical (in the direction of the positive y-axis) and a horizontal component (in the direction of the positive x-axis), and we throw at time $t = 0$ from the origin of the coordinate system at an angle θ to the positive x-axis, see Fig. 2.10.

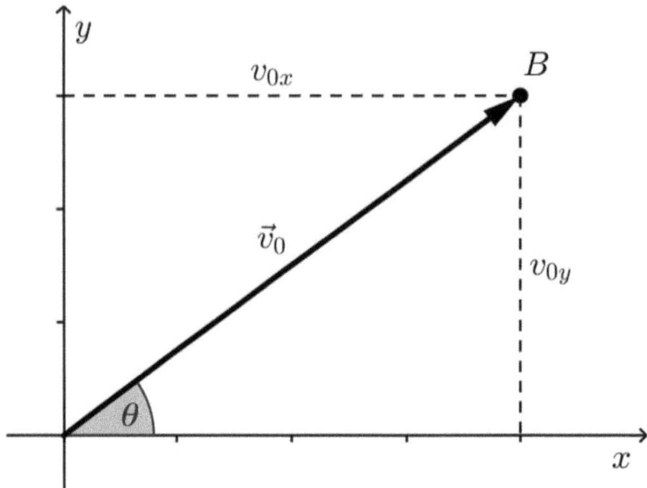

Figure 2.10.: Oblique throw

Then the components of the Earth's acceleration apply

$$\vec{a} = \begin{pmatrix} a_x \\ a_y \end{pmatrix} = \begin{pmatrix} 0 \\ -g \end{pmatrix}$$

and for the components of the velocity

$$\vec{v}_0 = \begin{pmatrix} v_{0x} \\ v_{0y} \end{pmatrix} = \begin{pmatrix} v_0 \cos\theta \\ v_0 \sin\theta \end{pmatrix}.$$

Since there is *no* acceleration in the x-direction, the x-component of the velocity is constant for all times

$$v_x(t) = v_{0x}.$$

The y-component changes with time

$$v_y(t) = v_{0y} - gt.$$

So it is

$$\vec{v}(t) = \begin{pmatrix} v_{0x} \\ v_{0y} - gt \end{pmatrix}.$$

By integrating component by component, we get the trajectory of the ball

$$\vec{r}(t) = \begin{pmatrix} x(t) \\ y(t) \end{pmatrix} = \begin{pmatrix} v_{0x}t + x_0 \\ v_{0y}t - \dfrac{1}{2}gt^2 + y_0 \end{pmatrix}$$

with the integration constants x_0 and y_0. If we consider, that we throw from the origin, i.e.

$$\vec{r}(0) = \begin{pmatrix} 0 \\ 0 \end{pmatrix},$$

then $x_0 = 0$ and $y_0 = 0$ and thus

$$\vec{r}(t) = \begin{pmatrix} x(t) \\ y(t) \end{pmatrix} = \begin{pmatrix} v_{0x}\, t \\ v_{0y}\, t - \dfrac{1}{2} g t^2 \end{pmatrix}.$$

Example 2.5.
We want to determine the range of a ball that is thrown into the air with a speed of $50\,\mathrm{m/s}$ and an angle of $36.87°$ to the horizontal. The components of the initial velocity are

$$\vec{v}_0 = \begin{pmatrix} v_{0x} \\ v_{0y} \end{pmatrix} = \begin{pmatrix} v_0 \cos\theta \\ v_0 \sin\theta \end{pmatrix} = \begin{pmatrix} 50\cos(36.87°) \\ 50\sin(36.87°) \end{pmatrix} \approx \begin{pmatrix} 40\,\mathrm{m/s} \\ 30\,\mathrm{m/s} \end{pmatrix},$$

where the symbol $\approx$ means "approximately equal". First, we determine the time T_2 that the ball flies before it falls back to the Earth. When the ball hits the Earth, the y-component of the trajectory is zero, so it follows

$$0 = y(T_2) = v_{0y}\, T_2 - \frac{1}{2} g T_2^2.$$

We know that at the start time $t = 0$ the ball was at the origin, and we assume that $T_2 \neq 0$, divide the equation by T_2 and solve for T_2:

$$T_2 = \frac{2 v_{0y}}{g} \tag{2.9}$$

the total flight time is thus twice as long as the ascent time, see (2.4). If we insert the numbers, we get

$$T_2 = \frac{2 v_{0y}}{g} = \frac{30 \cdot 2}{9.81} = 6.1\,\mathrm{s}.$$

Since the ball moves in the horizontal direction at a constant speed of $40\,\mathrm{m/s}$, the distance R that the ball has covered in the positive x-direction is:

$$R = v_{0x}\, T_2 = 40 \cdot 6.1 = 244\,\mathrm{m} \;\square$$

If one wants to answer the question of how the launch angle must be chosen so that the range is maximized, one first inserts the general formula for T_2 (2.9) into the last equation:

$$R = v_{0x}\, T_2 = v_{0x}\, \frac{2 v_{0y}}{g} = \frac{2 v_0 \cos\theta\, v_0 \sin\theta}{g} = \frac{2 v_0^2 \cos\theta \sin\theta}{g}$$

This equation can be simplified by using the trigonometric equation for the double angle:

$$\sin(2\theta) = 2\cos\theta\sin\theta$$

(see Chap. 29 on page 577). This gives

$$R = \frac{v_0^2}{g}\sin 2\theta.$$

The sine becomes maximal when $2\theta = 90°$, so $\theta = 45°$, i.e., the maximum range is

$$R_{max} = \frac{v_0^2}{g} \tag{2.10}$$

and the optimal launch angle is 45°. If the launch speed v_0 is so large at the optimal launch angle that R_{max} exceeds the Earth's radius, i.e.,

$$\frac{v_0^2}{g} > 6378\,\text{km}$$

(see Chap. 30 on page 581), the ball no longer falls back to Earth, but "falls past it". If one wants to put it into an orbit that does not collide with the Earth, it must, however, receive another boost in the direction of that orbit.

Circular Motion

We examine the motion of a satellite moving at a constant speed v on a circular orbit with radius r around the Earth. Since the direction of the satellite's speed is always changing, the satellite experiences an acceleration called the **centripetal acceleration**. To determine the centripetal acceleration, we remember that the velocity $\vec{v}$ of a moving particle always points along the tangent to the path curve at the current location of the particle. In the case of a circular path, this means that the velocity vector $\vec{v}$ is perpendicular to the radius.

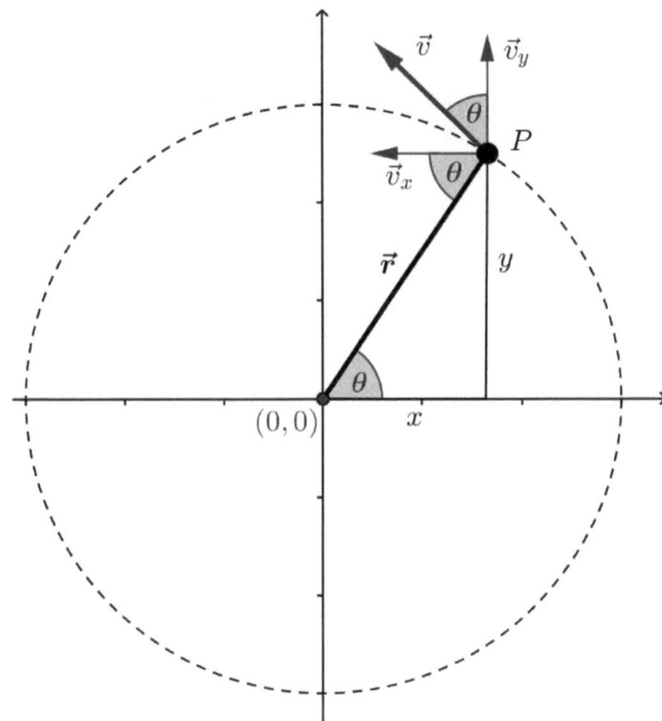

Figure 2.11.: Circular satellite orbit

Fig. 2.11 shows a satellite, which is currently located at point P and moves counterclockwise. The angle θ, which is enclosed by the position vector $\vec{r}$ and the positive x-axis, is also plotted. The angle θ also appears between the vectors $\vec{v}_x$ and $\vec{r}$ as well as between the vectors $\vec{v}$ and $\vec{v}_y$. If we consider that the vector $\vec{v}_x$ with

$$\vec{v}_x = \begin{pmatrix} v_x \\ 0 \end{pmatrix} = \begin{pmatrix} -v\sin\theta \\ 0 \end{pmatrix}$$

points in the negative x-direction and the vector $\vec{v}_y$ with

$$\vec{v}_y = \begin{pmatrix} 0 \\ v\cos\theta \end{pmatrix}$$

points in the positive y-direction, it follows that

$$\vec{v} = v_x\vec{e}_x + v_y\vec{e}_y = -v\sin\theta\,\vec{e}_x + v\cos\theta\,\vec{e}_y.$$

We now use the fact that $\sin\theta = y/r$ and $\cos\theta = x/r$, and obtain

$$\vec{v} = \left(\frac{-vy}{r}\right)\vec{e}_x + \left(\frac{vx}{r}\right)\vec{e}_y.$$

To obtain the acceleration, we need to differentiate this equation with respect to time. Since we have assumed a constant speed v and a constant radius r, these quantities do not change over time. Therefore, for the centripetal acceleration we get

$$\vec{a} = \frac{d\vec{v}}{dt} = \left(\frac{-v}{r} \frac{dy}{dt} \right) \vec{e}_x + \left(\frac{v}{r} \frac{dx}{dt} \right) \vec{e}_y.$$

Because $dy/dt = v_y = v \cos\theta$ and $dx/dt = v_x = -v \sin\theta$, we get

$$\vec{a} = \left(-\frac{v^2 \cos\theta}{r} \right) \vec{e}_x + \left(-\frac{v^2 \sin\theta}{r} \right) \vec{e}_y.$$

If we consider Eq. (2.7):

$$\cos^2\theta + \sin^2\theta = 1,$$

we find with Eq. (2.5) for the magnitude of the acceleration

$$a = \sqrt{a_x^2 + a_y^2} = \frac{v^2}{r} \sqrt{\cos^2\theta + \sin^2\theta} = \frac{v^2}{r}. \tag{2.11}$$

For the direction φ of $\vec{a}$, we get from Fig. 2.12:

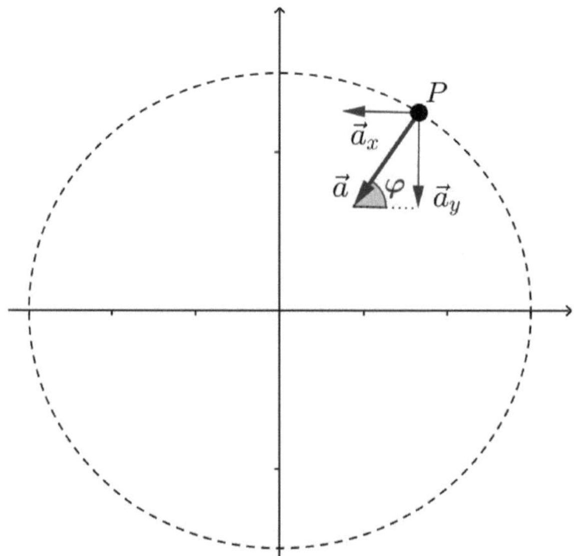

Figure 2.12.: Direction of the centripetal acceleration

$$\tan \varphi = \frac{a_y}{a_x} = \frac{\left(-\dfrac{v^2 \sin \theta}{r}\right)}{\left(-\dfrac{v^2 \cos \theta}{r}\right)} = \frac{\sin \theta}{\cos \theta} = \tan \theta,$$

so $\varphi = \theta$, i.e., the vector $\vec{a}$ points towards the center of the circle. Since the gravitational acceleration also points towards the center of the Earth, we can conclude that in the (near) Earth's field, the centripetal acceleration on a circular satellite orbit is proportional to the gravitational acceleration. Since the satellite flies at a constant speed v, the **orbital period** T is

$$v = \frac{2\pi r}{T}, \text{ thus } T = \frac{2\pi r}{v}. \tag{2.12}$$

The quantity $2\pi/T$ is called **angular frequency** and is denoted by the Greek letter Omega ω:

$$\omega = \frac{2\pi}{T} = \frac{v}{r} \tag{2.13}$$

We want to apply the results to the satellite orbit in a specific example.

Example 2.6.
A satellite is moving 200 km above the Earth's surface at a constant speed on a circular orbit around the Earth's center. What speed does it have, assuming that the centripetal acceleration is still 9.81 m/s^2 at this height, and how long does it take for one orbit?
The radius of the orbit is equal to the Earth's radius plus 200 km:

$$r = 6378 + 200 = 6578 \text{ km}$$

We calculate the speed of the satellite using Eq. (2.11):

$$v^2 = r \cdot a = 6.578.000 \cdot 9.81 = 64.530.180,$$

so

$$v = 8.033 \text{ m/s} \approx 8 \text{ km/s}.$$

For the orbital period we get

$$T = \frac{2\pi r}{v} \approx \frac{2\pi \cdot 6578}{8} = 5166 \text{ s} \approx 86 \text{ min}. \ \square$$

2.3. Generalization to Three Dimensions

So far, we have focused on movements in two dimensions, since the graphical representations are much simpler and therefore more understandable than in three dimensions. However, the general laws can easily be generalized to the three-dimensional case. When dealing with particles moving in three-dimensional space, there are now three coordinates instead of two that describe the path of the particle:

$$\vec{r}(t) = x(t)\,\vec{e}_x + y(t)\,\vec{e}_y + z(t)\,\vec{e}_z = \begin{pmatrix} x(t) \\ y(t) \\ z(t) \end{pmatrix},$$

where the unit vectors

$$\vec{e}_x = \begin{pmatrix} 1 \\ 0 \\ 0 \end{pmatrix}, \vec{e}_y = \begin{pmatrix} 0 \\ 1 \\ 0 \end{pmatrix}, \vec{e}_z = \begin{pmatrix} 0 \\ 0 \\ 1 \end{pmatrix}$$

are given. The other calculation rules of vector calculus are transferred in a meaningful way, e.g. for the magnitude of $\vec{r}(t)$

$$r(t) = \sqrt{x^2(t) + y^2(t) + z^2(t)}. \tag{2.14}$$

Velocity and acceleration are obtained analogously to

$$\begin{aligned} \vec{v}(t) &= \lim_{\Delta t \to 0} \frac{\Delta x \vec{e}_x + \Delta y \vec{e}_y + \Delta z \vec{e}_z}{\Delta t} = \frac{dx}{dt}\vec{e}_x + \frac{dy}{dt}\vec{e}_y + \frac{dz}{dt}\vec{e}_z \\ &= \begin{pmatrix} \dot{x}(t) \\ \dot{y}(t) \\ \dot{z}(t) \end{pmatrix} = \begin{pmatrix} v_x(t) \\ v_y(t) \\ v_z(t) \end{pmatrix}, \end{aligned}$$

as well as

$$\begin{aligned} \vec{a}(t) &= \frac{dv_x}{dt}\vec{e}_x + \frac{dv_y}{dt}\vec{e}_y + \frac{dv_z}{dt}\vec{e}_z = \frac{d^2x}{dt^2}\vec{e}_x + \frac{d^2y}{dt^2}\vec{e}_y + \frac{d^2z}{dt^2}\vec{e}_z \\ &= \begin{pmatrix} \ddot{x}(t) \\ \ddot{y}(t) \\ \ddot{z}(t) \end{pmatrix} = \begin{pmatrix} a_x(t) \\ a_y(t) \\ a_z(t) \end{pmatrix}. \end{aligned}$$

3. Newton's Laws

An interaction that can cause an *acceleration* of a body is called a **force**. Isaac Newton (1642 - 1727) was the first to formulate the relationship between forces and accelerations in laws, which constitute the so-called **Newtonian mechanics**. This was applicable without restrictions for any mechanical problem until the beginning of the 20th century. Then it was discovered that Newtonian mechanics must be replaced by Einstein's **Special Theory of Relativity**, when the velocities of the interacting bodies approach the speed of light. If the interacting bodies are of the order of atomic structures, then **quantum mechanics** replaces Newtonian mechanics. Nevertheless, Newtonian mechanics still has great relevance today, as it can be applied to the motion of objects whose size ranges from small to astronomical scales (e.g., **galaxy clusters**).

Newton broke away from the centuries-old notion that a force is required to keep a body in a **uniform motion**. He postulated, like Galileo before him, that a body moves at a constant speed or remains at rest if *no* forces act on it. He also recognized that the acceleration of a body depends on its mass and the force exerted on it. The greater the force, the greater the acceleration; and the greater the mass, the smaller the acceleration. The direction of acceleration is the direction of the force. The force is therefore itself a vector and is denoted by $\vec{F}$. If n forces $\vec{F}_i$ act on the body at the same time, then the acceleration $\vec{a}$ is proportional to the vector sum of all forces

$$\vec{F} = \vec{F}_1 + \vec{F}_2 + \cdots + \vec{F}_n.$$

From these considerations, two laws can be derived:

First Newton's Law (Principle of Inertia): If no forces act on a body ($\vec{F} = \vec{0}$), then its velocity (magnitude and/or direction) cannot change.

Second Newton's Law (Principle of Action): The acceleration $\vec{a}$ of a body is inversely proportional to its mass m and directly proportional to the resulting total force $\vec{F}$, that acts on it:

$$\vec{a} = \frac{\vec{F}}{m} \quad \text{or} \quad \vec{F} = m\vec{a}$$

M. Ruhrländer, *Ascent to the Einstein Equations*,
https://doi.org/10.1007/978-3-662-72672-3_3

The unit of force is 1 Newton (N) and corresponds to the force that must be applied to accelerate a body with the mass $1\,kg$ at $1\,m/s^2$. Newton formulated a third law:

Third Newton's Law (Principle of Reaction): Forces always occur in pairs. If body A exerts a force on body B, then an equal but opposite force from body B acts on body A.

If n forces $\vec{F}_i$ occur in a system, there is therefore a counterforce $\vec{F}_{i;opp}$ for each of these forces with

$$\vec{F}_i + \vec{F}_{i;opp} = \vec{0},$$

so that for the total force $\vec{F}$, which also takes into account all counterforces,

$$\vec{F} = \left(\vec{F}_1 + \vec{F}_{1;opp}\right) + \cdots + \left(\vec{F}_n + \vec{F}_{n;opp}\right) = \vec{0}$$

applies. The total force in a system is zero. From the third law of Newton, an important physical law can be derived, namely the conservation of momentum.

3.1. Conservation of Momentum

The **momentum** $\vec{p}$ of a particle is defined as the product of its mass m and its velocity $\vec{v}$

$$\vec{p} = m\vec{v}.$$

Momentum, like velocity, is a vector quantity. One can think of momentum as a measure of the difficulty of bringing the particle to rest. A heavy truck moving at a certain speed has a higher momentum than a car moving at a lower speed. It takes more braking force to stop the truck than the car. The second law of Newton can be formulated using momentum, assuming that the mass is constant, i.e., does not change over time:

$$\vec{F} = m\vec{a} = m\,\frac{d\vec{v}}{dt} = \frac{d\,(m\vec{v})}{dt} = \frac{d\vec{p}}{dt} \tag{3.1}$$

Let's consider two particles that each exert an equal but opposite force on each other. If $\vec{F}_{12}$ is the force from the first particle on the second and $\vec{F}_{21}$ is the force from the second on the first, then

$$\vec{F}_{12} = \frac{d\vec{p}_1}{dt}, \quad \vec{F}_{21} = \frac{d\vec{p}_2}{dt}.$$

If we add these two equations and take into account the third law of Newton ($\vec{F}_{12} = -\vec{F}_{21}$), we get

$$0 = \frac{d\vec{p}_1}{dt} + \frac{d\vec{p}_2}{dt} = \frac{d\left(\vec{p}_1 + \vec{p}_2\right)}{dt},$$

so

$$\vec{p}_1 + \vec{p}_2 = constant,$$

since a function whose derivative is zero can only be a constant. We can extend this statement to a closed system with n particles with momenta $\vec{p}_i = m_i \vec{v}_i$. If we define the **total momentum** of the system as

$$\vec{p}_{all} = \vec{p}_1 + \cdots + \vec{p}_n,$$

it follows, since the total force of the system is equal to zero, analogously

$$\vec{p}_{all} = \vec{p}_1 + \cdots + \vec{p}_n = m_1 \vec{v}_1 + \cdots + m_n \vec{v}_n = constant.$$

> The total momentum of a closed system of (possibly interacting) particles remains constant over time.

With the help of Newton's laws, we can solve a multitude of motion problems. All applications of Newton's laws boil down to two principal possibilities:

1. If all the forces acting on a particle are known, the acceleration of the particle can be determined. Because

$$\vec{F} = m\vec{a} = m\ddot{\vec{r}}(t)$$

the trajectory of the particle can in principle be derived from the acceleration; in principle means that often a considerable mathematical effort is required to solve the equation of motion or that one must be satisfied with approximate solutions.

2. Conversely, if we know the acceleration of a particle, we can determine the forces acting on the particle.

In the following, we present some applications of Newton's laws.

3.2. The Gravitational Force in the Earth's Environment

When we talk about the gravitational force or weight $\vec{F}_g$ in this section, we mean the force that "pulls" a body towards the center of the Earth, i.e., acts perpendicularly downwards towards the ground. In the following chapters, we will generalize this simplified approach.

Let's assume that a body of mass m is in free fall under the influence of the Earth's acceleration with magnitude $g = 9.81\,\text{m/s}^2$. Neglecting air resistance and the like, the only force acting on the body is the gravitational force. This downward-directed force and the corresponding downward-directed acceleration can be linked using the second law of Newton. For this, we again set up the coordinate system so that the body falls downwards along the upward-directed y-axis. Then follows

$$\vec{F}_g = m\vec{a} = m(-g)\,\vec{e}_y = \begin{pmatrix} 0 \\ -mg \\ 0 \end{pmatrix}$$

and from this for the magnitude of the gravitational force

$$F_g = mg.$$

The **weight** G of a body corresponds to the magnitude of the total force measured by an observer on the ground, which must be exerted to prevent the body from falling freely. For example, to hold a ball still in your hand, you have to apply an upward force that compensates for the gravitational force exerted by the Earth. That is, the weight G of a body is equal to the magnitude F_g of the gravitational force exerted on this body:

$$G = F_g = mg$$

The weight is thus proportional to the mass of a body, but is not equal to the mass of a body. If we move the body, for example, to the Moon, its weight is only about one-sixth as large as on Earth due to the lower gravitational acceleration on the Moon, but its mass does not change.

Example 3.1.
As an example, we want to calculate the acceleration of a body of mass m that slides down a frictionless inclined plane, which is inclined at an angle θ against the horizontal, see Fig. 3.1.

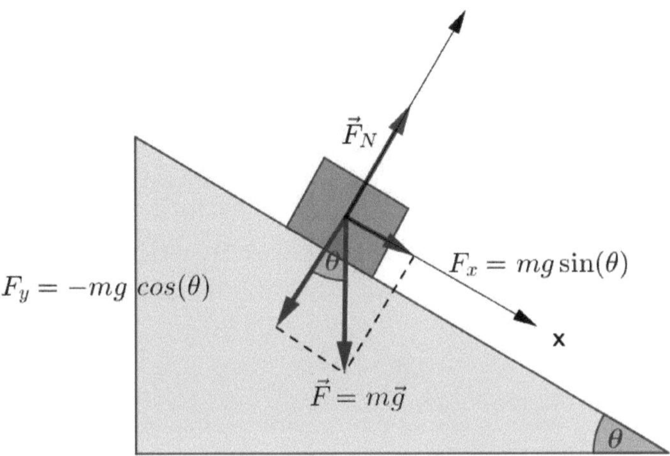

Figure 3.1.: Acceleration on inclined plane

The weight force $\vec{F}$ and the **normal force** $\vec{F}_N$, which is exerted by the inclined plane on the body and is the counterforce to $F_y\,\vec{e}_y$, act on the body. We choose a coordinate system whose coordinate axes are parallel (x-axis) or perpendicular (y-axis) to the inclined plane. Then the acceleration $\vec{a}$ acting on the body only has the component a_x. Because the component F_y is just offset by the normal force, which acts in the positive y direction, according to the third law of Newton. So for the x-component of $\vec{F}$

$$ma_x = F_x = mg\sin\theta$$

follows and from this

$$a_x = g\sin\theta.$$

The acceleration on the inclined plane is constant and has the value $g\sin\theta$. At $\theta = 0$, the plane is horizontal and the weight force only has a y-component, which is offset by the normal force. The acceleration is zero:

$$a_x = g\sin 0 = 0$$

In the other extreme case, $\theta = 90°$, the inclined plane vertical. Then the weight force only has an x-component and the normal force is equal to zero:

$$F_N = mg\cos(90°) = 0$$

The acceleration is then

$$a_x = g\sin(90°) = g.$$

The body is in free fall. $\square$

Example 3.2.
As another example, we want to derive the **centripetal force**, which acts on
a satellite in an Earth orbit. We have shown in the section on satellite orbits
that the *centripetal acceleration* $\vec{a}_z$ points from the satellite to the center of
the Earth, i.e., if $\vec{r}$ is the position vector of the satellite (with the center of the
Earth as the origin), then with the **unit vector in r-direction** $\vec{e}_r = \vec{r}/r$ and
the fact that the centripetal acceleration has the magnitude

$$|\vec{a}_z| = \frac{v^2}{r}$$

(see Eq. (2.11)), the relationship

$$\vec{a} = -\frac{v^2}{r}\,\vec{e}_r$$

follows, where the minus sign indicates the direction towards the center of the
Earth. It follows from the second law of Newton that, if m_s denotes the mass
of the satellite, the centripetal force acting on the satellite is

$$\vec{F}_z = m_s\,\vec{a}_z = -\frac{m_s v^2}{r}\,\vec{e}_r.$$

The centripetal force thus acts in the direction of the position vector $\vec{r}$ (which
is called **radial**), and its strength depends on the distance r from the origin
at the respective position (of course, the strength of the force also depends
on the mass and the orbital speed, but these do not change from position to
position). The centripetal force is an example of a so-called central force.

Forces whose strength depends only on the distance to the origin and whose
direction is radial are called **central forces**. They can be represented in the
form

$$\vec{F} = f(r) \cdot \vec{e}_r \tag{3.2}$$

with a function $f(r)$, which position depends only from the distance r to the
origin.

---

4. Work and Energy

4.1. Work in One Dimension in the Case of Constant Force

The **work**, which a force performs on a body/point mass, is defined as the product of this force and the local displacement of the point of application of the force. If the force and displacement point in different directions, only the component of the force that acts in the direction of the displacement does work on the body. For simplicity, we first consider the case where the force is constant and the body can only move along the x-axis. If θ is the angle between the force $\vec{F}$ and the displacement Δx, then the work W performed on the body is

$$W = F \cos \theta \Delta x = F_x \, \Delta x, \tag{4.1}$$

see Fig. 4.1.

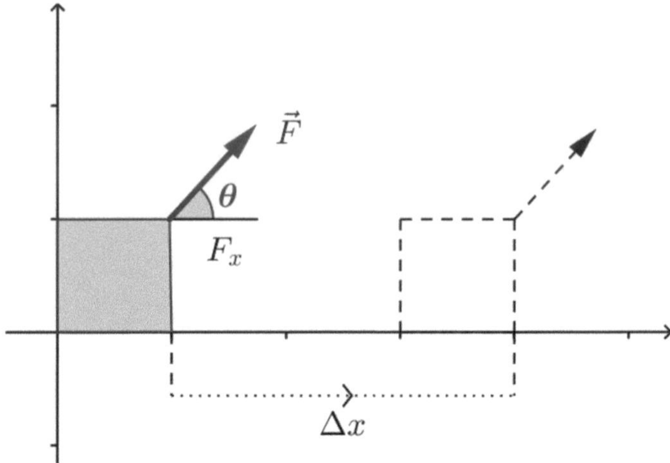

Figure 4.1.: Work in one dimension

Work is a scalar quantity (i.e., a number) that takes a positive value when $\vec{F}$ and Δx point in the same direction, and a negative one when they are opposite. The dimension of work is force times distance (or "force times path"), its unit is **Joule** (J):

$$J = N \cdot m,$$

see also Tab. 30.2. If multiple forces act on the body, the works add up. One can also first form the vector sum of the forces and then calculate the work of the resulting total force.

There is a relationship between the work on a point mass and its initial and final velocity. If F_x is the component of the force that acts in the direction of displacement, then $F_x = ma_x$. Since the force is constant, the acceleration a_x is also constant. If v_a and v_e denote the initial and final velocities and t_a and t_e the initial and final times of the displacement, then with Eq. (2.2)

$$v_e = a_x \left(t_e - t_a\right) + v_a,$$

from which

$$t_e - t_a = \frac{v_e - v_a}{a_x}$$

follows. Therefore, with Eq. (2.3) we can write the difference

$$\Delta x = x\left(t_e\right) - x\left(t_a\right)$$

as

$$\Delta x = \frac{1}{2} a_x \left(t_e - t_a\right)^2 + v_a \left(t_e - t_a\right).$$

If we replace the time difference $t_e - t_a$ by the penultimate expression, we get

$$\Delta x = \frac{1}{2} a_x \left(\frac{v_e - v_a}{a_x}\right)^2 + v_a \left(\frac{v_e - v_a}{a_x}\right).$$

Multiplying out and rearranging gives

$$2a_x \Delta x = v_e^2 - 2v_a v_e + v_a^2 + 2\left(v_a v_e - v_a^2\right) = v_e^2 - v_a^2.$$

This inserted into (4.1) yields

$$W = F_x \Delta x = ma_x \Delta x = \frac{1}{2} mv_e^2 - \frac{1}{2} mv_a^2.$$

The quantity $mv^2/2$ is called the **kinetic energy** of a point mass. It is a scalar quantity and depends on the mass and the velocity of the particle

$$E_{kin} = \frac{1}{2} mv^2. \tag{4.2}$$

So, work is the change in kinetic energy

$$W = \Delta E_{kin} = \frac{1}{2} mv_e^2 - \frac{1}{2} mv_a^2.$$

4.2. Work in the Case of Variable Force

In Fig. 4.2 the constant force F_x is shown as a function of the position x.

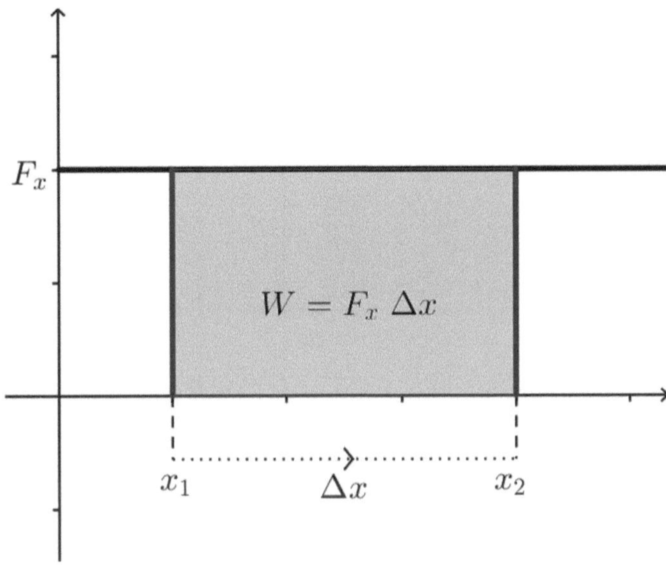

Figure 4.2.: Work in the case of constant force

The work W is represented by the marked area under the force-path curve. However, many forces encountered in nature are not constant, but depend on location. For example, when you stretch a spring, the **spring force** depends on how far the spring is already stretched, it is proportional to the deflection of the spring

$$F_{spring} = -k\,(x - x_0)$$

(„**Hooke's Law**"). Here, k denotes the spring constant, x_0 the rest position, and x the deflection; the minus sign indicates that the force wants to return the spring to its resting position.

In Fig. 4.3 we again use a graphical representation of work as the area under a curve to extend our definition of work to non-constant forces. The interval from $a = x_1$ to $b = x_{n+1}$ is divided into n small intervals $\Delta x_i = x_{i+1} - x_i$. For each small interval, the force is approximately constant equal to F_{x_i}, and we consider the shaded rectangle $F_{x_i}\Delta x_i$ in the figure as an approximation for the area under the force curve in the interval Δx_i:

$$W_i \approx F_{x_i}\Delta x_i,$$

where $\approx$ again means "approximately equal".

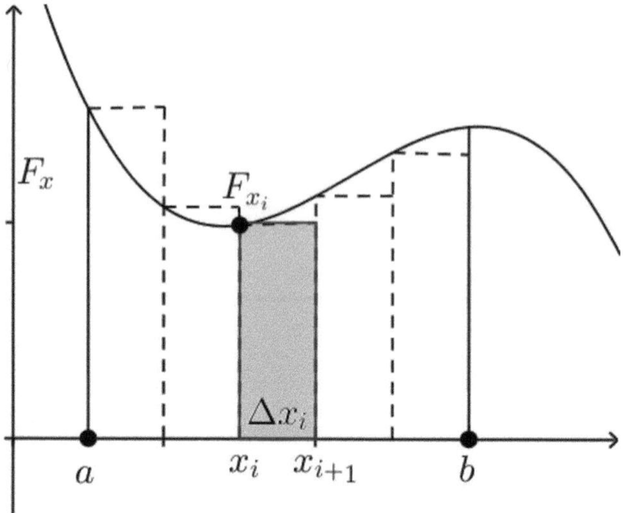

Figure 4.3.: Work in the case of non-constant force

The sum of all n rectangle areas approaches the actual area under the curve more and more, the smaller the intervals Δx_i become, i.e., the larger n becomes

$$W = \lim_{n \to \infty} \sum_{i=1}^{n} W_i = \lim_{n \to \infty} \sum_{i=1}^{n} F_{x_i} \Delta x_i.$$

This limit (if it exists) is referred to as the **definite integral** of F_x over the interval $[a, b]$ and is written as follows:

$$W = \lim_{n \to \infty} \sum_{i=1}^{n} F_{x_i} \Delta x_i = \int_{a}^{b} F_x \, dx$$

Since for each small interval $\Delta x_i = x_{i+1} - x_i$

$$W_i \approx \Delta E_{kin}^i = \frac{1}{2} m v_{i+1}^2 - \frac{1}{2} m v_i^2$$

and for any number n

$$W_1 + \cdots + W_n \approx \left(\frac{1}{2} m v_2^2 - \frac{1}{2} m v_1^2 \right) + \cdots + \left(\frac{1}{2} m v_{n+1}^2 - \frac{1}{2} m v_n^2 \right),$$

all intermediate terms cancel out in the summation and it remains

$$\lim_{n\to\infty} \sum_{i=1}^{n} W_i \;=\; \lim_{n\to\infty} \left(\frac{1}{2}mv_{n+1}^2 - \frac{1}{2}mv_1^2 \right)$$

$$= \lim_{n\to\infty} \left(\frac{1}{2}mv_b^2 - \frac{1}{2}mv_a^2 \right) = \frac{1}{2}mv_b^2 - \frac{1}{2}mv_a^2.$$

So, just like in the case of constant force, we get for the general case

$$W = \Delta E_{kin} = \frac{1}{2}mv_b^2 - \frac{1}{2}mv_a^2.$$

To calculate the work done on a body by a force in a specific situation, one must calculate the above definite integral. It has already been said that integration is the reverse of differentiation. We want to shed some more light on this.

Remark 4.1. **MF: Fundamental Theorem of Calculus**

One can calculate a definite integral over the interval $[a, b]$ using the following expression:

$$\int_a^b f(x)\, dx = [F(x)]_a^b = F(b) - F(a),$$

where $F(x)$ is an antiderivative of $f(x)$.

The choice of the antiderivative is arbitrary, because if $G(x)$ is another antiderivative of $f(x)$, then $G(x) = F(x) + konst$, so

$$G(b) - G(a) = F(b) + konst. - (F(a) + konst.) = F(b) - F(a).$$

The expression $[F(x)]_a^b$ is just a symbol, which can be pronounced as "F in the limits a and b", and helps to keep track during the actual calculation. $\square$

We want to calculate a simple example.

Example 4.1.
A body of mass $4\,\text{kg}$ is connected to a horizontal spring on a frictionless table, which obeys Hooke's law

$$F_x = -k(x - x_0)$$

with $k = 400\,\text{N/m}$. The spring, whose rest position should be the origin ($x_0 = 0$), is pulled to the point $x_1 = 5\,\text{cm}$. We want to calculate the work done by the spring on the body on the way from $0.05\,\text{m}$ to $0\,\text{m}$ (use the same units!), as well as the speed of the body at $0\,\mathrm{m}$. Since $x_0 = 0$, $F_x = -kx$. Furthermore,

$$G(x) = -k\,\frac{1}{2}x^2$$

is an antiderivative of F_x. So it follows

$$W = \int_{0.05}^{0} F_x\,dx = \int_{0.05}^{0} (-kx)\,dx = \left[-k\,\frac{1}{2}x^2\right]_{0.05}^{0}$$

$$= -\frac{k}{2}0^2 - \left(-\frac{k}{2}(0.05)^2\right) = \frac{400}{2}0.0025 = 0.5\,\mathrm{J}.$$

Since the initial speed at $0.05\,\mathrm{m}$ is zero $(v_a = 0)$, it follows that

$$\Delta E_{kin} = E_{kin} = \frac{1}{2}mv_e^2 = W = 0.5\,\mathrm{J}.$$

So it follows

$$v_e = \sqrt{\frac{2E_{kin}}{m}} = \sqrt{\frac{2\cdot 0.5}{4}} = 0.5\,\mathrm{m/s}.\ \square$$

4.3. Work and Energy in Three Dimensions

We consider a point mass that moves under the influence of a force $\vec{F}$ on a curve $\vec{r}(t)$ from a point P_1 to the point P_2 in space. If φ denotes the angle between the force vector $\vec{F}$ and a small displacement $\Delta\vec{r}_i$ along this curve, then the component of the force in the direction of $\Delta\vec{r}_i$ is equal to

$$F_r = F\cos\varphi,$$

see Fig. 4.4. The work done by the force during this small displacement is then

$$\Delta W_i = F_r\Delta r_i = F\cos\varphi\Delta r_i.$$

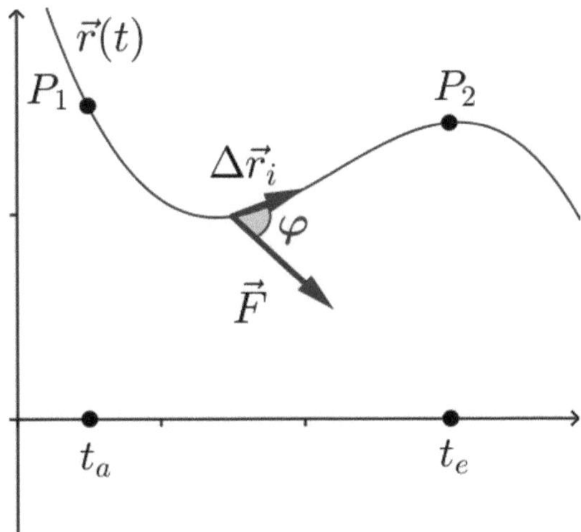

Figure 4.4.: Infinitesimal work

This link between $\vec{F}$ and $\Delta\vec{r}_i$ plays in physics an important role and is referred to as the **dot product** of the vectors $\vec{F}$ and $\Delta\vec{r}_i$.

Remark 4.2. **MF: Dot product of two vectors**
The dot product of two arbitrary vectors $\vec{a}$ and $\vec{b}$ is denoted by $\vec{a}\cdot\vec{b}$ and is defined by

$$\vec{a}\cdot\vec{b} = ab\cos\varphi = ab\cos\left(\angle\left(\vec{a},\vec{b}\right)\right),\tag{4.3}$$

i.e., φ is the angle between $\vec{a}$ and $\vec{b}$.

If $\vec{a}$ and $\vec{b}$ are non-zero, then $\vec{a}\cdot\vec{b} = 0$ means that the vectors $\vec{a}$ and $\vec{b}$ are perpendicular to each other (also said: **orthogonal** to each other), because $\cos\varphi = 0$ then implies

$$\varphi = \angle(\vec{a},\vec{b}) = 90°\,bzw.\,270°.$$

If, on the other hand, $\vec{a}$ and $\vec{b}$ are parallel, then $\varphi = 0°$ and

$$\vec{a}\cdot\vec{b} = ab\cos 0° = ab,$$

from which it generally follows:

$$\vec{a}\cdot\vec{a} = a^2$$

From the definition it also follows that the dot product is **commutative**, i.e. that

$$\vec{a}\cdot\vec{b} = \vec{b}\cdot\vec{a}.$$

In addition, the **distributive law** is fulfilled:

$$(\vec{a} + \vec{b}) \cdot \vec{c} = \vec{a} \cdot \vec{c} + \vec{b} \cdot \vec{c}$$

For the dot products of the unit vectors it holds

$$\vec{e}_x \cdot \vec{e}_x = \vec{e}_y \cdot \vec{e}_y = \vec{e}_z \cdot \vec{e}_z = 1,$$

$$\vec{e}_x \cdot \vec{e}_y = 0, \vec{e}_x \cdot \vec{e}_z = 0, \vec{e}_y \cdot \vec{e}_z = 0,$$

since the length of the unit vectors is 1 and they are perpendicular to each other.

For two arbitrary vectors

$$\vec{a} = \begin{pmatrix} a_x \\ a_y \\ a_z \end{pmatrix}, \vec{b} = \begin{pmatrix} b_x \\ b_y \\ b_z \end{pmatrix}$$

the dot product can be expressed by the components of the vectors:

$$\vec{a} \cdot \vec{b} = (a_x \vec{e}_x + a_y \vec{e}_y + a_z \vec{e}_z) \cdot (b_x \vec{e}_x + b_y \vec{e}_y + b_z \vec{e}_z)$$

With the distributive law we multiply out the two brackets, use the dot products of the unit vectors and obtain in turn

$$a_x \vec{e}_x (b_x \vec{e}_x + b_y \vec{e}_y + b_z \vec{e}_z) = a_x b_x \underbrace{\vec{e}_x \cdot \vec{e}_x}_{=1} + a_x b_y \underbrace{\vec{e}_x \cdot \vec{e}_y}_{=0} + a_x b_z \underbrace{\vec{e}_x \cdot \vec{e}_z}_{=0},$$

$$a_y \vec{e}_y (b_x \vec{e}_x + b_y \vec{e}_y + b_z \vec{e}_z) = a_y b_x \underbrace{\vec{e}_y \cdot \vec{e}_x}_{=0} + a_y b_y \underbrace{\vec{e}_y \cdot \vec{e}_y}_{=1} + a_y b_z \underbrace{\vec{e}_y \cdot \vec{e}_z}_{=0},$$

$$a_z \vec{e}_z (b_x \vec{e}_x + b_y \vec{e}_y + b_z \vec{e}_z) = a_z b_x \underbrace{\vec{e}_z \cdot \vec{e}_x}_{=0} + a_z b_y \underbrace{\vec{e}_z \cdot \vec{e}_y}_{=0} + a_z b_z \underbrace{\vec{e}_z \cdot \vec{e}_z}_{=1}.$$

So it follows in total

$$\vec{a} \cdot \vec{b} = a_x b_x + a_y b_y + a_z b_z. \qquad \Box \tag{4.4}$$

Let's return to the consideration of the point mass, which moves under the influence of a force $\vec{F}$, along a curve $\vec{r}(t)$ from a point P_1 to a point P_2. Using the scalar product, the work done by a force $\vec{F}$ during a small displacement $\Delta \vec{r}_i$ along the curve can be written as

$$\Delta W_i = F \cos \varphi \Delta r_i = \vec{F} \cdot \Delta \vec{r}_i.$$

To obtain the total work W, all ΔW_i are summed up and the displacement converges to zero or n to ∞:

$$W = \lim_{\Delta r_i \to 0, n \to \infty} \sum_{i=1}^{n} \Delta W_i = \lim_{\Delta r_i \to 0, n \to \infty} \sum_{i=1}^{n} \vec{F} \cdot \Delta \vec{r}_i.$$

This limit is referred to as the **line integral** (or **curve integral of $\vec{F}$ along the curve $\vec{r}(t)$** and is written:

$$W = \lim_{\Delta r_i \to 0, n \to \infty} \sum_{i=1}^{n} \vec{F} \cdot \Delta \vec{r}_i = \int_{P_1}^{P_2} \vec{F} \cdot d\vec{r} \tag{4.5}$$

In order to be able to calculate such a line integral concretely, we need some new mathematical tools.

Remark 4.3. **MT: Scalar and Vector Fields**
So far, we have considered functions whose variable (x or t) is a number. But there are also functions whose values depend, for example, on the location where these values are determined. The temperature, for example, is such a function, it provides (usually different) values for points in space. Since the location $\vec{r}$ is determined by its coordinates x, y, z, a function that depends on the location, i.e., on several variables (x, y, z), is called **multivariable**. A multivariable function

$$f = f(\vec{r}) = f(x, y, z),$$

whose values are *numbers*, is called a **scalar field**. If the function is *vector-valued* and multivariable, it is called a **vector field** and is written as follows

$$\vec{f} = \vec{f}(\vec{r}) = \vec{f}(x, y, z) = \begin{pmatrix} f_1(x, y, z) \\ f_2(x, y, z) \\ f_3(x, y, z) \end{pmatrix}, \tag{4.6}$$

where the f_1, f_2, f_3 are scalar fields and the **component functions** of $\vec{f}$. To return to our physical problem: In calculating the work done by a force along a path, we have not yet explicitly pointed out that the force can also depend on the position

$$\vec{r} = \begin{pmatrix} x \\ y \\ z \end{pmatrix}$$

of the particle. The function $\vec{F}$ is thus a vector field

$$\vec{F}(\vec{r}) = \vec{F}(x, y, z) = \begin{pmatrix} F_x(x, y, z) \\ F_y(x, y, z) \\ F_z(x, y, z) \end{pmatrix}$$

with the scalar fields F_x (= force in x-direction), F_y (= force in y-direction) and F_z (= force in z-direction). $\square$

Remark 4.4. **MT: Differential**

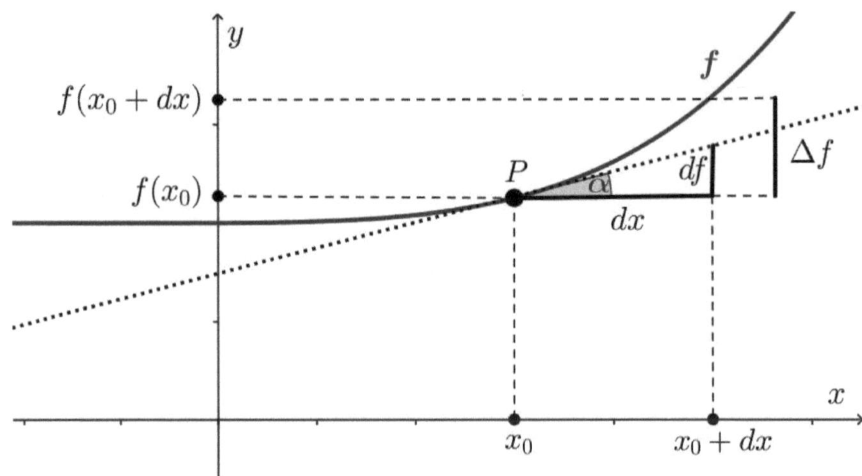

Figure 4.5.: Differential of a function

To determine the increase of a function $f(x)$ in a neighborhood of a point x_0, we calculate the function values at the position x_0 and at $x_0 + dx$. If the x-value changes by dx, then the function value changes by

$$\Delta f = f(x_0 + dx) - f(x_0).$$

The change of the tangent through the point $P = (x_0, f(x_0))$ between x_0 and $x_0 + dx$ is marked with df in Fig. 4.5 and is referred to as the **differential** of f. df is the change of the y-value, when one proceeds from $(x_0, f(x_0))$ along the local tangent by dx in x-direction. Because

$$\tan \alpha = f'(x_0) = \frac{df}{dx}$$

it follows

$$df = f'(x_0)\, dx. \tag{4.7}$$

For small dx it approximately holds

$$\Delta f = f(x_0 + dx) - f(x_0) \approx f'(x_0)\, dx.$$

This estimation is used in this book especially for simplifying complicated matters.

We want to extend the concept of differentials to vector-valued functions $\vec{r}(t)$. The first derivative of $\vec{r}(t)$ with respect to time was defined by

$$\dot{\vec{r}}(t) = \lim_{\Delta t \to 0} \frac{\Delta \vec{r}}{\Delta t} = \frac{d\vec{r}}{dt},$$

i.e.

$$d\vec{r} = \dot{\vec{r}}(t) \, dt, \tag{4.8}$$

and for small dt it approximately holds

$$\Delta \vec{r} \approx \dot{\vec{r}}(t) \, dt,$$

and as above, the quantity $d\vec{r}$ is called the **differential** of $\vec{r}(t)$. $\square$

Remark 4.5. **MT: Line integral**
To calculate a line integral between two points P_1 and P_2 concretely, the curve $\vec{r}$, along which one integrates, must be given in the form

$$\vec{r}(t) = \begin{pmatrix} x(t) \\ y(t) \\ z(t) \end{pmatrix}.$$

The points are defined by $P_1 = \vec{r}(t_a)$ and $P_2 = \vec{r}(t_e)$, where t_a and t_e denote the start and end times of the displacement. The force $\vec{F}$, which can vary from place to place, is given by

$$\vec{F}(\vec{r}(t)) = \begin{pmatrix} F_x(\vec{r}(t)) \\ F_y(\vec{r}(t)) \\ F_z(\vec{r}(t)) \end{pmatrix}.$$

The line integral can be rewritten with the differential (4.8):

$$\int_{P_1}^{P_2} \vec{F} \cdot d\vec{r} = \int_{t_a}^{t_e} \vec{F}(\vec{r}(t)) \cdot \dot{\vec{r}}(t) \, dt \tag{4.9}$$

This means that the line integral is now an ordinary definite integral with the integral limits t_a and t_e. In component notation the scalar product is

$$\vec{F} \cdot \dot{\vec{r}}(t) = F_x \, \dot{x}(t) + F_y \, \dot{y}(t) + F_z \, \dot{z}(t),$$

so that overall it follows

$$W = \int_{P_1}^{P_2} \vec{F} \cdot d\vec{r} = \int_{t_a}^{t_e} \vec{F} \cdot \dot{\vec{r}}(t) \, dt = \int_{t_a}^{t_e} \left(F_x \, \dot{x}(t) + F_y \, \dot{y}(t) + F_z \, \dot{z}(t) \right) dt. \, \square$$

For the following example we need some more derivative and integration rules.

Remark 4.6. **MF: Product Rule**
If $g(t)$ and $h(t)$ are two differentiable functions, then for the derivative of the product $g \cdot h$ of the two functions it follows

$$\frac{d}{dt}[g\,(t) \cdot h\,(t)] = h\,(t) \cdot \dot{g}\,(t) + g\,(t) \cdot \dot{h}\,(t) = h\,\frac{dg}{dt} + g\,\frac{dh}{dt}.\;\square$$

Remark 4.7. **MF: Derivation and Integration of Functions of the Form** $\mathbf{r\,(t) = t^a}$
As an example, we apply the product rule to the function $f\,(t) = t^3$. We already know that

$$\frac{d}{dt}\,(t) = 1 \;\; and \;\; \frac{d}{dt}\,(t^2) = 2t$$

so it follows:

$$\frac{d}{dt}\,(t^3) = \frac{d}{dt}\,(t \cdot t^2) \overset{product\,rule}{=} t\,\frac{d\,t^2}{dt} + t^2\frac{d\,t}{dt} = t \cdot 2t + t^2 \cdot 1 = 3t^2$$

From this we get for $f\,(t) = t^4$:

$$\frac{d}{dt}\,(t^4) = \frac{d}{dt}\,(t \cdot t^3) \overset{product\,rule}{=} t\,\frac{d\,t^3}{dt} + t^3\frac{d\,t}{dt} = t \cdot 3t^3 + t^3 \cdot 1 = 4t^3$$

In general, for a function $r\,(t) = t^a$, where a can be any real number

$$\dot{r}\,(t) = \frac{d}{dt}\,(t^a) = at^{a-1} \;\; and \;\; \int r\,(t)\,dt = \int t^a\,dt = \frac{1}{a+1}\,t^{a+1}. \qquad (4.10)$$

So when differentiating, the exponent a becomes a factor and at the same time the remaining exponent is reduced by 1; the second formula results from the fact that integration is the reverse of differentiation: If you differentiate the right-hand side, you get $r\,(t) = t^a$ again. $\square$

Example 4.2.
As an application example, we consider the force field

$$\vec{F}\,(\vec{r}) = \vec{F}\,(x,y,z) = \begin{pmatrix} 2x^2 - 3y \\ 4yz \\ 3x^2 z \end{pmatrix}$$

and calculate the work along two different paths $\vec{r}_1$ and $\vec{r}_2$ between the points $P_1 = (0, 0, 0)$ and $P_2 = (1, 1, 1)$:

$$\vec{r}_1(t) = \begin{pmatrix} t \\ t \\ t \end{pmatrix}, 0 \le t \le 1,$$

$$\vec{r}_2(t) = \begin{pmatrix} t \\ t^2 \\ t^3 \end{pmatrix}, 0 \le t \le 1,$$

$\vec{r}_1$ is the straight line between P_1 and P_2. It holds

$$\dot{\vec{r}}_1(t) = \begin{pmatrix} 1 \\ 1 \\ 1 \end{pmatrix} \quad \text{and} \quad \dot{\vec{r}}_2(t) = \begin{pmatrix} 1 \\ 2t \\ 3t^2 \end{pmatrix}.$$

It is

$$\vec{F}(\vec{r}_1(t)) = \begin{pmatrix} 2t^2 - 3t \\ 4t \cdot t \\ 3t^2 \cdot t \end{pmatrix} = \begin{pmatrix} 2t^2 - 3t \\ 4t^2 \\ 3t^3 \end{pmatrix}$$

and

$$\vec{F}(\vec{r}_2(t)) = \begin{pmatrix} 2t^2 - 3t^2 \\ 4t^2 \cdot t^3 \\ 3t^2 \cdot t^3 \end{pmatrix} = \begin{pmatrix} -t^2 \\ 4t^5 \\ 3t^5 \end{pmatrix}.$$

This results in

$$\vec{F}(\vec{r}_1(t)) \cdot \dot{\vec{r}}_1(t) = \begin{pmatrix} 2t^2 - 3t \\ 4t^2 \\ 3t^3 \end{pmatrix} \cdot \begin{pmatrix} 1 \\ 1 \\ 1 \end{pmatrix} = 2t^2 - 3t + 4t^2 + 3t^3 = 3t^3 + 6t^2 - 3t$$

and

$$\vec{F}(\vec{r}_2(t)) \cdot \dot{\vec{r}}_2(t) = \begin{pmatrix} -t^2 \\ 4t^5 \\ 3t^5 \end{pmatrix} \cdot \begin{pmatrix} 1 \\ 2t \\ 3t^2 \end{pmatrix} = -t^2 + 8t^6 + 9t^7.$$

With this, we can calculate the work done along both paths:

$$W_{r_1} = \int_0^1 (3t^3 + 6t^2 - 3t)\, dt = \left[3\frac{1}{4}t^4 + 6\frac{1}{3}t^3 - 3\frac{1}{2}t^2 \right]_0^1 = \frac{3}{4} + \frac{6}{3} - \frac{3}{2} = \frac{5}{4},$$

$$\begin{aligned} W_{r_2} &= \int_0^1 (-t^2 + 8t^6 + 9t^7)\, dt = \left[(-1)\frac{1}{3}t^3 + 8\frac{1}{7}t^7 + 9\frac{1}{8}t^8 \right]_0^1 \\ &= -\frac{1}{3} + \frac{8}{7} + \frac{9}{8} = \frac{325}{168} \end{aligned}$$

In this example, the work is therefore dependent on the path! $\square$

Work Equals Difference in Kinetic Energy

We now want to derive that in the general case, work is equal to the difference in kinetic energy. To do this, we apply the product rule to the function $(v(t))^2 = v^2(t)$ using the shorthand notation $(v(t))^2 = v^2(t)$, where

$$\vec{v}(t) = \dot{\vec{r}}(t) = \begin{pmatrix} \dot{x}(t) \\ \dot{y}(t) \\ \dot{z}(t) \end{pmatrix}.$$

It is

$$\frac{d}{dt} v^2(t) = \frac{d}{dt}(\vec{v}(t) \cdot \vec{v}(t)) = \frac{d}{dt}(\dot{x}(t) \cdot \dot{x}(t) + \dot{y}(t) \cdot \dot{y}(t) + \dot{z}(t) \cdot \dot{z}(t)).$$

Application of the product rule to each summand yields:

$$\begin{aligned}
\frac{d}{dt} v^2(t) &= 2(\dot{x}(t) \cdot \ddot{x}(t) + \dot{y}(t) \cdot \ddot{y}(t) + \dot{z}(t) \cdot \ddot{z}(t)) \\
&= 2 \begin{pmatrix} \dot{x}(t) \\ \dot{y}(t) \\ \dot{z}(t) \end{pmatrix} \cdot \begin{pmatrix} \ddot{x}(t) \\ \ddot{y}(t) \\ \ddot{z}(t) \end{pmatrix} = 2\dot{\vec{r}}(t) \cdot \ddot{\vec{r}}(t)
\end{aligned}$$

The general formula for work can be transformed using Newton's second law $\vec{F} = m\vec{a} = m\ddot{\vec{r}}$ to

$$W = \int_{P_1}^{P_2} \vec{F} \cdot d\vec{r} = \int_{t_a}^{t_e} \vec{F}(\vec{r}(t)) \cdot \dot{\vec{r}}(t)\, dt = \int_{t_a}^{t_e} m\ddot{\vec{r}}(t) \cdot \dot{\vec{r}}(t)\, dt.$$

The integrand $m\ddot{\vec{r}}(t) \cdot \dot{\vec{r}}(t)$ is, as just derived, the derivative of $\frac{1}{2}mv^2(t)$, so with the main theorem of differential and integral calculus, we get

$$W = \int_{t_a}^{t_e} m\ddot{\vec{r}}(t) \cdot \dot{\vec{r}}(t)\, dt = \frac{1}{2}mv^2(t_e) - \frac{1}{2}mv^2(t_a) = \Delta E_{kin}. \qquad (4.11)$$

We return to example 4.2, where we found that in general work depends on the path along which the force acts. We now define a **conservative force** by eliminating this path dependence:

> The work performed by a **conservative force** on a point mass is independent of the path the mass takes. It only depends on the starting and ending points of the path.

If $\vec{r}_1$ and $\vec{r}_2$ are two different paths between two points P_1 and P_2 and is $\vec{F}$ a conservative force, and considering that the sign of an integral reverses when the integral limits are swapped:

$$\int_a^b f(x)\, dx = -\int_b^a f(x)\, dx,$$

then, if W_1 denotes the work from P_1 to P_2 along $\vec{r}_1$ and W_2 the work from P_1 to P_2 along $\vec{r}_2$:

$$0 = W_1 - W_2 = \int_{P_1}^{P_2} \vec{F} \cdot d\vec{r}_1 - \int_{P_1}^{P_2} \vec{F} \cdot d\vec{r}_2 = \int_{P_1}^{P_2} \vec{F} \cdot d\vec{r}_1 + \int_{P_2}^{P_1} \vec{F} \cdot d\vec{r}_2,$$

i.e., a conservative force can also be defined by the fact that the total work along a closed path is zero, which is symbolically denoted with

$$\oint \vec{F} \cdot d\vec{r} = 0. \tag{4.12}$$

Example 4.3.
Homogeneous force fields with

$$\vec{F}(\vec{r}) = \begin{pmatrix} 0 \\ 0 \\ F_z \end{pmatrix}$$

with $F_z = const.$, i.e. for example the gravitational force in the vicinity of the Earth, are conservative, because $\vec{F} \cdot d\vec{r} = F_z\, dz$ implies:

$$\int_{P_1}^{P_2} \vec{F} \cdot d\vec{r} = \int_{z_1}^{z_2} F_z\, dz = -\int_{z_2}^{z_1} F_z\, dz,$$

thus

$$\oint \vec{F} \cdot d\vec{r} = \int_{z_1}^{z_2} F_z\, dz + \int_{z_2}^{z_1} F_z\, dz = 0 \square$$

4.4. Potential Energy

If you bring a stationary body from a fixed point P_0 to another point P in a *conservative* force field, the work expended (or gained) only depends on P, so it is a function of P. This function is called the **potential energy** $E_{pot}(P)$.

The change in potential energy corresponds in magnitude to the work done by the conservative force, its sign is opposite to that of the work:

$$\Delta E_{pot} = E_{pot}\left(P\right) - E_{pot}\left(P_0\right) = -W = -\int_{P_0}^{P} \vec{F} \cdot d\vec{r} \qquad (4.13)$$

The sign is chosen so that work done against the force $\vec{F}$ during a movement $(\vec{F} \cdot d\vec{r} < 0)$ is counted as negative. Energy must be supplied to the body, which leads to an increase in potential energy, while the work done *by* the body $(\vec{F} \cdot d\vec{r} > 0)$ reduces its potential energy and can thus be used for other systems (e.g., water that drives a turbine as it falls).

The zero point of potential energy is not determined by the definition (4.13) because only the difference ΔE_{pot} is defined. Two conventions are common:

1. One chooses $E_{pot} = 0$ for the zero point of the coordinate system; e.g. for experiments where gravity $\vec{F} = (0, 0, -mg)$ plays a role, one chooses $E_{pot}\left(P_0\right) = 0$ for $P_0 = (0, 0, 0)$.

2. In problems where the body can reach infinity, which is denoted by the symbol ∞, $E_{pot}(\infty) = 0$ is set. This results in

$$\int_{P}^{\infty} \vec{F} \cdot d\vec{r} = -\int_{\infty}^{P} \vec{F} \cdot d\vec{r} = E_{pot}\left(P\right) - E_{pot}\left(\infty\right) = E_{pot}\left(P\right),$$

   i.e., the potential energy at point P is then equal to the work that must be expended (or gained) when moving the mass point from P to infinity.

The work that must be expended is independent of the choice of the zero point of potential energy, as it is only determined by the difference ΔE_{pot}.

Example 4.4.
We calculate the potential energy for a body with mass m in a constant gravitational field $\vec{F} = (0, 0, -mg)$, which is lifted to height h. The required work is calculated with $P_1 = (0, 0, 0)$ and $P_2 = (0, 0, h)$ to

$$W = \int_{P_1}^{P_2} \vec{F} \cdot d\vec{r} = \int_{0}^{h} F_z \, dz = -\int_{0}^{h} mg \, dz = -mgh,$$

and thus

$$-W = mgh = E_{pot}\left(P_2\right) - E_{pot}\left(P_1\right).$$

If we set $E_{pot}(P_1) = 0$, then it follows

$$E_{pot}(0, 0, h) = mgh,$$

i.e., the work done on the body has led to an increase in its potential energy.
□

We can derive an important conclusion from the previous results. For a conservative force field $\vec{F}$, for the movement of a body from one point P_1 to P_2, on the one hand we have

$$W = \int_{P_1}^{P_2} \vec{F} \cdot d\vec{r} = E_{kin}(P_2) - E_{kin}(P_1)$$

and on the other hand

$$W = \int_{P_1}^{P_2} \vec{F} \cdot d\vec{r} = E_{pot}(P_1) - E_{pot}(P_2),$$

from which

$$E_{kin}(P_2) + E_{pot}(P_2) = E_{kin}(P_1) + E_{pot}(P_1)$$

follows. Since the two points were chosen arbitrarily, the following statement generally applies:

Energy theorem of mechanics: In a conservative force field the sum of kinetic and potential energy of a particle is constant at any point P. This constant sum is called **total mechanical energy** E. The total energy

$$E = E_{kin} + E_{pot}$$

is conserved in conservative forces.

Example 4.5.
As an example, we again consider the free fall of a body of mass m, which is dropped from height h with initial velocity $v(h) = 0$. Then, as derived above, with the specification $E_{pot}(0) = 0$ for any $y \leq h$:

$$E_{pot}(y) = -\int_0^y -mg \, dy = mgy$$

The fall time, which the body needs to reach height y, i.e., for the distance $h - y$, is obtained with

$$h - y = \frac{1}{2} g t^2$$

to

$$t = \sqrt{\frac{2\left(h - y\right)}{g}}.$$

So, the velocity v at y follows:

$$v\left(y\right) = g \cdot t = g\sqrt{\frac{2\left(h - y\right)}{g}} = \sqrt{2g\left(h - y\right)}$$

and for the kinetic energy

$$E_{kin}\left(y\right) = \frac{m}{2}v^2\left(y\right) = \frac{m}{2}\left(\sqrt{2g\left(h - y\right)}\right)^2 = mg\left(h - y\right).$$

The sum

$$E_{pot}\left(y\right) + E_{kin}\left(y\right) = mgh$$

is therefore independent of y and equal to the total energy. $\square$

To illuminate the connection between a conservative force and potential energy more precisely, we need a bit more mathematics.

Remark 4.8. **MT: Partial Derivative and Gradient**
The potential energy $E_{pot}(P)$ changes its value depending on which point in space $P = (x, y, z)$ it is measured. It is therefore dependent on the three variables x, y, z and is numerical. For such a scalar field $f = f\left(x, y, z\right)$ it is often interesting to see how it changes per unit length when only one variable, e.g. x, is changed, but the other two variables are left unchanged. The expression

$$\lim_{\Delta x \to 0} \frac{f\left(x + \Delta x, y, z\right) - f\left(x, y, z\right)}{\Delta x} = \frac{\partial f\left(x, y, z\right)}{\partial x} \tag{4.14}$$

is called the **partial derivative** of f with respect to x at the point (x, y, z), provided of course, that the limit exists. If the function f is partially differentiable with respect to x for all points (x, y, z), one also writes briefly

$$\frac{\partial f}{\partial x} \quad \text{for} \quad \frac{\partial f\left(x, y, z\right)}{\partial x}.$$

The partial derivative with respect to x is thus formed in the same way as the ordinary one, if one considers the other two variables as constants. f becomes a function that depends only on the variable x by keeping y and z constant. The notation $\partial f / \partial x$ instead of the "normal" df/dx should make clear, however, that the function f actually depends on several variables. The partial derivatives of f with respect to y and z, namely $\partial f / \partial y$ and $\partial f / \partial z$ are defined analogously.
$\square$

Example 4.6.
As an example, we calculate the partial derivatives of the function

$$f(x, y, z) = x^2 + y^2 + z^2 :$$

$$\frac{\partial f}{\partial x} = \frac{\partial}{\partial x}\left(x^2 + y^2 + z^2\right) = 2x + 0 + 0 = 2x, \ \textit{since } y, z \textit{ constant},$$

$$\frac{\partial f}{\partial y} = \frac{\partial}{\partial y}\left(x^2 + y^2 + z^2\right) = 0 + 2y + 0 = 2y, \ \textit{since } x, z \textit{ constant},$$

$$\frac{\partial f}{\partial z} = \frac{\partial}{\partial z}\left(x^2 + y^2 + z^2\right) = 0 + 0 + 2z = 2z, \ \textit{since } x, y \textit{ constant}\ \square$$

The vector field

$$grad\, f(x, y, z) = \begin{pmatrix} \partial f(x, y, z)/\partial x \\ \partial f(x, y, z)/\partial y \\ \partial f(x, y, z)/\partial z \end{pmatrix}, \tag{4.15}$$

whose component functions form the partial derivatives of $f(x, y, z)$, is called the **gradient** of f. The symbol **Nabla** ∇, an inverted Delta, is also used as a shorthand for the gradient

$$grad\, f = \nabla f.$$

The gradient is the multivariable counterpart to the 1st derivative. $\square$

Remark 4.9. **MT: Differential of Scalar Fields**
If one wants to determine the (infinitesimal) change of a scalar field $f(\vec{r})$ when moving from a point $P_1(\vec{r}) = P_1(x, y, z)$ in the direction

$$d\vec{r} = \begin{pmatrix} dx \\ dy \\ dz \end{pmatrix}$$

to the neighboring point

$$P_2(\vec{r} + d\vec{r}) = P_2(x + dx, y + dy, z + dz)$$

for small displacements $d\vec{r}$, we get

$$f(\vec{r} + d\vec{r}) - f(\vec{r}) \approx df(\vec{r}) = \nabla f(\vec{r}) \cdot d\vec{r} = \frac{\partial f(\vec{r})}{\partial x}\, dx + \frac{\partial f(\vec{r})}{\partial y}\, dy + \frac{\partial f(\vec{r})}{\partial z}\, dz.$$
$$\tag{4.16}$$

As in the one-dimensional case (4.8), $df(\vec{r})$ is called the **differential** of f at the point $\vec{r}$. In future applications, we will always write the equality sign in the above approximation for simplicity when we assume that the displacement $d\vec{r}$ is infinitesimal, i.e.

$$f(\vec{r} + d\vec{r}) - f(\vec{r}) = df(\vec{r}) = \nabla f(\vec{r}) \cdot d\vec{r}.\ \square$$

We can now proceed with the differential and from the defining equation (4.13) for the potential energy

$$W = \int_{P_1}^{P_2} \vec{F} \cdot d\vec{r} = -\Delta E_{pot} = -\left(E_{pot}\left(P_2\right) - E_{pot}\left(P_1\right)\right)$$

derive that the (infinitesimal) work between *neighboring* points P_1 and P_2 can be calculated by

$$dW = -\left(E_{pot}\left(P_2\right) - E_{pot}\left(P_1\right)\right) = -dE_{pot}\left(P_1\right)$$

On the other hand, for the (infinitesimal) work done by the force $\vec{F}$ between P_1 and P_2,

$$dW = \vec{F}\left(P_1\right) \cdot d\vec{r}$$

and thus

$$-dE_{pot} = \vec{F} \cdot d\vec{r}.$$

Now

$$\vec{F} \cdot d\vec{r} = F_x\, dx + F_y\, dy + F_z\, dz$$

and

$$-dE_{pot} = -\frac{\partial E_{pot}}{\partial x}\, dx - \frac{\partial E_{pot}}{\partial y}\, dy - \frac{\partial E_{pot}}{\partial z}\, dz,$$

from which, due to the arbitrarily selectable paths dx, dy, dz, it follows that

$$
\begin{aligned}
F_x &= -\frac{\partial E_{pot}}{\partial x} \\
F_y &= -\frac{\partial E_{pot}}{\partial y} \\
F_z &= -\frac{\partial E_{pot}}{\partial z}
\end{aligned}
$$

holds, i.e. in summary

$$\vec{F} = -\nabla E_{pot}. \tag{4.17}$$

We have thus shown that in a conservative force field the force $\vec{F}$ can always be represented as the (negative) gradient of the potential function E_{pot}. But the reverse also holds.

> If a force $\vec{F}$ can be represented as the gradient of a scalar function $\phi\left(x, y, z\right)$, i.e.
> $$\vec{F} = \nabla\phi, \tag{4.18}$$
> then the force field generated by $\vec{F}$ is conservative.

To prove this, we need another derivative rule.

Remark 4.10. **MF: Chain Rule, Inverse Function**
If $f = f(x)$ is a function, the first derivative of f with respect to the variable x is usually written as

$$f'(x) = \frac{df}{dx},$$

while the notation

$$\dot{f}(t) = \frac{df}{dt}$$

is chosen for the first derivative when the function depends on the time variable t.

To derive the chain rule, we consider the function $f(x) = x^3$. If

$$x = x(t) = t^2 + t + 5$$

is itself a function that depends on the time variable t, then

$$f(x(t)) = f(t^2 + t + 5) = \left(t^2 + t + 5\right)^3.$$

The function $f(x(t))$ is called a **composite function** and is also written as

$$f \circ x(t) = f(x(t)).$$

The 1st derivative of a composite function $f \circ x(t)$ with respect to time can be calculated according to the **chain rule** as follows:

$$\overbrace{f \circ x(t)}^{\cdot} = \frac{df(x(t))}{dt} = \frac{df(x(t))}{dx} \cdot \frac{dx}{dt} = f'(x(t)) \cdot \dot{x}(t). \tag{4.19}$$

So, we differentiate the **outer function** f with respect to x and evaluate this derivative at the point $x(t)$. Then we multiply the term by the derivative of the **interior function** x at the point t. For the above example, with $f'(x) = 3x^2$ and $\dot{x}(t) = 2t + 1$ we get the following result:

$$\overbrace{f \circ x(t)}^{\cdot} = \frac{d}{dt}\left(\left(t^2 + t + 5\right)^3\right) = f'(x(t)) \cdot \dot{x}(t) = 3\left(t^2 + t + 5\right)^2 \cdot (2t + 1)$$

We use the chain rule to obtain a useful formula for the **inverse function**. The inverse mapping f^{-1} of f is, if it exists, defined by the two equations

$$\begin{aligned} f^{-1} \circ f(x) &= x \\ f \circ f^{-1}(y) &= y. \end{aligned} \tag{4.20}$$

The inverse mapping f^{-1} thus undoes everything that the function f has done to the variable x. Here is a simple example

$$f(x) = 2x + 1 \Rightarrow f^{-1}(x) = \frac{x - 1}{2},$$

because:

$$f^{-1} \circ f\,(x) = f^{-1}(2x+1) = \frac{(2x+1)-1}{2} = x$$

If we differentiate the first equation in (4.20) using the chain rule, we get

$$1 = x' = \left(f^{-1} \circ f\,(x)\right)' = \left(f^{-1}\right)'\,(f\,(x)) \cdot f'\,(x)\,,$$

from which, with $y = f\,(x)$ and $x = f^{-1}\,(y)$, the formula for the **derivative of the inverse function** follows

$$\left(f^{-1}\right)'\,(y) = \frac{1}{f'\,(x)} = \frac{1}{f'\,(f^{-1}\,(y))}. \tag{4.21}$$

We generalize the chain rule to the case where the function

$$f = f\,(\vec{r}) = f\,(x,y,z)$$

can depend on the three spatial variables and each spatial variable, in turn, depends on time

$$\vec{r}\,(t) = \begin{pmatrix} x\,(t) \\ y\,(t) \\ z\,(t) \end{pmatrix}.$$

Then, the **multivariable chain rule** applies analogously

$$\frac{d\,(f \circ \vec{r}\,(t))}{dt} = f'\,(\vec{r}\,(t)) \cdot \dot{\vec{r}}\,(t)\,. \tag{4.22}$$

Here, $f'\,(\vec{r}\,(t))$ is the first derivative of f at the point $f'\,(\vec{r}\,(t))$, so

$$f'\,(\vec{r}\,(t)) = \begin{pmatrix} \partial f\,(\vec{r}\,(t))\,/\partial x \\ \partial f\,(\vec{r}\,(t))\,/\partial y \\ \partial f\,(\vec{r}\,(t))\,/\partial z \end{pmatrix} = grad\, f(\vec{r}\,(t)) = \nabla f(\vec{r}\,(t))$$

and

$$\dot{\vec{r}}\,(t) = \begin{pmatrix} \dot{x}\,(t) \\ \dot{y}\,(t) \\ \dot{z}\,(t) \end{pmatrix} = \begin{pmatrix} dx/dt \\ dy/dt \\ dz/dt \end{pmatrix}.$$

On the right side of (4.22) there are two vectors that are scalarly multiplied, so it follows

$$\begin{aligned}
\frac{df\,(\vec{r}\,(t))}{dt} &= f'\,(\vec{r}\,(t)) \cdot \dot{\vec{r}}\,(t) = \begin{pmatrix} \partial f\,(\vec{r}\,(t))\,/\partial x \\ \partial f\,(\vec{r}\,(t))\,/\partial y \\ \partial f\,(\vec{r}\,(t))\,/\partial z \end{pmatrix} \cdot \begin{pmatrix} dx/dt \\ dy/dt \\ dz/dt \end{pmatrix} = \\
&= \frac{\partial f\,(\vec{r}\,(t))}{\partial x}\frac{dx}{dt} + \frac{\partial f\,(\vec{r}\,(t))}{\partial y}\frac{dy}{dt} + \frac{\partial f\,(\vec{r}\,(t))}{\partial z}\frac{dz}{dt}.
\end{aligned} \tag{4.23}$$

If

$$\vec{f} = \vec{f}(\vec{r}) = \vec{f}(x, y, z) = \begin{pmatrix} f_1(x, y, z) \\ f_2(x, y, z) \\ f_3(x, y, z) \end{pmatrix}$$

is a vector field (4.6) and the unknowns x, y, z are functions that depend on the variables x', y', z', then the following **multivariable vector-valued chain rule** applies to the component function f_1:

$$\frac{\partial f_1}{\partial x'} = \frac{\partial f_1}{\partial x}\frac{\partial x}{\partial x'} + \frac{\partial f_1}{\partial y}\frac{\partial y}{\partial x'} + \frac{\partial f_1}{\partial z}\frac{\partial z}{\partial x'}, \tag{4.24}$$

and analogously for the other variables and component functions. $\square$

We now show that from

$$\vec{F} = \nabla\phi$$

it follows that the force $\vec{F}$ is conservative. First, it holds

$$\vec{F} = \begin{pmatrix} F_x \\ F_y \\ F_z \end{pmatrix} = \nabla\phi = \begin{pmatrix} \partial\phi/\partial x \\ \partial\phi/\partial y \\ \partial\phi/\partial z \end{pmatrix}.$$

For the differential of ϕ it follows

$$d\phi = \frac{\partial\phi}{\partial x}\,dx + \frac{\partial\phi}{\partial y}\,dy + \frac{\partial\phi}{\partial z}\,dz = F_x\,dx + F_y\,dy + F_z\,dz.$$

Let P_1 and P_2 be two arbitrary points and $\vec{r}(t)$ a curve with $\vec{r}(t_a) = P_1$ and $\vec{r}(t_e) = P_2$, then as in Eq. (4.9) for the work done by the force $\vec{F}$ from P_1 to P_2,

$$\begin{aligned} W &= \int_{P_1}^{P_2} \vec{F}\cdot d\vec{r} = \int_{t_a}^{t_e} \vec{F}\cdot\dot{\vec{r}}(t)\,dt = \int_{t_a}^{t_e} \left(F_x\,\dot{x}(t) + F_y\,\dot{y}(t) + F_z\,\dot{z}(t)\right)dt \\ &= \int_{t_a}^{t_e} \left(\frac{\partial\phi}{\partial x}\,\dot{x}(t) + \frac{\partial\phi}{\partial y}\,\dot{y}(t) + \frac{\partial\phi}{\partial z}\,\dot{z}(t)\right)dt \\ &= \int_{t_a}^{t_e} \left(\frac{\partial\phi}{\partial x}\frac{dx}{dt} + \frac{\partial\phi}{\partial y}\frac{dy}{dt} + \frac{\partial\phi}{\partial z}\frac{dz}{dt}\right)dt \\ &\overset{(4.23)}{=} \int_{t_a}^{t_e} \frac{d\phi}{dt}\,dt. \end{aligned}$$

Taking into account that the function ϕ depends on the location, i.e. $\phi = \phi\left(\vec{r}\left(t\right)\right)$, it follows from the fundamental theorem of calculus

$$\int_{t_a}^{t_e} \frac{d\phi}{dt}\, dt = \int_{t_a}^{t_e} \frac{d\phi\left(\vec{r}\left(t\right)\right)}{dt}\, dt = \phi\left(\vec{r}\left(t_e\right)\right) - \phi\left(\vec{r}\left(t_a\right)\right) = \phi\left(P_2\right) - \phi\left(P_1\right),$$

so in total

$$W = \int_{P_1}^{P_2} \vec{F} \cdot d\vec{r} = \phi\left(P_2\right) - \phi\left(P_1\right).$$

The work done is therefore only dependent on the two points P_1 and P_2 and not on the path between the points. The force field is conservative.

5. Rotations

We first deal with **rotations** which are characterized by the fact that the axis of rotation is firmly anchored in space and perpendicular to the plane of motion. The quantities to be investigated, such as angular velocity, torque or angular momentum, take on a particularly simple form under these conditions, which show many analogies to linear motions („translations") in one dimension. In the second part, the requirement of a fixed spatial axis is dropped, and the statements and laws are extended to the general case. In conclusion, we compare the physical quantities of translations and rotations in a table.

5.1. Angular Velocity and Angular Acceleration

We consider in Fig. 5.1 a disc that is rotatably mounted in its center. The axis of rotation is perpendicular to the disc, looking out of the book page, and is marked by a small black ring.

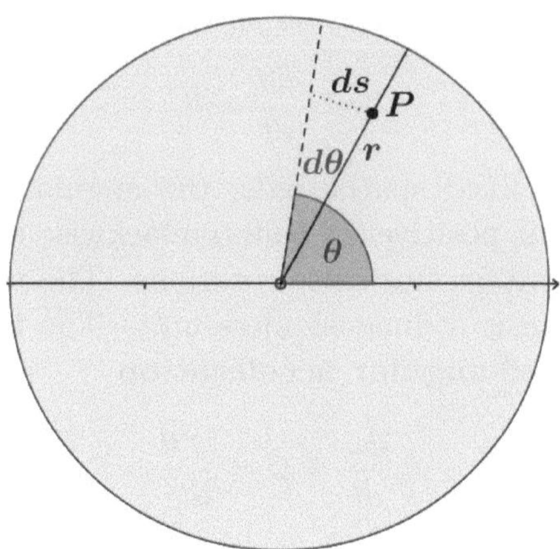

Figure 5.1.: Rotation angle

When the disc rotates, the points on the disc move at different speeds, a point on the edge rotates faster than a point near the axis. So it makes no sense to talk about the speed of the disc. However, when the disc has rotated

© The Author(s), under exclusive license to Springer-Verlag GmbH,
DE, part of Springer Nature 2026
M. Ruhrländer, *Ascent to the Einstein Equations*,
https://doi.org/10.1007/978-3-662-72672-3_5

completely once, all points on the disc have rotated once. A line between the axis of rotation and any point P moves by the same angle in a given time, regardless of the distance between the point and the axis of rotation.

We consider a particle of mass m at point P on the disc. P is defined by its distance r and the angle θ between a fixed reference axis (positive x-axis) and the line between the rotation point and the particle. During a short time interval dt, the particle moves along a circular arc by the distance ds, which is given by

$$ds = v\,dt,$$

where v is the magnitude of the velocity of the particle along the circular arc. During this time, the line between the rotation point and the particle sweeps the **rotation angle** $d\theta$. The angle $d\theta$ - measured in radians - is related to a full rotation (2π) as ds is to the circumference $(2\pi r)$:

$$\frac{d\theta}{2\pi} = \frac{ds}{2\pi r},$$

from which

$$d\theta = \frac{ds}{r}$$

follows. The **angular velocity** ω indicates how the angle $d\theta$ changes with time. It is the same for all particles on the disc

$$\omega = \frac{d\theta}{dt} = \dot{\theta}.$$

For rotations about a fixed spatial axis, the angular velocity has only two possible directions: It is positive for counterclockwise rotations (θ gets larger, so $d\theta > 0$) and negative for clockwise rotations. The unit of angular velocity is s^{-1}, since the radian is a dimensionless unit. The temporal change of the angular velocity is called **angular acceleration**

$$\alpha = \frac{d\omega}{dt} = \dot{\omega} = \frac{d^2\theta}{dt^2} = \ddot{\theta}.$$

The unit of angular acceleration is s^{-2}. One can link the **tangential velocity** ($=$ velocity along the circular arc) v of a particle on a disc with the angular velocity:

$$v = \frac{ds}{dt} = \frac{r\,d\theta}{dt} = r\omega$$

Similarly, there is a connection between the **tangential acceleration** a_T of a particle and the angular acceleration:

$$a_T = \frac{dv}{dt} = r\frac{d\omega}{dt} = r\alpha \tag{5.1}$$

In Eq. 2.11 on page 40, we calculated the magnitude of the centripetal acceleration $\vec{a}_Z$ directed towards the rotation point, which every particle on the disc experiences:

$$a_Z = \frac{v^2}{r}$$

This equation can be expressed with the angular velocity ω:

$$a_Z = \frac{v^2}{r} = \frac{(r\omega)^2}{r} = r\omega^2$$

There is an analogy between the angle of rotation, the angular velocity and the angular acceleration and the quantities displacement, velocity and acceleration in one-dimensional straight motion: If ω_0 and θ_0 are the initial values of angular velocity and position, then the rotational motions

$$
\begin{aligned}
\omega &= \omega_0 + \alpha t \\
\theta &= \theta_0 + \omega_0\, t + \frac{1}{2}\alpha t^2
\end{aligned}
$$

are analogous to the motion equations

$$
\begin{aligned}
v &= v_0 + at \\
x &= x_0 + v_0\, t + \frac{1}{2}at^2.
\end{aligned}
$$

5.2. Torque and Moment of Inertia

In Fig. 5.2, a force $\vec{F}$ acts at point P on a particle with mass m and sets the disc into rotation.

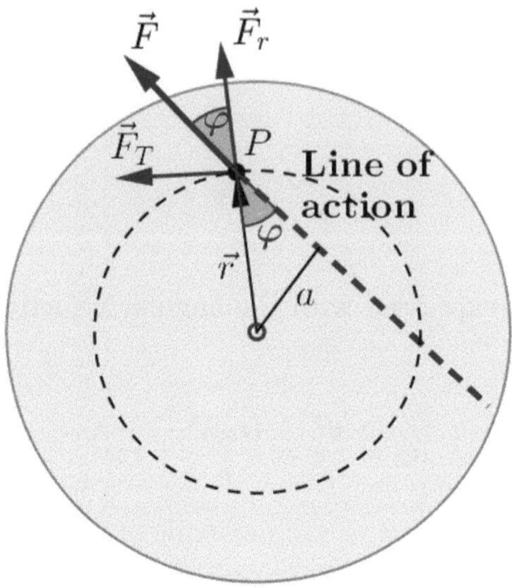

Figure 5.2.: Torque

The perpendicular distance between the line of action of the force (straight dashed line) and the axis of rotation, which should again go through the center of the disc, is called the **lever arm** a of the force. The product of force and lever arm is called **torque** (or **moment of force**) M. The torque determines the angular velocity of an object. The lever arm is calculated as

$$\frac{a}{r} = \sin\varphi \Rightarrow a = r\sin\varphi,$$

where φ is the angle between $\vec{F}$ and the position vector of the particle $\vec{r}$. For the torque M this results in

$$M = F \cdot a = F \cdot r \cdot \sin\varphi.$$

If the force is decomposed into the radial component $F_r = F \cdot \cos\varphi$ and the tangential component $F_T = F \cdot \sin\varphi$, then the radial component has no influence on the rotation of the disc. The torque can be written as a function of F_T:

$$M = F \cdot a = F \cdot r \cdot \sin\varphi = F_T \cdot r$$

The tangential component can be written according to the second law of Newton and with (5.1) as

$$F_T = ma_T = mr\alpha.$$

If both sides are multiplied by r, then one gets

$$M = F_T \cdot r = mr^2\alpha.$$

The expression mr^2 is called the **moment of inertia** I of the particle

$$I = mr^2.$$

The moment of inertia is a property of a body that depends on the axis of rotation (r is the distance of the particle from the axis of rotation) and represents a measure of the resistance that a body opposes to a change in its rotational motion. For the torque, therefore,

$$M = I \cdot \alpha.$$

This equation forms the analogue to the second law of Newton:

$$F = m \cdot a$$

If one has a rigid body with n particles (mass m_i, distance to the axis of rotation r_i), then the angular acceleration is the same for all particles. The resulting torque is obtained with the **extended moment of inertia**

$$I = m_1 r_1^2 + \cdots + m_n r_n^2 \tag{5.2}$$

to

$$M = m_1 r_1^2 \alpha + \cdots + m_n r_n^2 \alpha = \alpha \left(m_1 r_1^2 + \cdots + m_n r_n^2 \right) = I \cdot \alpha.$$

The kinetic energy of the particle's rotational motion can also be expressed by the moment of inertia:

$$E_{kin} = \frac{1}{2} mv^2 = \frac{1}{2} m(r\omega)^2 = \frac{1}{2} mr^2\omega^2 = \frac{1}{2} I\omega^2,$$

which corresponds to the equation

$$E_{kin} = \frac{1}{2} mv^2$$

for linear motion.

5.3. Angular Momentum

The **angular momentum** L of a particle with mass m, moving on a circular path with radius r at angular velocity ω, is defined as the product of momentum and radius:

$$L = pr = mvr = m(\omega r)r = mr^2\omega = I\omega,$$

where we again assume that the axis of rotation is perpendicular to the plane of motion. For arbitrary motions, the angular momentum of a particle relative to the origin is defined as

$$L = mvr_\perp = mvr \sin \theta.$$

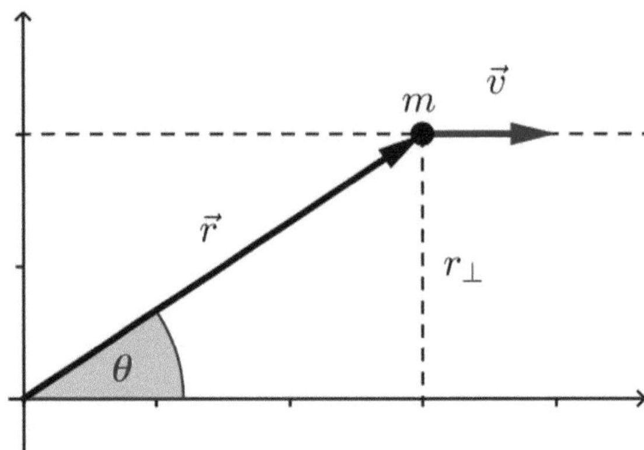

Figure 5.3.: Angular momentum

In Fig. 5.3, $\vec{v}$ is the particle velocity and $r_\perp = r \sin \theta$ is the perpendicular component of the position vector $\vec{r}$. Thus, the particle has *relative to the point* 0 an angular momentum, even though it is not moving on a circular path, but on a straight line at a distance $r_\perp$.

5.4. The General Case of Rotational Motions

We now want to drop the assumption of a fixed axis of rotation in space and generalize the statements about rotational motions. To represent the *vectorial* quantities associated with rotational motions, we need another mathematical concept.

Remark 5.1. **MT: Cross or Vector Product**
The **cross** or **vector product** of two vectors $\vec{a}$ and $\vec{b}$ is defined as *vector* $\vec{c} = \vec{a} \times \vec{b}$ with the following properties.

1. The magnitude of $\vec{c}$ is equal to the area of the parallelogram spanned by the vectors $\vec{a}$ and $\vec{b}$.

2. The vector $\vec{c}$ is perpendicular to both $\vec{a}$ and $\vec{b}$.

3. The vectors $\vec{a}$, $\vec{b}$ and $\vec{c}$ form a right-handed system.

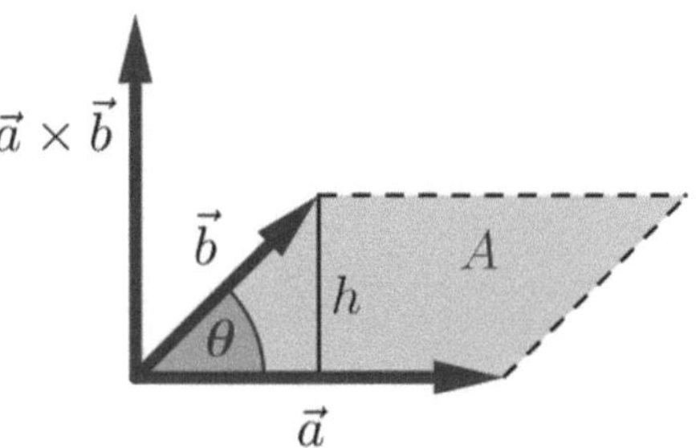

Figure 5.4.: Vector produkt

A right-handed system means that if I want to reach vector $\vec{b}$ from vector $\vec{a}$ by the shortest path, I have to rotate counterclockwise, and to reach vector $\vec{c} = \vec{a} \times \vec{b}$ from vector $\vec{b}$ by the shortest path, I again have to rotate counterclockwise. Or explained with another image: If you imagine that a slot is attached under the vector $\vec{a} \times \vec{b}$ into which I can insert a screwdriver to rotate vector $\vec{a}$ in the shortest way towards $\vec{b}$, I have to turn the screwdriver counterclockwise, etc.

The area A of the parallelogram in Fig. 5.4 is base a times height h:

$$A = a \cdot h = a \cdot b \sin \angle \left(\vec{a}, \vec{b} \right) = a \cdot b \sin \theta$$

So it follows

$$\left| \vec{a} \times \vec{b} \right| = A = a \cdot b \sin \theta.$$

From the definition, some calculation rules of the cross product can be derived.

1. $\vec{a} \times (\vec{b} + \vec{c}) = \vec{a} \times \vec{b} + \vec{a} \times \vec{c}$ and $(\vec{a} + \vec{b}) \times \vec{c} = (\vec{a} \times \vec{c}) + \left(\vec{b} \times \vec{c} \right)$ (**Distributive laws**)

2. $\vec{a} \times \vec{b} = -\vec{b} \times \vec{a}$ (**Antisymmetry law**)

3. $\lambda \cdot (\vec{a} \times \vec{b}) = (\lambda \cdot \vec{a}) \times \vec{b} = \vec{a} \times (\lambda \cdot \vec{b})$ (**Multiplication with scalar λ**)

If the two vectors $\vec{a}$ and $\vec{b}$ are parallel, i.e., there is a number k such that $\vec{a} = k \cdot \vec{b}$, then it follows

$$\vec{a} \times \vec{b} = \vec{0}, \text{ specifically } \vec{a} \times \vec{a} = \vec{0}, \tag{5.3}$$

because the area of the parallelogram spanned by the two vectors is equal to zero.

Next, we want to express the cross product by the components of the vectors and first calculate the cross products of the unit vectors $\vec{e}_x, \vec{e}_y, \vec{e}_z$. We note that the cross product of a vector with itself yields zero, that the unit vectors are perpendicular to each other, that the parallelogram spanned by two unit vectors is a square with area 1 and that $\vec{e}_x, \vec{e}_y, \vec{e}_z$ form a right-handed system. Then it follows:

$$
\begin{aligned}
\vec{e}_x \times \vec{e}_x &= 0 \\
\vec{e}_y \times \vec{e}_y &= 0 \\
\vec{e}_z \times \vec{e}_z &= 0 \\
\vec{e}_x \times \vec{e}_y &= \vec{e}_z \\
\vec{e}_y \times \vec{e}_z &= \vec{e}_x \\
\vec{e}_z \times \vec{e}_x &= \vec{e}_y
\end{aligned}
$$

If now

$$\vec{a} = a_x\,\vec{e}_x + a_y\,\vec{e}_y + a_z\,\vec{e}_z$$

and

$$\vec{b} = b_x\,\vec{e}_x + b_y\,\vec{e}_y + b_z\,\vec{e}_z$$

are two arbitrary vectors, then it follows with the distributive law

$$
\begin{aligned}
\vec{a} \times \vec{b} &= (a_x\,\vec{e}_x + a_y\,\vec{e}_y + a_z\,\vec{e}_z) \times (b_x\,\vec{e}_x + b_y\,\vec{e}_y + b_z\,\vec{e}_z) \\
&= a_x\,b_x\,\underbrace{(\vec{e}_x \times \vec{e}_x)}_{=0} + a_x\,b_y\,\underbrace{(\vec{e}_x \times \vec{e}_y)}_{=\vec{e}_z} + a_x\,b_z\,\underbrace{(\vec{e}_x \times \vec{e}_z)}_{=-\vec{e}_y} \\
&+ a_y\,b_x\,\underbrace{(\vec{e}_y \times \vec{e}_x)}_{=-\vec{e}_z} + a_y\,b_y\,\underbrace{(\vec{e}_y \times \vec{e}_y)}_{=0} + a_y\,b_z\,\underbrace{(\vec{e}_y \times \vec{e}_z)}_{=\vec{e}_x} \\
&+ a_z\,b_x\,\underbrace{(\vec{e}_z \times \vec{e}_x)}_{=\vec{e}_y} + a_z\,b_y\,\underbrace{(\vec{e}_z \times \vec{e}_y)}_{=-\vec{e}_x} + a_z\,b_z\,\underbrace{(\vec{e}_z \times \vec{e}_z)}_{=0} \\
&= (a_y\,b_z - a_z\,b_y)\,\vec{e}_x + (a_z\,b_x - a_x\,b_z)\,\vec{e}_y + (a_x\,b_y - a_y\,b_x)\,\vec{e}_z,
\end{aligned}
$$

i.e.

$$
\vec{a} \times \vec{b} = \begin{pmatrix} a_x \\ a_y \\ a_z \end{pmatrix} \times \begin{pmatrix} b_x \\ b_y \\ b_z \end{pmatrix} = \begin{pmatrix} a_y\,b_z - a_z\,b_y \\ a_z\,b_x - a_x\,b_z \\ a_x\,b_y - a_y\,b_x \end{pmatrix}.
$$

If $\vec{a}\,(t)$ and $\vec{b}\,(t)$ are vector-valued functions that depend on the variable t, then with the product rule for derivatives

$$
\frac{d}{dt}\left(\vec{a}\,(t) \times \vec{b}\,(t)\right) = \frac{d\vec{a}\,(t)}{dt} \times \vec{b}\,(t) + \vec{a}\,(t) \times \frac{d\vec{b}\,(t)}{dt}, \tag{5.4}
$$

so a similar product rule as in the multiplication of ordinary functions. We verify the formula exemplarily at the x-component of $\vec{a} \times \vec{b}$. It holds:

$$
\frac{d}{dt}\left(\vec{a}\,(t) \times \vec{b}\,(t)\right)_x = \frac{d}{dt}\left(a_y\,(t)\,b_z\,(t) - a_z\,(t)\,b_y\,(t)\right)
$$

$$
\begin{aligned}
&= \dot{a}_y(t)\, b_z(t) + a_y(t)\, \dot{b}_z(t) - \left(\dot{a}_z(t)\, b_y(t) + a_z(t)\, \dot{b}_y(t) \right) \\
&= \dot{a}_y(t)\, b_z(t) - \dot{a}_z(t)\, b_y(t) + \left(a_y(t)\, \dot{b}_z(t) - a_z(t)\, \dot{b}_y(t) \right) \\
&= \left(\dot{\vec{a}}(t) \times \vec{b}(t) \right)_x + \left(\vec{a}(t) \times \dot{\vec{b}}(t) \right)_x \quad \square
\end{aligned}
$$

After these explanations about the cross product, we now want to generalize the rotational movements. We consider a particle that moves on an arbitrary circle in space. Let $\vec{R}$ be the radius vector (vector from the center of the circle to the particle) and $\vec{v}$ the vector of the tangential velocity of the particle, which is perpendicular to $\vec{R}$. The two vectors $\vec{R}$ and $\vec{v}$ lie in the **plane of motion** of the particle, i.e., in the plane defined by the circle. We *define* the angular velocity as a vector $\vec{\omega}$, which is perpendicular to the circle, so that the vectors $\vec{R}$, $\vec{v}$ and $\vec{\omega}$ form a right-handed system, i.e.,

$$
\vec{\omega} = \text{constant} \cdot \left(\vec{R} \times \vec{v} \right).
$$

The constant is determined so that - as in the simple case - the magnitude of $\vec{\omega}$ is equal to

$$
\omega = d\theta/dt = v/R,
$$

i.e.,

$$
\vec{\omega} = \frac{1}{R^2} (\vec{R} \times \vec{v}).
$$

To check, we calculate again:

$$
\begin{aligned}
\omega &= |\vec{\omega}| = \left| \frac{1}{R^2} (\vec{R} \times \vec{v}) \right| = \frac{1}{R^2} \left| \vec{R} \times \vec{v} \right| \\
&= \frac{1}{R^2} R v \sin\left(\angle(\vec{R}, \vec{v}) \right) = \frac{v \sin(90°)}{R} = \frac{v}{R}.
\end{aligned}
$$

Fig. 5.5 shows a force $\vec{F}$, which acts on a particle at location P with the position vector $\vec{r}$. The torque exerted in this case is a *vector* $\vec{M}$ with the magnitude

$$
M = Fr \sin\varphi,
$$

which is perpendicular to the plane spanned by $\vec{F}$ and $\vec{r}$ (in the figure, this is the (x, y)-plane, but it can also be any plane lying arbitrarily in space). φ indicates the angle between $\vec{F}$ and $\vec{r}$.

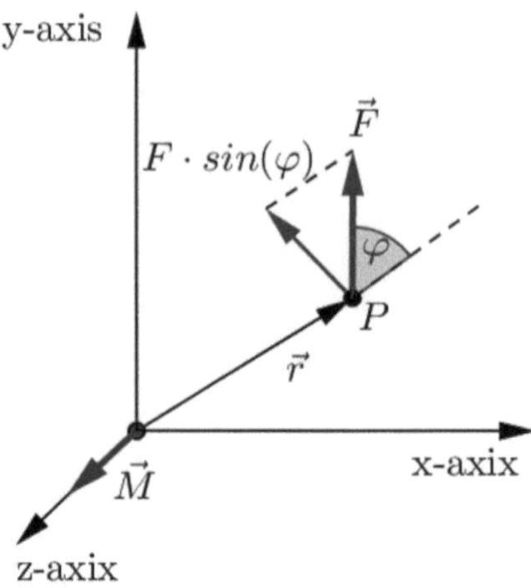

Figure 5.5.: Torque as a vector

This corresponds to the previous definition of torque, but now a direction is assigned to the torque:

$$\vec{M} = \vec{r} \times \vec{F}$$

In more detail, $\vec{M}$ is called the torque of the force acting on the particle m about the point 0. The angular momentum of a particle can also be written as a cross product:

$$\vec{L} = \vec{r} \times \vec{p} = \vec{r} \times m\vec{v} = m\,(\vec{r} \times \vec{v}) \tag{5.5}$$

Angular momentum and torque are always defined with respect to a fixed point (e.g., the origin of the coordinate system). If the origin of the coordinate system lies in the plane of rotation of the particle, i.e., if the position vector $\vec{r}$ is equal to the radius vector $\vec{R}$, then the angular momentum points in the same direction as the angular velocity

$$\vec{L} = \vec{r} \times \vec{p} = \vec{R} \times \vec{p} = \vec{R} \times m\vec{v} = m\left(\vec{R} \times \vec{v}\right) = mR^2\vec{\omega} = I\vec{\omega}.$$

If you derive the angular momentum using the derivative rule for cross products (5.4) with respect to time, you get with the second Newton's law (3.1):

$$\vec{F} = \frac{d\vec{p}}{dt}$$

the relationship

$$\frac{d\vec{L}}{dt} = \frac{d\,(\vec{r} \times \vec{p})}{dt} = \frac{d\vec{r}}{dt} \times \vec{p} + \vec{r} \times \frac{d\vec{p}}{dt} = \underbrace{\vec{v} \times m\vec{v}}_{=0} + \vec{r} \times \frac{d\vec{p}}{dt} = \vec{r} \times \frac{d\vec{p}}{dt} = \vec{r} \times \vec{F} = \vec{M}.$$

> The temporal change of the angular momentum is equal to the acting torque. This means that if the (total) torque is zero, the angular momentum is constant.

In central force fields, the following applies (see Eq. 3.2 on page 48)

$$\vec{F} = f\left(r\right) \vec{e}_r,$$

where

$$\vec{e}_r = \frac{\vec{r}}{r}$$

is the unit vector in the direction of $\vec{r}$. If $\vec{F}$ is a central force, then

$$\vec{M} = \vec{r} \times \vec{F} = \vec{r} \times f\left(r\right) \vec{e}_r = \vec{r} \times f\left(r\right) \frac{\vec{r}}{r} = \frac{f\left(r\right)}{r} \underbrace{\left(\vec{r} \times \vec{r}\right)}_{=0} = 0,$$

the torque is therefore zero, so also

$$\vec{M} = \frac{d}{dt} \vec{L} = 0, \tag{5.6}$$

i.e., we obtain an important result:

> In central force fields, the torque is zero, so the angular momentum is constant over time for all movements.

5.5. Comparison of Physical Quantities in Translations and Rotations

To conclude this chapter, we summarize the results obtained for rotational movements and compare them to the corresponding quantities in linear movements („**translations**") in a table.

Translations	Rotations
Length s	Angle θ
Mass m	Moment of inertia $I = mR^2$, $\vec{R}$ radius.
Velocity $\vec{v}$	Angular velocity $\vec{\omega} = \dfrac{1}{R^2}\left(\vec{R} \times \vec{v}\right)$
Momentum $\vec{p} = m\vec{v}$	Angular momentum $\vec{L} = \vec{r} \times m\vec{v}$, $\vec{r}$ position
Force $\vec{F}$	Torque $\vec{M} = \vec{r} \times \vec{F}$
$\vec{F} = \dfrac{d\vec{p}}{dt}$	$\vec{M} = \dfrac{d\vec{L}}{dt}$
Kin. Energy $E_{kin} = \dfrac{1}{2}mv^2$	Kin. Energy $E_{kin} = \dfrac{1}{2}I\omega^2$

Table 5.1.: Translations and Rotations

6. Newton's Law of Gravitation

After the preparatory statements in the last chapters, we now come to the core point of this first part, to Newton's law of gravitation. Kepler's laws for explaining the orbits of the planets were an important step towards understanding the processes in the solar system. They were rules established from the observational data of the time, which could not be derived from theoretical considerations. Only Newton realized that the Moon's orbit around the Earth and the orbits of the planets around the Sun must be attributed to forces that must also act over considerable distances. The Earth itself exerts its gravity on objects (falling apples etc.) that are not in direct contact with it. Newton postulated and proved that the planetary orbits and the Earth's attraction are due to a force that acts between two bodies and is smaller the further the two bodies are apart from each other. According to **Newton's law of gravitation** every body exerts an attractive force on every other body. This is proportional to the product of the masses of the two bodies and inversely proportional to the square of their distance. There is no reliable tradition on how Newton found his law of gravitation. One possible explanation is that he examined Kepler's third law for circular planetary orbits. This states that the ratio of the third power of the radii R of the orbits to the square of their orbital periods T yields the same value for all planets, i.e.

$$\frac{R^3}{T^2} = const.,$$

compare formula 1.5 on page 11. In Newton's time, the centripetal acceleration a for circular motions was also known (see Eq. 2.11 on page 40)

$$a = \frac{v^2}{R} \Rightarrow v^2 = aR,$$

where

$$v = \frac{2\pi R}{T} \Rightarrow T = \frac{2\pi R}{v}$$

is the average orbital speed. Then follows

$$const. = \frac{R^3}{T^2} = \frac{R^3}{\left(\frac{2\pi R}{v}\right)^2} = \frac{R^3}{\frac{4\pi^2 R^2}{v^2}} = \frac{R v^2}{4\pi^2} = \frac{R a R}{4\pi^2} = \frac{R^2 a}{4\pi^2}.$$

© The Author(s), under exclusive license to Springer-Verlag GmbH, DE, part of Springer Nature 2026

M. Ruhrländer, *Ascent to the Einstein Equations*,

https://doi.org/10.1007/978-3-662-72672-3_6

So if the expression $R^2 a/\left(4\pi^2\right)$ is constant, then the acceleration a must be proportional to $1/R^2$. This is how Newton might have stumbled upon his famous law of gravitation.

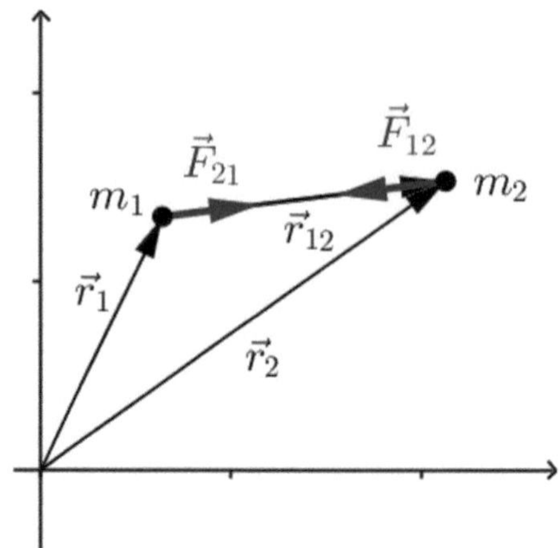

Figure 6.1.: Newton's law of gravitation

In Fig. 6.1 two masses m_1 and m_2 with their respective position vectors $\vec{r}_1$ and $\vec{r}_2$ are shown. The vector $\vec{r}_{12} = \vec{r}_2 - \vec{r}_1$ points from body m_1 to body m_2. The gravitational force $\vec{F}_{12}$, with which m_1 acts on m_2, is *defined* by

$$\vec{F}_{12} = -\frac{Gm_1 m_2}{r_{12}^2} \cdot \frac{\vec{r}_{12}}{r_{12}} = -\frac{Gm_1 m_2}{|\vec{r}_2 - \vec{r}_1|^2} \cdot \frac{\vec{r}_2 - \vec{r}_1}{|\vec{r}_2 - \vec{r}_1|}. \tag{6.1}$$

$\vec{r}_{12}/r_{12}$ is a unit vector pointing from m_1 to m_2, and the minus sign guarantees that $\vec{F}_{12}$ is an attractive force. G is the **gravitational constant**, which has the value

$$G = 6.67 \cdot 10^{-11}\,\mathsf{N} \cdot \mathsf{m}^2/\mathsf{kg}^2.$$

The force $\vec{F}_{21}$, which body m_2 exerts on m_1, is equal in magnitude to $\vec{F}_{12}$ according to Newton's third law, but points in the opposite direction. If r denotes the distance between two bodies, the magnitude of the gravitational force $\vec{F}$ is given by

$$F = \frac{Gm_1 m_2}{r^2}. \tag{6.2}$$

Example 6.1.
As an example, we calculate the gravitational force between two bodies with mass $1\,\mathsf{kg}$, which are $1\,\mathsf{m}$ apart from each other:

$$F = \frac{Gm_1 m_2}{r^2} = \frac{6.67 \cdot 10^{-11}\,\mathsf{N} \cdot \mathsf{m}^2/\mathsf{kg}^2 \cdot 1\,\mathsf{kg} \cdot 1\,\mathsf{kg}}{(1\,\mathsf{m})^2} = 6.674 \cdot 10^{-11}\,\mathsf{N}\,\square$$

The example shows that the gravitational force is extremely small. The weight force of the mass 1 kg is

$$F_G = mg = 9.81\,\text{N},$$

so more than 100 billion times as much. □

Example 6.2.
We show that as a consequence of Newton's law of gravitation, Galileo's law of falling near the Earth's surface results. If

$$M_E = 5.974 \cdot 10^{24}\,\text{kg}$$

is the mass of the Earth and

$$R_E = 6.378 \cdot 10^6\,\text{m}$$

is the average Earth radius (see also Tab. 30.3), then the gravitational force exerted by the Earth on any mass m at a distance $\approx R_E$ from the center of the Earth has the magnitude

$$F = \frac{GmM_E}{R_E^2} = m \cdot \frac{6.674 \cdot 10^{-11} \cdot 5.974 \cdot 10^{24}}{\left(6.378 \cdot 10^6\right)^2} \approx m \cdot 9.81 = mg.$$

Since both the Newtonian gravitational force and the Earth's gravitational force point towards the center of the Earth, there is thus a correspondence of both forces in the vicinity of the Earth. □

From the last equation it follows

$$g = \frac{GM_E}{R_E^2}. \tag{6.3}$$

Since the Earth's acceleration $g = 9.81\,\text{m/s}^2$ can be easily measured and the Earth's radius R_E is known, with (6.3) either the constant G or the Earth's mass M_E can be determined. Newton himself based his determination of the gravitational constant G on an estimate of the Earth's mass. Since the gravitational force is so small, the gravitational constant G is the least precisely known constant in physics, it is known only (with a small probability of error) to the fourth decimal place.

Example 6.3.
As a further test for the validity of the law of gravitation, we compare the acceleration of the Moon in its orbit with the acceleration of objects near the

Earth's surface. The distance to the Moon is about 60 times the Earth's radius, i.e., the Earth's acceleration should be $60^2 = 3600$ times as large as the Moon's acceleration towards the Earth. Assuming a circular orbit of the Moon around the Earth, the orbital speed according to Eq. 2.12 on page 41 is

$$v = 2\pi r/T,$$

where r denotes the distance Moon-Earth center and T the orbital period of the Moon. The acceleration of the Moon in its orbit a_m is according to formula 2.11 on page 40 (the Moon is an "Earth satellite") the centripetal acceleration

$$a_m = \frac{v^2}{r} = \frac{(2\pi r/T)^2}{r} = \frac{4\pi^2 r}{T^2}.$$

For $r = 3.84 \cdot 10^8\,\mathrm{m}$ and $T = 27.3\,\mathrm{days}$ it follows

$$a_m = \frac{4\pi^2 r}{T^2} = 2.72 \cdot 10^{-3}\,\mathrm{m/s^2}.$$

Comparing this number with the Earth's acceleration $g = 9.81\,\mathrm{m/s^2}$, it actually results in

$$\frac{g}{a_m} = \frac{9.81}{2.72 \cdot 10^{-3}} = 3607 \approx 60^2.\;\square$$

6.1. The Potential Energy of Newton's Gravitational Force

Assuming that the mass m_1 is located at the origin and that the mass m_2 has the position vector $\vec{r}$, Newton's law of gravitation can be written with the unit vector $\vec{e}_r = \vec{r}/r$ in the form

$$\vec{F} = -\frac{Gm_1 m_2}{r^2} \cdot \vec{e}_r \tag{6.4}$$

We now want to find a function that describes the potential energy of the gravitational force. To do this, we make some preliminary considerations. With the derivative rule

$$\frac{d}{dx}\,x^a = ax^{a-1}$$

(see Eq. (4.10)) and the chain rule (see Eq. ((4.22)) it follows with $r = |\vec{r}| = \sqrt{x^2 + y^2 + z^2}$:

$$\frac{\partial}{\partial x}\left(\frac{1}{\sqrt{x^2 + y^2 + z^2}}\right) = \frac{\partial}{\partial x}\left(x^2 + y^2 + z^2\right)^{-1/2}$$

$$= -\frac{1}{2}\left(x^2 + y^2 + z^2\right)^{-3/2} \cdot 2x$$

$$= \frac{-x}{\left(x^2 + y^2 + z^2\right)^{3/2}}$$

$$= \frac{-x}{\left(x^2 + y^2 + z^2\right)\sqrt{x^2 + y^2 + z^2}}$$

$$= -\frac{1}{r^2}\frac{x}{r}.$$

And similarly

$$\frac{\partial}{\partial y}\left(\frac{1}{\sqrt{x^2 + y^2 + z^2}}\right) = -\frac{1}{r^2}\frac{y}{r},$$

$$\frac{\partial}{\partial z}\left(\frac{1}{\sqrt{x^2 + y^2 + z^2}}\right) = -\frac{1}{r^2}\frac{z}{r}.$$

So for the gradient it holds

$$\nabla\frac{1}{r} = \nabla\left(\frac{1}{\sqrt{x^2 + y^2 + z^2}}\right) = -\frac{1}{r^2}\begin{pmatrix} x/r \\ y/r \\ z/r \end{pmatrix} = -\frac{1}{r^2}\frac{\vec{r}}{r} = -\frac{1}{r^2}\vec{e}_r.$$

It follows

$$\nabla\left(\frac{Gm_1m_2}{r}\right) = -\frac{Gm_1m_2}{r^2}\vec{e}_r = \vec{F},$$

so due to $-\nabla E\left(r\right) = \vec{F}$ (see Eq. (4.17)), the potential energy is given by the following expression:

$$E_{pot}\left(r\right) = -\frac{Gm_1m_2}{r}. \tag{6.5}$$

Thus, there exists a potential function for the gravitational force, and the gravitational field is conservative according to Eq. (4.18). The Newtonian equation of motion in a gravitational field, which is generated by a body with mass M, is therefore for a particle with mass m

$$m\ddot{\vec{r}} = -m\,\nabla\phi\left(\vec{r}\right) \tag{6.6}$$

with the potential function

$$\phi\left(\vec{r}\right) = -\frac{GM}{r}. \tag{6.7}$$

We want to find out how a body can overcome the binding to the Earth caused by the gravitational force. First, we consider a body with mass m near the

Earth's surface and assume that the gravitational force is constant mg. If we shoot the object vertically upwards with the velocity v_0, then it will reach a maximum height. We derive this from the conservation of energy, which applies in a conservative force field, with the following considerations. We set the potential energy of the Earth's gravitational field on the Earth's surface to zero, i.e., the body initially has no potential energy. The starting kinetic energy is $mv_0^2/2$. When the body reaches its maximum height h_{max}, its speed, and therefore its kinetic energy, is zero, and the potential energy is mgh_{max}. Therefore, from the conservation of energy, we have

$$\frac{1}{2}\,mv_0^2 = mgh_{max}$$

and from this

$$h_{max} = \frac{v_0^2}{2g}.$$

The equation states that the maximum height increases with increasing initial speed. But it also says that a body can never overcome its bond to the Earth. No matter how large the initial speed is chosen, the body would always reach a maximum height and then fall back to Earth. However, we know that it is indeed possible to overcome the Earth's bond, our mistake lies in the approach used. The assumption that the gravitational field is homogeneous is only practical for heights of up to $10\,\mathrm{km}$ above the ground. For our question, we must rather use Newton's law of gravitation, which is not homogeneous, but decreases like $1/r^2$.

In the following, let M_E denote the Earth's mass and R_E the Earth's radius. The change in potential energy when a particle of mass m is moved from point $\vec{r}_1$ to point $\vec{r}_2$ is given by

$$E_{pot}\left(r_2\right) - E_{pot}\left(r_1\right) = \frac{GM_E\,m}{r_1} - \frac{GM_E\,m}{r_2}.$$

If we set $r_1 = R_E$ and the potential energy on the Earth's surface equal to zero, then for any location $r\,(> R_E)$ we get

$$E_{pot}\left(r\right) = -\frac{GM_E\,m}{r},$$

and because

$$E_{pot}\left(R_E\right) = -\frac{GM_E\,m}{R_E} = 0$$

we obtain

$$E_{pot}\left(r\right) = \frac{GM_E\,m}{R_E} - \frac{GM_E\,m}{r}.$$

If we write the potential energy as a function of the distance

$$y = r - R_E$$

from the Earth's surface, then with the main denominator and using Eq. (6.3) we get

$$
\begin{aligned}
E_{pot}\left(r\right) &= \frac{GM_E\,m}{R_E} - \frac{GM_E\,m}{r} = \frac{GM_E\,m}{r R_E}\left(r - R_E\right)\\
&= m \cdot \left(\frac{GM_E}{R_E^2}\right) \cdot y \cdot \frac{R_E}{r} = mgy \cdot \frac{R_E}{r}.
\end{aligned}
$$

This equation once again shows that near the Earth's surface ($r \approx R_E$) the potential energy is approximately mgy, which we had already shown above under the assumption of a homogeneous gravitational force.

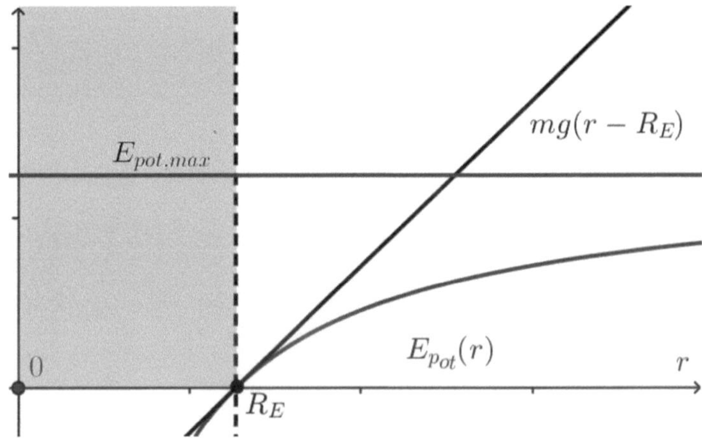

Figure 6.2.: Potential energy in the gravitational field of the Earth, $E_{pot}\left(R_E\right) = 0$

Fig. 6.2 shows the potential energy as a function of r. The straight line is the potential energy for the constant gravitational field $F = mg$. The function $E_{pot}(r)$ does not grow to infinity (like the straight line), but has its maximum at

$$E_{pot,max} = \frac{GM_E\,m}{R_E} = mgR_E,$$

which can be read directly from the function equation for the potential energy

$$E_{pot}(r) = \frac{GM_E\,m}{R_E} - \frac{GM_E\,m}{r}$$

since for every r the second term reduces the first.

We can now answer the question of how large the launch speed must be at least for a body to overcome the binding force of the Earth. Although it is also the case in the Newtonian gravitational field that a body increases its potential energy with increasing distance r from the center of the Earth, it cannot exceed the maximum mgR_E. Since we are in a conservative force field, the law of energy conservation applies, i.e., the kinetic energy at launch (where the potential energy is zero) can at most be reduced by the maximum value of the potential energy. If the body therefore has a higher kinetic energy than the maximum potential energy, the body overcomes the bond to the Earth and flies, if no other influences act on it, into infinity. The critical initial velocity that the body with mass m must have to escape is called **escape velocity** v_F and is calculated from

$$E_{kin} = \frac{1}{2} mv_F^2 = E_{pot,max} = \frac{GM_E\, m}{R_E} = mgR_E$$

to

$$
\begin{aligned}
v_F &= \sqrt{\frac{2GM_E}{R_E}} = \sqrt{2gR_E} = \sqrt{2 \cdot 9.81 \cdot 6.37 \cdot 10^6} \\
&= 11.2\,\mathrm{km/s} \approx 40.000\,\mathrm{km/h}.
\end{aligned}
\tag{6.8}
$$

In section 4.4 it was shown that one has free choice when determining the zero point of potential energy. We now want to set the zero point of potential energy, which was on the Earth's surface in the last explanations, so that the potential energy between two bodies becomes zero when the distance between them is infinite, i.e., with Eq. (6.5) follows

$$E_{pot}\left(r\right) = -\frac{GM_E\, m}{r}, \quad E_{pot}\left(r\right) = 0 \; \; f\ddot{u}r\, r \to \infty.$$

The potential energy is always negative by this choice, thus work is released when, for example, a body falls to the ground.

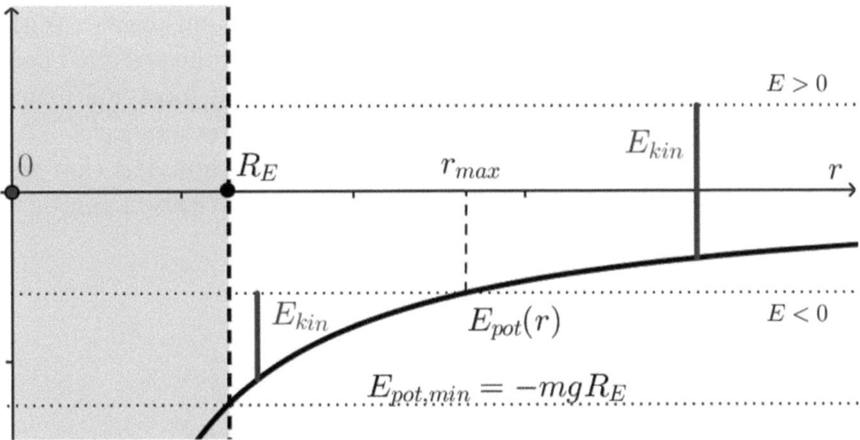

Figure 6.3.: Potent. energy in the gravitational field of the Earth, $E_{pot}\left(\infty\right) = 0$

Fig. 6.3 again shows the potential energy as a function of r. The minimum value of the potential energy is at

$$E_{pot,min} = E_{pot}\left(R_E\right) = -\frac{GM_E\,m}{R_E} = -mgR_E,$$

thus assumed at the Earth's surface. The function increases with increasing r and approaches zero as r tends to infinity, i.e., the potential energy can again grow at most by mgR_E, just like in the case when $E_{pot}(R_E) = 0$ was. It also follows that a body can overcome the Earth's bond if the kinetic energy is greater than or equal to mgR_E. Since the potential energy on the Earth's surface is equal to $-mgR_E$, the required total energy $E = E_{kin} + E_{pot}$ of the body must be greater than or equal to zero. In figure 6.3, two possible values for the total energy are entered ($E < 0$ and $E > 0$). A negative total energy means that the initial kinetic energy on the Earth's surface is less than mgR_E. In this case, the total energy and the potential energy intersect at a point r_{max}, which cannot be exceeded. The body is bound to the system. If, on the other hand, the total energy E is positive, then there is no limit to the distance r, the body is not bound to the system and can move away as far as it likes.

6.2. Derivation of Kepler's Laws from Newton's Law of Gravitation

Newton himself showed that his law of gravitation leads to the three Keplerian laws of planetary orbits around the Sun. We want to understand this (long) calculation in detail because we will later precisely demonstrate at which points the Einsteinian theory causes a change in the Newtonian orbits. With these orbits modified by Einstein, we will for example explain the observed perihelion precession of Mercury in reality. This section is the most mathematically demanding in part I and therefore requires a lot of perseverance!

Assuming that the mass of the Sun M_S is at the origin and that the mass of a planet m has the position vector $\vec{r}$, the law of gravitation can be written in the form

$$\vec{F} = -\frac{GM_S\, m}{r^2}\,\vec{e}_r = f\left(r\right)\vec{e}_r$$

and the potential energy in the form

$$E_{pot}\left(r\right) = -\frac{GM_S\, m}{r}, \quad E_{pot}\left(r\right) = 0\ \left(r \to \infty\right).$$

The gravitational field is a central force field, so according to (5.6) the torque is equal to zero and thus the angular momentum vector $\vec{L}\left(t\right) = \vec{L}$ is constant over time. Since the angular momentum vector $\vec{L}\left(t\right)$ at any time t due to

$$\vec{L}\left(t\right) = m\,\vec{r}\left(t\right) \times \vec{v}\left(t\right)$$

is perpendicular to the position and velocity vectors $\vec{r}\left(t\right)$ and $\vec{v}\left(t\right)$ of the planet, but its direction does not change over time, the Sun and the planet lie in a fixed plane at any time, which we want to consider as the (x, y)-plane. We want to use (time-dependent) polar coordinates (see Eq. 2.6 on page 30) for the derivation of the possible planetary orbits:

$$\vec{r}\left(t\right) = \begin{pmatrix} x\left(t\right) \\ y\left(t\right) \end{pmatrix} = \begin{pmatrix} r\left(t\right)\cos\varphi\left(t\right) \\ r\left(t\right)\sin\varphi\left(t\right) \end{pmatrix}$$

The orientation of the orbit is chosen so that the planets move counterclockwise around the Sun, i.e., for the angle φ within one orbital period

$$\varphi\left(t_2\right) > \varphi\left(t_1\right) \ \text{ for } t_2 > t_1. \tag{6.9}$$

For further treatment, we need the derivatives of the trigonometric functions.

Remark 6.1. **MF: Derivative of** $\sin\left(t\right)$, $\cos\left(t\right)$
In Fig. 6.4 the quantities

$$\sin\left(t\right), \sin\left(t + \Delta t\right), \cos\left(t\right) \text{ and } \cos\left(t + \Delta t\right)$$

are represented on the *unit circle*.

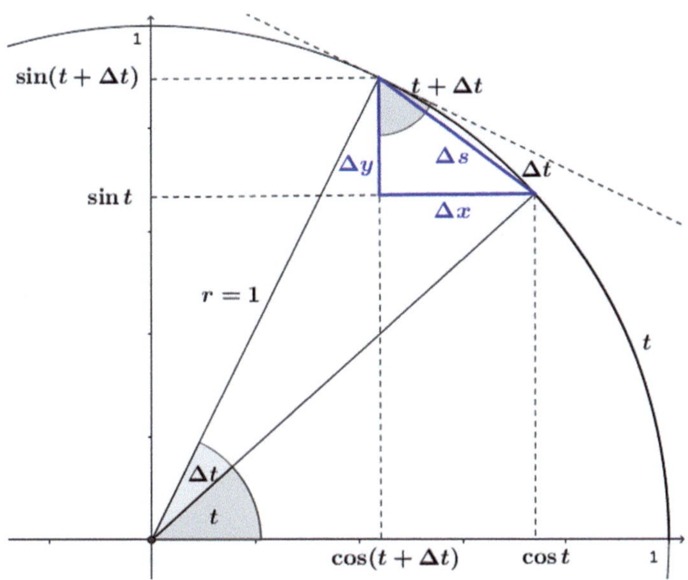

Figure 6.4.: Derivative of the sine and cosine function

First of all, it should be noted that for a circle with radius r for any arc t with the associated angle φ the ratio of the arc t to the total circumference $2\pi r$ is equal to the ratio of the angle φ to the full angle 2π. So it applies

$$\frac{t}{2\pi r} = \frac{\varphi}{2\pi} \Rightarrow t = \varphi r.$$

For the unit circle $r = 1$, the arc length t is therefore equal to the angle t. From Fig. 6.4, it can be seen that the triangle with sides $(\Delta x, \Delta y, \Delta s)$ is a good approximation for the arc triangle $(\Delta x, \Delta y, \Delta t)$ for small Δt. Similarly, the angle $t + \Delta t$ for small Δt is a good approximation for the angle between the legs Δy and Δs. Thus, it follows

$$\frac{\sin(t + \Delta t) - \sin t}{\Delta t} = \frac{\Delta y}{\Delta t} \approx \frac{\Delta y}{\Delta s} \approx \cos(t + \Delta t),$$

and

$$\frac{\cos(t + \Delta t) - \cos t}{\Delta t} = -\frac{\Delta x}{\Delta t} \approx -\frac{\Delta x}{\Delta s} \approx -\sin(t + \Delta t),$$

i.e.

$$\begin{aligned}
\frac{d}{dt}(\sin t) &= \lim_{\Delta t \to 0} \frac{\sin(t + \Delta t) - \sin t}{\Delta t} = \lim_{\Delta t \to 0} \frac{\Delta y}{\Delta t} \\
&= \lim_{\Delta t \to 0} \cos(t + \Delta t) = \cos t,
\end{aligned} \tag{6.10}$$

$$\frac{d}{dt}\left(\cos t\right) = \lim_{\Delta t \to 0} \frac{\cos\left(t + \Delta t\right) - \cos t}{\Delta t} = \lim_{\Delta t \to 0} -\frac{\Delta x}{\Delta t}$$
$$= \lim_{\Delta t \to 0} -\sin\left(t + \Delta t\right) = -\sin t. \;\square \qquad (6.11)$$

Trajectory of a particle in polar coordinates

With these results, we can calculate the derivative of the trajectory of a particle in polar coordinates. Note that with the chain rule (4.19), the derivative of $\sin\varphi\left(t\right)$ becomes

$$\frac{d}{dt}\left(\sin\varphi\left(t\right)\right) = \sin'\varphi\left(t\right) \cdot \frac{d}{dt}\left(\varphi\left(t\right)\right) = \cos\varphi\left(t\right) \cdot \dot{\varphi}\left(t\right)$$

and similarly

$$\frac{d}{dt}\left(\cos\varphi\left(t\right)\right) = \cos'\varphi\left(t\right) \cdot \frac{d}{dt}\left(\varphi\left(t\right)\right) = -\sin\varphi\left(t\right) \cdot \dot{\varphi}\left(t\right),$$

so with the product rule

$$\begin{aligned}
\dot{\vec{r}}\left(t\right) &= \begin{pmatrix} \dot{x}\left(t\right) \\ \dot{y}\left(t\right) \end{pmatrix} = \begin{pmatrix} \dfrac{d}{dt}\left[r\left(t\right)\cos\varphi\left(t\right)\right] \\ \dfrac{d}{dt}\left[r\left(t\right)\sin\varphi\left(t\right)\right] \end{pmatrix} \\[2ex]
&= \begin{pmatrix} \dot{r}\left(t\right)\cos\varphi\left(t\right) + r\left(t\right)\dfrac{d}{dt}\left(\cos\varphi\left(t\right)\right) \\ \dot{r}\left(t\right)\sin\varphi\left(t\right) + r\left(t\right)\dfrac{d}{dt}\left(\sin\varphi\left(t\right)\right) \end{pmatrix} \\[2ex]
&= \begin{pmatrix} \dot{r}\left(t\right)\cos\varphi\left(t\right) - r\left(t\right)\dot{\varphi}\left(t\right)\sin\varphi\left(t\right) \\ \dot{r}\left(t\right)\sin\varphi\left(t\right) + r\left(t\right)\dot{\varphi}\left(t\right)\cos\varphi\left(t\right) \end{pmatrix} \\[2ex]
&= \dot{r}\left(t\right)\begin{pmatrix} \cos\varphi\left(t\right) \\ \sin\varphi\left(t\right) \end{pmatrix} + r\left(t\right)\dot{\varphi}\left(t\right)\begin{pmatrix} -\sin\varphi\left(t\right) \\ \cos\varphi\left(t\right) \end{pmatrix} \\[2ex]
&= \dot{r}\left(t\right)\vec{e}_r + r\left(t\right)\dot{\varphi}\left(t\right)\vec{e}_\varphi, \qquad (6.12)
\end{aligned}$$

where

$$\vec{e}_r\left(t\right) = \frac{\vec{r}\left(t\right)}{r\left(t\right)} = \begin{pmatrix} \cos\varphi\left(t\right) \\ \sin\varphi\left(t\right) \end{pmatrix}$$

is defined as the unit vector in r-direction and

$$\vec{e}_\varphi\left(t\right) = \begin{pmatrix} -\sin\varphi\left(t\right) \\ \cos\varphi\left(t\right) \end{pmatrix}$$

is defined as the unit vector in φ-direction. Since

$$|\vec{e}_r(t)| = \left| \begin{pmatrix} \cos\varphi(t) \\ \sin\varphi(t) \end{pmatrix} \right| = \sqrt{\cos^2\varphi(t) + \sin^2\varphi(t)} = 1$$

and

$$|\vec{e}_\varphi(t)| = \left| \begin{pmatrix} -\sin\varphi(t) \\ \cos\varphi(t) \end{pmatrix} \right| = \sqrt{\sin^2\varphi(t) + \cos^2\varphi(t)} = 1,$$

hey are indeed unit vectors, and because

$$\begin{aligned}
\vec{e}_r(t) \cdot \vec{e}_\varphi(t) &= \begin{pmatrix} \cos\varphi(t) \\ \sin\varphi(t) \end{pmatrix} \cdot \begin{pmatrix} -\sin\varphi(t) \\ \cos\varphi(t) \end{pmatrix} \\
&= -\cos\varphi(t)\sin\varphi(t) + \cos\varphi(t)\sin\varphi(t) = 0
\end{aligned}$$

the two unit vectors are perpendicular to each other. However, note that these vectors are not constant, they depend on the respective angle $\varphi(t)$.

Determination of planetary orbits

We now want to use the conservation of angular momentum and energy to arrive at a determination equation for the planetary orbits. To make these determination equations as simple as possible, we again use the conservation of angular momentum and, for the sake of clarity, omit the time variable t in the following equations for the (vector) functions, i.e., we write e.g. $\vec{r}$ for $\vec{r}(t)$.

Since the cross product of two parallel vectors is zero, with (6.12)

$$\begin{aligned}
\vec{r} \times \dot{\vec{r}} &= r\vec{e}_r \times (\dot{r}\vec{e}_r + r\dot{\varphi}\vec{e}_\varphi) = \underbrace{r\vec{e}_r \times \dot{r}\vec{e}_r}_{=0} + r\vec{e}_r \times r\dot{\varphi}\vec{e}_\varphi \\
&= r\vec{e}_r \times r\dot{\varphi}\vec{e}_\varphi = r^2\dot{\varphi}(\vec{e}_r \times \vec{e}_\varphi) \\
&= r^2\dot{\varphi}\vec{e}_z,
\end{aligned} \qquad (6.13)$$

since the unit vector in z-direction $\vec{e}_z$ is perpendicular to the (x,y)-plane spanned by $\vec{e}_r$ and $\vec{e}_\varphi$. With (5.5) it follows from (6.13) for the angular momentum vector $\vec{L}$

$$\vec{L} = \vec{r} \times m\vec{v} = \vec{r} \times m\dot{\vec{r}} = m\vec{r} \times \dot{\vec{r}} = mr^2\dot{\varphi}\vec{e}_z,$$

so the magnitude of the angular momentum in polar coordinates is

$$L = \left|\vec{L}\right| = mr^2\,|\dot{\varphi}| = mr^2\dot{\varphi}. \qquad (6.14)$$

The last equation holds, since according to (6.9) $\varphi(t)$ is a monotonically increasing function and therefore the derivative of $\varphi(t)$ is everywhere greater than or equal to zero is: $\dot\varphi(t) \geq 0$. Because of

$$
\begin{aligned}
v^2 &= \left|\dot{\vec{r}}\right|^2 = (\dot{r}\vec{e}_r + r\dot\varphi\vec{e}_\varphi)\cdot(\dot{r}\vec{e}_r + r\dot\varphi\vec{e}_\varphi) \\
&= \dot{r}\vec{e}_r\cdot\dot{r}\vec{e}_r + \dot{r}\vec{e}_r\cdot r\dot\varphi\vec{e}_\varphi + r\dot\varphi\vec{e}_\varphi\cdot\dot{r}\vec{e}_r + r\dot\varphi\vec{e}_\varphi\cdot r\dot\varphi\vec{e}_\varphi \\
&= \dot{r}^2 \underbrace{\vec{e}_r\cdot\vec{e}_r}_{=1} + 2\dot{r}r\dot\varphi\underbrace{\vec{e}_\varphi\cdot\vec{e}_r}_{=0} + r^2\dot\varphi^2\underbrace{\vec{e}_\varphi\cdot\vec{e}_\varphi}_{=1} \\
&= \dot{r}^2 + r^2\dot\varphi^2
\end{aligned}
$$

the conservation of energy in the gravitational field is

$$
E = E_{kin} + E_{pot}(r) = \frac{1}{2}mv^2 + E_{pot}(r) = \frac{1}{2}m\left(\dot{r}^2 + r^2\dot\varphi^2\right) + E_{pot}(r).
$$

The quantity $\dot\varphi$ can be eliminated from the energy equation with Eq. (6.14), from which

$$
\dot\varphi^2 = \frac{L^2}{m^2 r^4}
$$

follows, and we obtain:

$$
E = \frac{1}{2}m\left(\dot{r}^2 + r^2\dot\varphi^2\right) + E_{pot}(r) = \frac{1}{2}m\dot{r}^2 + \frac{L^2}{2mr^2} + E_{pot}(r)
$$

We solve this equation for $\dot{r}(t)$

$$
\frac{dr}{dt} = \dot{r} = \sqrt{\frac{2\left(E - E_{pot}(r)\right)}{m} - \frac{L^2}{m^2 r^2}}.
$$

From Eq. (6.14) it follows

$$
\frac{d\varphi}{dt} = \frac{L}{mr^2}.
$$

Noting that with the chain rule

$$
\frac{dr}{dt} = \frac{dr}{d\varphi}\cdot\frac{d\varphi}{dt}
$$

holds, we get

$$
\frac{dr}{dt} = \frac{dr}{d\varphi}\cdot\frac{d\varphi}{dt} = \frac{dr}{d\varphi}\cdot\frac{L}{mr^2} = \sqrt{\frac{2\left(E - E_{pot}(r)\right)}{m} - \frac{L^2}{m^2 r^2}}
$$

and finally from this by division with L/m

$$
\frac{1}{r^2}\frac{dr}{d\varphi} = \sqrt{\frac{2m\left(E - E_{pot}(r)\right)}{L^2} - \frac{1}{r^2}}.
$$

We define a function $\sigma(\varphi)$ by $\sigma(\varphi) = 1/r(\varphi)$ and obtain with the derivative formula 4.10 on page 60

$$\frac{d\sigma}{dr} = \frac{d}{dr}\left(\frac{1}{r}\right) = \frac{d}{dr}\left(r^{-1}\right) = (-1)\,r^{-2} = -\frac{1}{r^2}$$

and from this with the chain rule

$$\frac{d\sigma}{d\varphi} = \frac{d\sigma}{dr}\frac{dr}{d\varphi} = -\frac{1}{r^2}\frac{dr}{d\varphi}.$$

Note that the potential energy can be written as

$$E_{pot} = -\frac{GM_sm}{r} = -GM_sm\sigma = -A\sigma$$

with $A = GM_sm$, so it follows

$$-\frac{d\sigma}{d\varphi} = \frac{1}{r^2}\frac{dr}{d\varphi} = \sqrt{\frac{2m\,(E+A\sigma)}{L^2} - \sigma^2}.\tag{6.15}$$

We want to simplify this equation and further transform the root term on the right side, where we define p by $p = \dfrac{L^2}{mA}$

$$\sqrt{\frac{2m\,(E+A\sigma)}{L^2} - \sigma^2} = \sqrt{-\sigma^2 + 2\sigma\,\frac{mA}{L^2} + \frac{2mE}{L^2}}$$

$$= \sqrt{-\sigma^2 + 2\sigma\,\frac{1}{p} + \frac{2mE}{L^2}}.$$

We make a **quadratic completion**, i.e., we subtract and add the term $1/p^2$ and obtain with the **Binomial formulas** (see Chap. 29)

$$\sqrt{\frac{2m\,(E+A\sigma)}{L^2} - \sigma^2} = \sqrt{-\left(\sigma^2 - 2\sigma\,\frac{1}{p} + \left(\frac{1}{p}\right)^2\right) + \frac{2mE}{L^2} + \left(\frac{1}{p}\right)^2}$$

$$= \sqrt{-\left(\sigma - \frac{1}{p}\right)^2 + \frac{2mE}{L^2} + \frac{1}{p^2}}$$

$$= \sqrt{-\left(\sigma - \frac{1}{p}\right)^2 + \frac{1}{p^2}\left(\frac{2mEp^2}{L^2} + 1\right)}$$

$$= \sqrt{-\left(\sigma - \frac{1}{p}\right)^2 + \frac{1}{p^2}\left(\frac{2EL^2}{mA^2} + 1\right)}.$$

If one defines the quantity ε by

$$\varepsilon = \sqrt{1 + \frac{2EL^2}{mA^2}} = \sqrt{1 + \frac{2EL^2}{m\left(GM_sm\right)^2}} = \sqrt{1 + \frac{2EL^2}{m^3G^2M_s^2}}, \qquad (6.16)$$

then it follows overall for the root term

$$\sqrt{\frac{2m\left(E + A\sigma\right)}{L^2} - \sigma^2} = \sqrt{\frac{\varepsilon^2}{p^2} - \left(\sigma - \frac{1}{p}\right)^2}.$$

With this result we obtain from Eq. (6.15)

$$-\frac{d\sigma}{d\varphi} = \sqrt{\frac{\varepsilon^2}{p^2} - \left(\sigma - \frac{1}{p}\right)^2}.$$

This equation is squared to

$$\left(\frac{d\sigma}{d\varphi}\right)^2 + \left(\sigma - \frac{1}{p}\right)^2 = \frac{\varepsilon^2}{p^2}.$$

If one sets

$$\tau = \frac{p}{\varepsilon}\left(\sigma - \frac{1}{p}\right),$$

then it follows

$$\frac{d\tau}{d\varphi} = \frac{d}{d\varphi}\left(\frac{p}{\varepsilon}\left(\sigma - \frac{1}{p}\right)\right) = \frac{p}{\varepsilon}\frac{d\sigma}{d\varphi}$$

and thus from the above equation

$$\left(\frac{d\tau}{d\varphi}\right)^2 + \tau^2 = 1.$$

This equation is satisfied by $\tau = \cos\varphi$, because it is

$$\left(\frac{d\tau}{d\varphi}\right)^2 + \tau^2 = (-\sin\varphi)^2 + \cos^2\varphi = \sin^2\varphi + \cos^2\varphi = 1.$$

Because

$$\cos\varphi = \tau = \frac{p}{\varepsilon}\left(\sigma - \frac{1}{p}\right)$$

it follows that

$$\sigma = \frac{\varepsilon\cos\varphi + 1}{p}$$

and from this finally

$$r(\varphi) = \frac{1}{\sigma(\varphi)} = \frac{p}{\varepsilon \cos \varphi + 1}. \tag{6.17}$$

This equation provides information about the possible orbits in the solar system that a celestial body can take in the polar coordinates (r, φ). To better understand how to interpret this equation, we return to the Cartesian coordinates (x, y) of the orbital plane. It applies

$$r^2 = x^2 + y^2 \ \text{ and } \ \cos \varphi = \frac{x}{r} = \frac{x}{\sqrt{x^2 + y^2}}.$$

These two expressions inserted into (6.17) results in

$$\sqrt{x^2 + y^2} = \frac{p}{\frac{\varepsilon x}{\sqrt{x^2 + y^2}} + 1} = \frac{p}{\frac{\varepsilon x + \sqrt{x^2 + y^2}}{\sqrt{x^2 + y^2}}} = \frac{p \sqrt{x^2 + y^2}}{\varepsilon x + \sqrt{x^2 + y^2}},$$

from which

$$1 = \frac{p}{\varepsilon x + \sqrt{x^2 + y^2}}$$

follows and from this

$$\sqrt{x^2 + y^2} = p - \varepsilon x.$$

This squared leads to

$$x^2 + y^2 = p^2 - 2p\varepsilon x + \varepsilon^2 x^2.$$

The terms with x^2 are combined

$$x^2 \left(1 - \varepsilon^2\right) + y^2 = p^2 - 2p\varepsilon x. \tag{6.18}$$

Division by $1 - \varepsilon^2$, where we assume $\varepsilon \neq 1$, and rearrangement results in

$$x^2 + \frac{2p\varepsilon x}{1 - \varepsilon^2} + \frac{y^2}{1 - \varepsilon^2} = \frac{p^2}{1 - \varepsilon^2}.$$

On both sides of the equation, the quadratic complement

$$\frac{p^2 \varepsilon^2}{\left(1 - \varepsilon^2\right)^2}$$

is added, i.e.

$$x^2 + \frac{2p\varepsilon x}{1 - \varepsilon^2} + \frac{p^2 \varepsilon^2}{\left(1 - \varepsilon^2\right)^2} + \frac{y^2}{1 - \varepsilon^2} = \frac{p^2}{1 - \varepsilon^2} + \frac{p^2 \varepsilon^2}{\left(1 - \varepsilon^2\right)^2},$$

so that

$$\left(x + \frac{p\varepsilon}{1-\varepsilon^2}\right)^2 + \frac{y^2}{1-\varepsilon^2} = \frac{p^2\left(1-\varepsilon^2\right) + p^2\varepsilon^2}{(1-\varepsilon^2)^2} = \frac{p^2}{(1-\varepsilon^2)^2}$$

results. If you set $a = \dfrac{p}{1-\varepsilon^2}$ and divide the equation by a^2, it follows

$$\frac{(x+a\varepsilon)^2}{a^2} + \frac{y^2}{a^2\left(1-\varepsilon^2\right)} = 1. \tag{6.19}$$

We interpret this equation by distinguishing the cases $\varepsilon < 1$ and $\varepsilon > 1$. We had excluded the case $\varepsilon = 1$ in the derivation of (6.19). This will be treated subsequently. We will see that the above equation describes orbital curves, which are referred to as **conic sections**.

Remark 6.2. **MF: Conic sections**
The **conic sections** are those plane curves that arise when a plane intersects a double cone, see Fig. 6.5. The following situations can arise.

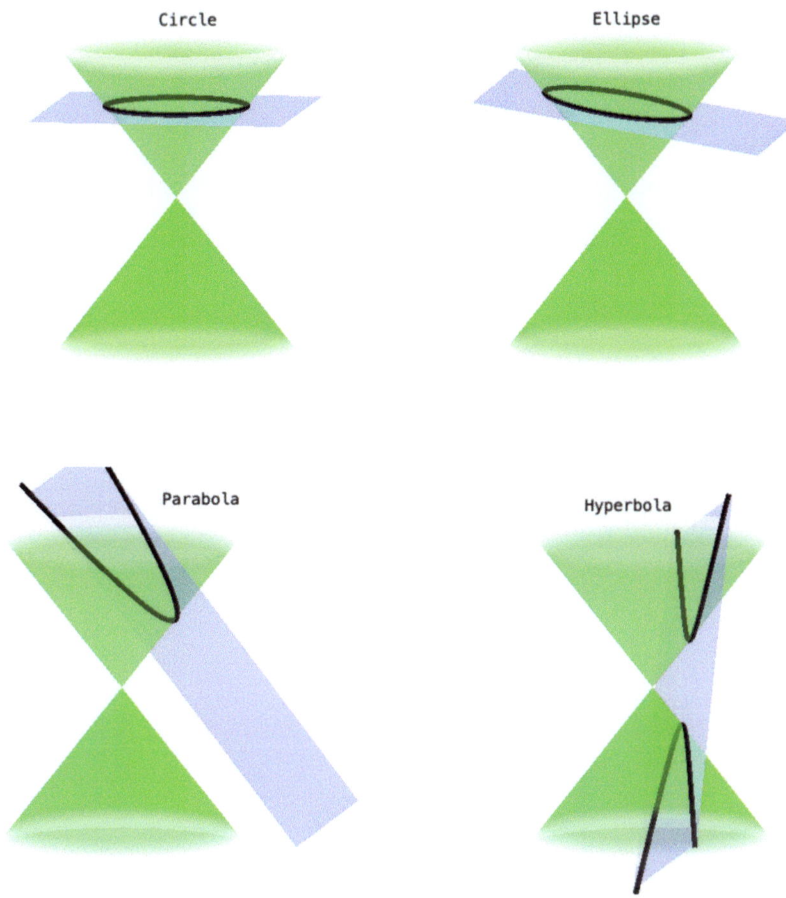

Figure 6.5.: Conic sections

If the cutting plane is parallel to the base of the cone, a **circle** is obtained
as the intersection line between the plane and the outer walls of the cone.
If the plane is tilted slightly, the intersection line results in an **ellipse**. If the
intersection line runs parallel to the outer walls, the curve is called a **parabola**.
If the cutting plane is even steeper, the intersection line results in a **hyperbola**.
The so-called **degenerate cases**, e.g. that the plane cuts the tip, we do not
consider further here. The four (non-degenerate) conic section functions in
Fig. 6.5 are characterized by the following properties:

- A *circle* consists of all points $P = (x, y)$ have the same distance $r > 0$
  from a given point $M = (x_0, y_0)$. M is called the **center** and r the **radius**

of the circle. The so-called **standard form** of the circle equation is

$$(x - x_0)^2 + (y - y_0)^2 = r^2.$$

- An *ellipse* consists of all points $P = (x, y)$ for which the sum of the distances from two given points F_1 and F_2 is constant. The standard form of the ellipse equation is

$$\frac{(x - x_0)^2}{a^2} + \frac{(y - y_0)^2}{b^2} = 1,$$

where $(x_0, y_0) = M$ denotes the center, a the major and b the minor semi-axis of the ellipse (see also Remark 1.1).

- A parabola consists of all points $P = (x, y)$ that have the same distance from a given line g and a given point F.

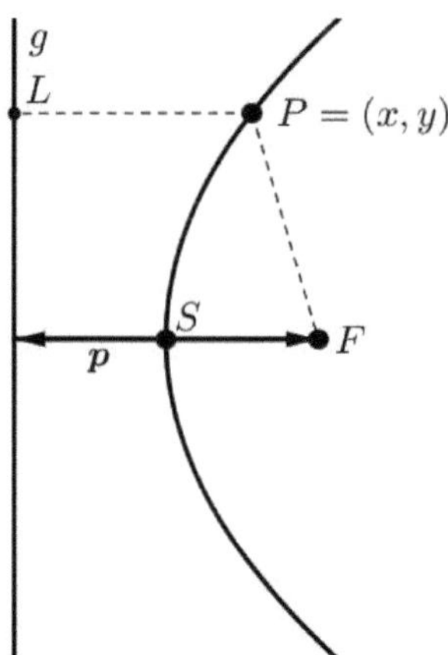

Figure 6.6.: Parabola with focus F and line g

In Fig. 6.6, the line g and the point F („**focus**") as well as the distance p („**parameter**") between the line and the focus are shown. The distances from point P to the focus and to the perpendicular point L are equal:

$$\overline{PF} = \overline{PL}.$$

The point $S = (x_0, y_0)$ with the smallest distance to the line g is called the **vertex** of the parabola. The main form of the parabola equation is

$$(y - y_0)^2 = 2p\,(x - x_0).$$

- A hyperbola consists of all points $P = (x, y)$, for which the difference of the distances from two given points F_1 and F_2 is constant.

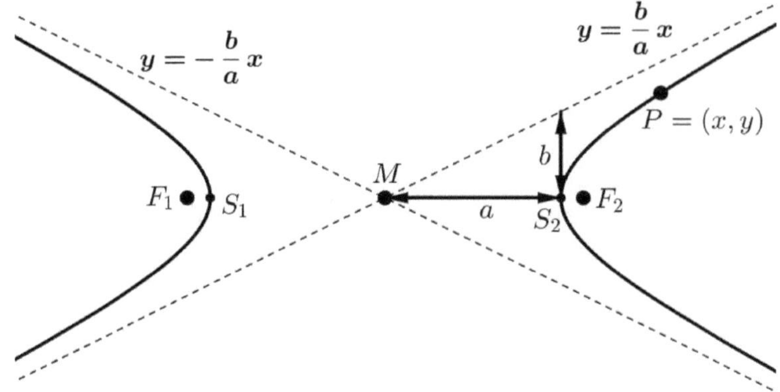

Figure 6.7.: Hyperbola with foci F_1 and F_2

In Fig. 6.7 the two foci F_1 and F_2 as well as the center $M = (x_0, y_0)$ between the foci are shown. The hyperbola consists of the two (bold) branches and has the property that for all points $P = (x, y)$ it holds

$$\left|\overline{PF_1} - \overline{PF_2}\right| = const.$$

In the figure, the two vertices S_1 and S_2, which mark the shortest distance of the two hyperbola branches, are shown. The distance a from the center M to the vertices is called the **major semi-axis**. The **minor semi-axis** b is defined by

$$b^2 = \overline{MF_2}^2 - a^2$$

The hyperbola approaches the two lines $y = \pm\dfrac{b}{a}\,x$ („**asymptotes**") for very large positive and negative x-values. The main form of the hyperbola equation is

$$\frac{(x - x_0)^2}{a^2} - \frac{(y - y_0)^2}{b^2} = 1.\,\square$$

The case $\varepsilon < 1$

The first Kepler's law

In Eq. (6.16) we had defined ε as

$$\varepsilon = \sqrt{1 + \frac{2EL^2}{mA^2}} = \sqrt{1 + \frac{2EL^2}{m^3 G^2 M_s^2}},$$

i.e., if $\varepsilon < 1$ is to be, this means that the total energy E must be negative. We have already worked out in Sect. 6.1 that a particle cannot escape from the gravitational field of the Sun into infinity if the total energy is negative. It is bound to the Sun, its trajectory is entirely finite. If $\varepsilon < 1$, then the quantity b is defined by

$$b \;=\; a\sqrt{1-\varepsilon^2} = \frac{p}{1-\varepsilon^2}\sqrt{1-\varepsilon^2} = \frac{p}{\sqrt{1-\varepsilon^2}}$$

$$= \frac{L^2}{mA\sqrt{1-\varepsilon^2}} = \frac{L^2}{m^2GM_s\sqrt{1-\varepsilon^2}} > 0.$$

So Eq. (6.19) can be written as

$$\frac{(x+a\varepsilon)^2}{a^2} + \frac{y^2}{a^2(1-\varepsilon^2)} = \frac{(x+a\varepsilon)^2}{a^2} + \frac{y^2}{b^2} = 1.$$

This equation describes an ellipse with semi-axes a and b. The difference to the ellipse equation 1.4 on page 10 is that the coordinate origin is now no longer in the center of the ellipse, but in one of the focal points at a distance $e = a\varepsilon$ from the center, as shown in Fig. 6.8.

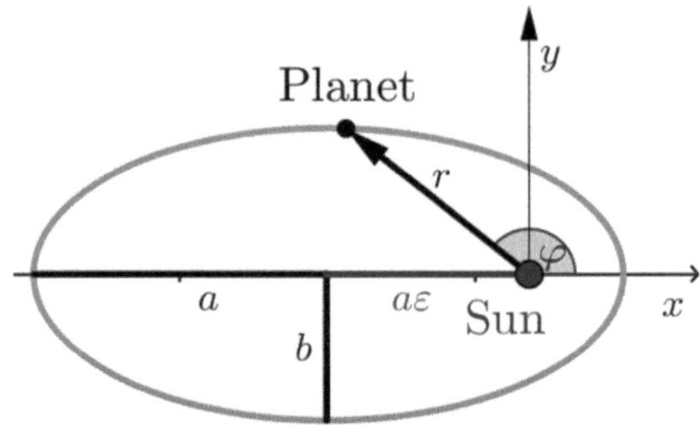

Figure 6.8.: Elliptical planetary orbit

We have thus derived the first Kepler's law from Newton's law of gravitation. The planets move on an elliptical orbit around the Sun. Here,

$$a = \frac{p}{1-\varepsilon^2} = \frac{L^2}{mA} \cdot \frac{1}{1 - \left(1 + \dfrac{2EL^2}{mA^2}\right)} = \frac{A}{2(-E)} = \frac{GM_s\, m}{2(-E)}$$

denotes the major semi-axis and

$$b \;=\; \frac{p}{\sqrt{1-\varepsilon^2}} = \frac{L^2}{m^2GM_s\sqrt{1-\varepsilon^2}}$$

$$= \frac{L^2}{m^2 G M_s \sqrt{1 - \left(1 + \dfrac{2EL^2}{mA^2}\right)}}$$

$$= \frac{L}{m^2 G M_s \sqrt{-\dfrac{2E}{mA^2}}}$$

$$= \frac{L}{m^2 G M_s \sqrt{-\dfrac{2E}{m^3 G^2 M_s^2}}}$$

$$= \frac{L}{\sqrt{2m(-E)}}$$

the minor semi-axis and

$$\varepsilon = \sqrt{1 + \frac{2EL^2}{mA^2}} = \sqrt{1 + \frac{2EL^2}{m^3 G^2 M_s^2}}$$

the eccentricity (see remark 1.3 on page 8) of the ellipse. Note that because $E < 0$, the above root expressions in which $-E$ occurs are well-defined.

As a special case, the circular orbit is included. For this, the two semi-axes a and b must be equal, so

$$a = \frac{GM_s m}{2(-E)} = \frac{L}{\sqrt{2m(-E)}} = b,$$

from which the condition

$$E = -\frac{G^2 M_s^2 m^3}{2L^2}$$

results by squaring and solving for E. If one considers

$$E = \frac{mv^2}{2} - \frac{GM_s m}{r}$$

and

$$L^2 = m^2 v^2 r^2,$$

then from the above equation

$$\frac{mv^2}{2} - \frac{GM_s m}{r} = -\frac{G^2 M_s^2 m^3}{2m^2 v^2 r^2}$$

and from this

$$\frac{v^4 r^2}{2v^2 r^2} - \frac{2GM_s r v^2}{2r^2 v^2} + \frac{G^2 M_s^2}{2v^2 r^2} = 0,$$

so

$$v^4 r^2 - 2GM_s r v^2 + G^2 M_s^2 = 0.$$

Division by r^2 results in

$$v^4 - \frac{2GM_s}{r} v^2 + \frac{G^2 M_s^2}{r^2} = \left(v^2 - \frac{GM_s}{r} \right)^2 = 0,$$

from which the relationship between the orbit radius r and the orbital velocity v follows:

$$v = \sqrt{\frac{GM_s}{r}}.$$

This formula looks similar to that of the escape velocity in Eq. 6.8 on page 92, but has in the denominator of the right side a radius of orbit that is always greater than the solar radius R_s. Also, the numerator is smaller by a factor of 2. Therefore, the orbital speed on a circular path is naturally smaller than the escape speed.

The second Kepler's law

The second Kepler's law follows directly from the fact that in the gravitational field the angular momentum is constant over time. Fig. 6.9 shows a planet orbiting the Sun. In the time interval dt, the planet moves by $d\vec{s} = \vec{v}\,dt$, the radius vector $\vec{r}$ sweeps over the area dA shown in this time interval.

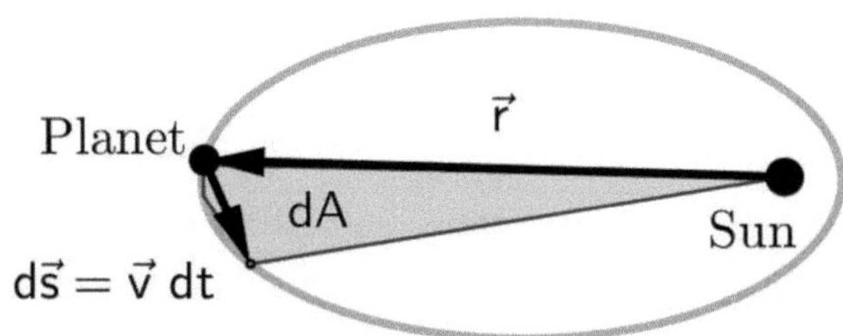

Figure 6.9.: Kepler's 2nd law, Area law

This is exactly half the size of the parallelogram spanned by the two vectors $\vec{r}$ and $d\vec{s}$ with the area $|\vec{r} \times d\vec{s}|$, so

$$dA = \frac{1}{2} |\vec{r} \times d\vec{s}| = \frac{1}{2} |\vec{r} \times \vec{v}\,dt| = \frac{dt}{2m} |\vec{r} \times m\vec{v}| = \frac{dt}{2m} \left| \vec{L} \right|,$$

where $\vec{L} = \vec{r} \times m\vec{v}$ is the angular momentum of the planet relative to the Sun. So it follows

$$\frac{dA}{dt} = \frac{1}{2m}\left|\vec{L}\right| = \frac{L}{2m},$$

i.e., the area swept per time interval is proportional to the magnitude of the angular momentum, hence constant. This is the statement of the second Kepler's law.

If T denotes the full orbital period, the area law states that for the area content A of the ellipse follows

$$A = \int_0^T \frac{dA}{dt}\, dt = \int_0^T \frac{L}{2m}\, dt = \frac{L}{2m}\int_0^T dt = \frac{TL}{2m}. \tag{6.20}$$

Kepler's 3rd Law

To derive the third Kepler's law, we still need to determine the area content of an arbitrary ellipse with the semi-axes a and b.

Remark 6.3. **MT: Area content of an ellipse**
To derive the area content of an ellipse, we first consider the unit circle with the origin as the center, which can be described by the equation

$$x^2 + y^2 = 1$$

From this it follows

$$y = y(x) = \pm\sqrt{1 - x^2},$$

where the functions $y(x) = \sqrt{1 - x^2}$ describe the upper half circle and $y(x) = -\sqrt{1 - x^2}$ the lower half circle. To determine the area content of the unit circle, it is sufficient to calculate the area content of the upper half circle and multiply it by 2. The area content of the upper half circle A_{uhc} is calculated by integrating over $y(x) = \sqrt{1 - x^2}$ in the limits from -1 to 1

$$A_{uhc} = \int_{-1}^{1} \sqrt{1 - x^2}\, dx.$$

To calculate this integral, we use the **substitution method**.

Remark 6.4. **MF: substitution method**
If $y = f(x)$ is a function that depends on the variable x and $x = g(t)$ is also a function that depends on the variable t , then the integral $\int f(x)\, dx$, which has x as the integration variable, can also be calculated by an integral with t as the integration variable. To do this, at each point of the original integral,

the variable x must be replaced by $g(t)$. This means: $f(x)$ becomes $f(g(t))$ and because of

$$\frac{dx}{dt} = \frac{g(t)}{dt} = \dot{g}(t)$$

the differential dx is replaced by

$$dx = \dot{g}(t)\,dt.$$

Therefore, the substitution formula follows

$$\int f(x)\,dx = \int f(g(t))\,\dot{g}(t)\,dt = \int f(g(t)) \cdot \frac{dx}{dt}\,dt. \qquad (6.21)$$

If you want to calculate a definite integral $\int_a^b f(x)\,dx$ using the substitution method, the integral limits change: If the variable x varies between a and b, then the variable t varies between $g^{-1}(a)$ and $g^{-1}(b)$ due to $t = g^{-1}(x)$. Here, $g^{-1}(x)$ is the **inverse function** of $g(t)$, which is defined by

$$t = g^{-1}(g(t)) \quad and \quad x = g\left(g^{-1}(x)\right).$$

So the substitution formula for definite integrals is:

$$\int_a^b f(x)\,dx = \int_{g^{-1}(a)}^{g^{-1}(b)} f(g(t)) \cdot \frac{dx}{dt}\,dt \;\square$$

We now apply the substitution formula to the integral

$$\int_{-1}^1 \sqrt{1 - x^2}\,dx$$

set

$$x = \sin t$$

and get

$$\frac{dx}{dt} = \cos t.$$

The inverse function of the sine is called **arcsine** ($\arcsin x$), i.e., it holds

$$t = \arcsin x$$

and the limits of the integral after substitution are

$$\arcsin(-1) = -\frac{\pi}{2},$$

since $\sin(-\pi/2) = -1$, and

$$\arcsin(1) = \frac{\pi}{2},$$

since $\sin(\pi/2) = 1$. Note that because

$$\sin^2 t + \cos^2 t = 1$$

it follows that for $-\pi/2 \leq t \leq \pi/2$

$$\cos t = \sqrt{1 - \sin^2 t}$$

is valid, so you get

$$\int_{-1}^{1} \sqrt{1 - x^2}\, dx = \int_{\arcsin(-1)}^{\arcsin(1)} \sqrt{1 - \sin^2(t)} \cdot \frac{dx}{dt}\, dt = \int_{-\frac{\pi}{2}}^{\frac{\pi}{2}} \cos t \cdot \cos t\, dt.$$

The integral on the right side is calculated using the formula of **partial integration**.

Remark 6.5. **MF: Formula of partial integration**
According to the product rule of differentiation (see Remark 4.6) the derivative of a product of two functions $u(x)$ and $v(x)$:

$$(u \cdot v)' = u' \cdot v + u \cdot v'$$

If you integrate both sides over the interval $[a, b]$, it follows

$$\int_a^b (u \cdot v)'\, dx = \int_a^b (u' \cdot v + u \cdot v')\, dx = \int_a^b u' \cdot v\, dx + \int_a^b u \cdot v'\, dx.$$

The left side is according to the main theorem of differential and integral calculus (see Remark 4.1 on page 53) but nothing else as

$$\int_a^b (u \cdot v)'\, dx = [u(x) \cdot v(x)]_a^b = u(b) \cdot v(b) - u(a) \cdot v(a).$$

So, after rearranging, we obtain the formula for partial integration

$$\int_a^b u' \cdot v\, dx = u(b) \cdot v(b) - u(a) \cdot v(a) - \int_a^b u \cdot v'\, dx. \qquad \Box \tag{6.22}$$

We apply this formula to the above integral and obtain with

$$u(x) = \sin x,\; v(x) = \cos x,\; u'(x) = \cos x,\; v'(x) = -\sin x$$

the result

$$\int_{-\frac{\pi}{2}}^{\frac{\pi}{2}} \cos t \cdot \cos t \, dt = [\sin(t)\cos(t)]_{-\frac{\pi}{2}}^{\frac{\pi}{2}} - \int_{-\frac{\pi}{2}}^{\frac{\pi}{2}} \sin t \, (-\sin t) \, dt.$$

Since

$$\cos\left(\frac{\pi}{2}\right) = \cos\left(-\frac{\pi}{2}\right) = 0$$

and

$$\sin^2(t) = 1 - \cos^2 t,$$

it follows

$$\begin{aligned}
\int_{-\frac{\pi}{2}}^{\frac{\pi}{2}} \cos^2 t \, dt &= \int_{-\frac{\pi}{2}}^{\frac{\pi}{2}} \sin^2 t \, dt = \int_{-\frac{\pi}{2}}^{\frac{\pi}{2}} 1 \, dt - \int_{-\frac{\pi}{2}}^{\frac{\pi}{2}} \cos^2 t \, dt \\
&= \frac{\pi}{2} - \left(-\frac{\pi}{2}\right) - \int_{-\frac{\pi}{2}}^{\frac{\pi}{2}} \cos^2 t \, dt.
\end{aligned}$$

Now we bring the integrals to one side and get

$$2\int_{-\frac{\pi}{2}}^{\frac{\pi}{2}} \cos^2 t \, dt = \pi.$$

Dividing by 2 finally yields

$$A_{uhc} = \int_{-1}^{1} \sqrt{1 - x^2} \, dx = \int_{-\frac{\pi}{2}}^{\frac{\pi}{2}} \cos^2 t \, dt = \frac{\pi}{2}$$

and thus for the area of the unit circle

$$A_{UC} = 2A_{uhc} = \pi.$$

We extend this result for the area A_{circle} of a circle with radius R, which has its center at the origin and is represented by the following equation

$$x^2 + y^2 = R^2.$$

With the substitution $x = t \cdot R$ and thus $\dfrac{dx}{dt} = R$

$$\begin{aligned}
A_{circle} &= 2\int_{-R}^{R} \sqrt{R^2 - x^2} \, dx = 2\int_{-1}^{1} \sqrt{R^2 - R^2 t^2} \cdot R \, dt \\
&= 2R^2 \int_{-1}^{1} \sqrt{1 - t^2} \, dt = 2R^2 \frac{\pi}{2} = \pi R^2.
\end{aligned}$$

We naturally obtain the well-known result.

To calculate the area of an arbitrary ellipse, we start from the defining equation

$$\frac{x^2}{a^2} + \frac{y^2}{b^2} = 1, a > 0, b > 0.$$

From this follows for the upper half ellipse

$$y = b\sqrt{1 - \frac{x^2}{a^2}}.$$

With the result of the circle, we obtain for the area A of the ellipse

$$\begin{aligned}
A &= 2\int_{-a}^{a} b\sqrt{1 - \frac{x^2}{a^2}}\, dx = 2b\int_{-a}^{a} \sqrt{\frac{a^2 - x^2}{a^2}}\, dx \\
&= \frac{2b}{a}\int_{-a}^{a} \sqrt{a^2 - x^2}\, dx = \frac{2b}{a}\frac{\pi a^2}{2} = \pi ab.\ \square
\end{aligned}$$

We combine this result with the above from Eq. (6.20) together, we get

$$A = \pi ab = \frac{TL}{2m}$$

and from that

$$T^2 = \frac{4m^2\pi^2 a^2 b^2}{L^2}$$

with the semi-axes derived further above

$$a = \frac{GM_s m}{2\left(-E\right)}$$

and

$$b = \frac{L}{\sqrt{2m\left(-E\right)}}.$$

We now check the third Kepler's law, i.e., whether the ratio of the third powers of the major semi-axis $\left(a^3\right)$ to the square of the orbital period $\left(T^2\right)$ is constant for all planets in the solar system, i.e., independent of the mass, the orbital period, the distance of the planets from the Sun or other planet-specific properties. It applies:

$$\frac{a^3}{T^2} = \frac{a^3}{\dfrac{4m^2\pi^2 a^2 b^2}{L^2}} = \frac{aL^2}{4m^2\pi^2 b^2} = \frac{\dfrac{GM_S m}{2(-E)}\cdot L^2}{4m^2\pi^2 \cdot \dfrac{L^2}{2m(-E)}} = \frac{GM_S}{4\pi^2}$$

The term on the right side contains only the universal gravitational constant G, the mass of the Sun M_S and the number $4\pi^2$, so no planet-specific peculiarities anymore. Thus, the third Kepler's law is also derived from the Newtonian law of gravitation.

However, Newton's approach goes beyond Kepler. It also includes solutions for orbital curves of celestial bodies that are not bound to the Sun. Before we turn to these cases, a note on the model used.

Real Planetary Orbits

In the derivation of the Kepler laws, for the sake of simplicity, we assumed that the Sun is stationary in one of the foci of the elliptical orbit. But the Sun actually also moves on an elliptical orbit opposite to the planet's orbit, with the center of mass of the Sun and planet in one of the common foci of the ellipse (for further details see [25]). Therefore, the statement of the first Kepler's law:

> „All planets move on elliptical orbits around the Sun. The Sun is in one of the two foci of the respective elliptical orbits."

is only approximately correct. We also assumed that we are dealing with a system consisting only of one planet and the Sun ("two-body problem"). This assumption is not true in reality. In particular, the gravitational forces of the planets among each other cause the planetary orbits to show (small) deviations from ellipses. But other effects such as e.g. the self-rotation (and associated flattening) of the Sun contribute to deviations from closed elliptical orbits. After each orbit, the line from the center to the point closest to the Sun, the **perihelion** of the orbit, is slightly rotated. This perihelion rotation is greatest for Mercury and of the order of one arcsecond per orbit; it is caused to about 90% by the other planets and to 10% by other effects. Already in the 19. century it was found that there is a (small) difference in the amount of about 43 arcseconds per century between the observed values of the perihelion rotation of Mercury and those calculated from the Newtonian theory of gravitation, which could not be explained. We will return to this phenomenon later and will show that the General Theory of Relativity exactly explains this difference between the calculation within the Newtonian theory and the astronomical observation.

The cases $\varepsilon > 1$ and $\varepsilon = 1$

We consider again Eq. 6.16 on page 100:

$$\varepsilon = \sqrt{1 + \frac{2EL^2}{mA^2}} = \sqrt{1 + \frac{2EL^2}{m^3 G^2 M_s^2}}$$

If $\varepsilon > 1$, this means that the total energy E must be positive. We showed in section 6.1 that a particle can escape from the gravitational field of the Sun into infinity if the total energy is positive. It is not bound to the Sun, its trajectory comes from infinity and tends back to infinity. If $\varepsilon > 1$, then the magnitude

$$b = a\sqrt{\varepsilon^2 - 1}$$

is greater than zero. We consider Eq. 6.19 on page 102:

$$\frac{(x + a\varepsilon)^2}{a^2} + \frac{y^2}{a^2 (1 - \varepsilon^2)} = 1$$

and can also write this as

$$\frac{(x + a\varepsilon)^2}{a^2} - \frac{y^2}{a^2 (\varepsilon^2 - 1)} = \frac{(x + a\varepsilon)^2}{a^2} - \frac{y^2}{b^2} = 1. \tag{6.23}$$

This equation represents a hyperbola with the semi-axes a and b and describes, for example, the path of a comet, whose direction is indeed changed by the gravitational field of the Sun, but due to its positive total energy, it moves away from the solar system again, as shown in Fig. 6.10.

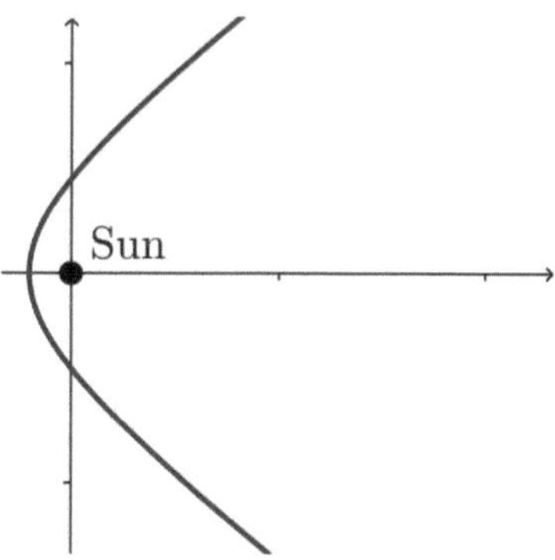

Figure 6.10.: Hyperbolic comet orbit

The case $\varepsilon = 1$ means that the energy E is zero, i.e., the particle does escape to infinity, but arrives there with vanishing kinetic energy. Since we divided by the term $1 - \varepsilon^2$ in the derivation of Eq. (6.19), we first set in Eq. (6.18):

$$x^2(1 - \varepsilon^2) + y^2 = p^2 - 2p\varepsilon x$$

With $\varepsilon = 1$, it follows that in this case

$$y^2 = p^2 - 2px$$

applies. The trajectory is therefore a parabola, see Fig. 6.11.

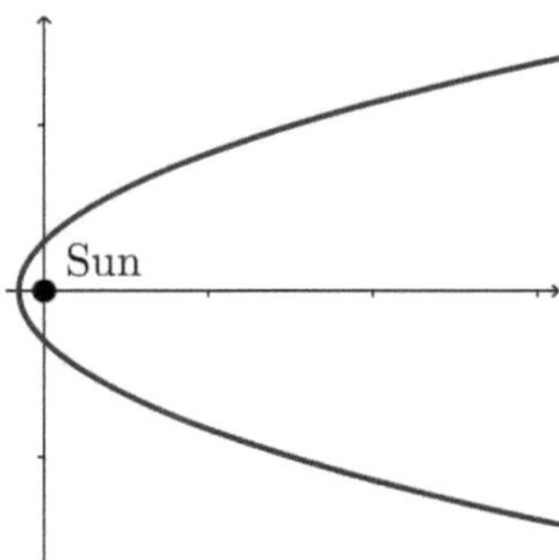

Figure 6.11.: Parabolic comet orbit

Overall, the conclusion from Newton's law of gravitation is that celestial bodies in the solar system (always in the ideal two-body situation) move on conic section curves. The derivation of the various possible trajectories of celestial bodies in the solar system has also shown that not only the three Kepler laws follow from Newton's law of gravitation, but that (among other phenomena) hyperbolic or parabolic orbits can also be predicted. Thus, Newton's theory of gravitation went beyond the reproduction of then already known knowledge and was able to explain almost all astronomical observations in the following nearly 250 years.

6.3. The Gravitational Field of Extended Bodies

So far, we have (silently) assumed that a body that generates a gravitational field is a point mass of total mass M at the center. Now we want to investigate the influence of the spatial mass distribution on the gravitational field and will find out that the gravitational force generated by a central mass at a point on the surface or outside of it in many situations is roughly the same as if the entire mass is localized in the center.

Hollow Sphere

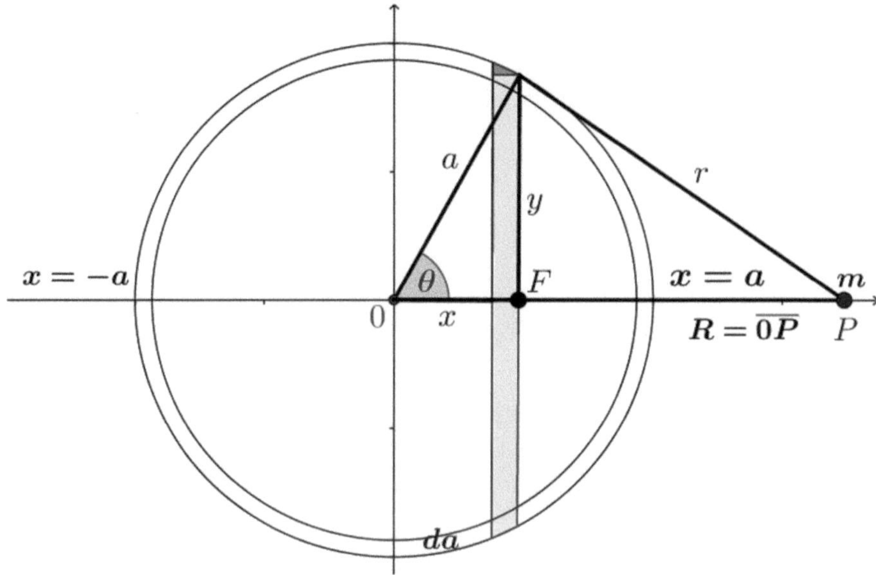

Figure 6.12.: Gravitational field of a spherical shell

We first calculate the gravitational field of a hollow sphere for an observer with mass m at point P outside the sphere.

The spherical shell may have the radius a and an infinitesimally small wall thickness $da \ll a$ („$\ll$" means „is very small compared to"). The mass M of the spherical shell is homogeneous, i.e. evenly distributed on the spherical shell. If ϱ is the **mass density**, which is constant at every location on the spherical shell, then ϱ is defined by the mass divided by the volume of the spherical shell

$$\varrho = \frac{M}{\text{Volume of the spherical shell}}.$$

A circular disc of (small) thickness dx cuts a circular ring (gray) from the spherical shell. We calculate its width ds by enlarging and interpreting the small triangle in Fig. 6.12. First, we assume that ds, which is actually a piece of circular arc, is approximately a straight line due to the "smallness" of the sizes da and dx, see Fig. 6.13. Since the angle between a and dx is equal to the angle θ, which is formed by the radius of the sphere a and the distance R between the origin and point P, the angle between dx and ds is equal to $90° - \theta$.

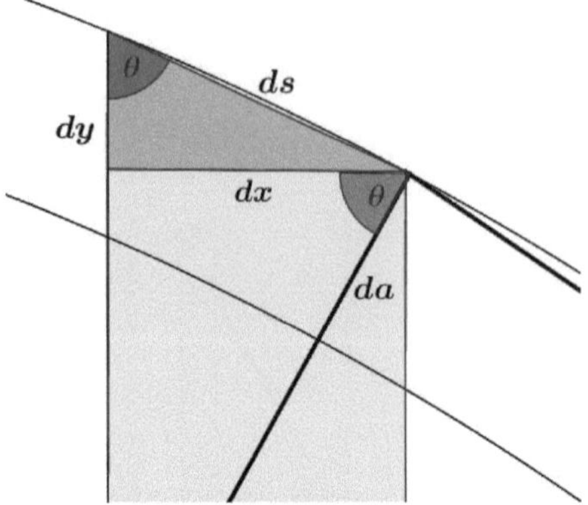

Figure 6.13.: Section of the spherical shell

   This means, one finds the angle θ also in the small triangle as the angle between dy and ds, so that for the width of the circular ring applies („$\Rightarrow$" means „it follows"):

$$\frac{dx}{ds} = \sin\theta \Rightarrow ds = \frac{dx}{\sin\theta} \Rightarrow dx = ds\,\sin\theta$$

The mass dM of the strip with the width ds and the thickness da and the circumference $2\pi y$ is calculated by

$$dM = \varrho \cdot Volume\,of\,the\,circular\,ring = \varrho \cdot 2\pi y \cdot ds \cdot da,$$

which can be written as

$$dM = \varrho \cdot 2\pi a \cdot ds \cdot \sin\theta \cdot da = \varrho \cdot 2\pi a \cdot dx \cdot da$$

due to $y = a\sin\theta$. All mass elements dM have the same distance r from point P, so the potential energy of a small test mass m in P in the gravitational field, which is generated by dM, is equal to

$$dE_{pot} = -\frac{G\,dM\,m}{r}.$$

The contribution of the entire hollow sphere to the gravitational field at point P is now obtained by summing all mass elements dM of the spherical shell, i.e. by integrating over all strips dx from $x = -a$ to $x = a$:

$$E_{pot} = \int_{-a}^{a} dE_{pot} = -Gm\int_{-a}^{a}\frac{dM}{r} = -Gm\varrho 2\pi a \cdot da\int_{-a}^{a}\frac{dx}{r}$$

In order to be able to calculate this integral, we still need to find a relationship between x and r. From the above graphic, we read with the *Pythagorean theorem*

$$r^2 = y^2 + (R - x)^2 = y^2 + x^2 + R^2 - 2Rx = a^2 + R^2 - 2Rx$$

from which

$$x = -\frac{r^2}{2R} + \frac{a^2}{2R} + \frac{R}{2}$$

follows. This gives us

$$\frac{dx}{dr} = \frac{d}{dr}\left(-\frac{r^2}{2R} + \frac{a^2}{2R} + \frac{R}{2}\right) = -\frac{r}{R}.$$

In order to apply the substitution formula $x \to r$ to the last integral, the changed integral limits must still be calculated. We set $x = -a$ and get

$$-a = -\frac{r^2}{2R} + \frac{a^2}{2R} + \frac{R}{2} \Rightarrow r^2 = R^2 + 2aR + a^2 \Rightarrow r^2 = (R + a)^2 \Rightarrow r = R + a.$$

Similarly for $x = a$:

$$a = -\frac{r^2}{2R} + \frac{a^2}{2R} + \frac{R}{2} \Rightarrow r^2 = R^2 - 2aR + a^2 \Rightarrow r^2 = (R - a)^2 \Rightarrow r = R - a > 0 \tag{6.24}$$

This results in

$$
\begin{aligned}
E_{pot} &= -Gm\varrho 2\pi a \cdot da \int_{-a}^{a} \frac{dx}{r} = -Gm\varrho 2\pi a \cdot da \int_{R+a}^{R-a} \frac{1}{r} \cdot \frac{dx}{dr}\, dr \\
&= -Gm\varrho 2\pi a \cdot da \int_{R+a}^{R-a} \frac{1}{r} \cdot \left(-\frac{r}{R}\right) dr = \frac{Gm\varrho 2\pi a \cdot da}{R} \int_{R+a}^{R-a} dr \\
&= \frac{Gm\varrho 2\pi a \cdot da}{R}\left[R - a - (R + a)\right] = -\frac{Gm\varrho 4\pi a^2 \cdot da}{R}.
\end{aligned}
\tag{6.25}
$$

We use the relationships again

$$\frac{dx}{ds} = \sin\theta \Rightarrow ds = \frac{dx}{\sin\theta} \Rightarrow dx = ds\sin\theta,$$

to calculate the surface of a sphere with radius a. With the same arguments as above, because $y = a\sin\theta$ for an infinitesimally small surface element dA on the sphere, we get

$$dA = 2\pi y \cdot ds = 2\pi a\sin\theta \cdot ds = 2\pi a \cdot dx.$$

If we sum over all surface elements dA, we get analogously

$$A = \int_{-a}^{a} 2\pi a \, dx = 2\pi a \int_{-a}^{a} dx = 2\pi a \left[a - (-a)\right] = 4\pi a^2.$$

If we consider that the volume V of the spherical shell is equal to the surface times the wall thickness, i.e.,

$$V = A \cdot da = 4\pi a^2 da,$$

then the mass M of the spherical shell follows as

$$M = \varrho V = \varrho 4\pi a^2 da,$$

so that the potential energy finally follows as

$$E_{pot} = -\frac{Gm\varrho 4\pi a^2 \cdot da}{R} = -\frac{GmM}{R}.$$

The gravitational force is obtained from

$$\vec{F} = -grad\, E_{pot} = -\frac{dE_{pot}}{dR}\, \vec{e}_R = -\frac{GmM}{R^2}\, \vec{e}_R.$$

This means:

> The gravitational force and potential energy in the gravitational field outside a homogeneous spherical shell of mass M are exactly the same as if the mass M were concentrated at the center of the hollow sphere.

If we are inside the hollow sphere, the derivation is similar. In such a case, the distance R of the point P from the center of the hollow sphere is less than the radius $R < a$, and the upper integration limit in Eq. (6.25) changes with (6.24) to

$$a = -\frac{r^2}{2R} + \frac{a^2}{2R} + \frac{R}{2} \Rightarrow r^2 = R^2 - 2aR + a^2 \Rightarrow r^2 = (a - R)^2 \Rightarrow r = a - R > 0.$$

Thus, the potential energy becomes

$$\begin{aligned}
E_{pot} &= \frac{Gm\varrho 2\pi a \cdot da}{R} \int_{R+a}^{a-R} dr = \frac{Gm\varrho 2\pi a \cdot da}{R}(-2R) \qquad (6.26)\\
&= -\frac{Gm\varrho 4\pi a^2 \cdot da}{a} = -\frac{GmM}{a} = const.
\end{aligned}$$

Therefore, for a particle inside the hollow sphere, the potential energy is always constant, regardless of where within the sphere the particle is located. Since the gravitational force results from the potential energy by derivation, in this case

$$\vec{F} = -grad\, E_{pot} = -grad\,(const.) = 0.$$

This means:

> The gravitational force is zero everywhere inside the hollow sphere. The contributions of the individual surface elements, which act in different directions, exactly cancel each other out.

Solid Sphere

We now move from the spherical shell to a homogeneous solid sphere. A homogeneous solid sphere with radius A is formed by summation/integration over concentric spherical shells, whose radii run from 0 to A. The mass of a homogeneous solid sphere is thus

$$M = \varrho V = \varrho \int_0^A 4\pi a^2 da = 4\pi\varrho \cdot \frac{A^3}{3}.$$

For a point P with mass m outside the solid sphere $(R > A)$ it follows from Eq. (6.25)

$$E_{pot} = \int_0^A -\frac{Gm\varrho 4\pi a^2}{R}\, da = -\frac{Gm\varrho 4\pi}{R} \int_0^A a^2\, da = -\frac{Gm\varrho 4\pi}{R}\frac{A^3}{3} = -\frac{GmM}{R}.$$

We thus obtain exactly the same result as for the spherical shell:

> Gravitational force and potential energy in the gravitational field outside a homogeneous solid sphere of mass M are exactly the same as in the case where the mass M would be concentrated at the center of the solid sphere.

For a point inside the solid sphere $(R < A)$ we divide the integration over a into two parts, namely one where the radius a is smaller than R $(0 \le a \le R$, where Eq. (6.25) is to be applied) and one part, where the radius a is larger than R $(R \le a \le A$, where Eq. (6.26) is to be used). It follows:

$$\begin{aligned}
E_{pot} &= \int_0^R -\frac{Gm\varrho 4\pi a^2}{R}\, da + \int_R^A -\frac{Gm\varrho 4\pi a^2 \cdot da}{a} \\
&= -Gm\varrho 4\pi \left(\int_0^R \frac{a^2}{R}\, da + \int_R^A a\, da \right)
\end{aligned}$$

With $M = 4\pi\varrho \cdot \dfrac{A^3}{3}$ it follows after calculation of the two integrals

$$
\begin{aligned}
E_{pot} &= -Gm\varrho 4\pi \left(\frac{R^2}{3} + \frac{A^2}{2} - \frac{R^2}{2} \right) = -Gm\varrho 4\pi \left(-\frac{R^2}{6} + \frac{A^2}{2} \right) \\
&= -\frac{GmM}{2A^3} \left(3A^2 - R^2 \right).
\end{aligned}
$$

Physically, the last equation means that only the mass elements of the solid sphere, whose distance to the center is less than R, contribute to the potential, the term $-3GmM/2A$ only provides a constant contribution to the potential energy and expresses that the mass elements of the solid sphere, which have a distance greater than R to the center, cancel each other out. The gravitational force $\vec{F_i}$ inside is given by

$$
\vec{F_i} = -grad\, E_{pot} = -\frac{d}{dR} \left(\frac{GmM}{2A^3} R^2 \right) \vec{e}_R = -\frac{GmM}{A^3} R\, \vec{e}_R.
$$

For the gravitational force F of a solid sphere with radius A it follows summarizing

$$
\vec{F} = \begin{cases} -\dfrac{GmM}{R^2}\, \vec{e}_R & f\ddot{u}r\ R \geq A \\[4mm] -\dfrac{GmM}{A^3}\, R\,\vec{e}_R & f\ddot{u}r\ R \leq A \end{cases}.
$$

We have shown that our assumption of a point mass in the derivation of the laws in the last chapters was justified at least for those central bodies that can be modeled approximately as hollow or solid spheres. It makes no difference whether the total mass of such bodies is distributed arbitrarily or concentrated at the center, as long as the test body is outside the central body.

6.4. The Poisson Equation

We derive another important equation that represents Newton's law of gravitation in a different form. Newton's equation of motion in a gravitational field, generated by a body with mass M, according to Eq. (6.6) for a (test) particle with mass m

$$
m\ddot{\vec{r}} = \vec{F} = -m\,\nabla\phi(\vec{r}) = -m\,\nabla \left(-\frac{GM}{r} \right) = m \left(-\frac{GM}{r^3}\,\vec{r} \right).
$$

If we imagine the mass M as the homogeneously distributed mass of a sphere with radius r and volume

$$
V = \frac{4}{3}\pi r^3
$$

and mass density ρ, then

$$M = \rho V = \rho \frac{4}{3}\pi r^3,$$

i.e., in such a case the gravitational force on the surface of the sphere with the distance r from the center of the sphere is:

$$\vec{F} = m\left(-\frac{GM}{r^3}\,\vec{r}\right) = m\left(-\frac{G\rho\frac{4}{3}\pi r^3}{r^3}\,\vec{r}\right) = -m\left(\frac{4}{3}\pi G\rho\,\vec{r}\right) \qquad (6.27)$$

Remark 6.6. **MT: Divergence**

We introduce another so-called **differential operator** (the *gradient* is also one), namely the **divergence**. The gradient is applied to a scalar function ϕ (e.g. $\phi(\vec{r}) = -GM/r$) and turns it into a vector (e.g. $\nabla\phi(\vec{r}) = (GM/r^3)\,\vec{r}$). The divergence is in a certain sense the reversal: If you have a vector field

$$\vec{A}(\vec{r}) = \begin{pmatrix} A_x(\vec{r}) \\ A_y(\vec{r}) \\ A_z(\vec{r}) \end{pmatrix},$$

then the divergence is defined as

$$div\,\vec{A} = \frac{\partial A_x}{\partial x} + \frac{\partial A_y}{\partial y} + \frac{\partial A_z}{\partial z},$$

where in the last equation we have again omitted the argument $\vec{r}$ for simplicity. Thus, the divergence turns a vector into a scalar, i.e., a number. $\square$

To calculate the divergence of the gravitational force on the sphere surface (6.27), we first calculate the divergence of the vector

$$\vec{r} = \begin{pmatrix} x \\ y \\ z \end{pmatrix} :$$

$$div\,\vec{r} = div\begin{pmatrix} x \\ y \\ z \end{pmatrix} = \frac{\partial x}{\partial x} + \frac{\partial y}{\partial y} + \frac{\partial z}{\partial z} = 1 + 1 + 1 = 3$$

So it follows

$$div\,\vec{F} = div\left(-m\left(\frac{4}{3}\pi G\rho\,\vec{r}\right)\right) = -m\frac{4}{3}\pi G\rho\,(div\,\vec{r})$$

$$= -m\frac{4}{3}\pi G\rho\, 3 = -m\, 4\pi G\rho.$$

Since the gravitational potential $\vec{F} = -m\,\nabla\phi(\vec{r})$ applies, it follows

$$div\,\nabla\phi(\vec{r}) = 4\pi G\rho.$$

If you write out the divergence of the gradient $\nabla\phi$, you get

$$
\begin{aligned}
div\,\nabla\phi(\vec{r}) &= div\left(\frac{\partial\phi}{\partial x}, \frac{\partial\phi}{\partial y}, \frac{\partial\phi}{\partial z}\right) = \frac{\partial}{\partial x}\left(\frac{\partial\phi}{\partial x}\right) + \frac{\partial}{\partial y}\left(\frac{\partial\phi}{\partial y}\right) + \frac{\partial}{\partial z}\left(\frac{\partial\phi}{\partial z}\right) \\
&= \frac{\partial^2\phi}{\partial x^2} + \frac{\partial^2\phi}{\partial y^2} + \frac{\partial^2\phi}{\partial z^2},
\end{aligned}
\tag{6.28}
$$

which is the sum of the second derivatives, also called the **Laplace operator**, abbreviated as Δ or ∇^2, so

$$\nabla^2\phi = \Delta\phi = 4\pi G\rho. \tag{6.29}$$

This equation is called the **Poisson equation** after the French mathematician Siméon Denis Poisson. It is also valid if there is no homogeneous mass distribution and no spherical volume. The mathematical means required to derive the general case, however, go beyond this introductory treatment of General Relativity. You can read the derivation of the general case in [25]. The Poisson equation can be interpreted as the gravitational field ϕ on the left side and its source, namely the mass density ρ, on the right side. It plays an important role in the derivation of the Einstein equations in part IV.

6.5. Gravitational and Inertial Mass, Equivalence Principle

We have called the property of an object that causes the gravitational force to act on another object mass and denoted it with m. We have used exactly the same term and notation to denote the resistance of a body to accelerations, without so far accounting for the fact that these properties of an object are the same. In fact, in physics, these two phenomena are distinguished and the first property is called **gravitational mass** and the second **inertial mass**. It is easy to imagine that the gravitational and inertial mass of an object may not be the same.

Let m_g denote the gravitational mass and m_i the inertial mass of a body, then the force that the Earth exerts on the body near the Earth's surface is given by:

$$F = \frac{GM_E\, m_g}{R_E^2},$$

where M_E denotes the gravitational mass of the Earth and R_E the Earth's radius. On the other hand, according to Newton's second law for any force

$$\vec{F} = m_i \vec{a},$$

from which the magnitude of the acceleration due to gravity is given by

$$a = \frac{F}{m_i} = \frac{GM_E}{R_E^2} \cdot \frac{m_g}{m_i}.$$

If the weight force were just another property of matter, like electrical charge or density, then the ratio m_g/m_i should depend on the chemical composition of the body, its temperature, or any other physical quantity, i.e., the acceleration due to gravity would be different for different bodies. However, experience shows that the acceleration due to gravity a is the same for all objects. Thus, the ratio m_g/m_i has the same value for every body, which is why we set

$$m_g = m_i = m$$

and thereby simultaneously determine the magnitude and units of Newton's gravitational constant G. The acceleration due to gravity is then given by

$$a = \frac{GM_E}{R_E^2} \tag{6.30}$$

and is no longer dependent on the mass of the body. Although the equivalence of gravitational and inertial mass has been tested with refined experiments to a relative error of 10^{-12} and is thus one of the best-secured physical laws, one must not forget that it is (only) an experimental finding. The equality of gravitational and inertial mass, which is called the **equivalence principle**, forms one of the most important foundations of Einstein's General Theory of Relativity, which we will discuss in part IV of this book.

7. Literature References and Further Reading for Part I

At the end of the first part, I would like to point out some books on Newtonian mechanics and the mathematics used in deriving the physical laws. There are about 2500 book offers on Newtonian mechanics alone on Amazon in the "Science and Technology" category. Therefore, literature recommendations are always very subjective, i.e., the small selection that I would like to present here consists of those books that were particularly helpful in the preparation of this book. In general, it is easier to acquire knowledge if you approach the subject you want to learn from different angles. This means that all the books presented here can help to better understand the contents of this first part.

In the category "popular science books", there are two recommendable books that go far beyond what we have discussed here in this first part. George Gamov describes in his thin booklet, Gravity [14], the Newtonian theory of gravitation and its implications mostly in everyday language and uses mathematical expressions at a level slightly below what has been used here so far.

Particularly noteworthy is the book by Bernard Schutz, Gravity from the ground up [28], which, much like this book, deals with both Newtonian and Einsteinian gravity theory, and manages with little mathematics (e.g., without calculus). The effects of the two theories are discussed. In the Earth-Moon system, in our solar system, in galaxies, with neutron stars and black holes, and for the entire universe, it is depicted in detail and visually attractive. So if you are looking for a more descriptive portrayal of the physical phenomena associated with gravity, Schutz is the place to be.

In the category "What should/could I bring to better understand the physics and mathematics up to this point" there are a number of physics and mathematics books. The school book, High School Physiks [31], should be mentioned, which deals with Newtonian mechanics at a similar level, but with many examples and vivid illustrations.

Among the books that teach basic mathematical knowledge, the school book, Essential Mathematics for Undergraduates [5], should be mentioned first. My (*German*) textbook [24], which aim to bridge the gap from school to university, serves a similar purpose. There, differential and integral calculus (in one dimension) as well as simple vector and matrix calculation are explained in

© The Author(s), under exclusive license to Springer-Verlag GmbH, DE, part of Springer Nature 2026

M. Ruhrländer, *Ascent to the Einstein Equations*,

https://doi.org/10.1007/978-3-662-72672-3_7

detail and with many examples.

In the category "books used/helpful for the creation of the first part", the very extensive textbook by Paul A. Tipler, Physics [35], should be mentioned first. In this companion text for the initial physics lectures that prospective scientists and engineers must attend, the first ten chapters lay the mathematical and physical foundations for the treatment of Newtonian mechanics very extensively and with many examples and graphic representations. The level of the book is slightly below what is described in the first part, but was a great help, especially in the drafting of the first chapters. For those also interested in physical theories besides Newtonian mechanics, Tipler's book can be highly recommended as an introduction to quantitative physics. A very similar concept is pursued by the (also very extensive) book, Physics by Halliday, Resnick and Walter [15], and it is almost a matter of taste which of the two books one prefers, as they are almost identical in structure and content.

As for the mathematics presented in the first part in eighteen "Mathematical Foundations" and eight "Mathematical Tools", the following three mathematical textbooks are well suited to anchor the new mathematical concepts and results more deeply in content. My (*German*) textbook, Lineare Algebra für Naturwissenschaftler und Ingenieure [23], just as (the english) Linear Algebra of Gilbert Strang [32], are easy to understand and cover the most importend aspects of linear algebra such as matrices and vectors. A very good introduction to calculus is provided by the textbook, Thomas' Calculus [33].

For the formulation of the somewhat more difficult passages in the first part of the book, such as the derivation of planetary orbits from Newton's law of gravitation, some "real" physics textbooks had to be used. Here, with increasing difficulty, the books by Feynman [9], Fliessbach [13] and Scheck [27] should be mentioned. These belong to the category of "further reading", as they go significantly beyond what is presented here in the first part, both in terms of mathematical representation and physical content.

Part II.

Vector and Tensor Calculus in the Euclidean Plane

In part I we have already seen that the physical phenomena discussed there can mostly be expressed in vector equations. The formulation of physical laws in the form of vector equations has - as we will see - the invaluable advantage that one has a free choice of the coordinate system, i.e., one can choose the coordinate system that is best suited for solving the given problem. We have already used this principle in part I at several points (for example, we calculated the orbits of planets using polar coordinates and placed the origin of coordinates at one focus of the ellipse) without each time pointing out this freedom of choice. The vectors or vector fields were also the most complicated quantities we had to deal with so far. In addition to scalars (numbers) and vectors, however, there are also the so-called **tensors**, which are characterized by the fact that they do not point in *no* direction (like scalars), not in *one* direction (like vectors), but in *several* directions at the same time. It will turn out that these quantities are suitable for describing the physical laws of relativity theory. Physical laws that are formulated as tensor equations have the same advantage as vector equations: they allow the choice of any coordinate system without changing their form.

In this part, we introduce tensor calculus for the simplest of all possible cases, namely for the two-dimensional Euclidean plane. We do this for one reason because the tensorial representations needed in later chapters can almost be transferred one-to-one from the two-dimensional, and for another reason because the derivations and formulas in two dimensions are simpler and more intuitive than in higher ones. Nevertheless, this second part is sometimes abstract in the first chapters and then formulated in a language ("index notation") that will require perseverance when working through it for the first time. But this effort is worthwhile because it already lays a large part of the (mathematical) foundations for relativity theory.

8. Vector Calculus in the Euclidean Plane

In this chapter, we expand our knowledge about vector calculus in the so-called **Euclidean plane**. Although the space surrounding us is three-dimensional, we are dealing with vector calculation in the plane in this chapter for *simplicity reasons*. The new concepts and terminologies of (two-dimensional) vector calculus can mostly be generalized one-to-one to the three- and later also to the four-dimensional case, so we refrain from the more complicated representations in three dimensions here.

8.1. Euclidean Plane, Distance, Scalar Product

We want to define in the following what exactly we mean by the Euclidean plane. The Euclidean plane consists of a set of points. We choose any of these points as the origin of a coordinate system, whose two perpendicular x- and y-axes intersect at this point. Such a coordinate system is named after the French mathematician René Descartes **Cartesian coordinate system**. The two axes of the coordinate system each represent the real numbers, for example, all positive real numbers are found on the x-axis to the right of the origin and all negative ones to the left. Each point P in the plane can then be uniquely identified by two numbers, namely its coordinates, i.e., the projections onto the coordinate axes, identify $P \to (P^x, P^y)$, as shown in Fig. 8.1.

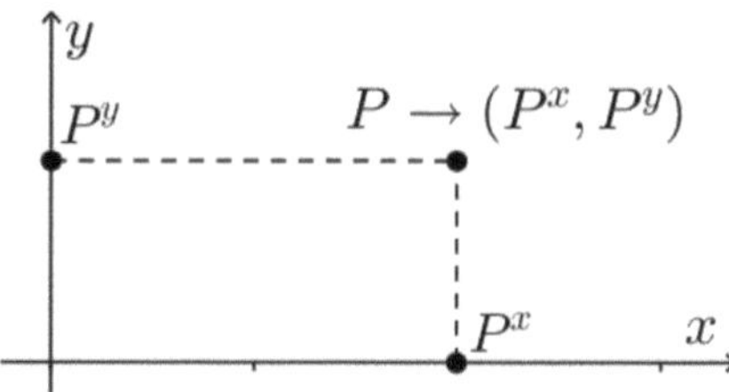

Figure 8.1.: Point in Euclidean plane

From now on, we use the symbol $\to$ to represent the coordinates of a point in order to distinguish between the geometric object "point of the plane" and its

© The Author(s), under exclusive license to Springer-Verlag GmbH, DE, part of Springer Nature 2026
M. Ruhrländer, *Ascent to the Einstein Equations*,
https://doi.org/10.1007/978-3-662-72672-3_8

coordinates in a specific (x, y) coordinate system K. The point as a geometric object always remains unchanged, however, its coordinates (P^x, P^y) change depending on how the coordinate system was chosen.

For example, if we choose a second (x', y') coordinate system K', whose origin is exactly the point P, then P in this coordinate system has the coordinates $(0,0)$. If it is important to document which coordinate system we are currently using, we write the coordinate system under the arrow $\rightarrow$, thus:

$$P \underset{K}{\rightarrow} (P^x, P^y) \;\; \text{bzw.} \;\; P \underset{K'}{\rightarrow} (0,0)$$

The shorthand $P \underset{K}{\rightarrow} (P^x, P^y)$ reads as follows: "The point P has the coordinates P^x and P^y in the coordinate system K".

Distance between two vectors

Let K be a Cartesian (x, y) coordinate system again. We *define* the **distance** d between two points $P \rightarrow (P^x, P^y)$ and $Q \rightarrow (Q^x, Q^y)$ as

$$d(P, Q) = \sqrt{(Q^x - P^x)^2 + (Q^y - P^y)^2}. \tag{8.1}$$

As can be seen in Fig. 8.2, the **Pythagorean theorem** (see Chap. 29 on page 577) was the basis for this formula for calculating the distance between two points. Since the two points P and Q are geometric objects, the distance between them is also a geometric object, i.e., it is always the same, no matter which coordinate system is chosen.

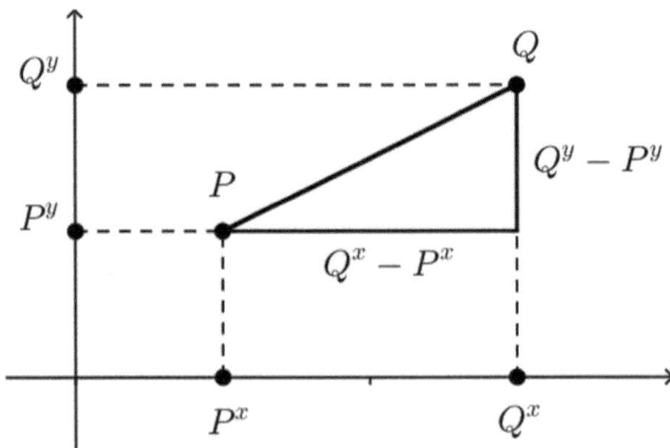

Figure 8.2.: Distance between two point in the Euclidian plane

This property is called **invariance under coordinate system change**. The two-dimensional space has received a coordinate system-independent (**invariant**)

structure by defining how distances are measured. This structure is called **Euclidean geometry**, and the set of all points is called the **Euclidean plane**. The Euclidean plane (and likewise the three-dimensional Euclidean space) is characterized by the way lengths are measured, and we will see that both in Special and General Relativity this natural method of measurement for us must be replaced by a completely different one.

We now want to show that from the invariance of length measurement also follows the invariance of angle measurement. To do this, we need to introduce the geometric object **vector** and first choose a geometric approach, which we will later replace with a *transformation-related* definition. Vectors are geometrically defined by two points P and Q. The point P is called the starting point and the point Q the endpoint of the vector. We graphically represent vectors as arrows and denote them with lowercase letters with an arrow above them $\vec{a}$. A vector can be represented - similar to a point in the plane by its coordinates - also by its **components**

$$\vec{a} \underset{K}{\rightarrow} \begin{pmatrix} a^x \\ a^y \end{pmatrix} = \begin{pmatrix} Q^x - P^x \\ Q^y - P^y \end{pmatrix},$$

which is graphically represented in Fig. 8.3.

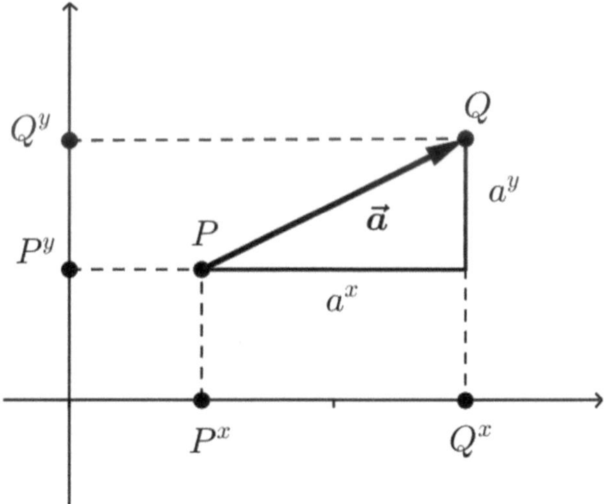

Figure 8.3.: Vector in the Euclidean plane

Here too, we have chosen a new notation. On the one hand, we want to make clear by using the arrow $\rightarrow$ that the components of a vector a^x, a^y depend on the chosen coordinate system (not the vector itself!), and on the other hand, we have moved the indices x, y upwards. The reasons for this will be explained later in Sect. 9.2. The length of a vector $\vec{a}$ is *defined* as the distance between its starting and ending points, we write for this

$$a = \sqrt{(a^x)^2 + (a^y)^2} = \sqrt{(Q^x - P^x)^2 + (Q^y - P^y)^2},$$

i.e., the length of a vector is also an invariant quantity. As already shown in Remark 2.4, the sum of two vectors is again a vector and also the product of a vector with a number.

Scalar product of two vectors

If we for two vectors

$$\vec{a} \to \begin{pmatrix} a^x \\ a^y \end{pmatrix}, \vec{b} \to \begin{pmatrix} b^x \\ b^y \end{pmatrix}$$

define a scalar product $\vec{a} \cdot \vec{b}$ by

$$\vec{a} \cdot \vec{b} = a^x b^x + a^y b^y \tag{8.2}$$

then the calculation rules apply as we have derived them in Remark 4.2 on page 55. In particular, it follows that the length of $\vec{a}$ can be expressed by the scalar product:

$$a = \sqrt{(a^x)^2 + (a^y)^2} = \sqrt{\vec{a} \cdot \vec{a}},$$

so

$$a^2 = \vec{a} \cdot \vec{a}.$$

Since the scalar product defined in (8.2) is calculated by the components of the two vectors and these depend on the choice of the coordinate system, it is not immediately clear whether the same result always comes out in any arbitrary coordinate system. That the scalar product is indeed an invariant quantity can be seen as follows. First, with the calculation rules for the scalar product, that with $\vec{c} = \vec{a} + \vec{b}$

$$c^2 = \left(\vec{a} + \vec{b}\right) \cdot \left(\vec{a} + \vec{b}\right) = \vec{a} \cdot \vec{a} + 2\vec{a} \cdot \vec{b} + \vec{b} \cdot \vec{b} = a^2 + 2\vec{a} \cdot \vec{b} + b^2$$

applies. By rearranging after the scalar product, one obtains from this

$$\vec{a} \cdot \vec{b} = \frac{1}{2}\left(c^2 - a^2 - b^2\right).$$

Since the right side of the equation is the same number in all coordinate systems, this also applies to the left side. The scalar product of two vectors is therefore an invariant quantity.

Example 8.1.
As an example, we calculate the scalar products of the two **Cartesian basis vectors**, which are defined by the stipulation that they should point in the direction of the positive coordinate axes and have the length 1, so

$$\vec{e}_x \;\rightarrow\; \begin{pmatrix} 1 \\ 0 \end{pmatrix},$$

$$\vec{e}_y \;\rightarrow\; \begin{pmatrix} 0 \\ 1 \end{pmatrix}.$$

Note that the basis vectors are *coordinate system dependent* as they take on the directions of the coordinate axes. It follows

$$\begin{aligned}
\vec{e}_x \cdot \vec{e}_x &= \vec{e}_y \cdot \vec{e}_y = 1, \\
\vec{e}_x \cdot \vec{e}_y &= \vec{e}_y \cdot \vec{e}_x = 0. \;\square
\end{aligned}$$

Angle measurement and linear combination

With the scalar product, we can now also define angle measurement invariantly. To do this, we assign two vectors $\vec{a}$ and $\vec{b}$ the angle $\varphi\left(\vec{a}, \vec{b}\right)$ formed between them by

$$\cos \varphi \left(\vec{a}, \vec{b}\right) = \frac{\vec{a} \cdot \vec{b}}{ab}.$$

Since only invariant quantities are on the right side, the left side is also invariant. Every vector

$$\vec{a} \rightarrow \begin{pmatrix} a^x \\ a^y \end{pmatrix}$$

can be written as a **linear combination** of the basis vectors:

$$\vec{a} = a^x \vec{e}_x + a^y \vec{e}_y,$$

because for the components of the vector on the right side it holds

$$a^x \begin{pmatrix} 1 \\ 0 \end{pmatrix} + a^y \begin{pmatrix} 0 \\ 1 \end{pmatrix} = \begin{pmatrix} a^x \\ a^y \end{pmatrix}.$$

Once a coordinate system with the corresponding basis vectors has been selected, the components of a vector

$$\vec{a} \rightarrow \begin{pmatrix} a^x \\ a^y \end{pmatrix}$$

in this coordinate system are given by

$$\begin{aligned}
\vec{a} \cdot \vec{e}_x &= a^x \cdot 1 + a^y \cdot 0 = a^x, \\
\vec{a} \cdot \vec{e}_y &= a^x \cdot 0 + a^y \cdot 1 = a^y.
\end{aligned}$$

8.2. Change of Basis

Firstly, it should be noted that the choice of a Cartesian coordinate system also uniquely determines the two basis vectors $\vec{e}_x, \vec{e}_y$ described above (in short: **basis**) and that conversely, the coordinate system is uniquely defined by the choice of a basis. Now, it is the case that there is not only *one* basis for the (two-dimensional) space, but arbitrarily (even infinitely) many. To illustrate this, imagine that we choose a Cartesian coordinate system and thus a basis, and then rotate the coordinate system by any angle around the origin. We then obtain a new coordinate system and thus a new basis for each angle. We will examine in detail what this looks like further below. Here we want to define the concept of a basis as broadly as possible. To do this, we use the following definition.

Two (arbitrary) vectors $\vec{e}_x, \vec{e}_y$ form a **basis of the two-dimensional space**, if *every* vector $\vec{v}$ can be represented as a *unique* linear combination of the vectors $\vec{e}_x, \vec{e}_y$, i.e.

$$\vec{v} = v^1 \vec{e}_x + v^2 \vec{e}_y,$$

the numbers v^1, v^2 are thus uniquely determined, they are called the **components** of $\vec{v}$ with respect to the basis $\vec{e}_x, \vec{e}_y$. As shown above, the Cartesian basis vectors $\vec{e}_x, \vec{e}_y$ fulfill this condition, they form a basis.

We now want to find out how to convert the basis vectors of two bases into each other. Let a second basis $\vec{e}_{1'}, \vec{e}_{2'}$ be given. Then each of these vectors can be represented as a linear combination of the basis vectors $\vec{e}_x, \vec{e}_y$:

$$\begin{aligned}
\vec{e}_{1'} &= a^1{}_1\, \vec{e}_x + a^2{}_1\, \vec{e}_y \\
\vec{e}_{2'} &= a^1{}_2\, \vec{e}_x + a^2{}_2\, \vec{e}_y
\end{aligned} \tag{8.3}$$

with four uniquely determined numbers $a^i{}_j, (i, j = 1, 2)$. Consider an arbitrary vector

$$\vec{v} = v^1 \vec{e}_x + v^2 \vec{e}_y,$$

which can also be written as a linear combination of the basis vectors $\vec{e}_{1'}, \vec{e}_{2'}$:

$$\vec{v} = v^{1'} \vec{e}_{1'} + v^{2'} \vec{e}_{2'},$$

then with Eq. (8.3)

$$\begin{aligned}
\vec{v} &= v^{1'} \vec{e}_{1'} + v^{2'} \vec{e}_{2'} \\
&= v^{1'} \left(a^1{}_1 \vec{e}_x + a^2{}_1 \vec{e}_y \right) + v^{2'} \left(a^1{}_2 \vec{e}_x + a^2{}_2 \vec{e}_y \right).
\end{aligned}$$

We combine the terms that belong to the basis vectors $\vec{e}_x, \vec{e}_y$ and obtain

$$\vec{v} = \left(v^{1'} a^1{}_1 + v^{2'} a^1{}_2 \right) \vec{e}_x + \left(v^{1'} a^2{}_1 + v^{2'} a^2{}_2 \right) \vec{e}_y.$$

Because of the *uniqueness* of the linear combinations we read off the relationship between the vector components in the two bases. It holds

$$v^1 = v^{1'}a^1_1 + v^{2'}a^1_2,$$
$$v^2 = v^{1'}a^2_1 + v^{2'}a^2_2. \tag{8.4}$$

Comparing (8.3) with (8.4), we find that the basis vectors and the associated vector components transform *differently*. The transformation behavior of the vector components is also called **contravariant** (to the behavior of the basis vectors). This difference ensures that the vector $\vec{v}$ can be represented equally as a linear combination in both bases. A change of the basis vectors is neutralized by a counteracting (contravariant) change in the vector components.

Of particular importance in physics are basis changes that result from shifts of the original coordinate system, i.e. shifts of the origin, or by rotations of the original coordinate system. We know from experience that the space surrounding us has no distinguished places or directions. This means that the form of the physical laws should not change if we rotate or shift the coordinate system. We want to consider some examples and calculate the transformations of basis vectors or vector components in a basis change.

Rotated and shifted coordinate systems

The simplest coordinate system change consists in shifting the coordinate system K by a certain vector $\vec{b}$.

Example 8.2.
We consider a two-dimensional vector $\vec{a}$, which is uniquely determined by a starting point P and an end point Q.

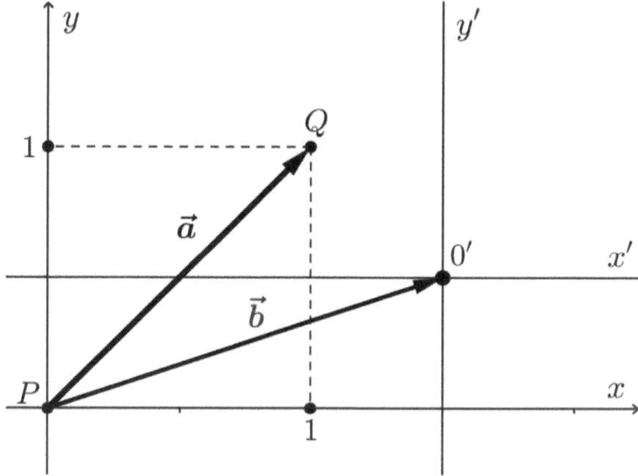

Figure 8.4.: Shift of the coordinate system

In Fig. 8.4 the vector $\vec{a}$ is drawn. It has the components

$$\vec{a} \underset{K}{\rightarrow} \begin{pmatrix} 1 \\ 1 \end{pmatrix}$$

in the (x, y) coordinate system K. The (x', y') coordinate system K' is obtained by shifting by the vector

$$\vec{b} \underset{K}{\rightarrow} \begin{pmatrix} b^x \\ b^y \end{pmatrix} = \begin{pmatrix} 1,5 \\ 0,5 \end{pmatrix}$$

from K. We want to calculate the components of the vector $\vec{a}$ in K'. For the coordinates of the points P and Q in K' it follows

$$P \underset{K'}{\rightarrow} \begin{pmatrix} P^{x'} \\ P^{y'} \end{pmatrix} = \begin{pmatrix} 0 - b^x \\ 0 - b^y \end{pmatrix} = \begin{pmatrix} 0 - 1,5 \\ 0 - 0,5 \end{pmatrix} = \begin{pmatrix} -1,5 \\ -0,5 \end{pmatrix}$$

and

$$Q \underset{K'}{\rightarrow} \begin{pmatrix} Q^{x'} \\ Q^{y'} \end{pmatrix} = \begin{pmatrix} 1 - b^x \\ 1 - b^y \end{pmatrix} = \begin{pmatrix} 1 - 1,5 \\ 1 - 0,5 \end{pmatrix} = \begin{pmatrix} -0,5 \\ 0,5 \end{pmatrix},$$

so it follows

$$\vec{a} \underset{K'}{\rightarrow} \begin{pmatrix} Q^{x'} - P^{x'} \\ Q^{y'} - P^{y'} \end{pmatrix} = \begin{pmatrix} -0,5 - (-1,5) \\ 0,5 - (-0,5) \end{pmatrix} = \begin{pmatrix} 1 \\ 1 \end{pmatrix}.$$

The vector $\vec{a}$ thus has the same components in the shifted coordinate system K' as in K and therefore also the same length. And the calculation has shown that this applies to any arbitrary vector $\vec{a}$. $\square$

Now we want to rotate the coordinate system as a second example, but without shifting the origin.

Example 8.3.

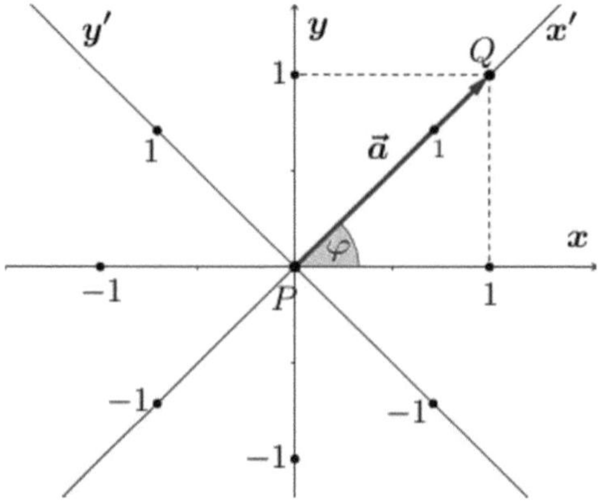

Figure 8.5.: Rotated coordinate system

If you rotate the (x, y) coordinate system by the angle φ, you get another coordinate system K' with the axes x' and y'. In this rotated coordinate system (in Fig. 8.5: $\varphi = 45°$) the *same* vector

$$\vec{a} \underset{K}{\to} \begin{pmatrix} 1 \\ 1 \end{pmatrix}$$

has the components $\sqrt{2}$ and 0, since $\vec{a}$ now lies on the x'-axis (i.e., $y' = 0$) and the length of $\vec{a}$ according to the Pythagorean theorem is $\sqrt{2}$, so

$$\vec{a} \underset{K'}{\to} \begin{pmatrix} \sqrt{2} \\ 0 \end{pmatrix}. \ \square$$

We consider in Fig. 8.6 the basis vectors in the two coordinate systems K and K'.

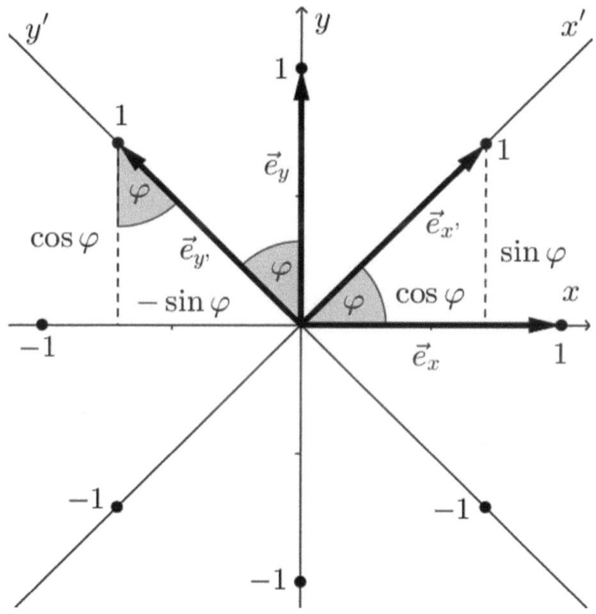

Figure 8.6.: Basis vectors in the rotated coordinate system

In the (x', y') coordinate system rotated by the angle φ, it holds that

$$\vec{e}_{x'} \underset{K'}{\to} \begin{pmatrix} 1 \\ 0 \end{pmatrix} \quad \text{and} \quad \vec{e}_{y'} \underset{K'}{\to} \begin{pmatrix} 0 \\ 1 \end{pmatrix}.$$

What components do these two vectors have in the coordinate system K? In Fig. 8.6 we read off:

$$\vec{e}_{x'} \underset{K}{\to} \begin{pmatrix} \cos \varphi \\ \sin \varphi \end{pmatrix}$$

$$\vec{e}_{y'} \quad \underset{K}{\to} \quad \begin{pmatrix} -\sin\varphi \\ \cos\varphi \end{pmatrix} \tag{8.5}$$

From this, we obtain the following representation for the basis vectors in K'

$$\begin{aligned} \vec{e}_{x'} &= \cos\varphi\,\vec{e}_x + \sin\varphi\,\vec{e}_y, \\ \vec{e}_{y'} &= -\sin\varphi\,\vec{e}_x + \cos\varphi\,\vec{e}_y. \end{aligned} \tag{8.6}$$

We can now use the transformation formula (8.4) to calculate the coordinates of a point P in the coordinate system K by the coordinates in the coordinate system K'. For this, we set

$$v^{1'} = x',\, v^{2'} = y',\, a^1_{\ 1} = \cos\varphi,\, a^2_{\ 1} = \sin\varphi,\, a^1_{\ 2} = -\sin\varphi,\, a^2_{\ 2} = \cos\varphi$$

and obtain

$$\begin{aligned} x &= x'\cos\varphi - y'\sin\varphi, \\ y &= x'\sin\varphi + y'\cos\varphi. \end{aligned} \tag{8.7}$$

Comparing this transformation with that from Eq. (8.6), it again results that the coordinates of a point transform contravariantly to the basis vectors.

We also want to derive the inverse transformation, i.e., the conversion of the (x, y) coordinates into (x', y') coordinates, and note that to do this, we have to rotate the K' coordinate system by $-\varphi$ to obtain the coordinate system K. Now, the cosine is an **even function**, i.e.

$$\cos\varphi = \cos\left(-\varphi\right),$$

and the sine is an **odd**, i.e.

$$-\sin\varphi = \sin\left(-\varphi\right),$$

i.e., it follows

$$\begin{aligned} x' &= x\cos(-\varphi) - y\sin(-\varphi) = x\cos\varphi + y\sin\varphi, \\ y' &= x\sin(-\varphi) + y\cos(-\varphi) = -x\sin\varphi + y\cos\varphi. \end{aligned} \tag{8.8}$$

Let's consider the above example 8.3 again. There, $(x, y) = (1, 1)$ and $\varphi = 45°$, i.e.

$$\cos\varphi = \sin\varphi = \frac{\sqrt{2}}{2}.$$

This results in

$$x' = x\cos(45°) + y\sin(45°) = \frac{\sqrt{2}}{2} + \frac{\sqrt{2}}{2} = \sqrt{2},$$

$$y' = -x\sin(45°) + y\cos(45°) = -\frac{\sqrt{2}}{2} + \frac{\sqrt{2}}{2} = 0,$$

so the same result as in example 8.3.

In Euclidean geometry, the length of a vector (the distance between two points) is an invariant, i.e., in all coordinate systems you get the same value. We consider an arbitrary two-dimensional position vector

$$\vec{r} \underset{K}{\to} \begin{pmatrix} x \\ y \end{pmatrix},$$

then the length of the vector in the coordinate system K

$$r = |\vec{r}| = \sqrt{x^2 + y^2}.$$

If K' is a coordinate system rotated by the angle φ, then with (8.8) for the length in the coordinate system K'

$$\begin{aligned}
r &= \sqrt{x'^2 + y'^2} = \sqrt{(x\cos\varphi + y\sin\varphi)^2 + (-x\sin\varphi + y\cos\varphi)^2} \\
&= \sqrt{x^2(\cos^2\varphi + \sin^2\varphi) + y^2(\sin^2\varphi + \cos^2\varphi)} \\
&= \sqrt{x^2 + y^2},
\end{aligned}$$

so you get the same value. For the vector

$$\vec{r} \underset{K'}{\to} \begin{pmatrix} x' \\ y' \end{pmatrix}$$

it holds

$$\vec{r} = x'\vec{e}_{x'} + y'\vec{e}_{y'}.$$

If you substitute the found transformation equations (8.8) and (8.6) for the components and the unit vectors, you get

$$\begin{aligned}
\vec{r} &= x'\vec{e}_{x'} + y'\vec{e}_{y'} \\
&= (x\cos\varphi + y\sin\varphi)(\cos\varphi\,\vec{e}_x + \sin\varphi\,\vec{e}_y) \\
&\quad + (-x\sin\varphi + y\cos\varphi)(-\sin\varphi\,\vec{e}_x + \cos\varphi\,\vec{e}_y) \\
&= x\cos^2\varphi\,\vec{e}_x + y\sin\varphi\cos\varphi\,\vec{e}_x + x\cos\varphi\sin\varphi\,\vec{e}_y + y\sin^2\varphi\,\vec{e}_y \\
&\quad + x\sin^2\varphi\,\vec{e}_x - y\cos\varphi\sin\varphi\,\vec{e}_x - x\sin\varphi\cos\varphi\,\vec{e}_y + y\cos^2\varphi\,\vec{e}_y \\
&= x(\cos^2\varphi + \sin^2\varphi)\,\vec{e}_x + y(\cos^2\varphi + \sin^2\varphi)\,\vec{e}_y \\
&= x\vec{e}_x + y\vec{e}_y.
\end{aligned}$$

So you get the expected result

$$\vec{r} = x'\vec{e}_{x'} + y'\vec{e}_{y'} = x\vec{e}_x + y\vec{e}_y.$$

Matrix Calculus

We want to further investigate the transformation equations found using the **matrix calculus**, which is presented in detail in [32].

Remark 8.1. **MT: Matrices**

A **matrix** is a rectangular array of numbers, in which the numbers are arranged in rows and columns. In this book, we deal with matrices for two-, three-, and four-dimensional spaces. We want to first deal with **square** matrices, which are matrices with the same number of rows and columns. Two-dimensional (square) matrices consist of four numbers, arranged in two rows and two columns. Three-dimensional matrices consist of nine numbers, arranged in three rows and columns. In the theory of relativity, time is added as a fourth dimension, the required four-dimensional square matrices have 16 numbers and four rows and columns. We want to explain or derive the properties and calculation rules of matrices always in the simplest case, namely the two-dimensional one. The "beautiful" thing about matrix calculations is that the two-dimensional properties and calculation rules can be transferred one-to-one to the other cases. We denote matrices with capital letters and write for a matrix $A = \left(a^i{}_j \right)$. The upper index i denotes the row number and the lower j the column number. The row and column numbers run from 1 to 2:

$$A = (a^i{}_j) = \begin{pmatrix} a^1{}_1 & a^1{}_2 \\ a^2{}_1 & a^2{}_2 \end{pmatrix}$$

The numbers $a^i{}_j$ are called **matrix elements**. A matrix can be divided into row vectors and column vectors. The **row vectors** are defined by

$$\begin{aligned} \vec{a}^1 &\rightarrow (a^1{}_1, a^1{}_2) \\ \vec{a}^2 &\rightarrow (a^2{}_1, a^2{}_2) \end{aligned}$$

and the **column vectors** by

$$\vec{a}_1 \rightarrow \begin{pmatrix} a^1{}_1 \\ a^2{}_1 \end{pmatrix}, \vec{a}_2 \rightarrow \begin{pmatrix} a^1{}_2 \\ a^2{}_2 \end{pmatrix}.$$

We now define some properties or operations, that can be performed with matrices.

1. Two matrices $A = \left(a^i{}_j \right)$ and $B = \left(b^i{}_j \right)$ are equal if

$$a^i{}_j = b^i{}_j$$

for all i and j.

2. If you swap the rows and columns of a matrix

$$A = \left(a^i_{\ j}\right) = \begin{pmatrix} a^1_{\ 1} & a^1_{\ 2} \\ a^2_{\ 1} & a^2_{\ 2} \end{pmatrix},$$

you get the **transposed matrix** of A

$$A^T = \left(a^j_{\ i}\right) = \begin{pmatrix} a^1_{\ 1} & a^2_{\ 1} \\ a^1_{\ 2} & a^2_{\ 2} \end{pmatrix}.$$

3. Two matrices A and B are added by adding their matrix elements:

$$A + B = \begin{pmatrix} a^1_{\ 1} + b^1_{\ 1} & a^1_{\ 2} + b^1_{\ 2} \\ a^2_{\ 1} + b^2_{\ 1} & a^2_{\ 2} + b^2_{\ 2} \end{pmatrix}$$

4. A matrix is multiplied by a number c by multiplying all matrix elements with the number c:

$$cA = \begin{pmatrix} ca^1_{\ 1} & ca^1_{\ 2} \\ ca^2_{\ 1} & ca^2_{\ 2} \end{pmatrix}$$

5. A matrix $A = \left(a^i_{\ j}\right)$ can be multiplied with a vector $\vec{v} \to \begin{pmatrix} v^1 \\ v^2 \end{pmatrix}$, the result is again a vector

$$A \cdot \vec{v} = \vec{w} \to \begin{pmatrix} w^1 \\ w^2 \end{pmatrix},$$

whose components are calculated as follows:

$$\begin{aligned} w^1 &= a^1_{\ 1} v^1 + a^1_{\ 2} v^2 \\ w^2 &= a^2_{\ 1} v^1 + a^2_{\ 2} v^2 \end{aligned}$$

If we assume that the scalar product of a row vector with a column vector is calculated in the same way as between two column vectors, the components of $\vec{w}$ can also be written as follows

$$\begin{aligned} w^1 &= \vec{a}^1 \cdot \vec{v} \\ w^2 &= \vec{a}^2 \cdot \vec{v}. \end{aligned}$$

6. A matrix $A = \left(a^i_{\ j}\right)$ can be multiplied with another matrix $B = \left(b^i_{\ j}\right)$, the result is again a matrix

$$A \cdot B = C = \left(c^i_{\ j}\right),$$

whose matrix elements are calculated as follows:

$$
\begin{aligned}
c^1{}_1 &= a^1{}_1 b^1{}_1 + a^1{}_2 b^2{}_1 \\
c^1{}_2 &= a^1{}_1 b^1{}_2 + a^1{}_2 b^2{}_2 \\
c^2{}_1 &= a^2{}_1 b^1{}_1 + a^2{}_2 b^2{}_1 \\
c^2{}_2 &= a^2{}_1 b^1{}_2 + a^2{}_2 b^2{}_2
\end{aligned}
$$

Here again, the scalar product calculation can be applied, it holds:

$$
\begin{aligned}
c^1{}_1 &= \vec{a}^1 \cdot \vec{b}_1 \\
c^1{}_2 &= \vec{a}^1 \cdot \vec{b}_2 \\
c^2{}_1 &= \vec{a}^2 \cdot \vec{b}_1 \\
c^2{}_2 &= \vec{a}^2 \cdot \vec{b}_2
\end{aligned}
$$

Note that the superscript and subscript indices on both sides of the equations correspond.

7. The concrete calculation of a product of two matrices A and B can be further clarified with the so-called **Falk scheme**:

$$
\begin{array}{cc|c}
 & & \begin{matrix} b^1{}_1 & b^1{}_2 \\ b^2{}_1 & b^2{}_2 \end{matrix} \\
\hline
\begin{matrix} a^1{}_1 & a^1{}_2 \\ a^2{}_1 & a^2{}_2 \end{matrix} & & c^2{}_2
\end{array}
$$

To calculate the element $c^2{}_2$, imagine taking the second column vector of B and placing it over the second row vector of A, then multiply the overlapping elements and subsequently add the two products thus formed. As an example, consider

$$
\begin{array}{cc|c}
 & & \begin{matrix} b^1{}_1 & 1 \\ b^2{}_1 & 2 \end{matrix} \\
\hline
\begin{matrix} a^1{}_1 & a^1{}_2 \\ 5 & 6 \end{matrix} & & c^2{}_2 = 17
\end{array},
$$

then $c^2{}_2$ is calculated by

$$
c^2{}_2 = 1 \cdot 5 + 2 \cdot 6 = 17.
$$

The Falk scheme is particularly helpful in the three- and four-dimensional case for the multiplication of matrices.

8. Unlike the multiplication of numbers, the order of the matrices matters in the multiplication of matrices, i.e., in general

$$A \cdot B \neq B \cdot A.$$

We say that the multiplication of matrices is **non-commutative**. We demonstrate this fact with a simple example. Let

$$A = \begin{pmatrix} 1 & 0 \\ 0 & 0 \end{pmatrix} \text{ and } B = \begin{pmatrix} 0 & 1 \\ 0 & 0 \end{pmatrix}$$

be two matrices, then it follows

$$A \cdot B = \begin{pmatrix} 1 & 0 \\ 0 & 0 \end{pmatrix} \begin{pmatrix} 0 & 1 \\ 0 & 0 \end{pmatrix} = \begin{pmatrix} 0 & 1 \\ 0 & 0 \end{pmatrix},$$

but

$$B \cdot A = \begin{pmatrix} 0 & 1 \\ 0 & 0 \end{pmatrix} \begin{pmatrix} 1 & 0 \\ 0 & 0 \end{pmatrix} = \begin{pmatrix} 0 & 0 \\ 0 & 0 \end{pmatrix}.$$

9. If three matrices A, B, C are given, you may multiply the first two together and then the result with the third, or first the last two and then the result with the first, i.e.,

$$A \cdot B \cdot C = (A \cdot B) \cdot C = A \cdot (B \cdot C),$$

however, the order of the matrices must be maintained.

10. Row and column vectors can be considered as special matrices. Row vectors are matrices with one row and two columns (so-called 1×2-matrix), column vectors are matrices with one column and two rows (2×1-matrix). If you consider that the transpose of a 2×1-matrix is a 1×2-matrix, i.e.

$$\begin{pmatrix} a \\ b \end{pmatrix}^T = \begin{pmatrix} a & b \end{pmatrix},$$

then you can calculate the scalar product of two vectors

$$\vec{a} \to \begin{pmatrix} a^x \\ a^y \end{pmatrix}, \vec{b} \to \begin{pmatrix} b^x \\ b^y \end{pmatrix}$$

also by matrix multiplication

$$\vec{a} \cdot \vec{b} = \begin{pmatrix} a^x \\ a^y \end{pmatrix}^T \begin{pmatrix} b^x \\ b^y \end{pmatrix} = \begin{pmatrix} a^x & a^y \end{pmatrix} \begin{pmatrix} b^x \\ b^y \end{pmatrix} = a^x b^x + a^y b^y.$$

11. If A, B are two matrices, then

$$(A \cdot B)^T = B^T \cdot A^T.$$

The changed order must be taken into account!

12. The matrix

$$E = \begin{pmatrix} 1 & 0 \\ 0 & 1 \end{pmatrix}$$

is called the **identity matrix**. The elements of the identity matrix are also written as

$$E = \left(\delta^i{}_j \right),$$

where the quantity

$$\delta^i{}_j = \begin{cases} 1 & for \ i = j \\ 0 & for \ i \neq j \end{cases}$$

is called **Kronecker-Delta**.

For a vector $\vec{v} \to \begin{pmatrix} v^1 \\ v^2 \end{pmatrix}$

$$E \cdot \vec{v} \to \begin{pmatrix} 1 & 0 \\ 0 & 1 \end{pmatrix} \begin{pmatrix} v^1 \\ v^2 \end{pmatrix} = \begin{pmatrix} 1 \cdot v^1 + 0 \cdot v^2 \\ 0 \cdot v^1 + 1 \cdot v^2 \end{pmatrix} = \begin{pmatrix} v^1 \\ v^2 \end{pmatrix},$$

so

$$E \cdot \vec{v} = \vec{v}.$$

For a matrix $A = \left(a^i{}_j \right)$ the same applies

$$A \cdot E = \begin{pmatrix} a^1{}_1 & a^1{}_2 \\ a^2{}_1 & a^2{}_2 \end{pmatrix} \begin{pmatrix} 1 & 0 \\ 0 & 1 \end{pmatrix} = \begin{pmatrix} a^1{}_1 & a^1{}_2 \\ a^2{}_1 & a^2{}_2 \end{pmatrix} = A.$$

Multiplication with the identity matrix does not change vectors and matrices, it is the analogue of multiplication with 1 for numbers.

13. If there exists a matrix X such that

$$A \cdot X = X \cdot A = E,$$

then X is called the **inverse matrix** of A and is written as $X = A^{-1}$. Not every matrix has an inverse matrix, for example, the matrix

$$A = \begin{pmatrix} 1 & 0 \\ 0 & 0 \end{pmatrix}$$

does not have an inverse matrix. Because if $X = (x^i{}_j)$ is any matrix, then

$$A \cdot X = \begin{pmatrix} 1 & 0 \\ 0 & 0 \end{pmatrix} \begin{pmatrix} x^1{}_1 & x^1{}_2 \\ x^2{}_1 & x^2{}_2 \end{pmatrix} = \begin{pmatrix} x^1{}_1 & x^1{}_2 \\ 0 & 0 \end{pmatrix},$$

i.e., one can never obtain the identity matrix E as a result.

14. The sum of the diagonal elements of a matrix $A = \left(a^i{}_j\right)$ is called the **trace** of A:

$$Trace\,(A) = a^1{}_1 + a^2{}_2 \,\square \tag{8.9}$$

In the following, it will often be referred back to the fact that matrices can also be viewed as **linear mappings** or **linear functions**.

Remark 8.2. **MT: Linear Mappings**
A vector field $\vec{\phi}$ is called **linear**, if for all vectors $\vec{x}, \vec{y}$ and all numbers α

$$\begin{aligned} \vec{\phi}\,(\vec{x} + \vec{y}) &= \vec{\phi}\,(\vec{x}) + \vec{\phi}\,(\vec{y}), \tag{8.10} \\ \vec{\phi}\,(\alpha\vec{x}) &= \alpha\vec{\phi}\,(\vec{x}) \end{aligned}$$

applies. One can consider the multiplication of a matrix with a vector as a linear function. Because if you multiply a matrix M with a sum of two vectors $\vec{a} + \vec{b}$, then according to the above calculation rules

$$M \cdot \left(\vec{a} + \vec{b}\right) = M \cdot \vec{a} + M \cdot \vec{b}. \tag{8.11}$$

The same applies to multiplication with a number c

$$M \cdot (c\vec{a}) = c\,(M \cdot \vec{a}),$$

as can be easily verified. $\square$

Rotation Matrices

The two equations in the formula (8.8) can be written as one equation using matrix notation

$$\begin{pmatrix} x' \\ y' \end{pmatrix} = \begin{pmatrix} x\cos\varphi + y\sin\varphi \\ -x\sin\varphi + y\cos\varphi \end{pmatrix} = \begin{pmatrix} \cos\varphi & \sin\varphi \\ -\sin\varphi & \cos\varphi \end{pmatrix} \begin{pmatrix} x \\ y \end{pmatrix}. \tag{8.12}$$

The matrix

$$D = \begin{pmatrix} \cos\varphi & \sin\varphi \\ -\sin\varphi & \cos\varphi \end{pmatrix} \tag{8.13}$$

on the right side is called **rotation matrix**. We now ask ourselves whether we can also determine the reverse coordinate transformation (8.7)

$$\begin{pmatrix} x \\ y \end{pmatrix} = \begin{pmatrix} \cos\varphi & -\sin\varphi \\ \sin\varphi & \cos\varphi \end{pmatrix} \begin{pmatrix} x' \\ y' \end{pmatrix} \tag{8.14}$$

using matrix calculation. We already know that this rule is indeed the reverse transformation. Nevertheless, we want to show it again explicitly using matrix calculation. To do this, we perform both transformations one after the other and expect that we then end up where we started. We want to start with the last equation, in which we substitute for x', y' (8.12)

$$\begin{aligned} \begin{pmatrix} x \\ y \end{pmatrix} &= \begin{pmatrix} \cos\varphi & -\sin\varphi \\ \sin\varphi & \cos\varphi \end{pmatrix} \begin{pmatrix} x' \\ y' \end{pmatrix} \\ &= \begin{pmatrix} \cos\varphi & -\sin\varphi \\ \sin\varphi & \cos\varphi \end{pmatrix} \begin{pmatrix} \cos\varphi & \sin\varphi \\ -\sin\varphi & \cos\varphi \end{pmatrix} \begin{pmatrix} x \\ y \end{pmatrix}. \end{aligned}$$

Since according to Remark 8.1 for three matrices $A\,(BC) = (AB)\,C$ applies, we first calculate the product of the two matrices on the right side

$$\begin{pmatrix} \cos\varphi & -\sin\varphi \\ \sin\varphi & \cos\varphi \end{pmatrix} \begin{pmatrix} \cos\varphi & \sin\varphi \\ -\sin\varphi & \cos\varphi \end{pmatrix} =$$

$$\begin{pmatrix} \cos^2\varphi + \sin^2\varphi & \sin\varphi\cos\varphi - \sin\varphi\cos\varphi \\ \sin\varphi\cos\varphi - \sin\varphi\cos\varphi & \cos^2\varphi + \sin^2\varphi \end{pmatrix} = \begin{pmatrix} 1 & 0 \\ 0 & 1 \end{pmatrix} = E,$$

i.e., the matrix in Eq. (8.14) is the inverse matrix to the rotation matrix D, and we obtain

$$\begin{pmatrix} x \\ y \end{pmatrix} = \begin{pmatrix} 1 & 0 \\ 0 & 1 \end{pmatrix} \begin{pmatrix} x \\ y \end{pmatrix},$$

thus an always valid equation.

General (Curvilinear) Coordinate Systems

We want to extend the concept of the coordinate system and also allow coordinate systems whose axes are not straight lines. First, however, we again consider a Cartesian (x, y) coordinate system K and call a vector-valued mapping

$$\vec{r}(s) \underset{K}{\rightarrow} \begin{pmatrix} x\,(s) \\ y\,(s) \end{pmatrix},$$

where the variable s is a real number and $x\,(s), y\,(s)$ are the component functions of $\vec{r}(s)$, a **curve**. It is said that the curve is **parametrized** by the real number s. Curves have already been encountered as trajectory curves of particles in section 6.2 in the derivation of the Kepler orbits of the planets; there the time t was the parameter. If we now assume that the mapping $\vec{r}(s)$ is differentiable, then we can form the derivative with respect to the parameter s. If $\vec{r}(s_0)$ is any point on the curve, then $d\vec{r}(s_0)/ds$ defines the **tangent vector** to the curve at the point $\vec{r}(s_0)$. Since the derivative is formed as the limit of a difference, the result is always a vector, because with the unit vectors $\vec{e}_x, \vec{e}_y$ it follows

$$
\begin{aligned}
\frac{d\vec{r}(s_0)}{ds} &= \lim_{h \to 0} \frac{\vec{r}(s_0 + h) - \vec{r}(s_0)}{h} \\
&= \lim_{h \to 0} \frac{x(s_0 + h) - x(s_0)}{h}\, \vec{e}_x + \lim_{h \to 0} \frac{y(s_0 + h) - y(s_0)}{h}\, \vec{e}_y \\
&= \frac{dx(s_0)}{ds}\, \vec{e}_x + \frac{dy(s_0)}{ds}\, \vec{e}_y.
\end{aligned}
$$

The tangent vector thus has the components

$$
\frac{d\vec{r}}{ds} \underset{K}{\to} \begin{pmatrix} dx/ds \\ dy/ds \end{pmatrix}
$$

and it points in every point *in the direction of the curve* (see also section 2.1). With these definitions, we can consider the basis vectors corresponding to a Cartesian coordinate system also as tangent vectors of the two coordinate axes. The coordinate axes are considered as curves, i.e., the x-axis is the curve

$$
\vec{K}_x(s) \underset{K}{\to} \begin{pmatrix} s \\ 0 \end{pmatrix}
$$

and correspondingly the y-axis

$$
\vec{K}_y(s) \underset{K}{\to} \begin{pmatrix} 0 \\ s \end{pmatrix}.
$$

If you form the tangent vectors of these two curves, you get

$$
\frac{d\vec{K}_x(s)}{ds} \underset{K}{\to} \begin{pmatrix} 1 \\ 0 \end{pmatrix}, \quad \frac{d\vec{K}_y(s)}{ds} \underset{K}{\to} \begin{pmatrix} 0 \\ 1 \end{pmatrix},
$$

thus

$$
\frac{d\vec{K}_x(s)}{ds} = \vec{e}_x, \quad \frac{d\vec{K}_y(s)}{ds} = \vec{e}_y.
$$

This definition of the basis vectors is very practical because it can be generalized to the case of curvilinear coordinate systems. But it can also be used to represent the tangent vector of a curve $\vec{r}(s)$ using the chain rule. With Eq. 4.23 on page 70 it follows

$$\frac{d\vec{r}}{ds} = \frac{\partial\vec{r}}{\partial x}\frac{dx}{ds} + \frac{\partial\vec{r}}{\partial y}\frac{dy}{ds} = \frac{dx}{ds}\,\vec{e}_x + \frac{dy}{ds}\,\vec{e}_y,$$

since

$$\frac{\partial\vec{r}}{\partial x} \rightarrow \begin{pmatrix} \partial x/\partial x \\ \partial y/\partial x \end{pmatrix} = \begin{pmatrix} 1 \\ 0 \end{pmatrix}$$

and analogously for the other basis vector

$$\frac{\partial\vec{r}}{\partial y} \rightarrow \begin{pmatrix} \partial y/\partial x \\ \partial y/\partial y \end{pmatrix} = \begin{pmatrix} 0 \\ 1 \end{pmatrix}$$

applies. Overall, it follows

$$\frac{\partial\vec{r}}{\partial x} = \vec{e}_x, \quad \frac{\partial\vec{r}}{\partial y} = \vec{e}_y.$$

We define a **general (curvilinear) coordinate system** K' for the Euclidean plane as a set of pairs of numbers (x', y'), which uniquely determine the position of each point P in the plane, and call the numbers x', y' the coordinates of the point P in the coordinate system K'. A **general coordinate transformation** in $\mathbb{R}^2$ is a rule $\vec{\phi}$, which maps the coordinates (x', y') to the Cartesian coordinates (x, y),

$$\vec{\phi}(x', y') = \begin{pmatrix} x \\ y \end{pmatrix} = \begin{pmatrix} x(x', y') \\ y(x', y') \end{pmatrix}.$$

We also write the mapping $\vec{\phi}$ memorably in the form

$$\begin{aligned} x &= x(x', y') \\ y &= y(x', y'), \end{aligned} \tag{8.15}$$

i.e., we denote the *component functions* of $\vec{\phi}$ also with x, y. However, it should be clear from the respective context whether we mean the Cartesian coordinates of a point or the component functions of $\vec{\phi}$ when we talk about x, y. We require that the function $\vec{\phi}$ be **differentiable** and **invertible**, i.e., in addition to the transformation $(x', y') \rightarrow (x, y)$ the reverse transformation $(x, y) \rightarrow (x', y')$ should also be possible. Both terms will be explained in the following. The first requirement we make of the function $\vec{\phi}$ is that it should be differentiable. For this, we need to define the derivative of a vector field (a multivariable vector-valued function) $\vec{\phi}$.

Remark 8.3. **MT: Derivative of a Vector Field**
Let a vector field

$$\vec{\phi}\,(x',y') \to \left(\begin{array}{c} x\,(x',y') \\ y\,(x',y') \end{array} \right)$$

be given. The two component functions x, y are multivariable *real-valued* functions, whose derivative according to remark 4.8 on page 66 is given by their respective gradients, i.e.

$$grad\,x \;\to\; \left(\frac{\partial x}{\partial x'}, \frac{\partial x}{\partial y'} \right),$$

$$grad\,y \;\to\; \left(\frac{\partial y}{\partial x'}, \frac{\partial y}{\partial y'} \right).$$

The components of the gradients are written as *row vectors*. The 1st derivative of $\vec{\phi}$ at a point (x_0', y_0') is defined as the 2×2-matrix ϕ' whose rows are the gradients of the component functions of $\vec{\phi}$, i.e.

$$\phi'(x_0', y_0') = \left(\begin{array}{cc} \dfrac{\partial x\,(x_0', y_0')}{\partial x'} & \dfrac{\partial x\,(x_0', y_0')}{\partial y'} \\[3ex] \dfrac{\partial y\,(x_0', y_0')}{\partial x'} & \dfrac{\partial y\,(x_0', y_0')}{\partial y'} \end{array} \right).$$

This matrix is called the **Jacobian matrix** (after the German mathematician Karl Gustav Jacob Jacobi). The first derivative of a vector-valued function at a specific point is thus a matrix and therefore a linear function. If you calculate the first derivative at another point, you get another matrix, but generally not the same as before, since the matrix elements are partial derivatives that vary with the points in space. However, for simplicity, we will omit the argument (x_0', y_0') in the following, when it is clear which point is meant or when all points are to be represented simultaneously. So we write briefly

$$\phi' = \left(\begin{array}{cc} \partial x/\partial x' & \partial x/\partial y' \\ \partial y/\partial x' & \partial y/\partial y' \end{array} \right). \;\square$$

Example 8.4.
As an example, we calculate the Jacobian matrix of the mapping

$$\vec{\phi}(x',y') \;\to\; \left(\begin{array}{c} x \\ y \end{array} \right) = \left(\begin{array}{c} x(x',y') \\ y(x',y') \end{array} \right) = \left(\begin{array}{cc} \cos\varphi & -\sin\varphi \\ \sin\varphi & \cos\varphi \end{array} \right) \left(\begin{array}{c} x' \\ y' \end{array} \right)$$

$$= \left(\begin{array}{c} x'\cos\varphi - y'\sin\varphi \\ x'\sin\varphi + y'\cos\varphi \end{array} \right)$$

from Eq. (8.7). For fixed φ we get

$$
\begin{aligned}
\frac{\partial x}{\partial x'} &= \frac{\partial}{\partial x'}\left(x'\cos\varphi - y'\sin\varphi\right) = \cos\varphi \\
\frac{\partial x}{\partial y'} &= \frac{\partial}{\partial y'}\left(x'\cos\varphi - y'\sin\varphi\right) = -\sin\varphi \\
\frac{\partial y}{\partial x'} &= \frac{\partial}{\partial x'}\left(x'\sin\varphi + y'\cos\varphi\right) = \sin\varphi \\
\frac{\partial y}{\partial y'} &= \frac{\partial}{\partial y'}\left(x'\sin\varphi + y'\cos\varphi\right) = \cos\varphi,
\end{aligned}
$$

so it follows

$$
\phi' = \begin{pmatrix} \partial x/\partial x' & \partial x/\partial y' \\ \partial y/\partial x' & \partial y/\partial y' \end{pmatrix} = \begin{pmatrix} \cos\varphi & -\sin\varphi \\ \sin\varphi & \cos\varphi \end{pmatrix}. \tag{8.16}
$$

The Jacobian matrix is therefore the same for all points (x', y'), i.e. constant.
□

Remark 8.4. **MT: Inverse function of a vector field**
We generalize the term *inverse function* (see remark 4.10 on page 69) to a vector field $\vec{\phi}(\vec{u})$. If there exists another vector field $\vec{\psi}(\vec{u})$ with

$$
\vec{\psi}\left(\vec{\phi}(\vec{u})\right) = \vec{\phi}\left(\vec{\psi}(\vec{u})\right) = \vec{u}, \tag{8.17}
$$

then $\vec{\psi}$ is referred to as the (unique) **inverse function** of $\vec{\phi}$ and is written as

$$
\vec{\psi} = \left(\vec{\phi}\right)^{-1} \square.
$$

Example 8.5.
As an example, let's consider the function $\vec{\phi}$ from Eq. (8.7). We check exemplarily whether the inverse function $\left(\vec{\phi}\right)^{-1}$ is defined by Eq. (8.14). With the rotation matrix (8.13) we get

$$
\begin{aligned}
\vec{\phi}\begin{pmatrix} x \\ y \end{pmatrix} = \vec{\phi}\begin{pmatrix} x'\cos\varphi - y'\sin\varphi \\ x'\sin\varphi + y'\cos\varphi \end{pmatrix} &= \\
\begin{pmatrix} \cos\varphi & \sin\varphi \\ -\sin\varphi & \cos\varphi \end{pmatrix}\begin{pmatrix} x'\cos\varphi - y'\sin\varphi \\ x'\sin\varphi + y'\cos\varphi \end{pmatrix} &= \\
\begin{pmatrix} x'\cos^2\varphi - y'\sin\varphi\cos\varphi + x'\sin^2\varphi + y'\cos\varphi\sin\varphi \\ -x'\left(\sin\varphi\cos\varphi - y'\sin^2\varphi\right) + x'\sin\varphi\cos\varphi + y'\cos^2\varphi \end{pmatrix} &= \\
\begin{pmatrix} x'\cos^2\varphi + x'\sin^2\varphi \\ y'\sin^2\varphi + y'\cos^2\varphi \end{pmatrix} &= \begin{pmatrix} x' \\ y' \end{pmatrix}.
\end{aligned}
$$

If we define the inverse mapping by

$$\left(\vec{\phi}\right)^{-1}(x',y') := \begin{pmatrix} x'\cos\varphi - y'\sin\varphi \\ x'\sin\varphi + y'\cos\varphi \end{pmatrix},$$

then

$$\vec{\phi}\left(\left(\vec{\phi}\right)^{-1}(x',y')\right) = \begin{pmatrix} x' \\ y' \end{pmatrix}.\;\square$$

It is often not immediately apparent whether an inverse function exists for a given function. However, there is the following test criterion. A multivariable vector-valued function

$$\vec{\phi}(x',y') \to \begin{pmatrix} x\,(x',y') \\ y\,(x',y') \end{pmatrix}$$

is exactly then (in a neighborhood of a point) invertible, when the Jacobian matrix

$$\phi' = \begin{pmatrix} \partial x/\partial x' & \partial x/\partial y' \\ \partial y/\partial x' & \partial y/\partial y' \end{pmatrix}$$

is invertible at this point, i.e., when the inverse matrix $(\phi')^{-1}$ exists. Another criterion for checking whether a matrix is invertible is provided by the so-called **determinant calculus**.

Remark 8.5. **MT: Determinant** of a matrix, **inverse matrix**
Let A be a 2×2 matrix

$$A = \begin{pmatrix} a & b \\ c & d \end{pmatrix},$$

then the expression

$$\det A = \det \begin{pmatrix} a & b \\ c & d \end{pmatrix} = a \cdot d - c \cdot b \tag{8.18}$$

is called the **determinant** of A. This formula only applies to 2×2 matrices, for matrices of higher order there are other calculation algorithms (see [32]), which we do not want to discuss here. The following statement now applies: The matrix A is invertible if and only if the determinant is not zero:

$$A \text{ invertierbar} \Leftrightarrow \det A \neq 0\;\square$$

Example 8.6.
As an example, we calculate the determinant of the Jacobian matrix for rotations (8.16)

$$\det \phi' \;=\; \det \begin{pmatrix} \cos\varphi & -\sin\varphi \\ \sin\varphi & \cos\varphi \end{pmatrix} = \cos^2\varphi - \sin\varphi(-\sin\varphi)$$

$$= \cos^2\varphi + \sin^2\varphi = 1$$

and again obtain the result that this matrix is invertible. $\square$

If the determinant of a matrix is not zero, the inverse of the matrix

$$A = \begin{pmatrix} a & b \\ c & d \end{pmatrix}$$

can be calculated using the following formula

$$A^{-1} = \frac{1}{\det A} \begin{pmatrix} d & -b \\ -c & a \end{pmatrix}, \tag{8.19}$$

because it holds

$$
\begin{aligned}
A \cdot A^{-1} &= \begin{pmatrix} a & b \\ c & d \end{pmatrix} \frac{1}{\det A} \begin{pmatrix} d & -b \\ -c & a \end{pmatrix} \\
&= \frac{1}{ad - cb} \begin{pmatrix} ad - cb & -ab + ba \\ cd - cd & -bc + ad \end{pmatrix} = \begin{pmatrix} 1 & 0 \\ 0 & 1 \end{pmatrix}.
\end{aligned}
$$

Example 8.7.
As an example, we calculate the inverse matrix of the Jacobian matrix for rotations

$$(\phi')^{-1} = \frac{1}{\det\phi'} \begin{pmatrix} \cos\varphi & \sin\varphi \\ -\sin\varphi & \cos\varphi \end{pmatrix} = \begin{pmatrix} \cos\varphi & \sin\varphi \\ -\sin\varphi & \cos\varphi \end{pmatrix},$$

since $\det\phi' = 1$. As expected, we thus obtain the rotation matrix D (8.13). $\square$

Coordinate lines

The curves described by

$$
\begin{aligned}
\vec{K}_1(s) &= \vec{\phi}(s, y') \\
\vec{K}_2(s) &= \vec{\phi}(x', s)
\end{aligned}
$$

where the x', y' are arbitrary but fixed and the parameter s varies, are called **coordinate lines**. Exactly two coordinate lines run through each point $(x, y) = \vec{\phi}(x', y')$.

As an example, we consider the coordinate lines of the polar coordinates in the Euclidean plane.

Example 8.8.
We have already introduced the polar coordinates in Eq. 2.6 on page 30 and used them intensively in the derivation of the Kepler orbits. They are the most important example for curvilinear coordinates in the plane. The polar coordinates P allow a (two-dimensional) vector $\vec{r}$ to be represented by its distance r from the origin and the angle φ between $\vec{r}$ and the positive x-axis are expressed, where the relationship between the Cartesian coordinates (x, y) and the polar coordinates of $\vec{r}$ is given by

$$\vec{\phi}(r, \varphi) \underset{K}{\rightarrow} \begin{pmatrix} r\cos\varphi \\ r\sin\varphi \end{pmatrix} = \begin{pmatrix} x(r, \varphi) \\ y(r, \varphi) \end{pmatrix}. \tag{8.20}$$

As usual, we have set $x' = r$ and $y' = \varphi$ where $r \geq 0$ and $-\pi < \varphi \leq \pi$. The curve $\vec{K}_1(r) = \vec{\phi}(r, \varphi)$ with a fixed angle φ is a straight line through the origin, the curve $\vec{K}_2(\varphi) = \vec{\phi}(r, \varphi)$ with a fixed r is a circle with radius r around the origin. Fig. 8.7 shows some coordinate lines.

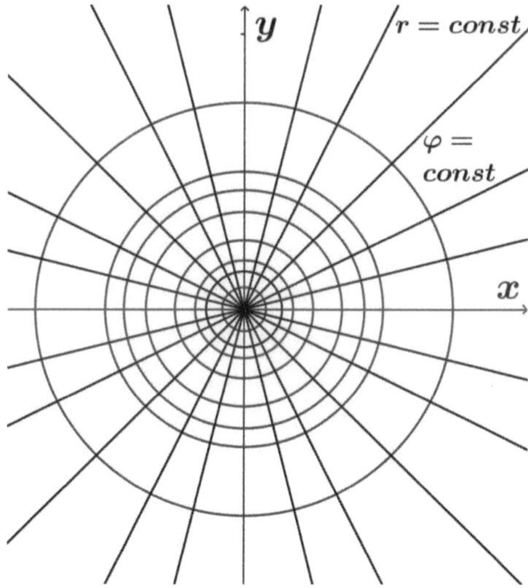

Figure 8.7.: Coordinate lines of the polar coordinates

Since the polar coordinates for $r = 0$ are not uniquely determined (the angle φ could be chosen arbitrarily), the inverse function of the polar coordinates is only defined for $(x, y) \neq (0, 0)$. It results in

$$\left(\vec{\phi}\right)^{-1}(x, y) \underset{P}{\rightarrow} \begin{pmatrix} \sqrt{x^2 + y^2} \\ arc\,(x, y) \end{pmatrix} = \begin{pmatrix} r\,(x, y) \\ \varphi\,(x, y) \end{pmatrix}, \tag{8.21}$$

where *arc* denotes the **arc function**. $\square$

Remark 8.6. **MT: The Arc Function**

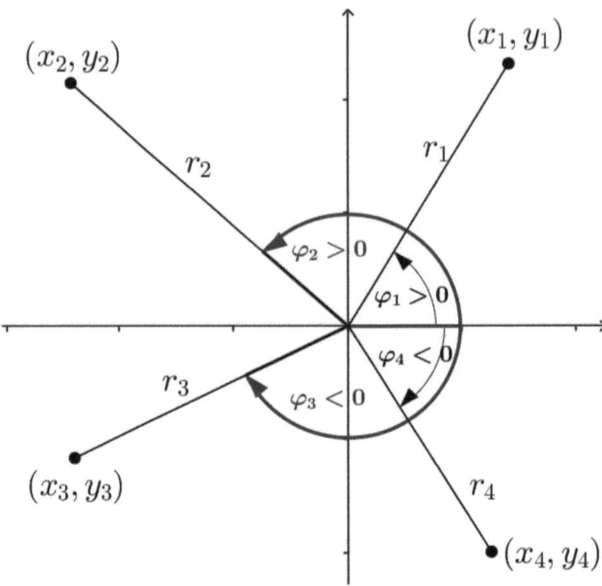

Figure 8.8.: Arc Function

The arc function provides the angle φ measured in radians between the line connecting the origin with a point $(x, y) \neq (0, 0)$ and the *positive* x-axis. The absolute smaller angle is always chosen. In Fig. 8.8, an arbitrary point (x, y), the connecting line to the origin with the length $r = \sqrt{x^2 + y^2} > 0$ and the angle φ to the positive x-axis are drawn in each quadrant. For $y \geq 0$ is $\varphi \geq 0$, for negative y values, φ is negative. For $y = 0, x < 0$ is $\varphi = \pi$ and for $y = 0, x > 0$ is $\varphi = 0$. Analytically, the arc function is defined by

$$\varphi = arc\,(x, y) = \begin{cases} \arccos \dfrac{x}{r} = \arccos \dfrac{x}{\sqrt{x^2 + y^2}} & y \geq 0 \\[2ex] -\arccos \dfrac{x}{r} = -\arccos \dfrac{x}{\sqrt{x^2 + y^2}} & y < 0 \end{cases}$$

where $\arccos x$ denotes the inverse function of the cosine **arc cosine**. $\square$

To determine the basis vectors of the polar coordinates, which we want to call $\vec{e}_r$ and $\vec{e}_\varphi$, we calculate the tangent vectors of the coordinate lines. To do this, we first differentiate the curve $\vec{K}_1(r) = \vec{\phi}(r, \varphi)$ with respect to r and obtain

$$\vec{e}_r = \frac{d\vec{K}_1}{dr} \rightarrow \begin{pmatrix} \partial x\,(r, \varphi)/\partial r \\ \partial y\,(r, \varphi)/\partial r \end{pmatrix} = \begin{pmatrix} \partial\,(r \cos \varphi)/\partial r \\ \partial\,(r \sin \varphi)/\partial r \end{pmatrix} = \begin{pmatrix} \cos \varphi \\ \sin \varphi \end{pmatrix}. \tag{8.22}$$

For the derivative of the curve $\vec{K}_2(\varphi) = \vec{\phi}(r, \varphi)$ with respect to φ we get

$$\vec{e}_\varphi = \frac{d\vec{K}_2}{d\varphi} \rightarrow \left(\begin{array}{c} \partial x\,(r, \varphi)\,/\partial\varphi \\ \partial y\,(r, \varphi)\,/\partial\varphi \end{array} \right) = \left(\begin{array}{c} \partial\,(r\cos\varphi)\,/\partial\varphi \\ \partial\,(r\sin\varphi)\,/\partial\varphi \end{array} \right) = \left(\begin{array}{c} -r\sin\varphi \\ r\cos\varphi \end{array} \right).$$

$$(8.23)$$

We first note that the basis vectors of the polar coordinates for each point $P = (x, y) = (r, \varphi)$ have a different value. Unlike in linear coordinate systems, where the basis vectors are constant, the basis vectors of the polar coordinates only apply to a specific point and are therefore also called **local basis**. If one calculates the scalar product of the two basis vectors (at the same point), one gets

$$\vec{e}_r \cdot \vec{e}_\varphi = -r\cos\varphi\sin\varphi + r\cos\varphi\sin\varphi = 0,$$

so the two basis vectors are perpendicular to each other at every point. The polar coordinates thus belong to the so-called **orthogonal** coordinate systems. The length of the two basis vectors is given by

$$\begin{aligned} |\vec{e}_r| &= \sqrt{\cos^2\varphi + \sin^2\varphi} = 1 \\ |\vec{e}_\varphi| &= \sqrt{r^2\sin^2\varphi + r^2\cos^2\varphi} = r, \end{aligned}$$

$$(8.24)$$

so they are *not* unit vectors.

We now want to investigate how the coordinates of an arbitrary point as well as the basis vectors change in general coordinate transformations. If we denote a small displacement of a two-dimensional (location) vector $\vec{r}$ with $\Delta\vec{r}$, then for the components of $\Delta\vec{r}$ in the Cartesian coordinate system K

$$\Delta\vec{r} \underset{K}{\rightarrow} (\Delta x, \Delta y)$$

and in a coordinate system K'

$$\Delta\vec{r} \underset{K'}{\rightarrow} (\Delta x', \Delta y').$$

To calculate the transformation/conversion between $\Delta x', \Delta y'$ and $\Delta x, \Delta y$, we use the formula for the total differential 4.16 on page 67:

$$\begin{aligned} \Delta x &= \frac{\partial x}{\partial x'}\Delta x' + \frac{\partial x}{\partial y'}\Delta y' \\ \Delta y &= \frac{\partial y}{\partial x'}\Delta x' + \frac{\partial y}{\partial y'}\Delta y' \end{aligned}$$

or in matrix notation

$$\left(\begin{array}{c} \Delta x \\ \Delta y \end{array} \right) = \left(\begin{array}{cc} \partial x/\partial x' & \partial x/\partial y' \\ \partial y/\partial x' & \partial y/\partial y' \end{array} \right) \left(\begin{array}{c} \Delta x' \\ \Delta y' \end{array} \right).$$

$$(8.25)$$

Of course, the inverse transformation also applies

$$\begin{pmatrix} \Delta x' \\ \Delta y' \end{pmatrix} = \begin{pmatrix} \partial x'/\partial x & \partial x'/\partial y \\ \partial y'/\partial x & \partial y'/\partial y \end{pmatrix} \begin{pmatrix} \Delta x \\ \Delta y \end{pmatrix}. \tag{8.26}$$

For the transformation of the basis vectors, the following applies analogously

$$\begin{aligned} \vec{e}_{x'} &= \frac{\partial x}{\partial x'}\,\vec{e}_x + \frac{\partial y}{\partial x'}\,\vec{e}_y \\ \vec{e}_{y'} &= \frac{\partial x}{\partial y'}\,\vec{e}_x + \frac{\partial y}{\partial y'}\,\vec{e}_y \end{aligned} \tag{8.27}$$

or

$$\begin{aligned} \vec{e}_x &= \frac{\partial x'}{\partial x}\,\vec{e}_{x'} + \frac{\partial y'}{\partial x}\,\vec{e}_{y'} \\ \vec{e}_y &= \frac{\partial x'}{\partial y}\,\vec{e}_{x'} + \frac{\partial y'}{\partial y}\,\vec{e}_{y'}. \end{aligned} \tag{8.28}$$

If you multiply the two transformation matrices, you get

$$\begin{pmatrix} \dfrac{\partial x}{\partial x'} & \dfrac{\partial x}{\partial y'} \\[2mm] \dfrac{\partial y}{\partial x'} & \dfrac{\partial y}{\partial y'} \end{pmatrix} \begin{pmatrix} \dfrac{\partial x'}{\partial x} & \dfrac{\partial x'}{\partial y} \\[2mm] \dfrac{\partial y'}{\partial x} & \dfrac{\partial y'}{\partial y} \end{pmatrix} = \begin{pmatrix} \dfrac{\partial x}{\partial x'}\dfrac{\partial x'}{\partial x} + \dfrac{\partial x}{\partial y'}\dfrac{\partial y'}{\partial x} & \dfrac{\partial x}{\partial x'}\dfrac{\partial x'}{\partial y} + \dfrac{\partial x}{\partial y'}\dfrac{\partial y'}{\partial y} \\[3mm] \dfrac{\partial y}{\partial x'}\dfrac{\partial x'}{\partial x} + \dfrac{\partial y}{\partial y'}\dfrac{\partial y'}{\partial x} & \dfrac{\partial y}{\partial x'}\dfrac{\partial x'}{\partial y} + \dfrac{\partial y}{\partial y'}\dfrac{\partial y'}{\partial y} \end{pmatrix}.$$

For the matrix elements on the right side, we apply the chain rule 4.24 on page 71 and get

$$\begin{aligned} \frac{\partial x}{\partial x'}\frac{\partial x'}{\partial x} + \frac{\partial x}{\partial y'}\frac{\partial y'}{\partial x} &= \frac{\partial x}{\partial x} = 1 \\[2mm] \frac{\partial x}{\partial x'}\frac{\partial x'}{\partial y} + \frac{\partial x}{\partial y'}\frac{\partial y'}{\partial y} &= \frac{\partial x}{\partial y} = 0 \\[2mm] \frac{\partial y}{\partial x'}\frac{\partial x'}{\partial x} + \frac{\partial y}{\partial y'}\frac{\partial y'}{\partial x} &= \frac{\partial y}{\partial x} = 0 \\[2mm] \frac{\partial y}{\partial x'}\frac{\partial x'}{\partial y} + \frac{\partial y}{\partial y'}\frac{\partial y'}{\partial y} &= \frac{\partial y}{\partial y} = 1, \end{aligned}$$

where we did not simply shorten the differentials, but calculated the partial derivatives. For example, $\partial x/\partial x$ denotes the derivative of the function $f(x,y) = x$ with respect to the variable x! In total we get

$$\begin{pmatrix} \partial x/\partial x' & \partial x/\partial y' \\ \partial y/\partial x' & \partial y/\partial y' \end{pmatrix} \begin{pmatrix} \partial x'/\partial x & \partial x'/\partial y \\ \partial y'/\partial x & \partial y'/\partial y \end{pmatrix} = \begin{pmatrix} 1 & 0 \\ 0 & 1 \end{pmatrix}. \tag{8.29}$$

The transformation matrices are, as one could also expect, thus inverse to each other. We want to clarify these relationships again using the example of polar coordinates.

Example 8.9.
In (8.20) it was shown that one can consider the Cartesian coordinates x and y as functions of r and φ:

$$\begin{aligned} x &= x(r, \varphi) = r \cos \varphi \\ y &= y(r, \varphi) = r \sin \varphi \end{aligned}$$

and conversely r, φ as functions of x, y

$$\begin{aligned} r &= r(x, y) = \sqrt{x^2 + y^2} \\ \varphi &= \varphi(x, y) = \pm \arccos \frac{x}{\sqrt{x^2 + y^2}}. \end{aligned}$$

For the Jacobian matrix, with $x' = r$ and $y' = \varphi$, we get

$$\phi' = \begin{pmatrix} \partial x/\partial r & \partial x/\partial \varphi \\ \partial y/\partial r & \partial y/\partial \varphi \end{pmatrix} = \begin{pmatrix} \cos \varphi & -r \sin \varphi \\ \sin \varphi & r \cos \varphi \end{pmatrix}.$$

To determine the matrix of the inverse transformation, we use the formula from Remark 8.5 on page 155. First, we calculate the determinant of ϕ'

$$\det \phi' = \det \begin{pmatrix} \cos \varphi & -r \sin \varphi \\ \sin \varphi & r \cos \varphi \end{pmatrix} = r \cos^2 \varphi + r \sin^2 \varphi = r,$$

i.e., for $r \neq 0$ the determinant is not equal to zero. For the Jacobian matrix of the inverse transformation, we then use the formula (8.19)

$$(\phi')^{-1} = \frac{1}{\det \phi'} \begin{pmatrix} r \cos \varphi & r \sin \varphi \\ -\sin \varphi & \cos \varphi \end{pmatrix} = \begin{pmatrix} \cos \varphi & \sin \varphi \\ -\sin \varphi/r & \cos \varphi/r \end{pmatrix}.$$

Of course, one can also directly calculate the Jacobian matrix of the inverse function by calculating the partial derivatives of the component functions, but the effort for this is considerably higher. For the determinant of the Jacobian matrix of the inverse transformation, we get

$$\det \begin{pmatrix} \cos \varphi & \sin \varphi \\ -\sin \varphi/r & \cos \varphi/r \end{pmatrix} = \frac{1}{r} \cos^2 \varphi + \frac{1}{r} \sin^2 \varphi = \frac{1}{r},$$

it is also defined for all $r \neq 0$ and not equal to zero. $\square$

The coordinate transformations of the polar coordinates are obtained with the formulas (8.25) and (8.26) to

$$\begin{pmatrix} \Delta x \\ \Delta y \end{pmatrix} = \begin{pmatrix} \partial x/\partial r & \partial x/\partial \varphi \\ \partial x/\partial r & \partial x/\partial \varphi \end{pmatrix} \begin{pmatrix} \Delta r \\ \Delta \varphi \end{pmatrix} = \begin{pmatrix} \cos\varphi & -r\sin\varphi \\ \sin\varphi & r\cos\varphi \end{pmatrix} \begin{pmatrix} \Delta r \\ \Delta \varphi \end{pmatrix} \tag{8.30}$$

and

$$\begin{pmatrix} \Delta r \\ \Delta \varphi \end{pmatrix} = \begin{pmatrix} \partial r/\partial x & \partial r/\partial y \\ \partial \varphi/\partial x & \partial \varphi/\partial y \end{pmatrix} \begin{pmatrix} \Delta x \\ \Delta y \end{pmatrix} = \begin{pmatrix} \cos\varphi & \sin\varphi \\ -\sin\varphi/r & \cos\varphi/r \end{pmatrix} \begin{pmatrix} \Delta x \\ \Delta y \end{pmatrix}. \tag{8.31}$$

The transformation to polar coordinates does not change the length of the displacement vector $\Delta\vec{r}$, because on the one hand it holds

$$\begin{aligned} |\Delta\vec{r}|^2 &= |\Delta x \vec{e}_x + \Delta y \vec{e}_y|^2 \\ &= \Delta x^2 \underbrace{\vec{e}_x \cdot \vec{e}_x}_{=1} + 2\Delta x \Delta y \underbrace{\vec{e}_x \cdot \vec{e}_y}_{=0} + \Delta y^2 \underbrace{\vec{e}_y \cdot \vec{e}_y}_{=1} = \Delta x^2 + \Delta y^2 \end{aligned}$$

and on the other hand with (8.24) and (8.31)

$$\begin{aligned} |\Delta\vec{r}|^2 &= |\Delta r\, \vec{e}_r + \Delta\varphi\, \vec{e}_\varphi|^2 = \Delta r^2 |\vec{e}_r|^2 + 2|\Delta r \Delta\varphi|\, \vec{e}_r \cdot \vec{e}_\varphi + \Delta\varphi^2 |\vec{e}_\varphi|^2 \\ &\underset{\vec{e}_r \cdot \vec{e}_\varphi = 0}{=} \Delta r^2 |\vec{e}_r|^2 + \Delta\varphi^2 |\vec{e}_\varphi|^2 \underset{|\vec{e}_r|=1;\,|\vec{e}_\varphi|=r}{=} \Delta r^2 + r^2 \Delta\varphi^2 \\ &= (\cos\varphi\, \Delta x + \sin\varphi\, \Delta y)^2 + r^2 \left(\frac{-\sin\varphi}{r}\Delta x + \frac{\cos\varphi}{r}\Delta y \right)^2 \\ &= \cos^2\varphi\, \Delta x^2 + 2\cos\varphi\sin\varphi\, \Delta x \Delta y + \sin^2\varphi\, \Delta y^2 \\ &\quad + r^2 \left(\frac{\sin^2\varphi}{r^2}\Delta x^2 - \frac{2\cos\varphi\sin\varphi}{r^2}\Delta x \Delta y + \frac{\cos^2\varphi}{r^2}\Delta y^2 \right) \\ &= \Delta x^2 \left(\cos^2\varphi + \sin^2\varphi\right) + \Delta y^2 \left(\cos^2\varphi + \sin^2\varphi\right) \\ &= \Delta x^2 + \Delta y^2, \end{aligned}$$

so there is a match.

We now derive a transformation formula between the components of a vector $\vec{v}$ in an (arbitrary) (x', y') coordinate system K' and those in the Cartesian (x, y) coordinate system K. The vector can be written in both coordinate systems as a linear combination of the basis vectors

$$\vec{v} = v^x \vec{e}_x + v^y \vec{e}_y = v^{x'} \vec{e}_{x'} + v^{y'} \vec{e}_{y'}.$$

We now replace the basis vectors in K' with Eq. (8.27)

$$\vec{v} = v^{x'} \vec{e}_{x'} + v^{y'} \vec{e}_{y'}$$

$$= v^{x'} \left(\frac{\partial x}{\partial x'} \vec{e}_x + \frac{\partial y}{\partial x'} \vec{e}_y \right) + v^{y'} \left(\frac{\partial x}{\partial y'} \vec{e}_x + \frac{\partial y}{\partial y'} \vec{e}_y \right).$$

We combine the terms that belong to the base vectors $\vec{e}_x, \vec{e}_y$ and obtain

$$\vec{v} = \left(v^{x'} \frac{\partial x}{\partial x'} + v^{y'} \frac{\partial x}{\partial y'} \right) \vec{e}_x + \left(v^{x'} \frac{\partial y}{\partial x'} + v^{y'} \frac{\partial y}{\partial y'} \right) \vec{e}_y.$$

From this, due to the uniqueness of the linear combinations, we read the relationship between the vector components in the two bases. It applies

$$\begin{pmatrix} v^x \\ v^y \end{pmatrix} = \begin{pmatrix} v^{x'} \partial x/\partial x' + v^{y'} \partial x/\partial y' \\ v^{x'} \partial y/\partial x' + v^{y'} \partial y/\partial y' \end{pmatrix} = \begin{pmatrix} \partial x/\partial x' & \partial x/\partial y' \\ \partial y/\partial x' & \partial y/\partial y' \end{pmatrix} \begin{pmatrix} v^{x'} \\ v^{y'} \end{pmatrix},$$
$$(8.32)$$

i.e., the components of a vector transform during a change of coordinate system just like the coordinate shifts. This was also to be expected, because the components of a vector are nothing more than the difference of the coordinates of the starting and ending points of the vector. As in section 8.2, the components of a vector also transform contravariantly to the base vectors during general coordinate changes. Of course, the reverse transformation also applies,

$$\begin{pmatrix} v^{x'} \\ v^{y'} \end{pmatrix} = \begin{pmatrix} \partial x'/\partial x & \partial x'/\partial y \\ \partial y'/\partial x & \partial y'/\partial y \end{pmatrix} \begin{pmatrix} v^x \\ v^y \end{pmatrix}. \qquad (8.33)$$

We want to derive an important principle/law from the two transformation formulas for vector components.

Vector equations in the Euclidean plane

A vector, whose components in a coordinate system are all zero, is also the zero vector in all other coordinate systems. Because let $\vec{v}$ be a vector, whose components in an arbitrary (x, y) coordinate system K are equal to zero,

$$\vec{v} \xrightarrow[K]{} \begin{pmatrix} 0 \\ 0 \end{pmatrix},$$

and let (x', y') be any other coordinate system K', then according to the transformation formula (8.33) for the components of $\vec{v}$ in K'

$$\begin{pmatrix} v^{x'} \\ v^{y'} \end{pmatrix} = \begin{pmatrix} \partial x'/\partial x & \partial x'/\partial y \\ \partial y'/\partial x & \partial y'/\partial y \end{pmatrix} \begin{pmatrix} 0 \\ 0 \end{pmatrix} = \begin{pmatrix} 0 \\ 0 \end{pmatrix}.$$

From this, one can deduce that the components of two vectors, which are equal in one (specific) coordinate system, also coincide in all other coordinate systems. Because let $\begin{pmatrix} v^x \\ v^y \end{pmatrix}$ be the components of a vector $\vec{v}$ and $\begin{pmatrix} w^x \\ w^y \end{pmatrix}$ are the components of a vector $\vec{w}$ in any (x, y) coordinate system K and it holds

$$\begin{pmatrix} v^x \\ v^y \end{pmatrix} = \begin{pmatrix} w^x \\ w^y \end{pmatrix}.$$

Then it follows

$$\begin{pmatrix} v^x \\ v^y \end{pmatrix} - \begin{pmatrix} w^x \\ w^y \end{pmatrix} = \begin{pmatrix} v^x - w^x \\ v^y - w^y \end{pmatrix} = \begin{pmatrix} 0 \\ 0 \end{pmatrix}$$

and thus in any other coordinate system K'

$$\begin{pmatrix} v^{x'} - w^{x'} \\ v^{y'} - w^{y'} \end{pmatrix} = \begin{pmatrix} 0 \\ 0 \end{pmatrix},$$

so

$$\begin{pmatrix} v^{x'} \\ v^{y'} \end{pmatrix} = \begin{pmatrix} w^{x'} \\ w^{y'} \end{pmatrix}.$$

Since physical laws often come in the form of a vector equation, for example, Newton's second law is given as

$$\vec{F} = m\vec{a},$$

this vector equation can also be written as an equation between the components of the vectors valid in any coordinate system, as

$$\begin{pmatrix} F^x \\ F^y \\ F^z \end{pmatrix} = m \begin{pmatrix} a^x \\ a^y \\ a^z \end{pmatrix},$$

without the *form of the equation* changing when moving to another coordinate system.

8.3. Vector Analysis in General Coordinate Systems

In this section, we want to deal with the derivatives of vector fields and allow for arbitrary coordinate systems for the representation of the component functions of the vectors. We have seen that the basis vectors of polar coordinates

$$\begin{aligned} \vec{e}_r &= \cos\varphi\,\vec{e}_x + \sin\varphi\,\vec{e}_y \\ \vec{e}_\varphi &= -r\sin\varphi\,\vec{e}_x + r\cos\varphi\,\vec{e}_y \end{aligned}$$

depend on r and φ from the perspective of the Cartesian coordinate system, i.e., they are not constant. The unit vector $\vec{e}_x$ is a constant vector field, it is the same size at every point in the Euclidean plane and points in the same direction. In polar coordinates P, it has the components according to (8.28) and (8.31)

$$\vec{e}_x = \cos\varphi\,\vec{e}_r - \frac{\sin\varphi}{r}\,\vec{e}_\varphi \;\underset{P}{\rightarrow}\; \begin{pmatrix} \cos\varphi \\ -\dfrac{\sin\varphi}{r} \end{pmatrix}.$$

Similarly,

$$\vec{e}_y = \sin\varphi\,\vec{e}_r + \frac{\cos\varphi}{r}\,\vec{e}_\varphi \;\underset{P}{\rightarrow}\; \begin{pmatrix} \sin\varphi \\ \dfrac{\cos\varphi}{r} \end{pmatrix}.$$

These are therefore not constant, although the vectors themselves are constant. The reason for this lies in the fact that the basis vectors in polar coordinates are also not constant. If we consider the vector $\vec{e}_x$ as a constant vector field and differentiate, we expect that its Jacobian matrix is the zero matrix. In Cartesian coordinates K we also get this result

$$(\vec{e}_x)' \;\underset{K}{\rightarrow}\; \begin{pmatrix} \partial(1)/\partial x & \partial(1)/\partial y \\ \partial(0)/\partial x & \partial(0)/\partial y \end{pmatrix} = \begin{pmatrix} 0 & 0 \\ 0 & 0 \end{pmatrix}.$$

If we consider the two column vectors of the Jacobian matrix, we can write this result in the form of two equations

$$\frac{\partial}{\partial x}\vec{e}_x \;\underset{K}{\rightarrow}\; \begin{pmatrix} 0 \\ 0 \end{pmatrix}$$

$$\frac{\partial}{\partial y}\vec{e}_x \;\underset{K}{\rightarrow}\; \begin{pmatrix} 0 \\ 0 \end{pmatrix}.$$

Similarly, we obtain

$$\frac{\partial}{\partial x}\vec{e}_y \;\underset{K}{\rightarrow}\; \begin{pmatrix} 0 \\ 0 \end{pmatrix}$$

$$\frac{\partial}{\partial y}\vec{e}_y \;\underset{K}{\rightarrow}\; \begin{pmatrix} 0 \\ 0 \end{pmatrix}. \tag{8.34}$$

In polar coordinates P, however, we get

$$(\vec{e}_x)' \;\underset{P}{\rightarrow}\; \begin{pmatrix} \partial(\cos\varphi)/\partial r & \partial(\cos\varphi)/\partial\varphi \\ \partial\left(-\dfrac{\sin\varphi}{r}\right)/\partial r & \partial\left(-\dfrac{\sin\varphi}{r}\right)/\partial\varphi \end{pmatrix} = \begin{pmatrix} 0 & -\sin\varphi \\ \sin\varphi/r^2 & -\cos\varphi/r \end{pmatrix},$$

so not the zero matrix. In curvilinear coordinates, it is therefore not sufficient to differentiate the components of a vector to obtain the derivative of the vector. Rather, the (usually) non-constant basis vectors themselves must also be differentiated.

Derivative of the basis vectors

Since $\vec{e}_x$ and $\vec{e}_y$ have constant components in the Cartesian coordinate system, similar to above

$$\frac{\partial}{\partial r}\vec{e}_x \underset{K}{\to} \begin{pmatrix} 0 \\ 0 \end{pmatrix}, \frac{\partial}{\partial \varphi}\vec{e}_x \underset{K}{\to} \begin{pmatrix} 0 \\ 0 \end{pmatrix}, \frac{\partial}{\partial r}\vec{e}_y \underset{K}{\to} \begin{pmatrix} 0 \\ 0 \end{pmatrix}, \frac{\partial}{\partial \varphi}\vec{e}_y \underset{K}{\to} \begin{pmatrix} 0 \\ 0 \end{pmatrix}.$$

This implies for the derivatives of the basis vectors in polar coordinates with (8.22), (8.23) and the product rule

$$\begin{aligned}
\frac{\partial}{\partial r}\vec{e}_r &= \frac{\partial}{\partial r}(\cos\varphi\,\vec{e}_x + \sin\varphi\,\vec{e}_y) \\
&= \left(\frac{\partial}{\partial r}\cos\varphi\right)\vec{e}_x + \cos\varphi\left(\frac{\partial}{\partial r}\vec{e}_x\right) \\
&\quad + \left(\frac{\partial}{\partial r}\sin\varphi\right)\vec{e}_y + \sin\varphi\left(\frac{\partial}{\partial r}\vec{e}_y\right) \\
&= 0\cdot\vec{e}_x + \cos\varphi\cdot\vec{0} + 0\cdot\vec{e}_y + \sin\varphi\cdot\vec{0} = \vec{0},
\end{aligned}$$

where we have again denoted the zero vector by $\vec{0}$. Furthermore, it holds

$$\begin{aligned}
\frac{\partial}{\partial \varphi}\vec{e}_r &= \frac{\partial}{\partial \varphi}(\cos\varphi\,\vec{e}_x + \sin\varphi\,\vec{e}_y) \\
&= \left(\frac{\partial}{\partial \varphi}\cos\varphi\right)\vec{e}_x + \cos\varphi\left(\frac{\partial}{\partial \varphi}\vec{e}_x\right) \\
&\quad + \left(\frac{\partial}{\partial \varphi}\sin\varphi\right)\vec{e}_y + \sin\varphi\left(\frac{\partial}{\partial \varphi}\vec{e}_y\right) \\
&= -\sin\varphi\cdot\vec{e}_x + \cos\varphi\cdot\vec{0} + \cos\varphi\cdot\vec{e}_y + \sin\varphi\cdot\vec{0} \\
&= -\sin\varphi\,\vec{e}_x + \cos\varphi\,\vec{e}_y \\
&= \frac{1}{r}\vec{e}_\varphi.
\end{aligned}$$

Similarly, using (8.23)

$$\begin{aligned}
\frac{\partial}{\partial r}\vec{e}_\varphi &= \frac{\partial}{\partial r}(-r\sin\varphi\,\vec{e}_x + r\cos\varphi\,\vec{e}_y) \\
&= \left(\frac{\partial}{\partial r}(-r\sin\varphi)\right)\vec{e}_x - r\sin\varphi\left(\frac{\partial}{\partial r}\vec{e}_x\right) \\
&\quad + \left(\frac{\partial}{\partial r}(r\cos\varphi)\right)\vec{e}_y + r\cos\varphi\left(\frac{\partial}{\partial r}\vec{e}_y\right) \\
&= -\sin\varphi\,\vec{e}_x - r\sin\varphi\cdot\vec{0} + \cos\varphi\,\vec{e}_y + r\cos\varphi\cdot\vec{0} \\
&= -\sin\varphi\,\vec{e}_x + \cos\varphi\,\vec{e}_y
\end{aligned}$$

$$= \frac{1}{r}\,\vec{e}_\varphi$$

and

$$
\begin{aligned}
\frac{\partial}{\partial\varphi}\,\vec{e}_\varphi &= \frac{\partial}{\partial\varphi}\left(-r\sin\varphi\,\vec{e}_x + r\cos\varphi\,\vec{e}_y\right)\\[2mm]
&= \left(\frac{\partial}{\partial\varphi}(-r\sin\varphi)\right)\vec{e}_x - r\sin\varphi\left(\frac{\partial}{\partial\varphi}\,\vec{e}_x\right)\\[2mm]
&\quad + \left(\frac{\partial}{\partial\varphi}(r\cos\varphi)\right)\vec{e}_y + r\cos\varphi\left(\frac{\partial}{\partial\varphi}\,\vec{e}_y\right)\\[2mm]
&= -r\cos\varphi\cdot\vec{e}_x - r\sin\varphi\cdot\vec{0} - r\sin\varphi\cdot\vec{e}_y + r\cos\varphi\cdot\vec{0}\\[2mm]
&= -r\cos\varphi\,\vec{e}_x - r\sin\varphi\,\vec{e}_y\\[2mm]
&= -r\,\vec{e}_r.
\end{aligned}
$$

Although we used the constancy of the Cartesian unit vectors in the derivation, no terms appear in the derivatives of the basis vectors of the polar coordinates that originate from the Cartesian coordinates. All derivatives can be represented as a linear combination of the basis vectors $\vec{e}_r$ and $\vec{e}_\varphi$, which is also natural, since the partial derivatives of the basis vectors themselves are again vectors. If we write the linear combinations of the partial derivatives of the basis vectors in a slightly different way, we obtain

$$
\begin{aligned}
\frac{\partial}{\partial r}\,\vec{e}_r &= 0\cdot\vec{e}_r + 0\cdot\vec{e}_\varphi = \Gamma^r_{rr}\,\vec{e}_r + \Gamma^\varphi_{rr}\,\vec{e}_\varphi\\[2mm]
\frac{\partial}{\partial\varphi}\,\vec{e}_r &= 0\cdot\vec{e}_r + \frac{1}{r}\cdot\vec{e}_\varphi = \Gamma^r_{r\varphi}\,\vec{e}_r + \Gamma^\varphi_{r\varphi}\,\vec{e}_\varphi\\[2mm]
\frac{\partial}{\partial r}\,\vec{e}_\varphi &= 0\cdot\vec{e}_r + \frac{1}{r}\cdot\vec{e}_\varphi = \Gamma^r_{\varphi r}\,\vec{e}_r + \Gamma^\varphi_{\varphi r}\,\vec{e}_\varphi\\[2mm]
\frac{\partial}{\partial\varphi}\,\vec{e}_\varphi &= -r\cdot\vec{e}_r + 0\cdot\vec{e}_\varphi = \Gamma^r_{\varphi\varphi}\,\vec{e}_r + \Gamma^\varphi_{\varphi\varphi}\,\vec{e}_\varphi.
\end{aligned}
\tag{8.35}
$$

From this we read that for the $2^3 = 8$ numbers, which we have written with the big gamma and three indices, the following applies:

$$
\begin{aligned}
\Gamma^r_{rr} &= 0\\[2mm]
\Gamma^\varphi_{rr} &= 0\\[2mm]
\Gamma^r_{r\varphi} &= 0\\[2mm]
\Gamma^\varphi_{r\varphi} &= \frac{1}{r}\\[2mm]
\Gamma^r_{\varphi r} &= 0
\end{aligned}
\tag{8.36}
$$

$$\Gamma^{\varphi}_{\varphi r} = \frac{1}{r}$$
$$\Gamma^{r}_{\varphi\varphi} = -r$$
$$\Gamma^{\varphi}_{\varphi\varphi} = 0$$

These numbers are called **Christoffel symbols** and are the components of the derivatives of the basis vectors in polar coordinates. The first lower index denotes the basis vector that is derived. The second lower index defines the variable by which (partially) is derived. The upper index is like the index of a vector component; it indicates to which basis vector the Christoffel symbol belongs. The Christoffel symbols depend on the chosen coordinate system. In Cartesian coordinates, all Christoffel symbols are equal zero, because according to Eq. (8.34) all derivatives of the basis vectors are zero. If you have a *random* (curvilinear) $\left(x',y'\right)$ coordinate system with basis vectors $\vec{e}_{x'}$ and $\vec{e}_{y'}$ present, then for the derivative vectors of the basis vectors

$$\frac{\partial}{\partial x'}\,\vec{e}_{x'} = \Gamma^{x'}_{x'x'}\,\vec{e}_{x'} + \Gamma^{y'}_{x'x'}\,\vec{e}_{y'}$$
$$\frac{\partial}{\partial y'}\,\vec{e}_{x'} = \Gamma^{x'}_{x'y'}\,\vec{e}_{x'} + \Gamma^{y'}_{x'y'}\,\vec{e}_{y'}$$
$$\frac{\partial}{\partial x'}\,\vec{e}_{y'} = \Gamma^{x'}_{y'x'}\,\vec{e}_{x'} + \Gamma^{y'}_{y'x'}\,\vec{e}_{y'}$$
$$\frac{\partial}{\partial y'}\,\vec{e}_{y'} = \Gamma^{x'}_{y'y'}\,\vec{e}_{x'} + \Gamma^{y'}_{y'y'}\,\vec{e}_{y'}.$$

We look again at the Cartesian unit vectors and their representation in polar coordinates:

$$\vec{e}_x = \cos\varphi\,\vec{e}_r - \frac{\sin\varphi}{r}\,\vec{e}_\varphi$$
$$\vec{e}_y = \sin\varphi\,\vec{e}_r + \frac{\cos\varphi}{r}\,\vec{e}_\varphi$$

and want to show that the derivatives each yield the zero vector. We again apply the product rule and the above results for the derivatives of $\vec{e}_r$ and $\vec{e}_\varphi$. Then follows

$$\frac{\partial}{\partial r}\,\vec{e}_x = \frac{\partial\left(\cos\varphi\right)}{\partial r}\,\vec{e}_r + \cos\varphi\left(\frac{\partial}{\partial r}\,\vec{e}_r\right) - \frac{\partial}{\partial r}\left(\frac{\sin\varphi}{r}\right)\vec{e}_\varphi - \frac{\sin\varphi}{r}\left(\frac{\partial}{\partial r}\,\vec{e}_\varphi\right)$$
$$= 0\cdot\vec{e}_r + \cos\varphi\cdot\vec{0} + \left(\frac{\sin\varphi}{r^2}\right)\vec{e}_\varphi - \frac{\sin\varphi}{r}\left(\frac{1}{r}\,\vec{e}_\varphi\right) = \vec{0}$$

and

$$\frac{\partial}{\partial \varphi}\, \vec{e}_x \;=\; \frac{\partial \left(\cos \varphi\right)}{\partial \varphi}\, \vec{e}_r + \cos \varphi \left(\frac{\partial}{\partial \varphi}\, \vec{e}_r\right) - \frac{\partial}{\partial \varphi}\left(\frac{\sin \varphi}{r}\right) \vec{e}_\varphi - \frac{\sin \varphi}{r}\left(\frac{\partial}{\partial \varphi}\, \vec{e}_\varphi\right)$$

$$\;=\; -\sin \varphi \cdot \vec{e}_r + \cos \varphi \cdot \frac{1}{r}\, \vec{e}_\varphi - \left(\frac{\cos \varphi}{r}\right) \vec{e}_\varphi - \frac{\sin \varphi}{r}\left(-r\, \vec{e}_r\right) = \vec{0}.$$

So we obtain the desired result. In the above calculations, both the components of $\vec{e}_x$ in polar coordinates (1. and 3. term) as well as the basis vectors themselves (2. and 4. term) were differentiated. Only in this way do all terms cancel out. In the same way, one obtains

$$\frac{\partial}{\partial r}\, \vec{e}_y \;=\; \vec{0}$$

$$\frac{\partial}{\partial \varphi}\, \vec{e}_y \;=\; \vec{0}.$$

Derivation of general vectors

We now consider a general vector field $\vec{v}$ with the components (v^r, v^φ) in polar coordinates

$$\vec{v} = v^r \vec{e}_r + v^\varphi \vec{e}_\varphi.$$

Then the derivatives are calculated analogously to

$$\frac{\partial \vec{v}}{\partial r} \;=\; \frac{\partial}{\partial r}\left(v^r \vec{e}_r + v^\varphi \vec{e}_\varphi\right)$$

$$\;=\; \frac{\partial v^r}{\partial r}\, \vec{e}_r + v^r \frac{\partial \vec{e}_r}{\partial r} + \frac{\partial v^\varphi}{\partial r}\, \vec{e}_\varphi + v^\varphi \frac{\partial \vec{e}_\varphi}{\partial r}$$

$$\;=\; \frac{\partial v^r}{\partial r}\, \vec{e}_r + \left(\frac{\partial v^\varphi}{\partial r} + \frac{v^\varphi}{r}\right) \vec{e}_\varphi,$$

since, as shown above:

$$\frac{\partial \vec{e}_r}{\partial r} = \vec{0}, \qquad \frac{\partial \vec{e}_\varphi}{\partial r} = \frac{1}{r}\, \vec{e}_\varphi$$

Similarly, it follows

$$\frac{\partial \vec{v}}{\partial \varphi} \;=\; \frac{\partial}{\partial \varphi}\left(v^r \vec{e}_r + v^\varphi \vec{e}_\varphi\right)$$

$$\;=\; \frac{\partial v^r}{\partial \varphi}\, \vec{e}_r + \frac{\partial v^\varphi}{\partial \varphi}\, \vec{e}_\varphi + v^r \frac{\partial \vec{e}_r}{\partial \varphi} + v^\varphi \frac{\partial \vec{e}_\varphi}{\partial \varphi}.$$

$$\;=\; \left(\frac{\partial v^r}{\partial \varphi} - r v^\varphi\right) \vec{e}_r + \left(\frac{\partial v^\varphi}{\partial \varphi} + \frac{v^r}{r}\right) \vec{e}_\varphi,$$

since, as shown above:

$$\frac{\partial \vec{e}_r}{\partial \varphi} = \frac{1}{r}\,\vec{e}_\varphi, \quad \frac{\partial \vec{e}_\varphi}{\partial \varphi} = -r\,\vec{e}_r$$

This explicitly shows that in general, the derivative of a vector is more than the partial derivative of its components

$$\begin{pmatrix} \partial v^r/\partial r \\ \partial v^\varphi/\partial r \end{pmatrix} \quad \text{bzw.} \quad \begin{pmatrix} \partial v^r/\partial \varphi \\ \partial v^\varphi/\partial \varphi \end{pmatrix},$$

which we also want to write in the following as

$$v^r{}_{,r} := \frac{\partial v^r}{\partial r}$$

$$v^r{}_{,\varphi} := \frac{\partial v^r}{\partial \varphi}$$

$$v^\varphi{}_{,r} := \frac{\partial v^\varphi}{\partial r}$$

$$v^\varphi{}_{,\varphi} := \frac{\partial v^\varphi}{\partial \varphi}$$

If we replace the derivatives of the basis vectors in the above equations with the representation using the Christoffel symbols (8.35), we get

$$\begin{aligned}
\frac{\partial \vec{v}}{\partial r} &= \frac{\partial v^r}{\partial r}\,\vec{e}_r + \frac{\partial v^\varphi}{\partial r}\,\vec{e}_\varphi + v^r\,\frac{\partial \vec{e}_r}{\partial r} + v^\varphi\,\frac{\partial \vec{e}_\varphi}{\partial r} \\
&= \frac{\partial v^r}{\partial r}\,\vec{e}_r + \frac{\partial v^\varphi}{\partial r}\,\vec{e}_\varphi + v^r\left(\Gamma^r_{rr}\,\vec{e}_r + \Gamma^\varphi_{rr}\,\vec{e}_\varphi\right) + v^\varphi\left(\Gamma^r_{\varphi r}\,\vec{e}_r + \Gamma^\varphi_{\varphi r}\,\vec{e}_\varphi\right).
\end{aligned}$$

We combine all terms that belong to $\vec{e}_r$ or $\vec{e}_\varphi$, and with the new notation for the partial derivatives we get

$$\frac{\partial \vec{v}}{\partial r} = \left(v^r{}_{,r} + v^r\,\Gamma^r_{rr} + v^\varphi\,\Gamma^r_{\varphi r}\right)\vec{e}_r + \left(v^\varphi{}_{,r} + v^r\,\Gamma^\varphi_{rr} + v^\varphi\,\Gamma^\varphi_{\varphi r}\right)\vec{e}_\varphi.$$

Similarly, we get

$$\frac{\partial \vec{v}}{\partial \varphi} = \left(v^r{}_{,\varphi} + v^r\,\Gamma^r_{r\varphi} + v^\varphi\,\Gamma^r_{\varphi\varphi}\right)\vec{e}_r + \left(v^\varphi{}_{,\varphi} + v^r\,\Gamma^\varphi_{r\varphi} + v^\varphi\,\Gamma^\varphi_{\varphi\varphi}\right)\vec{e}_\varphi.$$

For the expressions in the brackets, we also write compactly

$$\begin{aligned}
v^r{}_{;r} &:= v^r{}_{,r} + v^r\,\Gamma^r_{rr} + v^\varphi\,\Gamma^r_{\varphi r} \\
v^\varphi{}_{;r} &:= v^\varphi{}_{,r} + v^r\,\Gamma^\varphi_{rr} + v^\varphi\,\Gamma^\varphi_{\varphi r} \\
v^r{}_{;\varphi} &:= v^r{}_{,\varphi} + v^r\,\Gamma^r_{r\varphi} + v^\varphi\,\Gamma^r_{\varphi\varphi}
\end{aligned}$$

$$v^{\varphi}_{\ ;\varphi} \ := \ v^{\varphi}_{\ ,\varphi} + v^r\,\Gamma^{\varphi}_{r\varphi} + v^{\varphi}\,\Gamma^{\varphi}_{\varphi\varphi} \tag{8.37}$$

and calls these quantities the **components of the covariant derivative** of the vector $\vec{v}$. The semicolon in the lower index thus denotes the components of the covariant, the comma in the lower index the components of the partial derivative. In summary, the following representation results for the covariant derivatives of $\vec{v}$ in polar coordinates

$$\frac{\partial \vec{v}}{\partial r} = v^r_{\ ;r}\,\vec{e}_r + v^{\varphi}_{\ ;r}\,\vec{e}_\varphi$$
$$\frac{\partial \vec{v}}{\partial \varphi} = v^r_{\ ;\varphi}\,\vec{e}_r + v^{\varphi}_{\ ;\varphi}\,\vec{e}_\varphi.$$

If one generally has an *arbitrary* (curvilinear) (x', y') coordinate system with basis vectors $\vec{e}_{x'}$ and $\vec{e}_{y'}$ present, then the derivatives follow accordingly

$$\frac{\partial \vec{v}}{\partial x'} = v^{x'}_{\ ;x'}\,\vec{e}_{x'} + v^{y'}_{\ ;x'}\,\vec{e}_{y'}$$
$$\frac{\partial \vec{v}}{\partial y'} = v^{x'}_{\ ;y'}\,\vec{e}_{x'} + v^{y'}_{\ ;y'}\,\vec{e}_{y'},$$

where the components $v^{i'}_{\ ;j'}$ are defined analogously to the polar coordinates. Since in the Cartesian (x, y) coordinate system the derivatives of the basis vectors according to Eq. (8.34) are equal to zero and thus all Christoffel symbols vanish, it applies there that the components of the covariant derivative of a vector coincide with the partial ones

$$v^x_{\ ;x} = v^x_{\ ,x}$$
$$v^x_{\ ;y} = v^x_{\ ,y}$$
$$v^y_{\ ;x} = v^y_{\ ,x}$$
$$v^y_{\ ;y} = v^y_{\ ,y}.$$

This also makes it clear that the components of the covariant derivative of a scalar field ϕ, whose value at a fixed point does not depend on the basis vectors, but always takes the same real number independent of the coordinate system, are equal to those of the partial derivative, i.e.

$$\phi_{\,;x} = \frac{\partial \phi}{\partial x} = \phi_{\,,x}$$
$$\phi_{\,;y} = \frac{\partial \phi}{\partial y} = \phi_{\,,y}. \tag{8.38}$$

Example 8.10.

We want to illustrate the new definitions and terms using a concrete example and consider the vector field

$$\vec{v} = y\,\vec{e}_x + x\,\vec{e}_y,$$

which in the Cartesian (x, y) coordinate system K has the component function

$$\begin{pmatrix} v^x(x, y) \\ v^y(x, y) \end{pmatrix} = \begin{pmatrix} y \\ x \end{pmatrix}$$

For this vector field, we want to calculate the components of the covariant derivative in Cartesian and in polar coordinates. Since in Cartesian coordinates the components of the covariant derivative are equal to the partial derivatives, it follows

$$
\begin{aligned}
v^x{}_{;x} &= v^x{}_{,x} = \frac{\partial v^x}{\partial x} = \frac{\partial y}{\partial x} = 0 \\
v^x{}_{;y} &= v^x{}_{,y} = \frac{\partial v^x}{\partial y} = \frac{\partial y}{\partial y} = 1 \\
v^y{}_{;x} &= v^y{}_{,x} = \frac{\partial v^y}{\partial x} = \frac{\partial x}{\partial x} = 1 \\
v^y{}_{;y} &= v^y{}_{,y} = \frac{\partial v^y}{\partial y} = \frac{\partial x}{\partial y} = 0.
\end{aligned}
$$

To calculate the derivative of $\vec{v}$ in polar coordinates, we first derive the components v^r, v^φ of $\vec{v}$ in polar coordinates. For this, we use the transformation equation (8.31)

$$\begin{pmatrix} v^r \\ v^\varphi \end{pmatrix} = \begin{pmatrix} \partial r/\partial x & \partial r/\partial y \\ \partial\varphi/\partial x & \partial\varphi/\partial y \end{pmatrix} \begin{pmatrix} v^x \\ v^y \end{pmatrix} = \begin{pmatrix} \cos\varphi & \sin\varphi \\ -\sin\varphi/r & \cos\varphi/r \end{pmatrix} \begin{pmatrix} y \\ x \end{pmatrix}.$$

Because $y = r\sin\varphi$ and $x = r\cos\varphi$ we get from this

$$
\begin{aligned}
\begin{pmatrix} v^r \\ v^\varphi \end{pmatrix} &= \begin{pmatrix} \cos\varphi & \sin\varphi \\ -\sin\varphi/r & \cos\varphi/r \end{pmatrix} \begin{pmatrix} r\sin\varphi \\ r\cos\varphi \end{pmatrix} \\
&= \begin{pmatrix} r\sin\varphi\cos\varphi + r\sin\varphi\cos\varphi \\ (-\sin\varphi/r)(r\sin\varphi) + (\cos\varphi/r)(r\cos\varphi) \end{pmatrix},
\end{aligned}
$$

also

$$\begin{pmatrix} v^r \\ v^\varphi \end{pmatrix} = \begin{pmatrix} 2r\sin\varphi\cos\varphi \\ \cos^2\varphi - \sin^2\varphi \end{pmatrix}.$$

To calculate the components of the covariant derivatives of $\vec{v}$ in polar coordinates, the partial derivatives of the vector components v^r, v^φ must first be determined. It follows

$$
\begin{aligned}
v^r{}_{,r} &= \frac{\partial v^r}{\partial r} = \frac{\partial \left(2r \sin \varphi \cos \varphi\right)}{\partial r} = 2 \sin \varphi \cos \varphi \\
v^\varphi{}_{,r} &= \frac{\partial v^\varphi}{\partial r} = \frac{\partial \left(\cos^2 \varphi - \sin^2 \varphi\right)}{\partial r} = 0 \\
v^r{}_{,\varphi} &= \frac{\partial v^r}{\partial \varphi} = \frac{\partial \left(2r \sin \varphi \cos \varphi\right)}{\partial \varphi} = 2r \left(\cos^2 \varphi - \sin^2 \varphi\right) \\
v^\varphi{}_{,\varphi} &= \frac{\partial v^\varphi}{\partial \varphi} = \frac{\partial \left(\cos^2 \varphi - \sin^2 \varphi\right)}{\partial \varphi} = -2 \sin \varphi \cos \varphi - 2 \sin \varphi \cos \varphi \\
&= -4 \sin \varphi \cos \varphi.
\end{aligned}
$$

Now we use the formula (8.37) and the Christoffel symbols from (8.36), to calculate the components of the covariant derivative of $\vec{v}$

$$
\begin{aligned}
v^r{}_{;r} &= v^r{}_{,r} + v^r \underbrace{\Gamma^r_{rr}}_{=0} + v^\varphi \underbrace{\Gamma^r_{\varphi r}}_{=0} = 2 \sin \varphi \cos \varphi \\
v^\varphi{}_{;r} &= \underbrace{v^\varphi{}_{,r}}_{=0} + v^r \underbrace{\Gamma^\varphi_{rr}}_{=0} + v^\varphi \underbrace{\Gamma^\varphi_{\varphi r}}_{=\frac{1}{r}} = \frac{1}{r}\left(\cos^2 \varphi - \sin^2 \varphi\right) \\
v^r{}_{;\varphi} &= v^r{}_{,\varphi} + v^r \underbrace{\Gamma^r_{r\varphi}}_{=0} + v^\varphi \underbrace{\Gamma^r_{\varphi\varphi}}_{=-r} = 2r \left(\cos^2 \varphi - \sin^2 \varphi\right) - r \left(\cos^2 \varphi - \sin^2 \varphi\right) \\
&= r \left(\cos^2 \varphi - \sin^2 \varphi\right) \\
v^\varphi{}_{;\varphi} &= v^\varphi{}_{,\varphi} + v^r \underbrace{\Gamma^\varphi_{r\varphi}}_{=\frac{1}{r}} + v^\varphi \underbrace{\Gamma^\varphi_{\varphi\varphi}}_{=0} = -4 \sin \varphi \cos \varphi + \frac{1}{r}\left(2r \sin \varphi \cos \varphi\right) \\
&= -2 \sin \varphi \cos \varphi. \ \square
\end{aligned}
$$

$$(8.39)$$

Covariant Divergence

For any (curvilinear) (x', y') coordinate system, we define the **covariant divergence** of a vector $\vec{v}$ by

$$
v^{x'}{}_{;x'} + v^{y'}{}_{;y'}, \tag{8.40}
$$

see also Remark 6.6 for the definition of normal divergence. If we look at the example 8.10, we see that the covariant divergence is zero in both Cartesian coordinates and polar coordinates. We will later see that the covariant divergence always takes the same value regardless of the coordinate system.

Example 8.11.

As another example, we calculate the covariant divergence of an arbitrary vector field $\vec{v}$ in polar coordinates, using the covariant derivatives from Eq. (8.39)

$$
v^r{}_{;r} + v^\varphi{}_{;\varphi} = \left(v^r{}_{,r} + v^r \underbrace{\Gamma^r_{rr}}_{=0} + v^\varphi \underbrace{\Gamma^r_{\varphi r}}_{=0} \right) + \left(v^\varphi{}_{,\varphi} + v^r \underbrace{\Gamma^\varphi_{r\varphi}}_{=\frac{1}{r}} + v^\varphi \underbrace{\Gamma^\varphi_{\varphi\varphi}}_{=0} \right)
$$

$$
= \frac{\partial v^r}{\partial r} + \frac{\partial v^\varphi}{\partial \varphi} + \frac{1}{r} v^r,
$$

i.e., in addition to the sum of the partial derivatives, the term v^r/r appears.
□

In this chapter, in deriving the transformation formulas for the vector components, we used the corresponding transformation rules of the basis vectors, i.e., we first had to select a basis, then represent the components of vectors and their properties. Conversely, one can also *define* a vector as an object whose components change in a coordinate system change $K \to K'$ according to the transformation formula (8.33). This definition has the advantage that it does not require an explicit choice of a basis, but only requires the existence of coordinate systems. Of course, as shown above, one can take the resulting tangent vectors from the coordinate curves as basis vectors by derivation. We want to use this fundamental, "basis-free" possibility in the next chapters to define other objects (namely arbitrary **tensors**) using the transformation properties of their components.

9. Tensor Calculus in the Euclidean Plane

In the last chapter we named the four expressions

$$v^r_{;r}, v^\varphi_{;r}, v^r_{;\varphi}, v^\varphi_{;\varphi}$$

components of the covariant derivative, without it being clear, what object the covariant derivative is. It cannot be a vector, it only has two components each with an upper index. Objects whose description requires arbitrary components or indices are called **tensors**.

The **tensor calculus** is an indispensable tool to understand the equations of General Relativity. Also for Special Relativity, many otherwise cumbersome formulations are simplified by tensor calculus. In general, one can say that physical quantities can be described as tensors, with scalars and vectors turning out to be special cases of tensors. Physical laws are therefore equations between tensors, and to set up and reformulate such equations, one needs tensor calculus. In extension of what we said in Chap. 8 about vector equations, it will turn out that all physical laws formulated as tensor equations remain form-identical when changing from one coordinate system to another and performing the corresponding coordinate transformations.

We deal here with tensor calculus in the simplest case, namely in the (two-dimensional) Euclidean plane. We are dealing with only two coordinates and with flat space, which we understand intuitively well due to our sensory perception. Special Relativity then extends to four dimensions in still flat space (what this is, will become clear later), General Relativity then requires tensor calculus in four dimensions and in curved spaces. Despite the later necessary extensions, one can already derive a large part of the phenomena of tensor calculus in the Euclidean plane in such a way that the transition to the more complicated structures of relativity theory will prove to be quite simple. We have already seen in the last chapter that one can indeed work with arbitrary curvilinear coordinate systems in the (flat) Euclidean plane. We will continue this in this chapter and again use polar coordinates as a standard example. The two-dimensional tensor calculus with polar coordinates allows to develop most of the mathematical concepts in such a way that the later step to the curved spaces is not so big anymore.

© The Author(s), under exclusive license to Springer-Verlag GmbH, DE, part of Springer Nature 2026
M. Ruhrländer, *Ascent to the Einstein Equations*,
https://doi.org/10.1007/978-3-662-72672-3_9

Similar to vector calculus, we will also take a two-fold approach in tensor calculus. On the one hand, we consider tensors as (abstract) objects, on the other hand, we also represent tensors by their components in arbitrary coordinate systems and determine their properties by the corresponding properties of the components. The first, more abstract approach to tensor calculus helps at one or another point, by simplicity and clarity to recognize structures and similarities, the second is necessary when one wants to calculate concrete cases. Because then one always has to choose a suitable coordinate system and represent the corresponding quantities in the selected coordinates.

As with vectors, we want to linguistically *not* distinguish between tensors (objects, whose components are constant) and tensor fields (the components are functions), but we use both terms synonymously.

9.1. One-forms, Row Vectors

We want to start with the **one-forms** or **row vectors**. We have already encountered row vectors when we introduced matrix calculation. We called the rows of a matrix row vectors there. In general, we define a row vector $\tilde{p}$ as an object that can be represented by two numbers, i.e. its two components, when choosing a (x, y) coordinate system K

$$\tilde{p} \underset{K}{\to} \begin{pmatrix} p_x & p_y \end{pmatrix},$$

where the components should transform according to a coordinate system change $K \to K'$ as follows

$$
\begin{aligned}
p_{x'} &= \frac{\partial x}{\partial x'} p_x + \frac{\partial y}{\partial x'} p_y \\
p_{y'} &= \frac{\partial x}{\partial y'} p_x + \frac{\partial y}{\partial y'} p_y
\end{aligned}
$$

The components of a row vector are indicated by subscripts and written as a row, we want to mark the row vector itself with the symbol "tilde" above the p and thus make it clear that a row vector is something different than a "normal" vector with an arrow. Row vectors are also called **one-forms**, **covectors** or **covariant vectors** (in contrast to normal vectors, which are then called **contravariant**). We use the term "one-form" when we want to make statements about the abstract properties of these tensors. The word "row vector" comes into play whenever we want to emphasize the proximity to matrix calculation. If you write the components of a vector $\vec{v}$ as a column vector

$$\vec{v} \to \begin{pmatrix} v^x \\ v^y \end{pmatrix},$$

you can also consider the column vector as a (2×1)-matrix (two rows and one column), and many calculation rules for vectors are transferred one to one from the calculation rules for matrices. Similarly, you can consider the components of a row vector as a (2×1)-matrix and also transfer the calculation rules for matrices here. If, for example, $\tilde{p}$ and $\tilde{q}$ are two row vectors, then the sum of these row vectors is defined by

$$\tilde{p} + \tilde{q} \underset{K}{\rightarrow} \left(\begin{array}{cc} p_x & p_y \end{array}\right) + \left(\begin{array}{cc} q_x & q_y \end{array}\right) = \left(\begin{array}{cc} p_x + q_x & p_y + q_y \end{array}\right).$$

Also the multiplication

$$a \cdot \tilde{p} \underset{K}{\rightarrow} \left(\begin{array}{cc} a p_x & a p_y \end{array}\right)$$

with any real number a. We check the transformation property of the first summand as an example:

$$
\begin{aligned}
p_{x'} + q_{x'} &= \left(\frac{\partial x}{\partial x'} p_x + \frac{\partial y}{\partial x'} p_y\right) + \left(\frac{\partial x}{\partial x'} q_x + \frac{\partial y}{\partial x'} q_y\right) \\
&= \frac{\partial x}{\partial x'} (p_x + q_x) + \frac{\partial y}{\partial x'} (p_y + q_y).
\end{aligned}
$$

The sum of two row vectors is therefore again a row vector, as is the product of a row vector with a real number. One also says that the set of row vectors forms a **vector space**, which is called the **dual vector space**, but we do not want to delve into this further.

One can multiply a row vector with a vector, but first we need to generalize the multiplication of matrices.

Remark 9.1. **MT: Multiplication of two arbitrary matrices**
If $A = \left(a^i{}_j\right)$ is a $(m \times n)$-matrix and $B = \left(b^j{}_k\right)$ is a $(n \times l)$-matrix, i.e.

$$i = 1, \cdots, m; j = 1, \cdots, n; k = 1, \cdots, l,$$

then the product

$$C = \left(c^i{}_k\right) = A \cdot B$$

is a $(m \times l)$- matrix with

$$c^i{}_k = a^i{}_1 \, b^1{}_k + a^i{}_2 \, b^2{}_k + a^i{}_3 \, b^3{}_k + \cdots + a^i{}_n \, b^n{}_k.$$

If you imagine the matrix row $\vec{a}^i$ as a column vector, then the element $c^i{}_k$ is the scalar product of $\vec{a}^i$ with the column vector $\vec{b}_k$. Note that the general matrix multiplication is only defined for matrices where the number of columns of A equals the number of rows of B. We want to demonstrate this abstract definition in the case that currently interests us. So let A be a (1×2)-matrix

$$A = \left(\begin{array}{cc} a_1 & a_2 \end{array}\right)$$

and B a (2×1)-Matrix

$$B = \begin{pmatrix} b^1 \\ b^2 \end{pmatrix},$$

then it follows

$$A \cdot B = \begin{pmatrix} a_1 & a_2 \end{pmatrix} \cdot \begin{pmatrix} b^1 \\ b^2 \end{pmatrix} = a_1\, b^1 + a_2\, b^2.$$

The result is a (1×1)-matrix, i.e., a number, and the same result is obtained as with the scalar product of two (normal) vectors, if you imagine that

$$\begin{pmatrix} a_1 & a_2 \end{pmatrix} = \begin{pmatrix} a_1 \\ a_2 \end{pmatrix}. \square$$

The multiplication of a row vector

$$\tilde{p} \underset{K}{\to} \begin{pmatrix} p_x & p_y \end{pmatrix}$$

with a vector

$$\vec{v} \underset{K}{\to} \begin{pmatrix} v^x \\ v^y \end{pmatrix}$$

is now defined analogously to matrix multiplication as

$$\tilde{p}\,(\vec{v}) \underset{K}{=} p_x\, v^x + p_y\, v^y. \tag{9.1}$$

Since the number on the right side is formed by components that depend on the chosen coordinate system K, a K is noted under the equality sign. However, we will see later that this number is always the same, no matter which coordinate system is chosen. Note that the components of the row vector and the vector must always come from *the same* coordinate system!

In Remark 8.2 we showed that the multiplication of a matrix with a vector can be considered as a linear mapping. We now want to transfer this to the row vectors. So if $\tilde{p}$ is a row vector, $\vec{v}$ and $\vec{w}$ are two vectors and a is a real number, so it follows

$$\begin{aligned} \tilde{p}\,(\vec{v} + \vec{w}) &= p_x\,(v^x + w^x) + p_y\,(v^y + w^y) \\ &= (p_x\, v^x + p_y\, v^y) + (p_x\, w^x + p_y\, w^y) \\ &= \tilde{p}\,(\vec{v}) + \tilde{p}\,(\vec{w}) \end{aligned}$$

and

$$\tilde{p}\,(a\vec{v}) = p_x\, av^x + p_y\, av^y = a\,(p_x\, v^x + p_y\, v^y) = a\,\tilde{p}\,(\vec{v}).$$

Now we are also able to define the object row vector or one-form "abstractly":

> A **one-form** is a linear mapping that, when applied to a vector, yields a real number.

Which number? Well, the one that is on the right side of Eq. (9.1) and which we still have to show that it does not depend on the chosen coordinate system, because otherwise the mapping would not be uniquely defined.

We now derive some properties for one-forms. If

$$\vec{e}_x \underset{K}{\rightarrow} \begin{pmatrix} 1 \\ 0 \end{pmatrix}, \vec{e}_y \underset{K}{\rightarrow} \begin{pmatrix} 0 \\ 1 \end{pmatrix}$$

are the basis vectors in the chosen (x, y) coordinate system K, then it follows

$$\begin{aligned} \tilde{p}(\vec{e}_x) &= 1\,p_x + 0\,p_y = p_x \\ \tilde{p}(\vec{e}_y) &= 0\,p_x + 1\,p_y = p_y, \end{aligned} \tag{9.2}$$

i.e., we can write the formula (9.1) in the following way due to the linearity of $\tilde{p}$

$$\tilde{p}(\vec{v}) = \tilde{p}(v^x \vec{e}_x + v^y \vec{e}_y) = \tilde{p}(v^x \vec{e}_x) + \tilde{p}(v^y \vec{e}_y) = v^x \tilde{p}(\vec{e}_x) + v^y \tilde{p}(\vec{e}_y) = v^x p_x + v^y p_y.$$

The transformation property of one-forms can be written compactly in matrix form

$$\begin{pmatrix} p_{x'} & p_{y'} \end{pmatrix} = \begin{pmatrix} p_x & p_y \end{pmatrix} \begin{pmatrix} \partial x/\partial x' & \partial x/\partial y' \\ \partial y/\partial x' & \partial y/\partial y' \end{pmatrix}, \tag{9.3}$$

where, according to the rules for multiplying matrices, the components of the row vector must be multiplied *from the right* with the transformation matrix; the row vector (a (1×2)-matrix) is multiplied with a (2×2)-matrix. The components of row vectors thus transform inversely to the vector components. According to Eq. (8.33) the vector components are

$$\begin{pmatrix} v^{x'} \\ v^{y'} \end{pmatrix} = \begin{pmatrix} \partial x'/\partial x & \partial x'/\partial y \\ \partial y'/\partial x & \partial y'/\partial y \end{pmatrix} \begin{pmatrix} v^x \\ v^y \end{pmatrix}.$$

This opposing transformation behavior now ensures that the number on the right side in formula (9.1) is the same in every coordinate system, because

$$v^{x'} p_{x'} + v^{y'} p_{y'} = \begin{pmatrix} p_{x'} & p_{y'} \end{pmatrix} \begin{pmatrix} v^{x'} \\ v^{y'} \end{pmatrix} =$$

$$\begin{pmatrix} p_x & p_y \end{pmatrix} \underbrace{\begin{pmatrix} \partial x/\partial x' & \partial x/\partial y' \\ \partial y/\partial x' & \partial y/\partial y' \end{pmatrix} \begin{pmatrix} \partial x'/\partial x & \partial x'/\partial y \\ \partial y'/\partial x & \partial y'/\partial y \end{pmatrix}}_{=E} \begin{pmatrix} v^x \\ v^y \end{pmatrix}.$$

The two transformation matrices are inverse to each other according to Eq. (8.29), i.e., the product results in the unit matrix E. So it follows

$$v^{x'} p_{x'} + v^{y'} p_{y'} = \begin{pmatrix} p_{x'} & p_{y'} \end{pmatrix} \begin{pmatrix} v^{x'} \\ v^{y'} \end{pmatrix} = \begin{pmatrix} p_x & p_y \end{pmatrix} \begin{pmatrix} v^x \\ v^y \end{pmatrix} = v^x p_x + v^y p_y,$$

and thus it is shown that the expression $\tilde{p}\,(\vec{v})$ always delivers the same number regardless of the chosen coordinate system and is therefore well-defined.

Gradient of a scalar field

We now want to show that the gradient of a scalar field $\phi\,(\vec{r})$ is a one-form and not a normal vector. We do this by examining the transformation behavior of its components. In a (x, y) coordinate system K, the gradient is defined by its partial derivatives and we write (already suggestively)

$$\tilde{d}\phi \underset{K}{\to} \begin{pmatrix} \dfrac{\partial \phi}{\partial x} & \dfrac{\partial \phi}{\partial y} \end{pmatrix}.$$

If a second (x', y') coordinate system is given, then with the chain rule 4.24 on page 71

$$\frac{\partial \phi}{\partial x'} = \frac{\partial \phi}{\partial x}\frac{\partial x}{\partial x'} + \frac{\partial \phi}{\partial y}\frac{\partial y}{\partial x'}$$

$$\frac{\partial \phi}{\partial y'} = \frac{\partial \phi}{\partial x}\frac{\partial x}{\partial y'} + \frac{\partial \phi}{\partial y}\frac{\partial y}{\partial y'}$$

or again compactly in matrix notation

$$\begin{pmatrix} \dfrac{\partial \phi}{\partial x'} & \dfrac{\partial \phi}{\partial y'} \end{pmatrix} = \begin{pmatrix} \dfrac{\partial \phi}{\partial x} & \dfrac{\partial \phi}{\partial y} \end{pmatrix} \begin{pmatrix} \partial x/\partial x' & \partial x/\partial y' \\ \partial y/\partial x' & \partial y/\partial y' \end{pmatrix}.$$

The components of $\tilde{d}\phi$ thus transform like a one-form and we *define* the gradient of a scalar field ϕ as the one-form $\tilde{d}\phi$. This also gives greater clarity to the shorthand notation for the partial derivatives of vector components introduced in the last chapter. There we defined, for example,

$$\frac{\partial v^r}{\partial \varphi} = v^r{}_{,\varphi}.$$

Since the components of row vectors have a lower index and partial derivatives are components of a row vector, the placement of the index φ downwards is now clear. For $\partial \phi / \partial x, \partial \phi / \partial y$ we also write briefly

$$\frac{\partial \phi}{\partial x} = \phi_{,x} \quad \text{bzw.} \quad \frac{\partial \phi}{\partial y} = \phi_{,y}.$$

Covariant derivation of one-forms

We now want to derive the derivative of one-forms. Here, the components of one-forms are real-valued functions that depend on the coordinates of a point $P = (x, y)$, i.e.

$$\tilde{p}(P) \rightarrow \begin{pmatrix} p_x(x, y) & p_y(x, y) \end{pmatrix}.$$

We have already seen above in (8.38) that the components of the covariant derivative of a scalar function ϕ correspond to the partial derivatives

$$\phi_{;x} = \phi_{,x} = \frac{\partial \phi}{\partial x}$$
$$\phi_{;y} = \phi_{,y} = \frac{\partial \phi}{\partial y}.$$

To calculate the covariant derivative of a one-form $\tilde{p}$, we use the property that a one-form applied to a vector is a scalar function ϕ, i.e., independent of the coordinate system, it provides a real number, which we want to call ϕ

$$\phi := \tilde{p}(\vec{v}) = p_x v^x + p_y v^y.$$

So with the product rule we get

$$\phi_{;x} = \frac{\partial \phi}{\partial x} = \frac{\partial}{\partial x}(p_x v^x + p_y v^y) = \frac{\partial p_x}{\partial x} v^x + p_x \frac{\partial v^x}{\partial x} + \frac{\partial p_y}{\partial x} v^y + p_y \frac{\partial v^y}{\partial x}.$$

We now replace the partial derivatives of v^x and v^y with the covariant ones, i.e., analogously to Eq. (8.37) follows

$$\frac{\partial v^x}{\partial x} = v^x_{;x} - v^x \Gamma^x_{xx} - v^y \Gamma^x_{yx}$$
$$\frac{\partial v^y}{\partial x} = v^y_{;x} - v^x \Gamma^y_{xx} - v^y \Gamma^y_{yx}$$
$$\frac{\partial v^x}{\partial y} = v^x_{;y} - v^x \Gamma^x_{xy} - v^y \Gamma^x_{yy} \tag{9.4}$$
$$\frac{\partial v^y}{\partial y} = v^y_{;y} - v^x \Gamma^y_{xy} - v^y \Gamma^y_{yy}.$$

This inserted into the above equation results in the (lengthy) expression

$$\phi_{;x} = \frac{\partial p_x}{\partial x} v^x + p_x \left(v^x_{;x} - v^x \Gamma^x_{xx} - v^y \Gamma^x_{yx} \right) + \frac{\partial p_y}{\partial x} v^y + p_y \left(v^y_{;x} - v^x \Gamma^y_{xx} - v^y \Gamma^y_{yx} \right).$$

We rearrange this equation a bit and get

$$\phi_{;x} = \left(\frac{\partial p_x}{\partial x} - p_x \Gamma^x_{xx} - p_y \Gamma^y_{xx} \right) v^x + \left(\frac{\partial p_y}{\partial x} - p_x \Gamma^x_{yx} - p_y \Gamma^y_{yx} \right) v^y$$

$$+ \quad p_x\, v^x_{;x} + p_y\, v^y_{;x}. \tag{9.5}$$

Analogously, for the derivative with respect to y, we obtain

$$\phi_{;y} = \left(\frac{\partial p_x}{\partial y} - p_x\, \Gamma^x_{xy} - p_y\, \Gamma^y_{xy}\right) v^x + \left(\frac{\partial p_y}{\partial y} - p_x\, \Gamma^x_{yy} - p_y\, \Gamma^y_{yy}\right) v^y$$

$$+ \quad p_x\, v^x_{;y} + p_y\, v^y_{;y}. \tag{9.6}$$

We now *define* the covariant derivative of $\tilde{p}$ by considering the expressions in the brackets as their components, i.e., we define

$$p_{x\,;x} = \frac{\partial p_x}{\partial x} - p_x\, \Gamma^x_{xx} - p_y\, \Gamma^y_{xx} \tag{9.7}$$

$$p_{y\,;x} = \frac{\partial p_y}{\partial x} - p_x\, \Gamma^x_{yx} - p_y\, \Gamma^y_{yx}$$

$$p_{x\,;y} = \frac{\partial p_x}{\partial y} - p_x\, \Gamma^x_{xy} - p_y\, \Gamma^y_{xy}$$

$$p_{y\,;y} = \frac{\partial p_y}{\partial y} - p_x\, \Gamma^x_{yy} - p_y\, \Gamma^y_{yy}.$$

The above equation for $\phi_{;x}$ can thus be written as

$$\phi_{;x} = \left(p_x v^x + p_y v^y\right)_{;x} = \left(p_{x\,;x} v^x + p_{y\,;x} v^y\right) + \left(p_x v^x_{;x} + p_y v^y_{;x}\right)$$

and thus we obtain a kind of product rule for the covariant derivative. Before we continue with tensor calculus, we introduce a very helpful abbreviation, the so-called **Einstein's summation convention**.

9.2. Einstein's Summation Convention

So far, we have written the indices of vector components upwards, those of one-forms downwards. This different position of the indices allows us to introduce a simplifying notation:

Einstein's Summation Convention: Whenever an expression contains an index that is superscripted for one variable and subscripted for another, a summation over the values that the index can take is performed. For example, the expression $a_i\, b^i$ is a shorthand for the sum

$$a_i\, b^i = a_x\, b^x + a_y\, b^y,$$

if the index i can take the values x and y.

This notation also allows for an immediate generalization to more than two dimensions. If the set of values that the index i can take consists of x, y, z, then the expression $a_i \, b^i$ means $a_x \, b^x + a_y \, b^y + a_z \, b^z$. The index i is also called a **summation index**, it is *arbitrarily selectable*, i.e., the expressions $a_i \, b^i$ and $a_j \, b^j$ are equal:

$$a_i \, b^i = a_x \, b^x + a_y \, b^y = a_j \, b^j$$

In this chapter, we want to use small *Latin* letters like i, j, k etc. for the summation indices, where i, j, k can each take the values x, y. In the following chapters on relativity theory, we use small *Greek* letters for the indices, which can then take on the values t, x, y, z, for example. If we want to denote components in the (x', y') coordinate system K', we use i', j', k' etc. as summation indices, i.e., we then sum over x' and y'.

Example 9.1.

- To get used to the new notation, we consider as another example the expression $\vec{v} = v^i \vec{e}_i$. Since the index i appears both as a superscript and a subscript, the detailed notation is

$$\vec{v} = v^i \vec{e}_i = v^x \vec{e}_x + v^y \vec{e}_y.$$

- Another example of the application of the summation convention is the multiplication of two matrices, see Remark 9.1. There, $A = \left(a^i{}_j \right)$ was an $(m \times n)$-matrix and $B = \left(b^j{}_k \right)$ was an $(n \times l)$-matrix, so $i = 1, \cdots, m; j = 1, \cdots, n; k = 1, \cdots, l$ and the product

$$C = \left(c^i{}_k \right) = A \cdot B$$

was defined as an $(m \times l)$-matrix with

$$c^i{}_k = a^i{}_1 \, b^1{}_k + a^i{}_2 \, b^2{}_k + a^i{}_3 \, b^3{}_k + \cdots + a^i{}_n \, b^n{}_k.$$

If you apply the summation convention to this expression, you get the compact form

$$c^i{}_k = a^i{}_j \, b^j{}_k.$$

The use of the shorthand notation is only possible because we have written the matrix elements with one index above and one index below.

- Expressions like $a_i \, b^j$, $c^k d_{ij}$ and $a_i \, a_i$, on the other hand, do *not* fall under the summation convention, as they do not contain any identical indices, both superscript and subscript. $\square$

You can also sum multiple times in an equation, but you have to make sure that the summation indices are chosen differently to avoid confusion about which terms should be summed. As an example, we consider Eq. 9.5:

$$\phi_{;x} = \left(\frac{\partial p_x}{\partial x} - p_x\,\Gamma^x_{xx} - p_y\,\Gamma^y_{xx} \right) v^x + \left(\frac{\partial p_y}{\partial x} - p_x\,\Gamma^x_{yx} - p_y\,\Gamma^y_{yx} \right) v^y$$
$$+ \; p_x\,v^x_{;x} + p_y\,v^y_{;x}$$

On the right side are several sums, which we shorten one after the other with Einstein's summation convention. First, we look at the last two terms and write

$$p_x\,v^x_{;x} + p_y\,v^y_{;x} = p_i\,v^i_{;x}.$$

The terms in the two brackets also contain sums, which we abbreviate as follows

$$\frac{\partial p_x}{\partial x} - p_x\,\Gamma^x_{xx} - p_y\,\Gamma^y_{xx} = \frac{\partial p_x}{\partial x} - p_j\,\Gamma^j_{xx}$$
$$\frac{\partial p_y}{\partial x} - p_x\,\Gamma^x_{yx} - p_y\,\Gamma^y_{yx} = \frac{\partial p_y}{\partial x} - p_k\,\Gamma^k_{yx}.$$

As an intermediate result, we get

$$\phi_{;x} = \left(\frac{\partial p_x}{\partial x} - p_j\,\Gamma^j_{xx} \right) v^x + \left(\frac{\partial p_y}{\partial x} - p_k\,\Gamma^k_{yx} \right) v^y + p_i\,v^i_{;x}.$$

Now we shorten the first two summands, while simultaneously in the second bracket we exchange the index k for j, and obtain

$$\phi_{;x} = \left(\frac{\partial p_l}{\partial x} - p_j\,\Gamma^j_{lx} \right) v^l + p_i\,v^i_{;x},$$

thus a very compact expression. The index x is the only index that is not summed over. It denotes the variable after which the derivative is taken. Compared to the long notation of $\phi_{;x}$ above, the short notation makes it clearer and immediately readable, at which three places x appears on the right side. The index x is a so-called **free index** and free indices are essentially used to combine several equations into one. Let's look at Eq. 9.6 for $\phi_{;y}$

$$\phi_{;y} = \left(\frac{\partial p_x}{\partial y} - p_x\,\Gamma^x_{xy} - p_y\,\Gamma^y_{xy} \right) v^x + \left(\frac{\partial p_y}{\partial y} - p_x\,\Gamma^x_{yy} - p_y\,\Gamma^y_{yy} \right) v^y$$
$$+ \; p_x\,v^x_{;y} + p_y\,v^y_{;y}.$$

This equation can also be shortened with Einstein's summation convention to

$$\phi_{;y} = \left(\frac{\partial p_l}{\partial y} - p_j\,\Gamma^j_{ly} \right) v^l + p_i\,v^i_{;y},$$

and we can use a free index m, which can take the two values x and y, to represent the two equations for $\phi_{;x}$ and $\phi_{;y}$ in one, namely

$$\phi_{;m} = \left(\frac{\partial p_l}{\partial m} - p_j\,\Gamma^j_{lm}\right) v^l + p_i\,v^i_{\;;m}.$$

Free indices in an equation always mean that there is actually a system of several equations. If the free index can take two values, there are two equations; if it can take three values, there are three equations, etc. In an equation, the free indices must always appear in the same place on both sides, i.e., superscript or subscript.

Of course, the notation with Einstein's summation convention and the free indices is initially unfamiliar and not necessarily necessary for the two-dimensional case considered here. However, it is a good preparation for the more complicated equations of relativity theory to practice these new notations in the simple two-dimensional case. And we will do this again and again in the next sections.

The free index m can also be replaced by any other, it is only necessary to make sure that the change is made at *every* place where m appears. The expression

$$\phi_{;n} = \left(\frac{\partial p_l}{\partial n} - p_j\,\Gamma^j_{ln}\right) v^l + p_i v^i_{\;;n}$$

is thus identical to

$$\phi_{;m} = \left(\frac{\partial p_l}{\partial m} - p_j\,\Gamma^j_{lm}\right) v^l + p_i v^i_{\;;m}.$$

Example 9.2.
The transformation formulas for the components of vectors and one-forms are shortened by Einstein's summation convention to

$$v^{i'} = \frac{\partial i'}{\partial i}\,v^i \tag{9.8}$$

for vectors or

$$p_{i'} = \frac{\partial i}{\partial i'}\,p_i \tag{9.9}$$

for one-forms. It can be seen that the free indices are in the same position on both sides of the equations. $\square$

9.3. $(0,2)$-Tensors

General properties of $(0,2)$-Tensors

One-forms are linear mappings that, when applied to a vector, yield a real number; $\tilde{p}\,(\vec{v})$ is a scalar, i.e., a number that takes the same value in every coordinate system. They are also called $(0, 1)$-tensors because they have *one* vector as an argument. $(0, 2)$-**tensors** are **bilinear** mappings that have *two* vectors as arguments and yield a real number.

Remark 9.2. **MT: Multilinear mappings**
A mapping f is called **multilinear** if it has more than one argument and is linear in *each* of its arguments. As an example, consider a function $f(x, y, z)$ that has three arguments. Linearity in each argument means

$$
\begin{aligned}
f(x_1 + x_2, y, z) &= f(x_1, y, z) + f(x_2, y, z) \\
f(x, y_1 + y_2, z) &= f(x, y_1, z) + f(x, y_2, z) \\
f(x, y, z_1 + z_2) &= f(x, y, z_1) + f(x, y, z_2)
\end{aligned}
$$

and

$$
\begin{aligned}
f(ax, y, z) &= a f(x, y, z) \\
f(x, ay, z) &= a f(x, y, z) \\
f(x, y, az) &= a f(x, y, z)
\end{aligned}
$$

for any real number a. Multilinear mappings with two arguments are also called **bilinear**. $\square$

A $(0, 2)$-tensor $\mathbf{f}$ is a bilinear mapping, which, when choosing a (x, y) coordinate system K, can be represented by *four* numbers f_{ij}, i.e., its components

$$
\mathbf{f} \underset{K}{\rightarrow} f_{ij},
$$

$(i = x, y; j = x, y)$, whereby the components should tansform according to a coordinate system change $K \rightarrow K'$ as

$$
\begin{aligned}
f_{x'x'} &= \frac{\partial x}{\partial x'}\frac{\partial x}{\partial x'} f_{xx} + \frac{\partial x}{\partial x'}\frac{\partial y}{\partial x'} f_{xy} + \frac{\partial y}{\partial x'}\frac{\partial x}{\partial x'} f_{yx} + \frac{\partial y}{\partial x'}\frac{\partial y}{\partial x'} f_{yy} = \frac{\partial i}{\partial x'}\frac{\partial j}{\partial x'} f_{ij} \\
f_{x'y'} &= \frac{\partial x}{\partial x'}\frac{\partial x}{\partial y'} f_{xx} + \frac{\partial x}{\partial x'}\frac{\partial y}{\partial y'} f_{xy} + \frac{\partial y}{\partial x'}\frac{\partial x}{\partial y'} f_{yx} + \frac{\partial y}{\partial x'}\frac{\partial y}{\partial y'} f_{yy} = \frac{\partial i}{\partial x'}\frac{\partial j}{\partial y'} f_{ij} \\
f_{y'x'} &= \frac{\partial x}{\partial y'}\frac{\partial x}{\partial x'} f_{xx} + \frac{\partial x}{\partial y'}\frac{\partial y}{\partial x'} f_{xy} + \frac{\partial y}{\partial y'}\frac{\partial x}{\partial x'} f_{yx} + \frac{\partial y}{\partial y'}\frac{\partial y}{\partial x'} f_{yy} = \frac{\partial i}{\partial y'}\frac{\partial j}{\partial x'} f_{ij} \\
f_{y'y'} &= \frac{\partial x}{\partial y'}\frac{\partial x}{\partial y'} f_{xx} + \frac{\partial x}{\partial y'}\frac{\partial y}{\partial y'} f_{xy} + \frac{\partial y}{\partial y'}\frac{\partial x}{\partial y'} f_{yx} + \frac{\partial y}{\partial y'}\frac{\partial y}{\partial y'} f_{yy} = \frac{\partial i}{\partial y'}\frac{\partial j}{\partial y'} f_{ij},
\end{aligned}
$$

i.e. with the free index notation, briefly according to

$$
f_{i'j'} = \frac{\partial i}{\partial i'}\frac{\partial j}{\partial j'} f_{ij}. \tag{9.10}
$$

On the right side, the summation indices are chosen differently so that it is clear what is added with whom. If you write the components of $\mathbf{f}$ as a (2×2)-matrix

$$(f_{ij}) = \begin{pmatrix} f_{xx} & f_{xy} \\ f_{yx} & f_{yy} \end{pmatrix},$$

then you can also write the transformation formula as a matrix equation

$$(f_{i'j'}) = \begin{pmatrix} \dfrac{\partial x}{\partial x'} & \dfrac{\partial x}{\partial y'} \\[2mm] \dfrac{\partial y}{\partial x'} & \dfrac{\partial y}{\partial y'} \end{pmatrix}^{T} \begin{pmatrix} f_{xx} & f_{xy} \\ f_{yx} & f_{yy} \end{pmatrix} \begin{pmatrix} \dfrac{\partial x}{\partial x'} & \dfrac{\partial x}{\partial y'} \\[2mm] \dfrac{\partial y}{\partial x'} & \dfrac{\partial y}{\partial y'} \end{pmatrix}. \tag{9.11}$$

So you have to transpose the left Jacobian matrix of the coordinate change (see remark 8.1) to reproduce the correct transformation formula. In the following we use the handy index notation ((9.10)), when we want to calculate transformations of tensor components. For two arbitrary vectors

$$\vec{v} = v^x \vec{e}_x + v^y \vec{e}_y = v^i \vec{e}_i \text{ and } \vec{w} = w^j \vec{e}_j$$

$\mathbf{f}$ is defined as a function by

$$\mathbf{f}(\vec{v}, \vec{w}) = v^i w^j f_{ij} \tag{9.12}$$

where again the double summation convention must be observed. We show that this definition always gives the same number regardless of the coordinate system used. Let K' be another arbitrary coordinate system, then with (9.8) and (9.10)

$$\begin{aligned} v^{i'} w^{j'} f_{i'j'} &= \left(\frac{\partial i'}{\partial i} v^i \right) \left(\frac{\partial j'}{\partial j} w^j \right) \left(\frac{\partial k}{\partial i'} \frac{\partial l}{\partial j'} f_{kl} \right) \\ &= v^i w^j f_{kl} \left(\frac{\partial i'}{\partial i} \frac{\partial k}{\partial i'} \right) \left(\frac{\partial j'}{\partial j} \frac{\partial l}{\partial j'} \right). \end{aligned}$$

Now the following applies to the matrix products in the two brackets

$$\frac{\partial i'}{\partial i} \frac{\partial k}{\partial i'} = \delta_i^k ; \quad \frac{\partial j'}{\partial j} \frac{\partial l}{\partial j'} = \delta_j^l$$

with the *Kronecker delta* δ (see remark 8.1). So we get with

$$v^{i'} w^{j'} f_{i'j'} = v^i w^j f_{kl} \, \delta_i^k \, \delta_j^l = v^i w^j f_{ij}$$

the desired independence from the coordinate system.

For a $(0,2)$-tensor, the order of its two arguments is important. By swapping the two arguments, one generally obtains different results:

$$\mathbf{f}\left(\vec{v},\vec{w}\right) = v^i w^j f_{ij} \neq w^i v^j f_{ij} = \mathbf{f}\left(\vec{w},\vec{v}\right)$$

Therefore, a $(0,2)$-tensor $\mathbf{f}$ is called **symmetric**, if for arbitrary vectors $\vec{v}$ and $\vec{w}$

$$\mathbf{f}\left(\vec{v},\vec{w}\right) = \mathbf{f}\left(\vec{w},\vec{v}\right)$$

holds. For the components this means that

$$f_{ij} = f_{ji}$$

holds, i.e., they are *symmetric* in the two lower indices. Similarly, a tensor is called **antisymmetric**, if for arbitrary vectors $\vec{v}$ and $\vec{w}$

$$\mathbf{f}\left(\vec{v},\vec{w}\right) = -\mathbf{f}\left(\vec{w},\vec{v}\right)$$

holds. The components are then *antisymmetric* in the two lower indices

$$f_{ij} = -f_{ji}.$$

The Metric Tensor

The most important $(0,2)$-tensor for us is the **metric tensor g**, which is also briefly referred to as **metric** and whose components g_{ij} in the Cartesian (x,y) coordinate system K are defined by

$$\mathbf{g} \underset{K}{\rightarrow} g_{ij} = \begin{cases} 1 & i = j \\ 0 & i \neq j \end{cases},$$

or in matrix form

$$(g_{ij}) = \begin{pmatrix} 1 & 0 \\ 0 & 1 \end{pmatrix} = (\delta_{ij}),$$

where we have also written the Kronecker delta on the right side with two lower indices. The components of the metric tensor thus correspond to the unit matrix in the Cartesian coordinate system.

Given two arbitrary vectors

$$\vec{v} = v^x \vec{e}_x + v^y \vec{e}_y = v^i \vec{e}_i$$

and

$$\vec{w} = w^x \vec{e}_x + w^y \vec{e}_y = w^j \vec{e}_j$$

it follows that

$$\mathbf{g}\left(\vec{v},\vec{w}\right) \;=\; v^i w^j\, g_{ij} = v^i w^j\, \delta_{ij} = v^x w^x + v^y w^y = \vec{v}\cdot\vec{w},$$

i.e., the metric tensor is defined by the scalar product of two vectors. We have already shown in Chap. 8 that the scalar product of two vectors is invariant under coordinate transformations and is also bilinear, so the metric tensor is well-defined and bilinear. Since the order of the vectors does not matter in scalar multiplication, the metric tensor is symmetric, i.e.

$$\mathbf{g}\left(\vec{v},\vec{w}\right) = \vec{v}\cdot\vec{w} = \vec{w}\cdot\vec{v} = \mathbf{g}\left(\vec{w},\vec{v}\right).$$

One-form corresponding to a vector

The metric tensor can be used to define a corresponding one-form $\tilde{v}$ for a fixed vector $\vec{v}$:

$$\tilde{v} \to v_i = g_{ij} v^j \tag{9.13}$$

We check whether the components fulfill the transformation rules for one-forms. To do this, another coordinate system K' is chosen, then follows with (9.10) and (9.8)

$$
\begin{aligned}
v_{i'} \;=&\; g_{i'j'}\, v^{j'} = \frac{\partial i}{\partial i'}\frac{\partial j}{\partial j'}\, g_{ij}\, \frac{\partial j'}{\partial k}\, v^k = \frac{\partial i}{\partial i'}\left(g_{ij}\, v^k\right)\underbrace{\left(\frac{\partial j}{\partial j'}\frac{\partial j'}{\partial k}\right)}_{=\delta_k^j} \\
=&\; \frac{\partial i}{\partial i'}\left(g_{ij}\, v^k \delta_k^j\right) = \frac{\partial i}{\partial i'}\left(g_{ij}\, v^j\right) = \frac{\partial i}{\partial i'}\, v_i,
\end{aligned}
$$

i.e., $\tilde{v}$ is a one-form. In the Cartesian coordinate system one has $g_{ij} = \delta_{ij}$, there (and only there!) it applies the relationship

$$v_i = v^i, \tag{9.14}$$

i.e., the components of one-form and normal vector coincide. Therefore, we did not make a mistake when we considered the gradient of a scalar function as a vector and not as a one-form in part I. If you look at the expression (9.13), the multiplication and summation (which is briefly called the **contraction** of the vector with the metric) of the components of a vector v^j with the components of the metric tensor g_{ij}, results in a component of a one-form. If you relate this fact to the position of the indices, an upper index (on the vector) becomes a lower index (on the one-form) by contraction with the metric. This change in the position of the indices from top to bottom is also called **lowering with the metric**.

Vector corresponding to a one-form

We now ask the reverse question: Can we find a corresponding vector $\vec{p}$ for any one-form $\tilde{p}$? To answer the question, we assume that the matrix g_{ij} is invertible. We write the inverse matrix as g^{ij}, it is also symmetric and must satisfy the equation

$$g_{ij}\,g^{jk} = g^{kj}g_{ji} = \delta_i^{\,k} = \begin{cases} 1 & i = k \\ 0 & i \neq k \end{cases}.$$

If p_i are the components of the one-form $\tilde{p}$, the same applies to $\vec{p}$:

$$\vec{p} \to p^i = g^{ij}p_j \tag{9.15}$$

Example 9.3.

As an example, we want to calculate the metric tensor for polar coordinates, i.e., we set $(x', y') = (r, \varphi)$ and obtain with Eq. (8.31) and with formula (9.11):

$$
\begin{aligned}
\left(g_{i'j'}\right) &= \begin{pmatrix} \cos\varphi & -r\sin\varphi \\ \sin\varphi & r\cos\varphi \end{pmatrix}^T \begin{pmatrix} 1 & 0 \\ 0 & 1 \end{pmatrix} \begin{pmatrix} \cos\varphi & -r\sin\varphi \\ \sin\varphi & r\cos\varphi \end{pmatrix} \\[2mm]
&= \begin{pmatrix} \cos\varphi & \sin\varphi \\ -r\sin\varphi & r\cos\varphi \end{pmatrix} \begin{pmatrix} \cos\varphi & -r\sin\varphi \\ \sin\varphi & r\cos\varphi \end{pmatrix} \\[2mm]
&= \begin{pmatrix} 1 & 0 \\ 0 & r^2 \end{pmatrix}
\end{aligned}
$$

So, for the components of the metric tensor in polar coordinates P

$$\mathbf{g} \underset{P}{\to} \left(g_{i'j'}\right) = \begin{pmatrix} g_{rr} & g_{r\varphi} \\ g_{\varphi r} & g_{\varphi\varphi} \end{pmatrix} = \begin{pmatrix} 1 & 0 \\ 0 & r^2 \end{pmatrix}. \tag{9.16}$$

The inverse matrix $\left(g^{i'j'}\right)$ is obtained from this with the formula (8.19):

$$A^{-1} = \frac{1}{\det A}\begin{pmatrix} d & -b \\ -c & a \end{pmatrix}$$

and with

$$\det\left(g_{i'j'}\right) = r^2$$

to

$$\left(g^{i'j'}\right) = \frac{1}{\det\left(g_{i'j'}\right)}\begin{pmatrix} r^2 & 0 \\ 0 & 1 \end{pmatrix} = \frac{1}{r^2}\begin{pmatrix} r^2 & 0 \\ 0 & 1 \end{pmatrix} = \begin{pmatrix} 1 & 0 \\ 0 & \frac{1}{r^2} \end{pmatrix}.\;\square \tag{9.17}$$

We now consider the gradient $\tilde{d}\phi$ of a scalar function ϕ and use the formula (9.15) to calculate the components of the associated vector $\vec{d}\phi$ in polar coordinates. It holds

$$
\begin{aligned}
\left(\vec{d}\phi\right)^r &= g^{rj}\phi_{,j} = g^{rr}\phi_{,r} + g^{r\varphi}\phi_{,\varphi} = \phi_{,r} = \frac{\partial\phi}{\partial r} \\
\left(\vec{d}\phi\right)^\varphi &= g^{\varphi j}\phi_{,j} = g^{\varphi r}\phi_{,r} + g^{\varphi\varphi}\phi_{,\varphi} = \frac{1}{r^2}\phi_{,\varphi} = \frac{1}{r^2}\frac{\partial\phi}{\partial\varphi},
\end{aligned}
$$

i.e., the components of the vector associated with the gradient *do not* coincide with the components of the gradient

$$
\tilde{d}\phi \underset{P}{\rightarrow} \left(\begin{array}{cc} \dfrac{\partial\phi}{\partial r} & \dfrac{\partial\phi}{\partial\varphi} \end{array} \right).
$$

Even though we are in the Euclidean plane, the vectors generally have different components than their associated one-forms. The Cartesian coordinate system is the *only* one in which the components coincide.

Length of a vector

The metric tensor can be used to determine the length of vectors (hence the name "metric"). In the Euclidean plane, the (infinitesimal) displacement vector $d\vec{s}$ has the Cartesian components

$$
d\vec{s} \underset{K}{\rightarrow} \left(\begin{array}{c} dx \\ dy \end{array} \right).
$$

The square of the length of $d\vec{s}$, which is the same in all coordinate systems, is obtained with the metric as follows

$$
\begin{aligned}
d\vec{s}\cdot d\vec{s} = ds^2 &= g_{xx}\,dx\,dx + g_{xy}\,dx\,dy + g_{yx}\,dy\,dx + g_{yy}\,dy\,dy \\
&= dx^2 + dy^2.
\end{aligned}
$$

In polar coordinates, we get

$$
d\vec{s} \underset{P}{\rightarrow} \left(\begin{array}{c} dr \\ d\varphi \end{array} \right) \tag{9.18}
$$

and for the length

$$
\begin{aligned}
ds^2 &= g_{rr}\,dr\,dr + g_{r\varphi}\,dr\,d\varphi + g_{\varphi r}\,d\varphi\,dr + g_{\varphi\varphi}\,d\varphi\,d\varphi \\
&= dr^2 + r^2\,d\varphi^2.
\end{aligned}
$$

The covariant derivative of a $(0, 2)$-Tensor

The following derivation of the covariant derivative contains a lot of "index gymnastics" and is therefore not easy to follow. Of course, it offers a good opportunity to further practice the summation convention.

To derive the covariant derivative of a $(0, 2)$-tensor, we use Eq. (9.12):

$$\mathbf{f}(\vec{v}, \vec{w}) = v^i w^j f_{ij} = \phi$$

We consider the right side, similar to the derivation of the covariant derivative of one-forms, as a scalar function ϕ, whose covariant derivative coincides with the partial one. Thus follows

$$\phi_{;k} = \phi_{,k} = \frac{\partial \phi}{\partial k} = \frac{\partial \left(v^i w^j f_{ij} \right)}{\partial k}.$$

Applying the product rule twice to the right side, we get

$$\frac{\partial \left(v^i w^j f_{ij} \right)}{\partial k} = \frac{\partial v^i}{\partial k} w^j f_{ij} + \frac{\partial w^j}{\partial k} v^i f_{ij} + \frac{\partial f_{ij}}{\partial k} v^i w^j.$$

Now we replace as in Eq. (9.4) the partial derivatives of the components of the two vectors with their covariant ones, which we can write briefly with the summation convention as

$$\frac{\partial v^i}{\partial k} = v^i{}_{;k} - v^l \, \Gamma^i_{lk}$$

and

$$\frac{\partial w^j}{\partial k} = w^j{}_{;k} - w^m \, \Gamma^j_{mk}$$

We get

$$\frac{\partial \left(v^i w^j f_{ij} \right)}{\partial k} = \left(v^i{}_{;k} - v^l \, \Gamma^i_{lk} \right) w^j f_{ij} + \left(w^j{}_{;k} - w^m \, \Gamma^j_{mk} \right) v^i f_{ij} + \frac{\partial f_{ij}}{\partial k} v^i w^j.$$

Rearranging the right side, we get

$$\frac{\partial \left(v^i w^j f_{ij} \right)}{\partial k} = \frac{\partial f_{ij}}{\partial k} v^i w^j - f_{ij} \, \Gamma^i_{lk} \, w^j v^l - f_{ij} \, \Gamma^j_{mk} \, v^i w^m + v^i{}_{;k} \, w^j f_{ij} + w^j{}_{;k} \, v^i f_{ij}.$$

We want to factor out the components of the two vectors in the first three terms, so we swap the summation indices $i \leftrightarrow l$ in the second term and $j \to l, m \to j$ in the third and get as an intermediate result

$$\frac{\partial \left(v^i w^j f_{ij} \right)}{\partial k} = \frac{\partial f_{ij}}{\partial k} v^i w^j - f_{lj} \, \Gamma^l_{ik} \, w^j v^i - f_{il} \, \Gamma^l_{jk} \, v^i w^j + v^i{}_{;k} \, w^j f_{ij} + w^j{}_{;k} \, v^i f_{ij}$$

and from this by factoring out

$$\frac{\partial \left(v^i w^j f_{ij} \right)}{\partial k} = \left(\frac{\partial f_{ij}}{\partial k} - f_{lj}\, \Gamma^l_{ik} - f_{il}\, \Gamma^l_{jk} \right) v^i w^j + v^i{}_{;k}\, w^j f_{ij} + w^j{}_{;k}\, v^i f_{ij}.$$

We now *define* the covariant derivative of **f** by considering the expressions in the bracket on the right side as their components, i.e., we define

$$f_{ij;k} = \frac{\partial f_{ij}}{\partial k} - f_{lj}\, \Gamma^l_{ik} - f_{il}\, \Gamma^l_{jk} = f_{ij,k} - f_{lj}\, \Gamma^l_{ik} - f_{il}\, \Gamma^l_{jk}, \tag{9.19}$$

in the last equation we have again used the shorthand for the partial derivative

$$f_{ij,k} = \frac{\partial f_{ij}}{\partial k}.$$

Covariant derivative of the metric tensor

We now want to calculate the covariant derivative of the metric tensor. To do this, we first note that if the components of a one-form or a $(0,2)$-tensor are zero in one coordinate system, they are also zero in all others. For example, with the transformation formula (9.10), it follows that the components of a $(0,2)$-tensor in a (x', y') coordinate system can be calculated by

$$f_{i'j'} = \frac{\partial j}{\partial i'}\, \frac{\partial k}{\partial j'}\, f_{jk}$$

i.e., if all $f_{jk} = 0$, then all $f_{i'j'}$ are also equal to zero. This property is now the reason why, if a so-called **tensor equation** is valid in one coordinate system, it is valid in the same form in any other coordinate system. A tensor equation is an equation that contains components of tensors of the same rank on the right and on the left side. If we bring all tensor components to the left side of the equation in the coordinate system, in which the equation is valid, so that zero is on the right, we have on the left side a tensor whose components are all zero, and then we know that this is also the case in all other coordinate systems. We then write the left side as a tensor in another coordinate system and then transfer the terms, which should be on the right side, to the right side.

With these preliminary remarks, we can now derive the covariant derivative of the metric tensor. In Cartesian coordinates, as shown above in (9.14), the components of a vector $\vec{v}$ and the corresponding one-form $\tilde{v}$ are equal

$$v_i = v^i,$$

and thus the partial derivatives are also equal

$$v^i{}_{,k} = v_{i,k}.$$

Since in Cartesian coordinates the Christoffel symbols are equal to zero, it follows

$$v^i{}_{;k} = v^i{}_{,k}$$

and

$$v_{i;k} = v_{i,k}.$$

So in Cartesian coordinates it follows

$$v_{i;k} = v^i{}_{;k}.$$

Now it holds

$$v^i{}_{;k} = \delta_{ij}\, v^j{}_{;k} = g_{ij}\, v^j{}_{;k}$$

in Cartesian coordinates, from which with the penultimate equation

$$v_{i;k} = g_{ij}\, v^j{}_{;k}$$

follows. On the left and right side are tensor components, so there is a tensor equation that is valid in all coordinate systems. That means, it also holds in arbitrary (x', y') coordinate systems

$$v_{i';k'} = g_{i'j'}\, v^{j'}{}_{;k'}. \tag{9.20}$$

We now want to show that the covariant derivative of the metric tensor is equal to zero. To do this, we apply the product rule to Eq. (9.13) and get

$$v_{i';k'} = \left(g_{i'j'}\, v^{j'}\right)_{;k'} = g_{i'j';k'}\, v^{j'} + g_{i'j'}\, v^{j'}{}_{;k'}.$$

Comparing this equation with (9.20), it follows for an arbitrary vector field $\vec{v}$

$$g_{i'j';k'}\, v^{j'} = 0$$

and thus again

$$g_{i'j';k'} = 0.$$

In the Cartesian coordinate system, the covariant derivative of the metric can also be calculated directly by using the formula (9.19)

$$g_{ij;k} = g_{ij,k} - g_{lj}\, \Gamma^l_{ik} - g_{il}\, \Gamma^l_{jk}.$$

The Christoffel symbols are zero in the Cartesian coordinate system, also the partial derivatives of the metric components 0 and 1 are zero, so the right side is zero.

9.4. (M, N)-Tensors

In this section we now deal with the general forms of tensors, but only as we need them in relativity theory. We largely dispense with detailed derivations. On the one hand because the treatment of vectors, one-forms and $(0, 2)$-tensors already used the same approaches and derivation paths as they also apply to more general tensors. On the other hand, the formulas and calculation steps (at least visually) become more confusing, since we have to deal with several lower and/or upper indices.

$(0, N)$-Tensors

We first generalize the definition of $(0, 2)$-tensors to $(0, N)$-tensors, where in relativity theory N is at most the number 4. $(0, N)$-tensors are *multilinear* mappings that have N vectors as arguments and deliver a real number. As a continuous example in this section we want to consider a $(0, 4)$-tensor $\mathbf{T}$. This tensor $\mathbf{T}$ thus maps four vectors to a real number, its 16 components have *four lower* indices, i.e.

$$\mathbf{T} \to T_{ijkl},$$

where in the (x, y) coordinate system each index i, j, k, l can take the values x, y. The tensor components must fulfill the following transformation formula:

$$T_{i'j'k'l'} = \frac{\partial i}{\partial i'} \frac{\partial j}{\partial j'} \frac{\partial k}{\partial k'} \frac{\partial l}{\partial l'} T_{ijkl}$$

So there is for *every* lower index a term of the form $\dfrac{\partial i}{\partial i'}$. Analogously to the $(0, 2)$-tensors, for four arbitrary vectors it results

$$\mathbf{T}(\vec{u}, \vec{v}, \vec{w}, \vec{z}) = u^i v^j w^k z^l \, T_{ijkl}.$$

The components of the covariant derivative of $\mathbf{T}$ are given by

$$T_{ijkl;m} = T_{ijkl,m} - T_{njkl} \, \Gamma^n_{im} - T_{inkl} \, \Gamma^n_{jm} - T_{ijnl} \, \Gamma^n_{km} - T_{ijkn} \, \Gamma^n_{lm}.$$

This formula may become somewhat more bearable when one realizes that for each lower index one gets a term with the Christoffel symbol, and the summation (over the summation index n) is done in order in the lower indices.

$(M, 0)$-Tensors

With a small "trick" we can also consider a vector $\vec{v}$ as a linear mapping, which has an arbitrary one-form $\tilde{p}$ as an argument and delivers a real number. The rule for this is

$$\vec{v}(\tilde{p}) = \tilde{p}(\vec{v}) = p_i \, v^i. \tag{9.21}$$

For this reason, a vector is also called a $(1,0)$-tensor. The components of such a tensor have *one upper* index. More generally, a $(M,0)$-tensor is a multilinear mapping, which has M one-forms ($M = 1,2,3,4$) as arguments and maps into the real numbers. The components of $(M,0)$-tensors have M *upper* indices. The statements of the last Sections on $(0,N)$ tensors can be analogously transferred to $(M,0)$-tensors. As a consistent example for this section, we consider the properties of a $(2,0)$ tensor

$$\mathbf{T} \to T^{ij},$$

whose components transform during a coordinate system change according to

$$T^{i'j'} = \frac{\partial i'}{\partial i}\,\frac{\partial j'}{\partial j}\,T^{ij}.$$

If $\tilde{p} \to p_i$ and $\tilde{q} \to q_j$ are two one-forms, then

$$\mathbf{T}\,(\tilde{p},\tilde{q}) = p_i\,q_j\,T^{ij}.$$

The components of the covariant derivative of $\mathbf{T}$ are given by

$$T^{ij}_{\;\;;k} = T^{ij}_{\;\;,k} + T^{lj}\,\Gamma^{i}_{lk} + T^{il}\,\Gamma^{j}_{lk}.$$

For each lower index, there is a positive term with the Christoffel symbol, and the summation (over the summation index l) is carried out in the upper indices in order.

A special $(2,0)$-tensor is the **metric tensor for one-forms**, which we want to denote by $\mathbf{G}$. This is defined by

$$\mathbf{G} \to g^{ij}, \tag{9.22}$$

where the components of the metric tensor for one-forms correspond to the elements of the inverse matrix to (g_{ij}).

(M,N)-Tensors

Finally, a (M,N)-tensor is a multilinear mapping that maps M one-forms and N vectors to real numbers, where M,N can again be the numbers 1 to 4. The components of a (M,N)-tensor thus have **M** *upper* and **N** *lower* indices. As a consistent example in this section, we will use a $(1,1)$ tensor. A $(1,1)$ tensor $\mathbf{T}$ requires a one-form $\tilde{p}$ and a vector $\vec{v}$ to produce the number $\mathbf{T}(\tilde{p},\vec{v})$. The components of

$$\mathbf{T} \to T^{i}_{\;j}$$

have a superscript (from the one-form) and a subscript (from the vector) index and transform during a coordinate system change according to

$$T^{i'}_{\ j'} \;=\; \frac{\partial i'}{\partial i}\,\frac{\partial j}{\partial j'}\,T^{i}_{\ j}.$$

If one follows the way of summations of the superscript and subscript indices, one sees that there can only be this one possibility for transformation. It arises quasi "automatically" due to the position of the indices. If $\vec{v}$ is a vector and $\tilde{p}$ is a one-form, then

$$\mathbf{T}(\tilde{p}, \vec{v}) = p_i\, v^j\, T^{i}_{\ j}.$$

The components of the covariant derivative of $\mathbf{T}$ are given by

$$T^{i}_{\ j;k} = T^{i}_{\ j,k} + T^{l}_{\ j}\,\Gamma^{i}_{lk} - T^{i}_{\ l}\,\Gamma^{l}_{jk}.$$

For the upper index, there is a positive term with the Christoffel symbol, for the lower one a negative one, and the summation (over the summation index l) is carried out in the indices in order.

Raising and Lowering Indices

In generalization of the correspondence between vectors and one-forms, one can use the metric tensor $\mathbf{g}$ to generate a corresponding $(M - 1, N + 1)$-tensor from a (M, N)-tensor. Similarly, one uses the tensor $\mathbf{G}$ to generate a corresponding $(M + 1, N - 1)$-tensor from a (M, N)-tensor. It is common to name these new tensors with the same name and only distinguish them by the index position of their components. For example, the corresponding tensor of a $(0, 2)$-tensor $\mathbf{T}$ is a $(1, 1)$-tensor, which is also denoted by $\mathbf{T}$. The components of the corresponding tensor $\mathbf{T}$ are derived from the components T_{ij} of the $(0, 2)$-tensor as well as from the components g^{ik} of $\mathbf{G}$ by

$$\mathbf{T} \to T^{i}_{\ j} = g^{ik}\, T_{kj}.$$

This means, one *"pulls the first index i of T_{ij} upwards with the inverse matrix g^{ik}"*. With exactly the same steps, one can also pull the second index of the components T_{ij} upwards and obtains

$$T^{j}_{\ i} = g^{jk} T_{ik}.$$

The components $T^{j}_{\ i}$ generally differ from $T^{i}_{\ j}$, see [25]. If we look at the corresponding tensor to a $(2, 0)$-tensor $\mathbf{T}$ as another example, we get just as well

$$T^{i}_{\ j} = T^{ik}\, g_{kj}.$$

In this case, one says, "*one pulls the second index j of T^{ij} downwards with the matrix g_{kj}*". With exactly the same steps, one can also pull the first index of the components T^{ij} downwards and obtains

$$T^j{}_i = g_{ik}T^{kj}.$$

If we choose the metric $\mathbf{g}$ with its components g_{ij} as a $(0,2)$-tensor, we get

$$g^i{}_j = g^{ik}g_{kj} = \delta^i{}_j,$$

since the two matrices (g_{ij}) and (g^{ij}) are inverse to each other. If we also pull up the other index, we get

$$g^{ij} = g^{jk}\delta^i{}_k = g^{ji} = g^{ij},$$

i.e., by pulling up the indices of the components of $\mathbf{g}$ twice, we get the components of $\mathbf{G}$. The metric is the *only* tensor with this property.

The Covariant Derivative as a Tensor

So far, we have derived the components of the covariant derivatives of various tensors without precisely stating what the covariant derivatives are. We want to catch up on this here and take the covariant derivative of a vector as a prime example. In the chapter on vectors, we argued that the expression $\partial\vec{v}/\partial j$ is a vector field. This assumes that the index j is a fixed quantity. But j can basically take two values, e.g. $j = x, y$, so we can also consider the numbers $v^i{}_{,j}$ as components of a $(1,1)$-tensor. This tensor field is called the **covariant derivative** of $\vec{v}$ and is denoted by $\nabla\vec{v}$. The equation

$$v^i{}_{;j} = v^i{}_{,j} + v^k\,\Gamma^i_{kj} \tag{9.23}$$

also includes the statement that the ordinary partial derivative is not a tensor, i.e., the expressions $v^i{}_{,j}$ are *not* components of a $(1,1)$-tensor, but the sum of the terms on the right side of (9.23) is. We will show this again with a different approach. If two coordinate systems (x,y) and (x',y') are given, then if the ordinary derivative were a $(1,1)$-tensor would be to transform the components $v^i{}_{,j}$ as follows

$$v^i{}_{,j} = \frac{\partial i}{\partial i'}\frac{\partial j'}{\partial j}\,v^{i'}{}_{,j'}.$$

However, with the transformation formula for vectors and the product rule, it applies

$$v^i{}_{,j} = \frac{\partial v^i}{\partial j} = \frac{\partial}{\partial j}\left(\frac{\partial i}{\partial i'}\,v^{i'}\right) = \frac{\partial}{\partial j}\left(\frac{\partial i}{\partial i'}\right)v^{i'} + \left(\frac{\partial i}{\partial i'}\right)\frac{\partial v^{i'}}{\partial j}.$$

If you apply the chain rule to both terms, you get

$$
\begin{aligned}
v^i{}_{,j} &= \frac{\partial j'}{\partial j} \cdot \frac{\partial}{\partial j'}\left(\frac{\partial i}{\partial i'}\right) v^{i'} + \left(\frac{\partial i}{\partial i'}\right) \frac{\partial j'}{\partial j} \frac{\partial v^{i'}}{\partial j'} \\
&= \frac{\partial j'}{\partial j} \cdot \frac{\partial^2 i}{\partial j' \partial i'} v^{i'} + \frac{\partial i}{\partial i'} \frac{\partial j'}{\partial j} v^{i'}{}_{,j'}.
\end{aligned}
$$

The right term on the right side in the last equation is exactly the one you would expect if a tensor transformation were present. Since the left term is usually different from zero in curvilinear coordinates, the ordinary derivative is therefore not a tensor.

We can use the last equation and the fact that the covariant derivative in the Cartesian coordinate system K is equal to the ordinary partial,

$$
v^i{}_{;j} = v^i{}_{,j},
$$

to arrive at a second representation of the covariant derivative in any coordinate system K'. Since the covariant derivative is a tensor, the components of the covariant derivative of a vector in K' can be calculated by

$$
v^{i'}{}_{;j'} = \frac{\partial i'}{\partial i} \frac{\partial j}{\partial j'} v^i{}_{;j} = \frac{\partial i'}{\partial i} \frac{\partial j}{\partial j'} v^i{}_{,j}.
$$

For $v^i{}_{,j}$ we substitute the right side of the penultimate equation with $i' \to k', j' \to l'$ and obtain with multiple application of the chain rule and considering that

$$
\frac{\partial i'}{\partial j'} = \delta^{i'}_{j'} = \begin{cases} 1 & i' = j' \\ 0 & i' \neq j' \end{cases}
$$

applies, for the components of the covariant derivative

$$
\begin{aligned}
v^{i'}{}_{;j'} &= \frac{\partial i'}{\partial i} \frac{\partial j}{\partial j'} v^i{}_{,j} = \frac{\partial i'}{\partial i} \frac{\partial j}{\partial j'}\left(\frac{\partial l'}{\partial j} \cdot \frac{\partial^2 i}{\partial l' \partial k'} v^{k'} + \frac{\partial i}{\partial k'} \frac{\partial l'}{\partial j} v^{k'}{}_{,l'}\right) \\
&= \frac{\partial i'}{\partial i} \underbrace{\left(\frac{\partial j}{\partial j'} \frac{\partial l'}{\partial j}\right)}_{=\delta^{l'}_{j'}} \cdot \frac{\partial^2 i}{\partial l' \partial k'} v^{k'} + \underbrace{\left(\frac{\partial i'}{\partial i} \frac{\partial i}{\partial k'}\right)}_{=\delta^{i'}_{k'}} \underbrace{\left(\frac{\partial j}{\partial j'} \frac{\partial l'}{\partial j}\right)}_{=\delta^{l'}_{j'}} v^{k'}{}_{,l'} \\
&= \frac{\partial i'}{\partial i} \cdot \frac{\partial^2 i}{\partial j' \partial k'} v^{k'} + v^{i'}{}_{,j'}.
\end{aligned}
$$

Through this derivation, we have found a new calculation method for the Christoffel symbols, because it applies with Eq. (9.23)

$$
v^{i'}{}_{;j'} = v^{i'}{}_{,j'} + v^{k'} \Gamma^{i'}_{k'j'} = v^{i'}{}_{,j'} + \frac{\partial i'}{\partial i} \cdot \frac{\partial^2 i}{\partial j' \partial k'} v^{k'},
$$

from which

$$\Gamma^{i'}_{k'j'} = \frac{\partial i'}{\partial i} \cdot \frac{\partial^2 i}{\partial j' \partial k'} \tag{9.24}$$

follows. Thus, the Christoffel symbols can be calculated in any coordinate system K' if the transformation from the Cartesian coordinate system K to K' is known. We will revisit this form of the Christoffel symbols in part IV about the basics of general relativity to define the covariant derivative of tensors in four-dimensional spacetime.

Example 9.4.
As an application example, we calculate a selected Christoffel symbol of the polar coordinates, (i.e. $i', j', k' = r, \varphi$), namely $\Gamma^r_{r\varphi}$, and note that for the polar coordinates

$$r = \sqrt{x^2 + y^2}, x = r \cos \varphi, y = r \sin \varphi$$

applies. This results in

$$
\begin{aligned}
\Gamma^r_{r\varphi} &= \frac{\partial r}{\partial x} \cdot \frac{\partial^2 x}{\partial r \partial \varphi} + \frac{\partial r}{\partial y} \cdot \frac{\partial^2 y}{\partial r \partial \varphi} \\
&= \frac{\partial \left(\sqrt{x^2 + y^2} \right)}{\partial x} \cdot \frac{\partial^2 (r \cos \varphi)}{\partial r \partial \varphi} + \frac{\partial \left(\sqrt{x^2 + y^2} \right)}{\partial y} \cdot \frac{\partial^2 (r \sin \varphi)}{\partial r \partial \varphi} \\
&= \frac{x}{\sqrt{x^2 + y^2}} (-\sin \varphi) + \frac{y}{\sqrt{x^2 + y^2}} \cos \varphi \\
&= \frac{x}{r} (-\sin \varphi) + \frac{y}{r} \cos \varphi \\
&= -\cos \varphi \sin \varphi + \sin \varphi \cos \varphi \\
&= 0,
\end{aligned}
$$

thus the same result as in Eq. 8.36 on page 167. $\square$

In generalization of our example about the covariant derivative of vectors, one can say that the covariant derivative of a (M, N)-tensor $\mathbf{T}$ results in a tensor that has one more lower index. $\nabla \mathbf{T}$ is thus a $(M, N + 1)$-tensor. In order for this system to also apply to the gradient of a scalar field ϕ, scalar fields, which do not have indices, are defined as $(0, 0)$-tensors. In more detail, the following rules apply to the covariant derivative:

- The covariant mapping of a scalar field, i.e., a $(0, 0)$-tensor is a $(0, 1)$-tensor (i.e., a one-form).

- The covariant mapping of a vector, i.e., a $(1, 0)$-tensor is a $(1, 1)$-tensor.

- The covariant derivative of a one-form, i.e., a $(0, 1)$-tensor is a $(0, 2)$-tensor.

- In general, the covariant derivative of a (M, N)-tensor is a $(M, N+1)$-tensor.

- For *every* raised index (vector index) in the covariant derivative, a term $+\Gamma$ appears, and for *every* lowered index (one-form index), a term $-\Gamma$ appears. As examples, we consider

$$
\begin{aligned}
\nabla_k T_{ij} &= T_{ij,k} - T_{lj}\,\Gamma^l_{ik} - T_{il}\,\Gamma^l_{jk} \\
\nabla_k T^{ij} &= T^{ij}_{\ \ ,k} + T^{lj}\,\Gamma^i_{lk} + T^{il}\,\Gamma^j_{lk} \\
\nabla_k T^i_{\ j} &= T^i_{\ j,k} + T^l_{\ j}\,\Gamma^i_{lk} - T^i_{\ l}\,\Gamma^l_{jk}.
\end{aligned}
\tag{9.25}
$$

Calculation of the Christoffel Symbols by the Metric

The fact that the covariant derivative of the metric is equal to zero provides another way to calculate the Christoffel symbols, namely using the metric and its ordinary derivatives. First, it should be noted that the equation

$$
g_{ij;k} = g_{ij,k} - \Gamma^l_{\ ik}\,g_{lj} - \Gamma^l_{\ jk}\,g_{il} = 0
\tag{9.26}
$$

allows us to calculate the ordinary derivative of the metric components $g_{ij,k}$ from the Christoffel symbols. To also show the reverse, we first prove that in any coordinate system the Christoffel symbols are symmetric in the two lower indices, i.e.

$$
\Gamma^l_{\ ij} = \Gamma^l_{\ ji}.
$$

Consider an arbitrary scalar field ϕ, then the covariant derivative $\nabla\phi$ is a one-form with the components

$$
\nabla\phi \rightarrow \frac{\partial\phi}{\partial i} = \phi_{,i},
$$

since the covariant derivative of a scalar function is equal to the normal partial. The second covariant derivative $\nabla(\nabla\phi)$ is a $(0,2)$-tensor and has the components

$$
\nabla\nabla\phi \rightarrow \phi_{,i;j}.
$$

In Cartesian coordinates, however,

$$
\phi_{,i;j} = \phi_{,i,j} = \frac{\partial}{\partial j}\left(\frac{\partial\phi}{\partial i}\right),
$$

since the Christoffel symbols are equal to zero.

For the further calculation, we use the **Schwarz's theorem**, which states that in the double partial derivative of a function of several variables the order of the derivatives can be swapped.

Remark 9.3. **MT: Schwarz's theorem**

If $\phi(x, y)$ is a real-valued function of two variables, then (in most cases)

$$\frac{\partial}{\partial y}\left(\frac{\partial \phi}{\partial x}\right) = \frac{\partial}{\partial x}\left(\frac{\partial \phi}{\partial y}\right). \square \tag{9.27}$$

With Schwarz's theorem it follows

$$\phi_{,i;j} = \phi_{,i,j} = \frac{\partial}{\partial j}\frac{\partial \phi}{\partial i} = \frac{\partial}{\partial i}\frac{\partial \phi}{\partial j} = \phi_{,j,i} = \phi_{,j;i}$$

in Cartesian coordinates. If (x', y') is another coordinate system, then with the transformation formula for the components of $(0, 2)$-tensors

$$\phi_{,i';j'} = \frac{\partial i}{\partial i'}\frac{\partial j}{\partial j'}\phi_{,i;j} = \frac{\partial j}{\partial j'}\frac{\partial i}{\partial i'}\phi_{,j;i} = \phi_{,j';i'}.$$

That is, if a tensor is symmetric in one coordinate system, then it is also symmetric in all other coordinate systems.

According to Eq. (9.7) the covariant derivative of the one-form $\nabla \phi$ is equal to

$$\nabla\nabla\phi \rightarrow \phi_{,i;j} = \frac{\partial \phi_{,i}}{\partial j} - \phi_{,l}\,\Gamma^l_{ij} = \phi_{,i,j} - \phi_{,l}\,\Gamma^l_{ij}.$$

Since due to the symmetry of the covariant derivative

$$\phi_{,i;j} = \phi_{,j;i} = \phi_{,j,i} - \phi_{,l}\,\Gamma^l_{ji}$$

and according to Schwarz's theorem

$$\phi_{,i,j} = \phi_{,j,i}$$

it follows from the last three equations that

$$\phi_{,l}\,\Gamma^l_{ij} = \phi_{,l}\,\Gamma^l_{ji}$$

for any scalar fields ϕ. If we specifically choose $\phi(x, y) = x$, it follows that

$$\phi_{,x} = \frac{\partial \phi}{\partial x} = 1,\ \phi_{,y} = \frac{\partial \phi}{\partial y} = 0,$$

and from this

$$\Gamma^x_{ij} = \underbrace{\phi_{,x}}_{=1}\,\Gamma^x_{ij} + \underbrace{\phi_{,y}}_{=0}\,\Gamma^y_{ij} = \phi_{,l}\,\Gamma^l_{ij} = \phi_{,l}\,\Gamma^l_{ji} = \phi_{,x}\,\Gamma^x_{ji} + \phi_{,y}\,\Gamma^y_{ji} = \Gamma^x_{ji}.$$

And if we choose $\phi(x, y) = y$, it follows as well that

$$\Gamma^y_{ij} = \Gamma^y_{ji},$$

so overall it holds that

$$\Gamma^l_{ij} = \Gamma^l_{ji}$$

in any coordinate system.

We now show how the Christoffel symbols can be calculated from the (ordinary) derivatives of the metric. For this, we write Eq. (9.26) three times with different indices

$$
\begin{aligned}
g_{ij,k} &= \Gamma^l_{ik}\, g_{lj} + \Gamma^l_{jk}\, g_{il} \\
g_{ik,j} &= \Gamma^l_{ij}\, g_{lk} + \Gamma^l_{kj}\, g_{il} \\
-g_{jk,i} &= -\Gamma^l_{ji}\, g_{lk} - \Gamma^l_{ki}\, g_{jl}.
\end{aligned}
$$

Now we add the three equations and use the symmetry of the metric tensor $(g_{ij} = g_{ji})$

$$g_{ij,k} + g_{ik,j} - g_{jk,i} = \left(\Gamma^l_{ik} - \Gamma^l_{ki}\right) g_{lj} + \left(\Gamma^l_{ij} - \Gamma^l_{ji}\right) g_{lk} + \left(\Gamma^l_{jk} + \Gamma^l_{kj}\right) g_{il}.$$

Due to the symmetry of the Christoffel symbols, the first two terms vanish and we get because $\Gamma^l_{jk} = \Gamma^l_{kj}$

$$g_{ij,k} + g_{ik,j} - g_{jk,i} = 2 g_{il}\, \Gamma^l_{jk}.$$

Now we divide by 2, multiply the equation with the inverse matrix $(g_{im})^{-1} = g^{im}$ and get

$$\frac{1}{2}\, g^{im}\left(g_{ij,k} + g_{ik,j} - g_{jk,i}\right) = g^{im}\, g_{il}\, \Gamma^l_{jk}.$$

Since

$$g^{im}\, g_{il} = \delta^m_l,$$

it follows that

$$g^{im}\, g_{il}\, \Gamma^l_{jk} = \delta^m_l \Gamma^l_{jk} = \Gamma^m_{jk}$$

and we finally get

$$\Gamma^m_{jk} = \frac{1}{2}\, g^{im}\left(g_{ij,k} + g_{ik,j} - g_{jk,i}\right). \tag{9.28}$$

Example 9.5.

As an example, we calculate the Christoffel symbol $\Gamma^\varphi_{r\varphi}$ of the polar coordinates again using the last formula. According to Eq. (9.17), the inverse metric tensor for polar coordinates is

$$\mathbf{G} \underset{P}{\to} \begin{pmatrix} g^{rr} & g^{r\varphi} \\ g^{\varphi r} & g^{\varphi\varphi} \end{pmatrix} = \begin{pmatrix} 1 & 0 \\ 0 & 1/r^2 \end{pmatrix},$$

and the derivatives of the metric were obtained above as

$$g_{i'j',r} = \begin{pmatrix} \dfrac{\partial g_{rr}}{\partial r} & \dfrac{\partial g_{r\varphi}}{\partial r} \\[2mm] \dfrac{\partial g_{\varphi r}}{\partial r} & \dfrac{\partial g_{\varphi\varphi}}{\partial r} \end{pmatrix} = \begin{pmatrix} 0 & 0 \\ 0 & 2r \end{pmatrix}$$

and

$$g_{i'j',\varphi} = \begin{pmatrix} \dfrac{\partial g_{rr}}{\partial \varphi} & \dfrac{\partial g_{r\varphi}}{\partial \varphi} \\[2mm] \dfrac{\partial g_{\varphi r}}{\partial \varphi} & \dfrac{\partial g_{\varphi\varphi}}{\partial \varphi} \end{pmatrix} = \begin{pmatrix} 0 & 0 \\ 0 & 0 \end{pmatrix}.$$

Thus it follows

$$\begin{aligned} \Gamma^{\varphi}_{r\varphi} &= \frac{1}{2} g^{r\varphi} \left(g_{rr,\varphi} + g_{r\varphi,r} - g_{r\varphi,r} \right) + \frac{1}{2} g^{\varphi\varphi} \left(g_{\varphi r,\varphi} + g_{\varphi\varphi,r} - g_{r\varphi,\varphi} \right) \\ &= \frac{1}{2} \cdot 0 \left(0 + 0 - 0 \right) + \frac{1}{2} \cdot \frac{1}{r^2} \left(0 + 2r - 0 \right) = \frac{1}{r} \end{aligned}$$

and of course we get the same result as in 8.36 on page 167. □

Tensor equations in the Euclidean plane

At the end of the chapter we would like to emphasize again the importance of tensor calculation for the formulation of physical laws. First of all, a tensor whose components are all zero in one coordinate system is also the zero tensor in all other coordinate systems. For example, let $\mathbf{T}$ be a $(1,2)$-tensor whose components in a (x,y) coordinate system K are zero

$$\mathbf{T} \underset{K}{\to} T^{i}{}_{jk} = 0,$$

and let (x',y') be any other coordinate system, then according to the transformation formula for the components of $\mathbf{T}$ in (x',y')

$$T^{i'}{}_{j'k'} = \frac{\partial i'}{\partial i}\frac{\partial j}{\partial j'}\frac{\partial k}{\partial k'} T^{i}{}_{jk} = 0,$$

since all $T^{i}{}_{jk}$ are equal to zero. We can now formulate:

> If a physical law applies in the Euclidean plane in any coordinate system and is formulated as a tensor equation, then this law applies in all coordinate systems.

Since vectors are special tensors, the statement also applies to physical laws that are formulated as vector equations.

10. The Inertia Tensor

In this chapter, after the more theoretical and abstract discussions of the last sections, we want to examine a concrete example of a tensor from Newtonian mechanics. However, in this example we do not stay in the Euclidean plane, but rather examine physical quantities in three-dimensional space. In more detail, we extend our statements about rotational movements from Chap. 5.

10.1. Extension of the Rotational Movements of Point Particles

To repeat, we look again at tab. 5.1 from part I, in which the physical quantities of translations and rotations are compared.

Translations	Rotations
Length s	Angle θ
Mass m	Moment of inertia $I = mR^2$, $\vec{R}$ radius.
Velocity $\vec{v}$	Angular velocity $\vec{\omega} = \dfrac{1}{R^2}\left(\vec{R} \times \vec{v}\right)$
Momentum $\vec{p} = m\vec{v}$	Angular momentum $\vec{L} = \vec{r} \times m\vec{v}$, $\vec{r}$ position
Force $\vec{F}$	Torque $\vec{M} = \vec{r} \times \vec{F}$
$\vec{F} = \dfrac{d\vec{p}}{dt}$	$\vec{M} = \dfrac{d\vec{L}}{dt}$
Kin. Energy $E_{kin} = \dfrac{1}{2}mv^2$	Kin. Energy $E_{kin} = \dfrac{1}{2}I\omega^2$

Table 10.1.: Comparison of translations and rotations

From the formula for the angular velocity $\vec{\omega}$, we can determine the tangential

M. Ruhrländer, *Ascent to the Einstein Equations*,
https://doi.org/10.1007/978-3-662-72672-3_10

velocity $\vec{v}$. For this, we need the following remark.

Remark 10.1. **MT: The "bac-cab" rule**

If $\vec{a}, \vec{b}, \vec{c}$ are three vectors in $\mathbb{R}^3$, then with the Cartesian components of the vectors

$$\vec{a} \times \left(\vec{b} \times \vec{c}\right) \;\rightarrow\; \begin{pmatrix} a^x \\ a^y \\ a^z \end{pmatrix} \times \left(\begin{pmatrix} b^x \\ b^y \\ b^z \end{pmatrix} \times \begin{pmatrix} c^x \\ c^y \\ c^z \end{pmatrix} \right)$$

$$= \begin{pmatrix} a^x \\ a^y \\ a^z \end{pmatrix} \times \left(\begin{pmatrix} b^y c^z - b^z c^y \\ b^z c^x - b^x c^z \\ b^x c^y - b^y c^x \end{pmatrix} \right)$$

$$= \begin{pmatrix} a^y \left(b^x c^y - b^y c^x\right) - a^z \left(b^z c^x - b^x c^z\right) \\ a^z \left(b^y c^z - b^z c^y\right) - a^x \left(b^x c^y - b^y c^x\right) \\ a^x \left(b^z c^x - b^x c^z\right) - a^y \left(b^y c^z - b^z c^y\right) \end{pmatrix}$$

$$= \begin{pmatrix} \left(b^x a^y c^y + b^x a^z c^z\right) - \left(a^y b^y c^x + a^z b^z c^x\right) \\ \left(b^y a^z c^z + b^y a^x c^x\right) - \left(a^z b^z c^y + a^x b^x c^y\right) \\ \left(b^z a^x c^x + b^z a^y c^y\right) - \left(a^x b^x c^z + a^y b^y c^z\right) \end{pmatrix}.$$

Now we do a small "trick" and add in each row in the first bracket a term $b^i a^i c^i$ and then subtract it again in the second bracket

$$\vec{a} \times \left(\vec{b} \times \vec{c}\right) \;\rightarrow\; \begin{pmatrix} b^x a^y c^y + b^x a^z c^z + b^x a^x c^x - \left(a^x b^x c^x + a^y b^y c^x + a^z b^z c^x\right) \\ b^y a^z c^z + b^y a^x c^x + b^y a^y c^y - \left(a^x b^x c^y + a^y b^y c^y + a^z b^z c^y\right) \\ b^z a^x c^x + b^z a^y c^y + b^z a^z c^z - \left(a^x b^x c^z + a^y b^y c^z + a^z b^z c^z\right) \end{pmatrix}$$

$$= \begin{pmatrix} b^x \left(a^y c^y + a^z c^z + a^x c^x\right) - \left(a^x b^x + a^y b^y + a^z b^z\right) c^x \\ b^y \left(a^z c^z + a^x c^x + a^y c^y\right) - \left(a^x b^x + a^y b^y + a^z b^z\right) c^y \\ b^z \left(a^x c^x + a^y c^y + a^z c^z\right) - \left(a^x b^x + a^y b^y + a^z b^z\right) c^z \end{pmatrix}.$$

In each row, the scalar products of $\vec{a} \cdot \vec{c}$ and $\vec{a} \cdot \vec{b}$ are listed, from which it follows

$$\vec{a} \times \left(\vec{b} \times \vec{c}\right) \rightarrow \begin{pmatrix} b^x \left(\vec{a} \cdot \vec{c}\right) - \left(\vec{a} \cdot \vec{b}\right) c^x \\ b^y \left(\vec{a} \cdot \vec{c}\right) - \left(\vec{a} \cdot \vec{b}\right) c^y \\ b^z \left(\vec{a} \cdot \vec{c}\right) - \left(\vec{a} \cdot \vec{b}\right) c^z \end{pmatrix} = \begin{pmatrix} b^x \\ b^y \\ b^z \end{pmatrix} \left(\vec{a} \cdot \vec{c}\right) - \begin{pmatrix} c^x \\ c^y \\ c^z \end{pmatrix} \left(\vec{a} \cdot \vec{b}\right).$$

Thus, we obtain the "bac-cab" formula:

$$\vec{a} \times \left(\vec{b} \times \vec{c}\right) = \vec{b}\left(\vec{a} \cdot \vec{c}\right) - \vec{c}\left(\vec{a} \cdot \vec{b}\right) \;\square$$

We use this formula to determine the tangential velocity $\vec{v}$. It holds

$$\vec{\omega} = \frac{1}{R^2}\left(\vec{R} \times \vec{v}\right) \Rightarrow R^2 \vec{\omega} = \vec{R} \times \vec{v}.$$

We multiply the last equation vectorially with $\vec{R}$ and obtain

$$R^2\vec{\omega}\times\vec{R} = \left(\vec{R}\times\vec{v}\right)\times\vec{R} = -\vec{R}\times\left(\vec{R}\times\vec{v}\right) = \vec{R}\times\left(\vec{v}\times\vec{R}\right).$$

On the right side, we apply the "bac-cab" formula

$$R^2\vec{\omega}\times\vec{R} = \vec{v}\underbrace{\left(\vec{R}\cdot\vec{R}\right)}_{=R^2} - \vec{R}\underbrace{\left(\vec{R}\cdot\vec{v}\right)}_{=0} = \vec{v}R^2,$$

since the tangential velocity is perpendicular to the radius vector. So, after division by R^2, we get

$$\vec{v} = \vec{\omega}\times\vec{R}.$$

In Chap. 5, we showed that the angular momentum vector $\vec{L}$ points in the same direction as the angular velocity $\vec{\omega}$, when the coordinate origin lies in the rotation plane of the particle. Then it holds

$$\vec{L} = mR^2\vec{\omega}.$$

We now consider the case where the origin of coordinates is located outside the plane of rotation of the particle, as illustrated in Fig. 10.1.

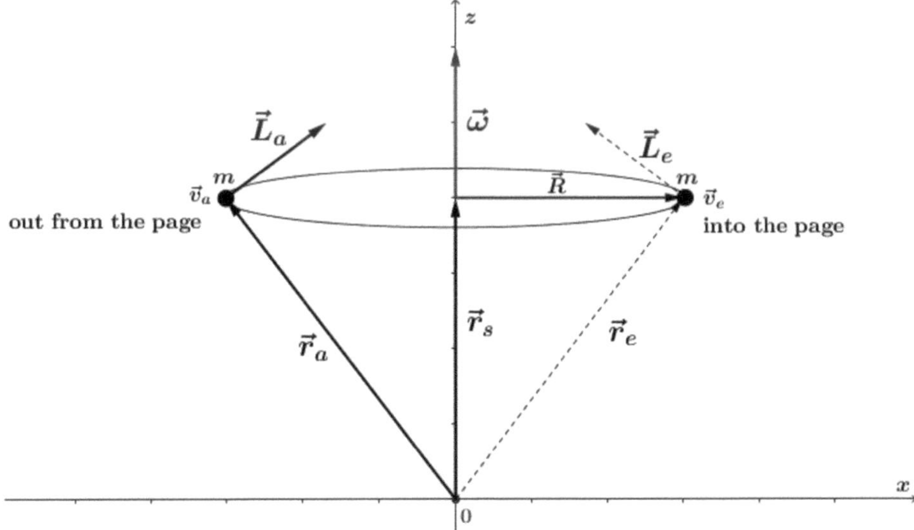

Figure 10.1.: Angular momentum not parallel to angular velocity

On the left side of Fig. 10.1 the starting point of a particle with mass m, the initial position vector $\vec{r}_a$ and the initial tangential velocity $\vec{v}_a$ are drawn, which are supposed to protrude from the side/the sheet. Also, the initial angular momentum $\vec{L}_a$ is entered, which is perpendicular to $\vec{r}_a$ and $\vec{v}_a$ and forms a right-handed system with these two vectors. The particle moves on a circular line with radius vector $\vec{R}$. After half a rotation, the particle has reached the point with position vector $\vec{r}_e$ and tangential velocity $\vec{v}_e$, which points into the side, as well as with the angular momentum vector $\vec{L}_e$. The vector of angular velocity $\vec{\omega}$ is perpendicular to $\vec{R}$ and $\vec{v}$ and points in the positive z-direction (axis of rotation). It can be seen that the angular momentum vector in this case is not parallel to the angular velocity and the direction of the angular momentum vector changes from point to point as the particle moves.

We now want to derive a relationship between the angular momentum and the angular velocity which in the general case cannot have the simple form

$$\vec{L} = mR^2\vec{\omega}$$

as we have just shown with Fig. 10.1. Rather, we need an object that will give us the angular momentum vector $\vec{L}$ when we apply the angular velocity $\vec{\omega}$ to it. Such an object is known to us from tensor calculus, because a $(1,1)$-tensor fulfills this requirement. Let

$$\mathbf{I} \to I_i^j,$$

$i, j, = x, y, z$, be such a tensor, then the expression

$$\mathbf{I}(\vec{\omega}) \to I_i^j\,\omega^i$$

is a vector (observe Einstein's summation convention). We are therefore looking for a tensor $\mathbf{I}$ that satisfies the equation

$$\vec{L} = \mathbf{I}(\vec{\omega})$$

or in component form

$$L^j = I_i^j\,\omega^i.$$

It can also be seen from Fig. 10.1 that the position vector of the particle $\vec{r}_e$ can be divided into a vector $\vec{r}_s$, which is parallel to the angular velocity $\vec{\omega}$ (i.e., to the axis of rotation), and into the radius vector $\vec{R}$, which is perpendicular to the angular velocity:

$$\vec{r}_e = \vec{r}_s + \vec{R}$$

We multiply this equation vectorially with the angular velocity

$$\vec{\omega} \times \vec{r}_e = \vec{\omega} \times \left(\vec{r}_s + \vec{R}\right) = \underbrace{\vec{\omega} \times \vec{r}_s}_{=0} + \vec{\omega} \times \vec{R} = \vec{\omega} \times \vec{R},$$

since $\vec{\omega}$ is parallel to $\vec{r}_s$. Now $\vec{\omega} \times \vec{R}$ is the tangential velocity $\vec{v}$, i.e., it holds for an arbitrary position vector $\vec{r}$ of the particle

$$\vec{v} = \vec{\omega} \times \vec{R} = \vec{\omega} \times \vec{r}.$$

To find the desired tensor, we consider the angular momentum and substitute for $\vec{v}$ the above expression

$$\vec{L} = \vec{r} \times m\vec{v} = m\left(\vec{r} \times \vec{v}\right) = m\left(\vec{r} \times \vec{\omega} \times \vec{r}\right).$$

On the right side, we apply the "bac-cab" formula and obtain

$$\vec{L} = m\left(\vec{\omega}(\vec{r} \cdot \vec{r}) - \vec{r}(\vec{r} \cdot \vec{\omega})\right) = m\left(r^2\vec{\omega} - \vec{r}(\vec{r} \cdot \vec{\omega})\right). \tag{10.1}$$

We calculate with

$$\vec{L} \to \begin{pmatrix} L^x \\ L^y \\ L^z \end{pmatrix}, \quad \vec{r} \to \begin{pmatrix} x \\ y \\ z \end{pmatrix}, \quad \vec{\omega} \to \begin{pmatrix} \omega^x \\ \omega^y \\ \omega^z \end{pmatrix}$$

in component form further

$$\begin{aligned} L^x &= m\left((x^2 + y^2 + z^2)\omega^x - x(x\omega^x + y\omega^y + z\omega^z)\right) \\ &= m\left((y^2 + z^2)\omega^x - xy\omega^y - xz\omega^z\right) \\ L^y &= m\left((x^2 + y^2 + z^2)\omega^y - y(x\omega^x + y\omega^y + z\omega^z)\right) \\ &= m\left((x^2 + z^2)\omega^y - yx\omega^x - yz\omega^z\right) \\ L^z &= m\left((x^2 + y^2 + z^2)\omega^z - z(x\omega^x + y\omega^y + z\omega^z)\right) \\ &= m\left((x^2 + y^2)\omega^z - zx\omega^x - zy\omega^y\right). \end{aligned}$$

These three equations can be summarized into a matrix equation

$$\begin{pmatrix} L^x \\ L^y \\ L^z \end{pmatrix} = m \begin{pmatrix} y^2 + z^2 & -xy & -xz \\ -yx & x^2 + z^2 & -yz \\ -zx & -zy & x^2 + y^2 \end{pmatrix} \begin{pmatrix} \omega^x \\ \omega^y \\ \omega^z \end{pmatrix}. \tag{10.2}$$

With this, we have found the desired components of the (symmetric) tensor **I**, which is called the **inertia tensor**. In short form, these can also be expressed by the Kronecker delta

$$\mathbf{I} \to I_i^j = m\left(\delta_i^j \, r^2 - ij\right), \quad i, j = x, y, z.$$

So for example

$$I_x^x = m\left(\delta_x^x r^2 - x \cdot x\right) = m\left((x^2 + y^2 + z^2) - x \cdot x\right) = m\left(y^2 + z^2\right)$$

or

$$I_z^y = m\left(\delta_z^y r^2 - z \cdot y\right) = m\left(-z \cdot y\right).$$

To understand the physical meaning of the components of the inertia tensor, we remember that the moment of inertia describes the resistance of the particle to rotational accelerations. This resistance depends not only on the mass of the particle, but also on the distance of the particle from the axis of rotation. The quantities I_i^j indicate how much angular momentum in the i direction is generated during rotation around the j axis. For example, I_x^x indicates how much angular momentum in the x direction is produced during rotations around the x axis. The mass m of the particle is multiplied with the square distance of the particle from the x axis:

$$I_x^x = m\left(y^2 + z^2\right)$$

So, this is the three-dimensional version of the equation $I = mr^2$, which is valid for planar rotations, where r denotes the distance of the particle from the axis of rotation. The other two terms on the diagonal of the momentum tensor matrix correspondingly represent the y- and z-components of the angular momentum during rotation around the y- and z-axis. The off-diagonal elements look a bit different, for example, in I_y^z the mass m is multiplied by the product of the y- and z-coordinates

$$I_y^z = -m\left(y \cdot z\right).$$

This matrix element describes the contribution of asymmetric distributions of mass with respect to the z-axis. To better understand this, we extend our model to a multi-particle system and demonstrate the various contributions of the inertia tensor to the angular momentum vector with some examples.

10.2. Rotational Movements in Multi-Particle Systems

Let's now consider N particles with masses m_i, which are supposed to form a rigid body, i.e., we imagine that the particles are connected with (massless) rods, which prevent the body formed by these particles from being deformed. It is rigid and its only possibilities of movement consist in that it can be moved linearly and rotated around any axis, which can also pass through the body. We choose a Cartesian coordinate system, place the center of gravity of the body at the origin of the coordinate system and denote the locations of the N particles with

$$\vec{r}_i \rightarrow \begin{pmatrix} x_i \\ y_i \\ z_i \end{pmatrix}.$$

Then the total angular momentum $\vec{L}$ of the body is the sum of the angular momenta of the individual particles, so with Eq. (10.1)

$$\vec{L} = \sum_{i=1}^{N} \vec{L}_i = \sum_{i=1}^{N} m_i \left(r_i^2 \vec{\omega} - \vec{r}_i \left(\vec{r}_i \cdot \vec{\omega} \right) \right),$$

where it should be noted that in a rigid body all particles rotate with the *same* angular velocity $\vec{\omega}$. If you perform exactly the same calculation as in the last section with the one-particle system, you get the equation analogous to (10.2)

$$\begin{pmatrix} L^x \\ L^y \\ L^z \end{pmatrix} = \begin{pmatrix} \sum m_i \left(y_i^2 + z_i^2 \right) & -\sum m_i\, x_i\, y_i & -\sum m_i\, x_i\, z_i \\ -\sum m_i\, y_i\, x_i & \sum m_i \left(x_i^2 + z_i^2 \right) & -\sum m_i\, y_i\, z_i \\ -\sum m_i\, z_i\, x_i & -\sum m_i\, z_i\, y_i & \sum m_i \left(x_i^2 + y_i^2 \right) \end{pmatrix} \begin{pmatrix} \omega^x \\ \omega^y \\ \omega^z \end{pmatrix},$$

where the summations each go from $i = 1$ to $i = N$. The inertia tensor thus looks very similar to the one-particle system. Here too, you can use the shorthand notation

$$\mathbf{I} \rightarrow I_j^k = \sum_{i=1}^{N} m_i \left(\delta_j^k\, r_i^2 - j_i\, k_i \right), \quad j_i, k_i = x_i, y_i, z_i.$$

The diagonal elements of the inertia tensor are called **moments of inertia** and the remaining elements are called **moments of deviation**. These cause a torque perpendicular to the axis of rotation due to centrifugal forces when rotating.

For further understanding and interpretation of the components of the inertia tensor, let's look at an example. Fig. 10.2 shows a rigid body in the shape of an octahedron (two pyramids glued together at the square base). The six masses m_i of the body are attached at the corners of the octahedron, and the connecting lines are assumed to be massless. The center of gravity of the body is located at the origin of the coordinates. We want to calculate the inertia tensor and assume that all masses m_i are equal, i.e. $m_i = m$. The height of the two pyramids is equal to the length of the sides of the square $(2a)$.

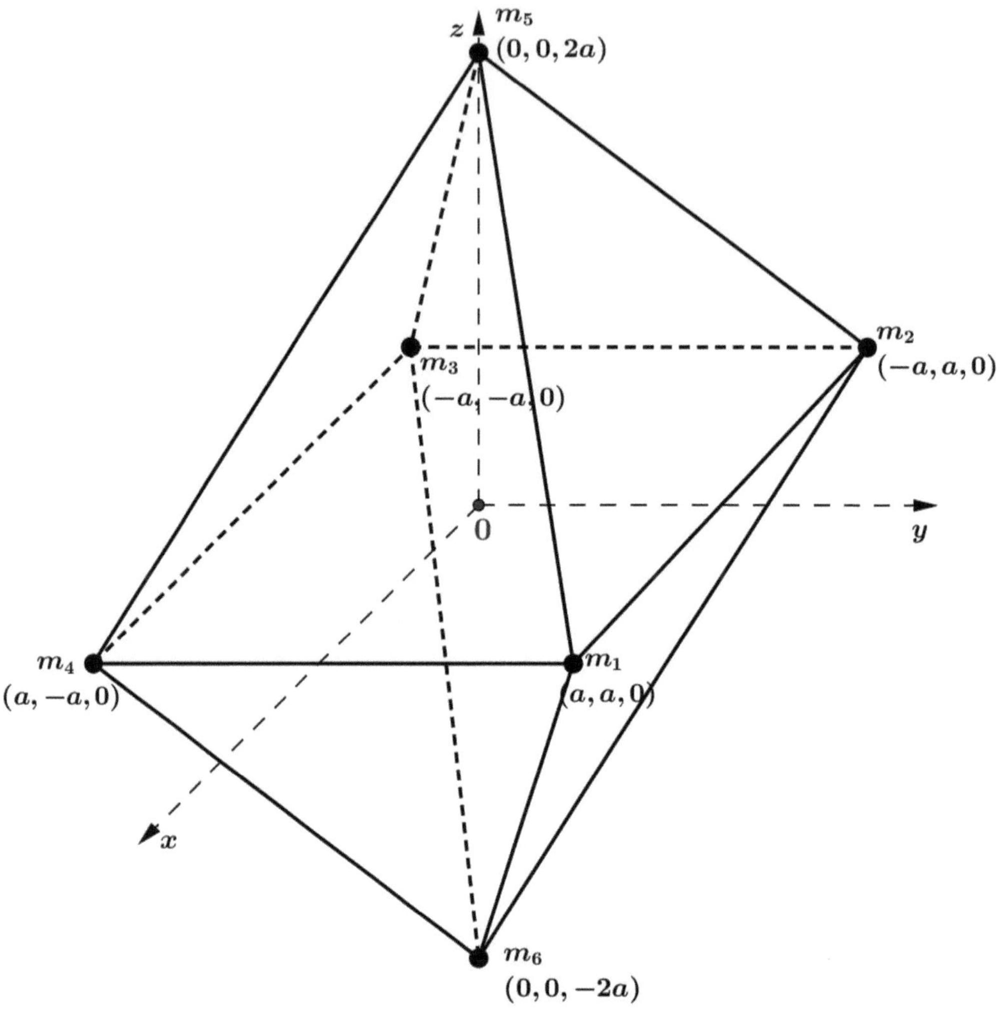

Figure 10.2.: Inertia tensor example octahedron

Then we get

$$
\begin{aligned}
I_x^x &= m \sum_{i=1}^{6} \left(y_i^2 + z_i^2 \right) \\
&= m \left[\left(y_1^2 + z_1^2 \right) + \left(y_2^2 + z_2^2 \right) + \cdots + \left(y_6^2 + z_6^2 \right) \right] \\
&= m \left[\left(a^2 \right) + \left(a^2 \right) + (-a)^2 + (-a)^2 + (2a)^2 + (-2a)^2 \right] \\
&= 12ma^2.
\end{aligned}
$$

Similarly, we get

$$I_y^y = 12ma^2.$$

For I_z^z, however, we get

$$
\begin{aligned}
I_z^z &= m \sum_{i=1}^{6} \left(x_i^2 + y_i^2 \right) \quad \bullet \\
&= m \left[\left(x_1^2 + y_1^2 \right) + \left(x_2^2 + y_2^2 \right) + \cdots + \left(x_6^2 + y_6^2 \right) \right] \\
&= m \left[a^2 + a^2 + (-a)^2 + a^2 + (-a)^2 + (-a)^2 + a^2 + (-a)^2 \right] \\
&= 8ma^2.
\end{aligned}
$$

The calculation of the off-diagonal elements yields

$$
\begin{aligned}
I_x^y &= -m \sum_{i=1}^{N} x_i\, y_i = -m \left[x_1\, y_1 + x_2\, y_2 + \cdots + x_6\, y_6 \right] \\
&= -m \left[aa + (-a)a + (-a)(-a) + a(-a) + 00 + 00 \right] \\
&= -m \left[2a^2 - 2a^2 \right] = 0
\end{aligned}
$$

$$
\begin{aligned}
I_x^z &= -m \sum_{i=1}^{N} x_i\, z_i = -m \left[x_1\, z_1 + x_2\, z_2 + \cdots + x_6\, z_6 \right] \\
&= -m \left[a0 + (-a)0 + (-a)0 + a0 + 02a + 0(-2a) \right] \\
&= 0
\end{aligned}
$$

$$
\begin{aligned}
I_y^z &= -m \sum_{i=1}^{N} y_i\, z_i = -m \left[y_1\, z_1 + y_2\, z_2 + \cdots + y_6\, z_6 \right] \\
&= -m \left[a0 + a0 + (-a)0 + (-a)0 + 02a + 0(-2a) \right] \\
&= 0.
\end{aligned}
$$

Overall, the inertia tensor looks like this

$$
\mathbf{I} \rightarrow \begin{pmatrix} 12ma^2 & 0 & 0 \\ 0 & 12ma^2 & 0 \\ 0 & 0 & 8ma^2 \end{pmatrix},
$$

i.e., it has a diagonal form, which is simultaneously due to

$$
\begin{pmatrix} L^x \\ L^y \\ L^z \end{pmatrix} = \begin{pmatrix} 12ma^2 & 0 & 0 \\ 0 & 12ma^2 & 0 \\ 0 & 0 & 8ma^2 \end{pmatrix} \begin{pmatrix} \omega^x \\ \omega^y \\ \omega^z \end{pmatrix} = \begin{pmatrix} 12ma^2\omega^x \\ 12ma^2\omega^y \\ 8ma^2\omega^z \end{pmatrix}
$$

means that the angular momentum vector is parallel to the axis of rotation, i.e., to the angular velocity $\vec{\omega}$. If the inertia tensor has diagonal form, we call the axes of rotation **principal axes of inertia** and the diagonal elements **principal moments of inertia**. In our example, the chosen x-, y-, and z-axes are principal axes of inertia. The principal axes of inertia indicate the symmetry of the body. The octahedron equipped with equal mass vertices is symmetric with respect to all three coordinate axes. The equality of the first two diagonal elements $\left(12ma^2\right)$ indicates that our example is a **symmetric gyroscope**, where a **gyroscope** is simply a rigid body that rotates around an axis. Other examples of symmetric gyroscopes are the sphere, the cube, and most toy tops, whose inertia tensors we cannot calculate here, as this requires advanced mathematical tools not covered in this book.

We want to calculate the components of the inertia tensor in another coordinate system K'. K' is to be created by rotating the x-axis by $\varphi = 30°$ from the Cartesian coordinate system K. So if x', y', z' are the coordinates of the coordinate system rotated around the x-axis, the coordinates y and z are rotated by $30°$, the coordinate x remains unchanged. Thus, with the rotation matrix 8.13 on page 149

$$
\begin{aligned}
x' &= x \\
\begin{pmatrix} y' \\ z' \end{pmatrix} &= \begin{pmatrix} \cos\varphi & \sin\varphi \\ -\sin\varphi & \cos\varphi \end{pmatrix} \begin{pmatrix} y \\ z \end{pmatrix},
\end{aligned}
$$

which can also be written as *one* matrix equation

$$
\begin{pmatrix} x' \\ y' \\ z' \end{pmatrix} = \begin{pmatrix} 1 & 0 & 0 \\ 0 & \cos\varphi & \sin\varphi \\ 0 & -\sin\varphi & \cos\varphi \end{pmatrix} \begin{pmatrix} x \\ y \\ z \end{pmatrix}.
$$

The rotation matrix

$$
R = \begin{pmatrix} 1 & 0 & 0 \\ 0 & \cos\varphi & \sin\varphi \\ 0 & -\sin\varphi & \cos\varphi \end{pmatrix}
$$

is, as in two dimensions, the Jacobian matrix of the coordinate change

$$
R = \left(\frac{\partial i'}{\partial i} \right)
$$

as well as **orthogonal**, i.e.

$$
R^{-1} = R^T = \begin{pmatrix} 1 & 0 & 0 \\ 0 & \cos\varphi & -\sin\varphi \\ 0 & \sin\varphi & \cos\varphi \end{pmatrix}.
$$

With the transformation rules for $(1,1)$-tensors, we obtain the components of **I** in K' by

$$I_{i'}^{j'} = \frac{\partial j'}{\partial j}\frac{\partial i}{\partial i'} I_i^j,$$

which can also be written as a matrix multiplication, i.e. $\left(I_{i'}^{j'}\right) =$

$$\begin{pmatrix} 1 & 0 & 0 \\ 0 & \cos\varphi & \sin\varphi \\ 0 & -\sin\varphi & \cos\varphi \end{pmatrix} \begin{pmatrix} 12ma^2 & 0 & 0 \\ 0 & 12ma^2 & 0 \\ 0 & 0 & 8ma^2 \end{pmatrix} \begin{pmatrix} 1 & 0 & 0 \\ 0 & \cos\varphi & -\sin\varphi \\ 0 & \sin\varphi & \cos\varphi \end{pmatrix}$$

$$= \begin{pmatrix} 1 & 0 & 0 \\ 0 & \cos\varphi & \sin\varphi \\ 0 & -\sin\varphi & \cos\varphi \end{pmatrix} \begin{pmatrix} 12ma^2 & 0 & 0 \\ 0 & 12ma^2\cos\varphi & -12ma^2\sin\varphi \\ 0 & 8ma^2\sin\varphi & 8ma^2\cos\varphi \end{pmatrix}$$

$$= \begin{pmatrix} 12ma^2 & 0 & 0 \\ 0 & 12ma^2\cos^2\varphi + 8ma^2\sin^2\varphi & -4ma^2\cos\varphi\sin\varphi \\ 0 & -4ma^2\cos\varphi\sin\varphi & 12ma^2\sin^2\varphi + 8ma^2\cos^2\varphi \end{pmatrix}.$$

By the coordinate change, i.e., by rotating the coordinate axes, the previously symmetrical gyroscope has become asymmetrical. The new coordinate axes are therefore *not* principal axes of inertia. If we also consider that $\cos 30° = \sqrt{3}/2$ and $\sin 30° = 1/2$ apply, we finally obtain

$$\left(I_{i'}^{j'}\right) = \begin{pmatrix} 12ma^2 & 0 & 0 \\ 0 & 12ma^2\frac{3}{4} + 8ma^2\frac{1}{4} & -4ma^2\frac{\sqrt{3}}{4} \\ 0 & -4ma^2\frac{\sqrt{3}}{4} & 12ma^2\frac{1}{4} + 8ma^2\frac{3}{4} \end{pmatrix}$$

$$= \begin{pmatrix} 12ma^2 & 0 & 0 \\ 0 & 11ma^2 & -\sqrt{3}ma^2 \\ 0 & -\sqrt{3}ma^2 & 9ma^2 \end{pmatrix}.$$

11. Literature References and Further Reading for Part II

The second part of this book is characterized by the fact that very little concrete physics is practiced, but rather the techniques and skills of vector and tensor calculation are presented and practiced. Accordingly, the (short) list of recommended books does not consist of physics textbooks, but of special treatises on these topics. The main literature source used in this part is the book by Bernard Schutz [29], in which, similar to here, the vector and tensor calculation is not only introduced via the transformation behavior of the components of tensors, but the tensors themselves are represented and treated as objects (i.e., linear mappings). This is - if you like - a somewhat "more modern" approach than in many other texts, which manage with coordinate transformations. These include the book Vectors and Tensors by Daniel Fleisch [11], which provides an easy introduction to vector and tensor calculation.

Among the advanced books in the field of tensor calculation are the works by Schutz, Geometrical methods of mathematical physics [30] and by Jaenich, Vektor Analysis [17], which, however, go far beyond what was considered in this part and also have an abstract representation.

M. Ruhrländer, *Ascent to the Einstein Equations*,
https://doi.org/10.1007/978-3-662-72672-3_11

Part III.

Special Theory of Relativity

The Special Theory of Relativity was formulated by Einstein in 1905, when he was dealing with the problem of how to reconcile the behavior of electromagnetic waves with the concepts of time and space known from classical (Newtonian) mechanics. As we will see, he did not succeed in this endeavor. Instead, he shook the traditional concept of space and time and introduced completely new concepts, which were also so difficult to grasp because they cannot be reconciled with our senses and our everyday experience. However, Einstein's revolutionary new ideas are only noticeable (i.e., significant in measurements), when objects are examined that move at high speeds, like atomic particles, which are accelerated to speeds close to the speed of light (approx. $300.000\,\mathrm{km/s}$) in particle accelerators like the LHC in Geneva. For objects that - compared to the speed of light - have rather moderate speeds, including the not insignificant average orbital speed of the Earth around the Sun, after all $\approx 30\,\mathrm{km/s} = 108.000\,\mathrm{km/h}$, the corrections to space, time, mass, etc., prescribed by the Special Theory of Relativity are so small that they do not weigh heavily in most calculations and can therefore be ignored.

In this part of the book, we proceed in two parts. First, the phenomena and consequences of the Special Theory of Relativity are presented in historical sequence. We use a mathematical description that can be largely understood with simple algebraic knowledge. In the later chapters, vector and tensor calculus are then introduced for the four-dimensional spacetime and a second formulation of the Special Theory of Relativity is presented using these tools. This second description is essential for the introduction and understanding of the General Theory of Relativity in the third part. It is tested here in generalization of the corresponding methods, which in part II were presented for the Euclidean plane, on a second "simple" object, namely the Special Theory of Relativity.

12. Principle of Relativity

12.1. The Galilean / Newtonian Principle of Relativity

In this chapter, we describe how space and time were imagined in the times of Galilei and Newton and which principles of physics can be derived from these ideas. The first Newtonian law states that a point mass does not change its state of motion - whether it is at rest or moving at a constant speed - if no external force acts on it. The first Newtonian law thus does not distinguish between a point mass at rest and one moving at a constant speed. Galilei already wrote about the fact that if someone is in an opaque box (e.g., below deck of a ship) that moves uniformly, i.e., without jerking or changing its direction, maintains a constant speed, it is impossible for him to determine whether he is moving or not. Everything inside the box happens just as if it were at rest. Even if you cut a hole in a side wall of the box and through this opening see a second box approaching your own, you still can't say anything about whether your own, the foreign, or both boxes are moving. The only thing you can reliably determine is that the two boxes are moving *relative* to each other. Therefore, it actually makes no sense to say, for example, "this object is moving at a speed of 1 km/h", without specifying relative to what (to the ground, to the conveyor belt?) it is moving. Of course, this "forbidden" expression is used constantly in everyday life, as in most cases the ground is meant as a reference system.

Under **reference system** we understand a coordinate system in which we can specify the position (i.e., the spatial coordinates) of objects at any point in time using scales and clocks. We imagine a dense network of observers in each reference system, equipped with identical clocks and scales. This dense network of observers is needed to achieve as accurate as possible results when measuring locations and times of events. For example, a time measurement of an event, which is some distance from an observer, is already distorted by the travel time of the information (such as a light signal from the event to the observer). Observers can avoid such difficulties if they only capture local events (events in their immediate surroundings) and leave other more distant events to observers for whom these are local.

© The Author(s), under exclusive license to Springer-Verlag GmbH, DE, part of Springer Nature 2026
M. Ruhrländer, *Ascent to the Einstein Equations*,
https://doi.org/10.1007/978-3-662-72672-3_12

Galilei and Newton generally assumed that space, time and also the mass of a body are absolute, i.e., that spatial distances and temporal differences as well as the mass of a body, measured in one reference system, also have the same measurements in any other reference system.

In an observer's reference system B_1, a stationary body, on which no external forces act, remains at rest according to the first Newtonian law (see Fig. 12.1). If the reference system of a second observer B_2 moves relative to that of B_1 at a constant speed, the force-free body in this reference system has a constant speed. Imagine you are standing on a platform and next to you is a suitcase. For an observer, who is passing by you in a train at a constant speed, your suitcase is moving away from him at (opposite) speed.

However, if the train accelerates while the observer is looking at your suitcase and the observer does not notice this acceleration, he concludes that your suitcase is accelerating, although no external forces are acting on it. The first Newtonian law does not apply in such (accelerated) reference systems. Reference systems in which the first Newtonian law applies are called **inertial systems**. According to the first Newtonian law, reference systems that move relative to an inertial system at a constant speed are also inertial systems.

We now want to investigate whether the other two Newtonian laws apply in the same way in all inertial systems, which is also referred to as **invariant**. To do this, we consider two reference systems S and S' with Cartesian coordinates x, y, z and origin 0 or x', y', z' and $0'$. The reference system S' is assumed to move at a constant speed v along the positive x-axis of the reference system S. Furthermore, we want to assume for simplicity that the origins of the two reference systems coincided at the time $t = 0$.

Fig. 12.1 shows the observer B_1 at rest at the origin of the reference system S and the observer B_2 also at rest at the origin of his reference system S'. The reference system S' moves at a constant speed $\vec{v} = (v, 0, 0)$ parallel to the x-axis and has moved a distance vt from S at time t.

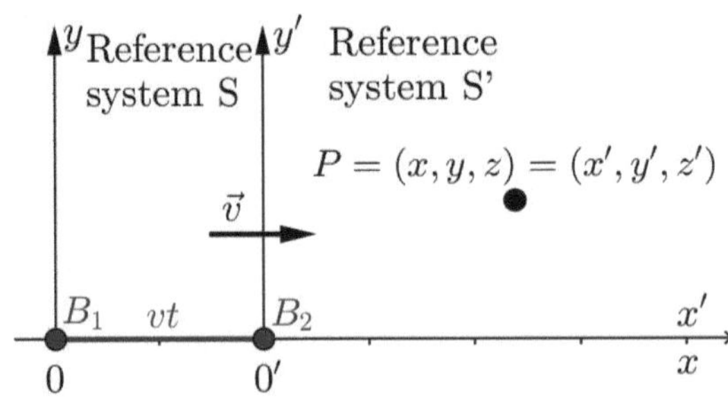

Figure 12.1.: Galilei-Transformation

Now consider the point P, P has the coordinates $P = (x, y, z)$ in system S and the coordinates $P = (x', y', z')$ in system S', where for simplicity the z- and z'-axes were omitted in the figure. These can be imagined as if they were coming out of the origin in S or in S' emerge from the paper. Let us denote the time in system S with t and in S' with t', so the assumption of an absolute time means that B_1 and B_2 each measure the same time, that is, t and t' are equal. Since the two reference systems coincided at the time $t = t' = 0$ and S' only moves along the x-axis, this movement does not change the y' and z' coordinates in the reference system S' over time. The relationships between the coordinates are therefore given by:

$$\begin{aligned}
x' &= x - vt \\
y' &= y \\
z' &= z \\
t' &= t
\end{aligned} \tag{12.1}$$

These four equations are referred to as the **Galilean transformation** and provide a rule for how to calculate the corresponding values in system S' by measurements in system S. Of course, the reverse also applies, i.e., with the equations

$$\begin{aligned}
x &= x' + vt \\
y &= y' \\
z &= z' \\
t &= t'
\end{aligned}$$

one obtains a rule for how to derive the corresponding values in system S from quantities in system S'. If a particle in system S has the velocity

$$\vec{u} \xrightarrow{S} \begin{pmatrix} u^x \\ u^y \\ u^z \end{pmatrix} = \begin{pmatrix} dx/dt \\ dy/dt \\ dz/dt \end{pmatrix},$$

then its velocity in the reference system S' is calculated due to $dt' = dt$ with the Galilean transformation to

$$\vec{u}\,' \xrightarrow{S'} \begin{pmatrix} dx'/dt' \\ dy'/dt' \\ dz'/dt' \end{pmatrix} = \begin{pmatrix} d(x - vt)/dt \\ dy/dt \\ dz/dt \end{pmatrix} = \begin{pmatrix} dx/dt - v \\ dy/dt \\ dz/dt \end{pmatrix} = \begin{pmatrix} u^x - v \\ u^y \\ u^z \end{pmatrix},$$

$$\tag{12.2}$$

so

$$\vec{u}\,' = \vec{u} - \vec{v}.$$

Differentiating again, taking into account that the velocity v of the two systems relative to each other is constant, yields a relationship between the accelerations in the two systems

$$\vec{a}\,' \rightarrow \begin{pmatrix} du^{x'}/dt \\ du^{y'}/dt \\ du^{z'}/dt \end{pmatrix} = \begin{pmatrix} d\,(u^x - v)\,/dt \\ du^y/dt \\ du^z/dt \end{pmatrix} = \begin{pmatrix} du^x/dt \\ du^y/dt \\ du^z/dt \end{pmatrix} = \begin{pmatrix} a^x \\ a^y \\ a^z \end{pmatrix},$$

so

$$\vec{a}\,' = \vec{a},$$

the accelerations are therefore identical. Since the mass of a particle in different inertial systems does not change ($m' = m$), the second law of Newton applies in both systems in the same way:

$$\vec{F} = m\vec{a} = m'\vec{a}\,' = \vec{F}\,'$$

From the equality of the second law, it follows directly that the third law also applies in the same way in system S'. We can therefore summarize the so-called **Newtonian principle of relativity** as follows:

> The Newtonian laws - and derived from them the entire Newtonian mechanics - are invariant in all reference systems that are linked by a Galilean transformation.

12.2. Light and Ether

Although every inertial system is suitable for the construction of mechanics, Newton assumed that there is an "absolute space" that is distinguished from all other inertial systems. In this absolute space the **ether** should rest, which Newton considered a thin and elastic medium that should fill the entire space. The obvious question: "Why then do the planets move unhindered through the resting ether?" Newton answered as follows:

> "If one assumes that the ether is $700,000$ times more elastic and at the same time $700,000$ times thinner than our air, its resistance would be over 600 million times less than that of water. Such a small resistance would not cause any noticeable change in the movement of the planets even in $10,000$ years."

In the 19*th* century, there were increasing indications that light, which Newton had considered a particle beam, has wave character. Thus, the English physicist Thomas Young showed that bundles of light can enhance each other just like water waves, but also cancel each other out. These so-called **interference phenomena** were also observed with sound, which was recognized as air vibration. And just as the water wave needed water and the sound wave needed air as carrier medium, light should be a vibration of the ether. Since light reaches us from the most distant stars, the carrier medium of light, like the ether, had to be present everywhere.

In the later development of electricity theory, the English physicist Michael Faraday saw a hint of "tensions in the ether" in the vicinity of the magnetic poles. He suspected that the magnetic force is transmitted from one pole to the other by pressure and tension in the ether. Finally, James C. Maxwell came to the realization that, if electricity and magnetism were indeed tension phenomena in the ether, then the tense ether - similar to a tense spring - must also be able to vibrate. He calculated that these vibrations spread at the speed of light. Light is therefore an electromagnetic wave. With these insights, the ether became a central concept of all physics, and physicists set about searching for the absolute space in which the ether should rest.

Because of the Earth's orbital motion around the Sun, it could not be assumed that the Earth rests in the ether. Therefore, attempts were made to determine the movement of the Earth through absolute space experimentally, assuming that electrical and magnetic forces are transmitted by tensions in the ether. A movement of the Earth through the ether should therefore lead to measurable changes in electromagnetic effects. But all experiments were negative, the desired changes did not occur. Among the experiments to detect the absolute space were also some that aimed to determine the speed of light relative to the Earth. According to the Galilean speed transformation (12.2) the light of a star, when the Earth moves towards the star with its orbital speed of $v = 30\,\mathrm{km/s}$, should be measured at the speed $c - v$. If the Earth later moves away from this star on its orbit, the speed of the starlight measured on Earth should then be $c + v$. So a speed difference of $2v$ should be measured. In fact, however, no difference could be measured at all. The light apparently always propagates at the same speed regardless of the speed of the Earth. In other words, it could not be proven that light and thus also the laws of electromagnetic phenomena set up by Maxwell comply with Newton's principle of relativity.

Subsequent attempts to modify the Maxwell equations to make them Galilei-invariant were also unsuccessful, because they predicted contradictory electromagnetic phenomena. Finally, Einstein solved the problem by concluding that the reason the speed of light is independent of the movement of the light source is always the same size is a fundamental error in our general conception of speed, i.e., since speed equals length divided by time, our conceptions of length/space and time are wrong.

12.3. The Einsteinian Principle of Relativity

The Special Theory of Relativity is based on two postulates proposed by Einstein in 1905:

1st Postulate (**Principle of Relativity**): There is no physically preferred inertial system. All laws of physics take the same form in all inertial systems.

2nd Postulate (**Principle of the Constancy of the Speed of Light**): The speed of light in a vacuum always has the value $c \approx 300000\,\text{km/s}$ in any inertial system.

Einstein suspected behind the failure of attempts to measure the movement of the Earth in the ether a general law of nature. If the ether does not exist at all, the idea of absolute motion or absolute rest is meaningless and only the relative motion of a body in relation to another can be of significance in physics. The first postulate does not directly reject the idea of an ether, but states that it plays no role in the formulation of the physical laws. So, why do we need it at all?

The precisely measured value for the speed of light in a vacuum is

$$c = 299792.458\,\text{km/s}.$$

If light propagates in a material medium such as water or glass, the *relative speed* of light in relation to the water or glass can also take other values. The 2nd postulate applies in such cases accordingly.

Neither the principle of the constancy of the speed of light nor the principle of relativity appear unusual at first glance. However, they involve fundamental changes in the concepts of space and time. Let's take for example a light signal that propagates in space. If we try to chase this light signal, it will - no matter how fast we run - always move away from us at the same speed c. We will now derive three fundamental conclusions from Einstein's postulates, namely that first, the simultaneity of two events loses its absolute character, that second, different observers do not measure the same duration for a process and that third, the length of an object is measured differently in different reference systems.

The Relativity of Simultaneity

Two events are to be simultaneous for an observer if light signals emitted from the locations of the events arrive at the observer at the same time. If we denote the location of the first event with A and that of the second with B, it follows from the constancy of the speed of light that the distance A - observer must be equal to the distance B - observer, the observer will therefore be in the middle between the two points A and B. It is again irrelevant due to the constancy of the speed of light whether the light sources are at rest or in uniform motion relative to the observer. Of course, this definition of simultaneity can also be extended to any number of further events and thus leads to a definition of "time" in physics. If we imagine that clocks of the same kind are set up at different points of a reference system S, whose hands position are simultaneously the same, then the time of an event is understood to be the time indication/hand position of the clock that is spatially adjacent to the event. To illustrate the relativity of simultaneity, we use an example that comes from Einstein himself (see [7]).

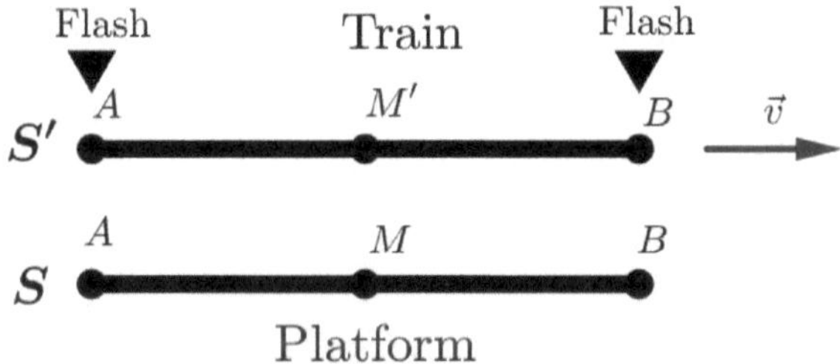

Figure 12.2.: Relativity of Simultaneity

In Fig. 12.2, we see a train passing by a railway platform at a constant speed $\vec{v}$. Let's assume that two lightning strikes occur at points A and B, which, from the perspective of an observer located in the middle between A and B at point M on the platform, happen simultaneously. Let M' be the midpoint of the line segment $\overline{AB}$ of the moving train. This point M' is at the same level as M at the time of the lightning strikes. If the train were not moving, an observer at M' would also perceive the lightning strikes as simultaneous. However, from the perspective of the platform, the train is moving to the right at a constant speed $\vec{v}$, i.e., towards the light signal from B and ahead of the light signal from A. Since the speed of the light flash emitted from A is equal to the speed of the light flash emitted from B, the observer at M' concludes that event B occurred earlier than event A. In general, we can make the following statement about the relativity of simultaneity.

> Relativity of Simultaneity: Events that are simultaneous in relation to an inertial system S are not simultaneous in relation to a second inertial system S' and vice versa. Each reference system has its own time. A time specification only makes sense if the reference system to which the time specification refers is also specified.

With this principle of the relativity of simultaneity, which directly follows from Einstein's postulates, the absolute time assumed by Newton has become obsolete.

Time Dilation

We now want to quantitatively determine the relationships between the different times in different reference systems. For this, we construct a simple clock. In this "light clock", the balance wheel is a back-and-forth moving light signal. In a cylinder, a flash of light travels from the bottom to the top to a mirror, is reflected there ("tick") and runs back to the ground. There it is registered ("tock"), immediately triggers a new flash, and the counter is increased by one. We imagine that we have set up and synchronized several such clocks along the x-axis of the (stationary) inertial system S at sufficiently small intervals so that we can read the time t in the reference system S at all interesting locations. Another identical clock moves at a constant speed $\vec{v} = (v, 0, 0)$ relative to system S in the positive x-axis direction. We want to answer the question of how much time t passes in the stationary system S when the time t' passes in the system S' of the moving clock.

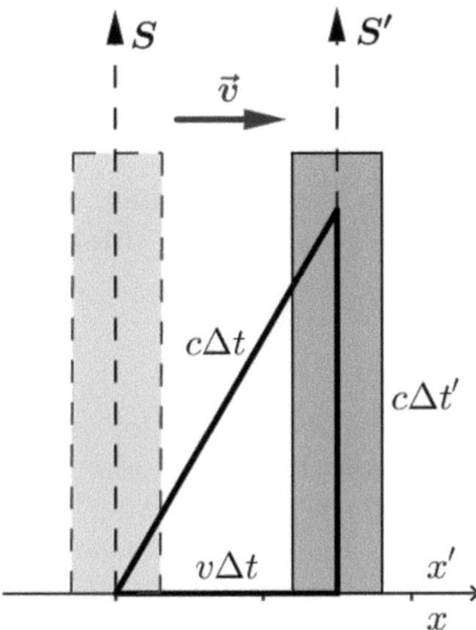

Figure 12.3.: Time dilation

In Fig. 12.3, the clock is at the dashed position at time t_1 in system S and t_1' in system S'. At this time, a flash of light is shot upwards. The clock moves within a time span $\Delta t = t_2 - t_1$ by the distance $v\Delta t$ in system S. The path covered by the light in the stationary system S is $c\Delta t$ due to the constancy of the speed of light. In the moving system S', the light beam covers the distance $c\Delta t' = c(t_2'-t_1')$. However, these two path lengths are not the same, so the time intervals Δt and $\Delta t'$ must differ! With the **Pythagorean theorem** (29.1), the relationship between the two measured values follows:

$$(v\Delta t)^2 + (c\Delta t')^2 = (c\Delta t)^2 \Rightarrow c^2(\Delta t')^2 = (\Delta t)^2(c^2 - v^2) \Rightarrow$$

$$(\Delta t')^2 = (\Delta t)^2 \left(1 - \frac{v^2}{c^2}\right)$$

and finally from this

$$\Delta t' = \Delta t \sqrt{1 - \frac{v^2}{c^2}}.$$

The time span $\Delta t'$ is therefore always smaller than Δt. The time between two events that occur in the same location in a reference system is called **proper time** Δt_E. The time interval $\Delta t_E = t_2' - t_1' = \Delta t'$, measured in the reference system S', is such a proper time. The time interval Δt, measured in any other reference system S, is therefore always larger by the factor

$$\gamma = \frac{1}{\sqrt{1 - v^2/c^2}} \tag{12.3}$$

(called **gamma factor**) than the proper time. This stretching of the time interval Δt compared to Δt_E is called **time dilation**:

$$\Delta t = \gamma \Delta t_E \tag{12.4}$$

Example 12.1.
To calculate a specific numerical example, let's imagine that a spaceship is flying away from Earth at a constant speed of $v = 0.6\,c$. The astronauts tell ground control that they will sleep for one hour. How long do they sleep in the Earth's reference system? Since the astronauts fall asleep and wake up in the same place in their reference system, their time is the proper time and the time interval on Earth is with (12.4)

$$\Delta t = \gamma \Delta t_E = \frac{1}{\sqrt{1 - v^2/c^2}} = \frac{1}{\sqrt{1 - (0.6c)^2/c^2}} = \frac{1}{\sqrt{0.64}} = 1.25,$$

so they sleep for one hour and 15 minutes in the Earth's time system! □

Length Contraction

From the time dilation it directly follows that length measurements in two inertial systems moving against each other at a speed v also differ. First, we note that if the reference system S' is moving to the right in the positive x-direction at speed v from the perspective of a (stationary) reference system S (the reference systems are again constructed so that the x- and x'-axes coincide), then the reverse also applies. That is, from the perspective of a (stationary) observer B_2 in S', the reference system S is moving to the left at the same speed v. If, for example, B_2 were to measure a lower speed u relative to S, i.e., $u < v$, this would be a physical law that, according to the principle of relativity, applies in every inertial system. This means that an observer B_1 in system S would also measure a smaller speed v relative to S', i.e., $v < u$, which is obviously a contradiction.

We now imagine that in system S two synchronized clocks are set up at two different points on the x-axis at a distance Δx, past which the observer B_2 in system S' drives by at a relative speed v. He measures the respective times at which he passes the two clocks and obtains a time interval $\Delta t'$ as the duration from one clock to the other. Exactly the same happens in system S. When observer B_2 passes the first clock, the time is taken and also when he reaches the second clock, this time difference is denoted by Δt. Since the same relative speed v is measured *in absolute terms* in both systems, it follows that

$$\frac{\Delta x}{\Delta t} = v = \frac{\Delta x'}{\Delta t'},$$

from which, with (12.4), it follows that

$$\Delta x' = \Delta x \,\frac{\Delta t'}{\Delta t} = \Delta x \,\sqrt{1 - \frac{v^2}{c^2}},$$

i.e., the length $\Delta x'$, measured in a moving reference system, is always shorter than the length Δx in the stationary reference system; this length is called **proper length** Δx_E and the shortening effect **length contraction**

$$\Delta x' = \Delta x_E \,\sqrt{1 - \frac{v^2}{c^2}} = \frac{1}{\gamma} \Delta x_E. \tag{12.5}$$

Now we consider whether there can also be a length distortion in directions perpendicular to the relative velocity $\vec{v}$. We again conduct a thought experiment and imagine that a train in a state of rest just barely fits into a narrow tunnel. If the width of the tunnel were to decrease from the train's perspective at a high relative speed, it would no longer fit into the tunnel, which, from its perspective, is approaching it. The entrance of the tunnel would either collide with the train or at least leave some scratches on the train. Now let's look at the situation from the other side. From the tunnel's perspective, the train is approaching at high speed. The lateral contraction of the tunnel, which the train has perceived, would now mean a lateral contraction of the train from the tunnel's perspective. The train would be narrower and could comfortably pass through the tunnel. This contradiction can only be resolved if the above assumption is rejected. So we can summarize:

> The length $\Delta x'$, measured in a moving reference system in the direction of motion, is always shorter by the factor γ than the length Δx_E in the stationary reference system: $\gamma \Delta x' = \Delta x_E$. Distances perpendicular to the direction of motion, on the other hand, do not shorten or lengthen.

The invariance of distances perpendicular to the direction of motion, by the way, we have already implicitly used in the derivation of the formula for time dilation, when we assumed that both observers agree that the light beam in the light clock actually covers the calculated distance perpendicular to the direction of motion.

In both phenomena discussed so far, namely time dilation and length contraction, the size

$$\gamma = \frac{1}{\sqrt{1 - v^2/c^2}}$$

appears. Fig. 12.4 shows the course γ as a function of the magnitude of the speed relation v/c.

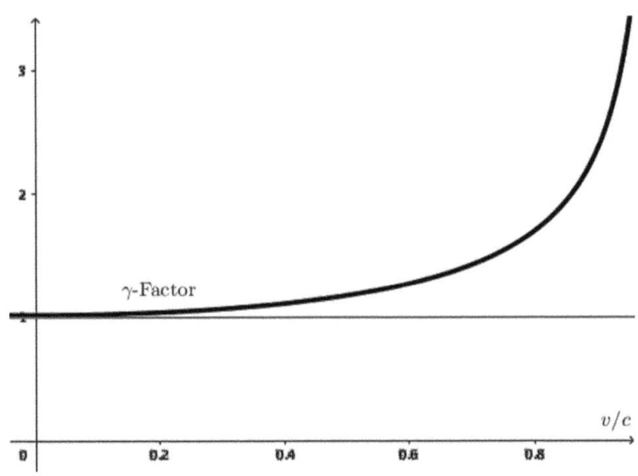

Figure 12.4.: Course of the gamma factor

One can see that the γ-factor (thick line) only deviates significantly from the line $\gamma = 1$ from a relative speed of approx. $0, 1\,c$, which is why time dilation and length contraction could only be practically demonstrated by experiments in particle accelerators. But also observations of very fast particles, which are produced from cosmic radiation by reactions with atomic nuclei and molecules of the atmosphere, have been able to impressively demonstrate time dilation and length contraction.

Example 12.2.
Muons are such secondary particles, they possess a speed of

$$v = 0.998\,c$$

and decay according to the law

$$N\,(t) = N_0 e^{-t/\tau}, \tag{12.6}$$

where N_0 is the number of muons at time $t = 0$, $N\,(t)$ is the number of muons at time t, and τ is the average lifetime, which for a muon at rest is approximately two microseconds $(= 2 \cdot 10^{-6}\,\text{s} = 2\,\mu\text{s})$. Since muons are created at an altitude of several thousand meters, only a few should reach the Earth's surface. A typical muon would only cover about 600 m in 2 μs. In the Earth's frame of reference, however, the muon's lifetime increases by the factor

$$\gamma = \frac{1}{\sqrt{1 - v^2/c^2}} = \frac{1}{\sqrt{1 - (0.998^2)\, c^2/c^2}} \approx 15.8,$$

i.e., the lifetime in the Earth's frame of reference is approximately 31 μs, the muon covers about 9500 m in this time. The same result is obtained if one moves into the muon's frame of reference. Here, the proper lifetime is still 2 μs, but a distance muon - Earth's surface of 9500 m contracts in the muon's frame of reference to

$$\Delta x' = \Delta x \sqrt{1 - \frac{v^2}{c^2}} = 9500 \sqrt{1 - \frac{v^2}{c^2}} = 9500 \sqrt{1 - \frac{(0.998^2)\, c^2}{c^2}} \approx 600\,\text{m}.$$

It is easy to check whether these predictions of the Special Theory of Relativity are correct. Suppose one observes at time $t = 0$ at 9000 m altitude 10^8 muons. According to the non-relativistic prediction, the muons need for the 9000 m to the Earth's surface the time

$$\frac{9000\,\text{m}}{0.998\,c} = \frac{9000\,\text{m}}{0.998 \cdot 300,000,000\,\text{m/s}} \approx 30\,\mu\text{s},$$

thus 15 times the average decay time τ. If one inserts this into the above formula (12.6), one obtains

$$N = 10^8 e^{-15} \approx 30.6,$$

i.e., of the original 100 million muons, only 31 would reach the Earth's surface. According to the Special Theory of Relativity, the Earth only has to cover 600 m in the muon's frame of reference, which takes about 2 μs $= 1\tau$. Thus, the number of muons expected at sea level is calculated to be

$$N = 10^8 e^{-1} \approx 3.68 \cdot 10^7.$$

Experiments have confirmed that the predictions of the Special Theory of Relativity are correct. In fact, approx. 36.8 million muons are observed on the Earth's surface. $\square$

Doppler Effect

As another example, we want to examine the so-called **Doppler effect**. The Doppler effect can be observed with any wave, it is responsible, for example, for the fact that the whistle of an approaching train is perceived as higher than that of a stationary train or one that is moving away. However, in this example we will limit ourselves to light waves in a vacuum. To do this, we first clarify what a wave is using simple means.

Remark 12.1. **MT: Oscillations and Waves**
A **wave** is, roughly speaking, an oscillation, that propagates through space. Which leads us to the question, what is an **oscillation**.

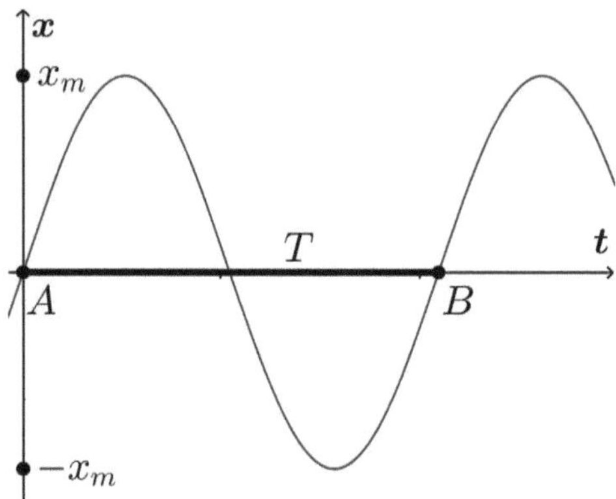

Figure 12.5.: Oscillation

An oscillation is a periodic (i.e., repeating) deflection of a particle from a resting position, e.g., the up and down of a boat on the water or the swinging of a clock pendulum. Oscillations thus refer to an object that is stimulated to oscillate around its rest point. The rest point itself does not move. Fig. 12.5 shows the *temporal* course of an oscillation with rest position zero and maximum deflection x_m.

At time $t = 0$, the particle is deflected from its rest position A, moves to the point x_m and then returns, crosses its rest position and moves to the point $-x_m$, turns there back, crosses its rest position again at time B etc. The time T, which passes until the particle has completed a full oscillation, is called **period**. The number of full oscillations per second is called the **frequency f** of the oscillation, which is calculated as

$$f = \frac{1}{T} \tag{12.7}$$

and is measured in Hertz (Hz). If the period of the oscillation is $T = 0.01\,$s, then the frequency is $f = 100\,$Hz.

A prerequisite for the creation of waves is that coupled oscillating elements are present, which can trigger wave movements in a chain reaction. Oscillations that trigger adjacent oscillations, which in turn trigger adjacent oscillations, etc., form a wave. The difference between oscillations and waves is that waves propagate, i.e., they cover distances, while oscillations occur locally. The famous stadium wave "La Ola" circles the field because many individual people move up and down, i.e., oscillate up and down. They do not leave their seats, but "La Ola" moves through the entire stadium. The coupled oscillating element in "La Ola" is the "willingness to participate". Because if some adjacent spectators remain seated, the wave subsides.

The graphical representation of a plane wave in Fig. 12.6 differs from that of an oscillation mainly in that the wave image is a *spatial* representation in the (x, y) coordinate system at a fixed point in time. Imagine that at each point on the x-axis there is a particle that performs an oscillation around the rest position $y = 0$ with the maximum alignment (**amplitude**) A_0. In Fig. 12.6 is a snapshot of a wave $A(x, t)$, which oscillates in the y-direction, at time $t = 0$. The wave has a propagation direction (arrow), which points in the positive x-direction.

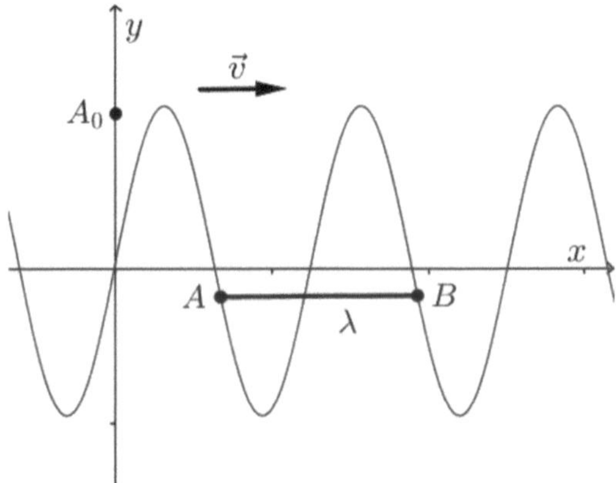

Figure 12.6.: Wave

The distance between two repetitions of the waveform (in the figure between A and B) is called **wavelength** λ. The wavelength indicates the periodicity of A on the x-axis. The magnitude of the **phase velocity** v, with which the wave propagates in the positive x-direction, is given by the wavelength λ divided by the time T, which is needed for a full oscillation

$$v = \frac{\lambda}{T} = \lambda f.$$

A plane wave is most easily described when the coordinate system is chosen so that one axis corresponds to its direction of propagation. In the directions perpendicular to the propagation no oscillation takes place. A (**harmonic homogeneous**) **plane wave** is defined by

$$A\left(x,t\right) = A_0 \sin\left(2\pi f\left(x/v - t\right) + \varphi\right), \tag{12.8}$$

where φ denotes the so-called **phase**. The phase indicates the wave shift from the zero point $(x = 0)$ at time $t = 0$. In Fig. 12.6, the phase of the drawn wave is zero.

In Eq. 2.13 on page 41 we have the *angular frequency* ω defined by

$$\omega = \frac{2\pi}{T}.$$

From (12.7) it follows that

$$\omega = \frac{2\pi}{T} = 2\pi f,$$

and we can write (12.8) as

$$A\left(x,t\right) = A_0 \sin\left(\omega\left(x/v - t\right) + \varphi\right). \tag{12.9}$$

We further define the so-called **wave number** k by

$$k = \frac{\omega}{v} = \frac{2\pi}{\lambda}$$

and obtain the usual representation of a plane wave in physics (with $\varphi = 0$):

$$A\left(x,t\right) = A_0 \sin\left(kx - \omega t\right) \ \text{ or } \ A\left(x,t\right) = A_0 \cos\left(kx - \omega t\right) \ \square \tag{12.10}$$

In Chap. 25 on gravitational waves, we will deal more intensively with the wave phenomenon.

Classical Doppler Effect

After these preliminary remarks, we now turn to the Doppler effect, which we initially treat classically, i.e. in the Newtonian world without relativity theory. Light, like other electromagnetic radiation, is subject to the so-called **wave-particle duality**, i.e., it appears in some experiments as a particle beam, in others as a wave phenomenon. We assume that a light wave with constant speed c and temporally unchanging frequency (color) f propagates in the positive x-direction. When two observers, who are moving relative to each other, look at this light beam, they see it in different colors, i.e., the perceived frequencies differ. This phenomenon is called **Doppler effect**, often simply referred to as **redshift** although this is only *one* possible manifestation of the Doppler effect, as we will see shortly.

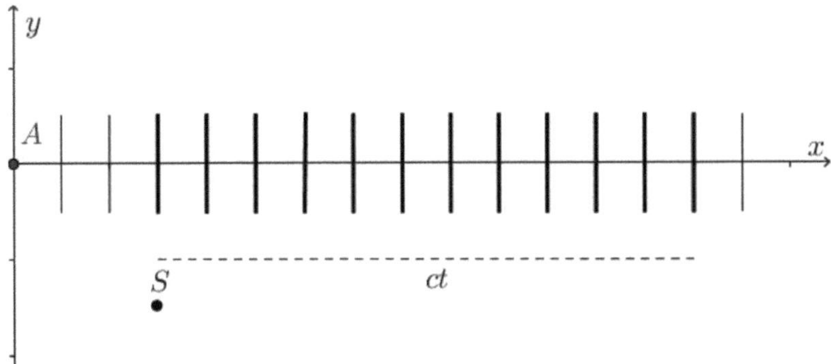

Figure 12.7.: Classical Doppler Effect-I

In Fig. 12.7, a light wave has been emitted from point A in the positive x direction. At time zero, an observer located at S has started to count the wave crests of the light wave. By the time t, the wave has covered the distance ct and the observer has seen the boldly marked wave crests pass by him. The bold wave crest on the far right was the first and the bold one on the far left was the last wave crest he counted during the time t. If the number of counted wave crests is equal to N, then the wave has the frequency from the perspective of S

$$f = \frac{N}{t}.$$

Now let's consider in Fig. 12.8 an observer S' moving relative to S, who is moving in the positive x-direction with the relative speed v.

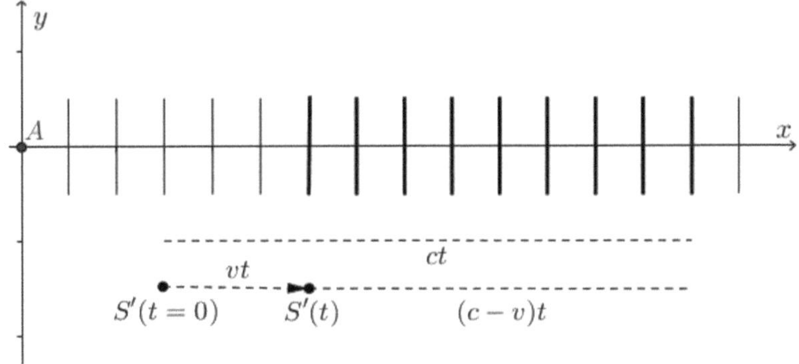

Figure 12.8.: Classical Doppler Effect II

The observer S' is at the position of S at time $t = 0$ and moves to the right by the distance vt in the time interval t. He counts the wave crests that pass him, which are again highlighted in bold in the figure. The number N'

he counts is less than that of S due to his movement in the direction of wave propagation. The ratio of N' to N is the same as that of the distances $(c - v)\,t$ to ct, so

$$\frac{N'}{N} = \frac{(c - v)\,t}{ct} = 1 - v/c.$$

From this, it follows for the frequencies

$$f' = \frac{N'}{t} = \frac{N}{t}(1 - v/c) = (1 - v/c)\,f. \tag{12.11}$$

f' is therefore smaller than f, the moving observer perceives the light at a lower frequency, i.e., for him, the color of the light is shifted towards the red end of the light spectrum. This is why we speak of *redshift*. However, if the observer S' moves towards the light source, v must be replaced by $-v$ in the above formula, and we get $f' > f$. This effect is called **blueshift**. Even in this classical consideration, it has become clear that the frequency of a wave is not an intrinsic property, but depends on the movement of the observer who measures the frequency.

Doppler Effect in Special Relativity

The formula (12.11) applies to speeds significantly below the speed of light. For astronomical applications (measurements of distance changes of objects, determination of planets in other solar systems), it is usually sufficient. But it is also used on Earth, for example in speed measurement via radar. For higher speeds, the formula must be adapted to the requirements of special relativity. According to time dilation, the time

$$t' = t \cdot \sqrt{1 - v^2/c^2}$$

is measured in the inertial system S', from which it follows for the frequencies

$$f' = \frac{N'}{t'} = \frac{N}{t\sqrt{1 - v^2/c^2}}(1 - v/c) = \frac{1 - v/c}{\sqrt{(1 - v/c)(1 + v/c)}}\,f$$

$$= \sqrt{\frac{1 - v/c}{1 + v/c}}\,f. \tag{12.12}$$

This relationship is called the **relativistic Doppler effect**. Also in this case, the formula "reverses" when the observer moves towards the light source or the light source approaches an observer, as the speed v must be replaced by $-v$. The Doppler effect means that when registering a light wave in space, its original frequency cannot be determined if the speed of the light source is unknown. Each observer measures a different frequency of the light wave, which is again a nice example of the "democracy" of inertial systems.

12.4. Clock Desynchronization

In this and the following sections, we want to derive further direct consequences from time dilation and length contraction. We start by calculating quantitatively how clocks in different reference systems desynchronize. We again consider the train example and imagine that in a train at positions A and B, which are at a distance L' from each other, two clocks are located, clock 2 at the beginning of the train and clock 1 at the end of the train, see Fig. 12.9. From the perspective of an accompanying observer M', who is exactly in the middle of A and B, these two clocks should be synchronized. The train passes a railway platform, on which an observer M stands, with the constant speed v to the right. M' ignites a flash of light when he is exactly level with M. In the system of the train S', the light spreads in both directions at the speed c and reaches both clocks at the same time after the light travel time $t' = L'/2c$. The question is whether M detects a time difference on the clocks and how large this is if applicable.

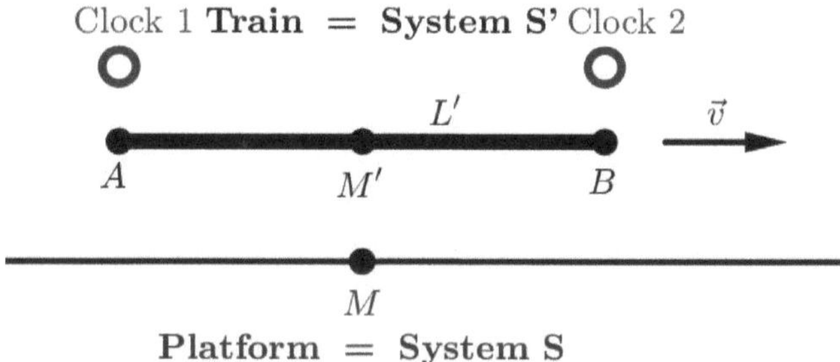

Figure 12.9.: Clock desynchronization

First, we note that the distance between the two clocks for the observer M is not L', but is due to the length contraction (12.5)

$$L = L' \sqrt{1 - \frac{v^2}{c^2}}. \tag{12.13}$$

From his point of view, the light also spreads in both directions at the speed c. However, clock 1 is approaching the light. In the time t_1, which the light needs to clock 1, it does not have to cover the distance $L/2$, but a path shortened by $v\,t_1$, so

$$c\,t_1 = \frac{L}{2} - v \cdot t_1.$$

Conversely, clock 2 is moving away from the light. The light has to cover not only the length $L/2$, but a path lengthened by $v\,t_2$ in the time t_2, so

$$c\,t_2 = \frac{L}{2} + v \cdot t_2.$$

Both equations are solved for t_1 and t_2 respectively

$$t_1 = \frac{L}{2\,(c+v)} \quad \text{and} \quad t_2 = \frac{L}{2\,(c-v)}$$

and subtracted from each other:

$$t_2 - t_1 = \frac{L}{2\,(c-v)} - \frac{L}{2\,(c+v)} = \frac{L\,(c+v) - L\,(c-v)}{2\,(c-v)\,(c+v)} = \frac{L}{2}\,\frac{2v}{c^2 - v^2} = \frac{Lv}{c^2 - v^2}$$

This is the time difference in system S. For the observer M, the light first reaches clock 1 and then clock 2, for him the two events are not simultaneous, in *his* system his clocks show a time difference of $\dfrac{Lv}{c^2 - v^2}$. Can the observer M also detect a time difference in system S'? First, he knows that according to the time dilation (12.4) the moving clocks run slower, i.e., the time difference $\Delta t' = t_2' - t_1'$ results in

$$
\begin{aligned}
\Delta t' &= (t_2 - t_1) \cdot \sqrt{1 - \frac{v^2}{c^2}} = \frac{Lv}{c^2 - v^2} \cdot \sqrt{1 - \frac{v^2}{c^2}} \\[2mm]
&= \frac{Lv}{c^2\left(1 - \dfrac{v^2}{c^2}\right)} \cdot \sqrt{1 - \frac{v^2}{c^2}} = \frac{Lv}{c^2} \cdot \frac{1}{\sqrt{1 - \dfrac{v^2}{c^2}}}.
\end{aligned}
$$

If we now insert the above relation (12.13)

$$L = L' \sqrt{1 - \frac{v^2}{c^2}}$$

between L and L', we get

$$\Delta t' = \frac{Lv}{c^2} \cdot \frac{1}{\sqrt{1 - \dfrac{v^2}{c^2}}} = \frac{L'v}{c^2}.$$

This time difference shows the discrepancy in the rate of the two clocks 1 and 2 in system S' as perceived by observer M. For observer M, the two clocks do not run at the same rate, rather they differ by the amount $L'v/c^2$, they are desynchronized. If two events are simultaneous and separated by the distance

L' for a stationary observer, then for another observer moving parallel to L' at speed v, the spatially leading event appears to be delayed by $L'v/c^2$ compared to the spatially trailing event. We illustrate these relationships again with an example.

Example 12.3.
A spaceship flies past Earth at a speed of $v = 0.8\,c$ to a planet eight light years away, whose clocks are synchronized with those on Earth. The clocks between Earth and the spaceship are synchronized and set to zero during the flyby.

1. How long does the spaceship's flight take from Earth's perspective?

   From Earth's perspective, the distance between Earth and the planet is eight light years
   ($1\,\text{light year} = 1\,\text{year} \cdot c$) , so the flight duration is

   $$\Delta t = \frac{\Delta x}{v} = \frac{8 \cdot c}{0.8 \cdot c} = 10\,\text{years}.$$

2. How much time passes in the moving system of the spaceship from Earth's perspective?

   Due to the (known on Earth) time dilation, a shorter flight time $\Delta t'$ is measured from Earth's perspective in the spaceship

   $$\Delta t' = \Delta t \cdot \sqrt{1 - \frac{v^2}{c^2}} = 10 \cdot \sqrt{1 - \frac{(0.8\,c)^2}{c^2}} = 10 \cdot 0.6 = 6\,\text{years}.$$

3. How large is the distance $\Delta x'$ Earth - Planet from the spaceship's perspective?

   Due to length contraction, we get

   $$\Delta x' = \Delta x \cdot \sqrt{1 - \frac{v^2}{c^2}} = 8 \cdot \sqrt{1 - \frac{(0.8\,c)^2}{c^2}} = 8 \cdot 0.6 = 4.8\,\text{light years}.$$

4. How long does it take for the spaceship, which is at rest in its system, until the planet flies past it?

   This is calculated as in 1. to

   $$\Delta t' = \frac{\Delta x'}{v} = \frac{4.8 \cdot c}{0.8 \cdot c} = 6\,\text{years}.$$

   As expected, the observers on Earth and the spaceship crew arrive at the same result for $\Delta t'$.

5. How much time has passed on the clocks on Earth from the perspective of the spaceship when the planet flies past?

From the perspective of the spaceship, the clocks on Earth run slower due to time dilation than its own. From the spaceship's perspective, there is a time duration of

$$\Delta t = \Delta t' \cdot \sqrt{1 - \frac{v^2}{c^2}} = 6 \cdot \sqrt{1 - \frac{(0.8\,c)^2}{c^2}} = 6 \cdot 0.6 = 3.6\,\text{years}.$$

6. How does a crew member explain the difference between the flight duration Δt, which she measures, and the one calculated on Earth?

On the spaceship, a clock reading of ten years was read off the planet during the flyby. A crew member explains the difference to the 3.6 years by the fact that the clocks in the Earth - Planet system are desynchronized, with the clock on the planet running ahead of the clocks on Earth (the spatially trailing system) by the amount

$$\frac{L \cdot v}{c^2} = \frac{8\,\text{Lichtjahre} \cdot 0.8 \cdot c}{c^2} = \frac{8 \cdot c \cdot 0.8 \cdot c}{c^2} = 8 \cdot 0.8 = 6.4\,\text{Jahre}.$$

If you add this number to his calculated flight time of 3.6 years, you also get ten years. $\square$

Twin Paradox

With example 12.3, we can also explain the so-called **Twin Paradox** and show that it is not a paradox at all. Let's imagine that the observer M', who has a twin M who stays on Earth, is sitting in the spaceship R_1. At the time of M' passing by the Earth, the two twins are the same age. The thought experiment is such that M' flies to the planet eight light years away and then flies back to Earth in a second spaceship R_2 at the same speed, i.e., we neglect for simplicity the procedure and time of decelerating the first rocket and accelerating the second rocket to the speed $0.8\,c$. Nevertheless, it is clear that M' changes his inertial system, because the spaceship R_1 flies at a constant speed always in the same direction away from the Earth, while the spaceship R_2 flies in the opposite direction. Let's now consider the times that have passed during the outward and return flights. As explained in point (1.) in example 12.3, the outward flight of M' takes ten years for his twin M on Earth. M' needs the same amount of time from M's perspective for the return flight, a total of twenty years. M', on the other hand, reads six years on his clock for the outward flight (see point (4.)), and he needs the same amount of time for the return flight, a total of twelve years. Therefore, when he returns to Earth, the age difference to his twin is eight years. M also comes to the same result, knowing that the aging $\Delta t'$ of his twin can be calculated from his aging Δt by time dilation:

$$\Delta t' = \Delta t \cdot \sqrt{1 - \frac{v^2}{c^2}} = 20 \cdot \sqrt{1 - \frac{(0.8\,c)^2}{c^2}} = 20 \cdot 0.6 = 12\,\text{years}$$

In this sense, there is agreement between the two twins and thus no paradox. The actual twin paradox is seen as the question of why M', for whom the Earth and the planet are moving, does not also experience a greater aging compared to his twin M. Well, this is because it is *not* a symmetrical situation. Because M' changes his reference system when he reaches the planet, while M is in his reference system Earth the whole time. But let's look at it more closely. When M' reaches the planet, according to point (6.), the local clocks show ten years, i.e., four years more than the flight time $\Delta t' = 6\,\text{years}$ measured in the spaceship R_1. When M' switches to the spaceship R_2, the local clocks are synchronized with those on the planet, causing a time jump of four years for M'. On the return journey, we then find exactly the same conditions as on the outward journey: The flight time in the spaceship R_2 is six years, while the clocks on Earth show four years more when the spaceship arrives. In total, this explains the age difference of eight years between the twins.

These different aging processes have been experimentally verified many times. For example, in the 1970s, two atomic clocks were synchronized and one of them was flown around the Earth in a commercial airplane. After landing, the clock readings of both clocks were compared, the moving clock showed a lower time according to the prediction of the Special Theory of Relativity than the one that remained on Earth.

12.5. Addition of Velocities

In the Special Theory of Relativity, the speed of light is the highest achievable speed. This upper limit has the consequence that speeds cannot simply be added. Let's imagine that in a spaceship flying away from Earth at $0.8\,c$, the second stage is ignited, which in turn flies away at a speed of $0.8\,c$ relative to the first stage, see Fig. 12.10. Relative to Earth, however, the speed of the second stage can be a maximum of c and *not* $0.8\,c + 0.8\,c = 1.6\,c$!

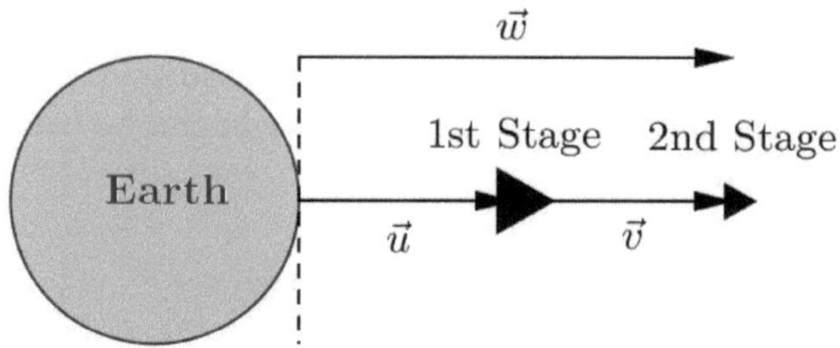

Figure 12.10.: Addition of velocities

We want to determine the speed w between the 2nd stage and the Earth. To do this, we imagine that there is a clock in the 2nd stage that "ticks" between two beats with a period T_2. How does an observer in the 1st stage see this clock ticking? First, the period for this observer is extended by the time dilation

$$T_2 \longrightarrow T_2 \cdot \gamma = \frac{T_2}{\sqrt{1 - v^2/c^2}}.$$

Since the 2nd stage moves away from the first at the speed v, and from the perspective of the first stage in the period $T_2\,\gamma$ by the path $T_2\,\gamma \cdot v$, for which a light signal from stage 1 to stage 2 takes the time $T_2\,\gamma \cdot v/c$, the total perceived period duration T_1 of the clock in stage 2 is

$$
\begin{aligned}
T_1 &= T_2 \cdot \gamma + \frac{T_2\,\gamma \cdot v}{c} = \frac{T_2}{\sqrt{1 - v^2/c^2}} + \frac{T_2\,v}{c\,\sqrt{1 - v^2/c^2}} \\
&= T_2\,\frac{1 + v/c}{\sqrt{1 - v^2/c^2}} = T_2\,\frac{\sqrt{1 + v/c}\,\sqrt{1 + v/c}}{\sqrt{(1 + v/c)\,(1 - v/c)}} \\
&= T_2\,\sqrt{\frac{1 + v/c}{1 - v/c}}.
\end{aligned}
$$

The period extension factor f_v observed in stage 1 for the clock in stage 2 is therefore

$$f_v = \sqrt{\frac{1 + v/c}{1 - v/c}},$$

which again confirms the relativistic Doppler formula (12.12). An observer on Earth measures a period for the signals of the clock in stage 2

$$T_3 = f_w T_2 = T_2\,\sqrt{\frac{1 + w/c}{1 - w/c}}.$$

If one considers stage 1 as a relay station that forwards signals coming from stage 2 without delay, an observer on Earth can also obtain the same period by

$$T_3 = f_u T_1 = f_u \cdot (f_v \cdot T_2),$$

i.e., it must hold

$$f_w = f_u \cdot f_v \Rightarrow \sqrt{\frac{1 + w/c}{1 - w/c}} = \sqrt{\frac{1 + u/c}{1 - u/c}} \cdot \sqrt{\frac{1 + v/c}{1 - v/c}}.$$

The last equation is squared and solved for w

$$\frac{1 + w/c}{1 - w/c} = \frac{1 + u/c}{1 - u/c} \cdot \frac{1 + v/c}{1 - v/c}$$

$$\Leftrightarrow \quad 1 + \frac{w}{c} = \frac{(1 + u/c)\,(1 + v/c)}{(1 - u/c)\,(1 - v/c)} - \frac{w}{c}\,\frac{(1 + u/c)\,(1 + v/c)}{(1 - u/c)\,(1 - v/c)}$$

$$\Leftrightarrow \quad \frac{w}{c} \cdot \left(1 + \frac{(1 + u/c)\,(1 + v/c)}{(1 - u/c)\,(1 - v/c)} \right) = \frac{(1 + u/c)\,(1 + v/c)}{(1 - u/c)\,(1 - v/c)} - 1$$

$$\Leftrightarrow \quad \frac{w}{c} \cdot \left(\frac{2\,(1 + uv/c^2)}{(1 - u/c)\,(1 - v/c)} \right) = \frac{2/c\,(u + v)}{(1 - u/c)\,(1 - v/c)}$$

$$\Leftrightarrow \quad w \cdot (1 + uv/c^2) = u + v,$$

from which the so-called **addition theorem** of Special Relativity Theory follows:

$$w = \frac{u + v}{1 + u \cdot v/c^2}. \tag{12.14}$$

If u and v are small compared to the speed of light, then the factor $u \cdot v/c^2$ is close to zero and with $w \approx u + v$ the addition formula for velocities in the Newtonian world is obtained. Let's consider a few special cases.

Example 12.4.

- If $u = v = 0$, then $w = 0$.

- If $u = v = c$, then

$$w = \frac{u + v}{1 + u \cdot v/c^2} = \frac{c + c}{1 + c \cdot c/c^2} = c.$$

- In our last example, $u = v = 0.8\,c$, so

$$w = \frac{u + v}{1 + u \cdot v/c^2} = \frac{0.8\,c + 0.8\,c}{1 + 0.8\,c \cdot 0.8\,c/c^2} = \frac{1.6\,c}{1 + 0.64} = 0.975\,c. \ \square$$

12.6. Momentum, Mass, Energy

After we have derived some consequences of the Einstein's postulates for the quantities time, space and velocity in the last sections, we will show in this section which modifications to the classical concepts of momentum, mass and energy must be made by the laws of Special Relativity Theory. In Newtonian mechanics, the momentum of a particle $\vec{p}$ is defined as the product $\vec{p} = m \cdot \vec{u}$ of the mass m and the velocity $\vec{u}$. We will show that this equation is only an approximate solution of a more general definition. To do this, we conduct the following thought experiment. We imagine a ball of mass m_0 being thrown at a certain speed u_0 against a wall, with the distance between the thrower and the wall being l, see Fig. 12.11. Thrower, ball and wall are on a train moving at speed v past an observer B on the railway platform.

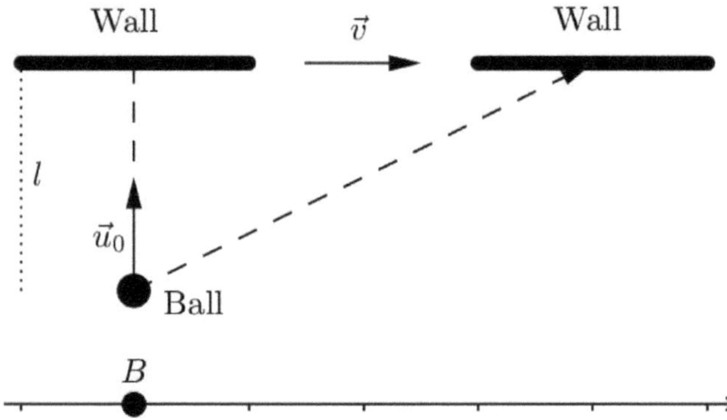

Figure 12.11.: Relativistic Momentum

When the thrower in the train and the observer are at the same height, the ball is thrown against the wall. From the thrower's point of view, the ball flies at the speed u_0 directly towards the wall. For the observer B on the railway embankment, the dashed arrow corresponds to the flight path of the ball. Regardless of the relative speed between the observer and the train, the wall always receives the same momentum perpendicular to the direction of the train's movement, which can be determined, for example, with a momentum measuring device attached to the wall. Let S be the inertial system of the train and S' the inertial system of the observer B. In the inertial system of the train, the magnitude of the momentum p is:

$$p = m_0 \cdot u_0 = m_0 \frac{l}{t},$$

since $u_0 = distance/time = l/t$, and similarly for the momentum p' in the inertial system of the observer:

$$p' = m' \cdot u' = m' \frac{l'}{t'}$$

Now, we know from section 12.3, that lengths perpendicular to the direction of motion do not differ, i.e., it is $l' = l$. Furthermore, t' is greater than t, according to time dilation it is $t' = \gamma \cdot t$. We thus obtain

$$p' = m' \cdot u' = m' \frac{l}{\gamma \cdot t}. \tag{12.15}$$

If we now use that the two impulses are equal $p = p'$, we get

$$m_0 \frac{l}{t} = m' \frac{l}{\gamma \cdot t}$$

$$\Rightarrow \quad m' = \gamma \cdot m_0 = \frac{m_0}{\sqrt{1 - v^2/c^2}}. \tag{12.16}$$

Masses thus appear in other inertial systems to be larger by the γ factor than in the rest system of the particle. If the relative velocity $v = 0$, no mass increase occurs: $m' = m_0$. At $v = c$ the mass m' would be infinite. Thus, no body with rest mass $m_0 > 0$ can be brought to the speed of light, the force required would be infinitely large. This is different for particles without rest mass such as **photons** (light particles). These can (and must) move at the speed of light c in a vacuum.

Let's look at the situation again from a different perspective and assume that the mass m_0 of a particle (the ball) is invariant, i.e., the same in all inertial systems: $m' = m_0$. Then the classical law of conservation of momentum no longer applies. Because from (12.15) we can derive that for the velocity u' in the y-direction measured by the observer on the platform

$$u' = \frac{l}{\gamma \cdot t} = \frac{u_0}{\gamma} = u_0 \sqrt{1 - v^2/c^2}$$

applies, i.e., the observer sees the ball flying against the wall at a lower speed, from which also

$$p = m_0 \cdot u_0 \neq m_0 \cdot u' = p'$$

follows. We now *define* the **relativistic momentum** $\vec{P}$ of a particle with mass m_0, moving with velocity $\vec{u}$, by

$$\vec{P} = \frac{m_0 \, \vec{u}}{\sqrt{1 - u^2/c^2}} \rightarrow \frac{m_0}{\sqrt{1 - u^2/c^2}} \begin{pmatrix} u_x \\ u_y \\ u_z \end{pmatrix} = \begin{pmatrix} P_x \\ P_y \\ P_z \end{pmatrix}, \tag{12.17}$$

where as usual

$$u^2 = |\vec{u}|^2 = u_x^2 + u_y^2 + u_z^2$$

applies. The relativistic momentum thus grows beyond all limits when the velocity u approaches the speed of light. If the velocity u is small compared to the speed of light, so $u^2/c^2 \approx 0$ and the relativistic momentum goes - as desired - into the classical: $\vec{P} \approx \vec{p}$. If one interprets in the above definition the quantity

$$\frac{m_0}{\sqrt{1 - u^2/c^2}}$$

as **relativistic mass** m_r, then one can also write the relativistic momentum as

$$\vec{P} = m_r\,\vec{u} = \frac{m_0\,\vec{u}}{\sqrt{1 - u^2/c^2}}.$$

We want to check whether with this definition the momenta of our example are invariant. To do this, we check the y-components of the relativistic momenta in the system S of the moving train and in the system S' of the observer on the embankment. For the thrower in the train the velocity of the ball, which is in his system S only moves in the y-direction, equal to

$$\vec{u} \rightarrow \begin{pmatrix} 0 \\ u_0 \\ 0 \end{pmatrix}$$

and thus $|\vec{u}|^2 = u_0^2$, so for the y-component of its momentum

$$P_y = \frac{m_0 \cdot u_0}{\sqrt{1 - u_0^2/c^2}}.$$

For the observer on the platform, the ball moves both in the x-direction (with speed v) and in the y-direction (with speed $u' = u_0/\gamma$). In his system S', the ball has a speed

$$\vec{w}' \rightarrow \begin{pmatrix} v \\ u_0/\gamma \\ 0 \end{pmatrix} = \begin{pmatrix} v \\ u_0\sqrt{1 - v^2/c^2} \\ 0 \end{pmatrix},$$

from which

$$(w')^2 = v^2 + u_0^2\left(1 - v^2/c^2\right)$$

follows, i.e., the y-component of its momentum is

$$P_y' = \frac{m_0 \cdot u_0/\gamma}{\sqrt{1 - (w')^2/c^2}}$$

$$= \frac{m_0 \cdot u_0/\gamma}{\sqrt{1 - \dfrac{v^2 + u_0^2\left(1 - v^2/c^2\right)}{c^2}}}$$

$$= \frac{m_0 \cdot u_0/\gamma}{\sqrt{1 - \dfrac{v^2}{c^2} - \dfrac{u_0^2}{c^2} + \dfrac{u_0^2 v^2}{c^4}}}$$

$$= \frac{m_0 \cdot u_0/\gamma}{\sqrt{\left(1 - \dfrac{v^2}{c^2}\right)\left(1 - \dfrac{u_0^2}{c^2}\right)}}$$

$$= \frac{m_0 \cdot u_0/\gamma}{\sqrt{1 - v^2/c^2}\,\sqrt{1 - u_0^2/c^2}}$$

$$= \frac{m_0 \cdot u_0/\gamma}{\dfrac{1}{\gamma}\sqrt{1 - u_0^2/c^2}}$$

$$= \frac{m_0 \cdot u_0}{\sqrt{1 - u_0^2/c^2}}$$

$$= P_y.$$

So the momentum is conserved, i.e., it is the same in both inertial systems.

The Most Famous Equation in Physics

We now derive the most famous equation in physics

$$E = mc^2.$$

There are two ways to do this, both of which yield the same result. The first, mathematically simpler method seems to be only an approximate solution at first, but is nevertheless exact, as we will see later. First, we recall that the difference between two values of a function f can be estimated by the differential df, i.e., for two values x and x_0, which are close to each other,

$$\Delta f = f(x) - f(x_0) \approx f'(x_0)(x - x_0) = df(x_0),$$

and the "approximately equal" sign $\approx$ turns into a real equality sign when we form the limit $\lim\limits_{x \to x_0}$ (see Eq. 4.8 on page 59). We now specifically consider the function

$$f(x) = \frac{1}{\sqrt{1 - x}}$$

and calculate its derivative using the chain rule 4.22 on page 70

$$
f'(x) \;=\; \left(\frac{1}{\sqrt{1-x}}\right)' = \left((1-x)^{-1/2}\right)' = -\frac{1}{2}(1-x)^{-3/2}(-1)
$$

$$
= \frac{1}{2(1-x)^{3/2}} = \frac{1}{2\sqrt{(1-x)^3}},
$$

i.e., we obtain $f'(0) = \dfrac{1}{2}$ and thus for small x-values

$$
f(x) - f(0) = \frac{1}{\sqrt{1-x}} - 1 \approx \frac{1}{2}x \implies \frac{1}{\sqrt{1-x}} \approx 1 + \frac{1}{2}x.
$$

We now apply this approximation formula for $x = \dfrac{v^2}{c^2}$, i.e., we assume that the speed v is very small compared to the speed of light c ($v \ll c$). We thus obtain as an approximate solution

$$
\frac{1}{\sqrt{1 - v^2/c^2}} \approx 1 + \frac{1}{2}\frac{v^2}{c^2}.
$$

Substituting this into the formula for mass increase (12.16) yields

$$
m' = \frac{m_0}{\sqrt{1 - v^2/c^2}} \approx m_0\left(1 + \frac{1}{2}\frac{v^2}{c^2}\right) = m_0 + \frac{1}{2}\frac{m_0\,v^2}{c^2}.
$$

Multiplying this equation on both sides by c^2, we get

$$
m'c^2 \approx m_0\,c^2 + \frac{1}{2}m_0\,v^2.
$$

The last expression is the kinetic energy of a particle with mass m_0 and speed v (see Eq. 4.2 on page 50). The quantity $m'c^2$ is called the **total energy** and m_0c^2 is the **rest energy**. If we denote the total energy as usual by E, then for small speeds v we have

$$
E \approx m_0\,c^2 + \frac{1}{2}m_0\,v^2.
$$

For $v = 0$ the equation is exact, and for a stationary particle with rest mass m_0 we obtain Einstein's most famous equation:

$$
E = m_0\,c^2
$$

The exact calculation requires a bit more mathematics. For simplicity, we consider here the one-dimensional case, i.e., we assume that only the x-component of the particle's speed is different from zero, i.e.

$$\vec{v} \to \begin{pmatrix} v \\ 0 \\ 0 \end{pmatrix},$$

and that the force $\vec{F}$ only acts in the x-direction, i.e.

$$\vec{F} \to \begin{pmatrix} F \\ 0 \\ 0 \end{pmatrix},$$

from which

$$\vec{F} \cdot d\vec{r} = \begin{pmatrix} F \\ 0 \\ 0 \end{pmatrix} \cdot \begin{pmatrix} dx \\ dy \\ dz \end{pmatrix} = F\,dx + 0 \cdot dy + 0 \cdot dz = F\,dx$$

follows. In Newtonian mechanics, the force acting on a particle with mass m_0 is defined as the time change of the (classical) momentum (see Eq. 3.1 on page 44):

$$F = \frac{dp}{dt}$$

Furthermore, the work done by the force on the particle is equal to the change in kinetic energy. Similarly, in relativistic mechanics, the force can be defined as the time change of the relativistic momentum and the work as the change in relativistic energy (see also Eq. 4.11 on page 62):

$$E_{kin} = \int_0^x \vec{F} \cdot d\vec{r} = \int_0^x F\,dx = \int_0^x \frac{dP}{dt}\,dx = \int_0^v \frac{dx}{dt}\,dP = \int_0^v u\,dP$$

$$\overset{(12.17)}{=} \int_0^v u\,d\left(\frac{m_0\,u}{\sqrt{1 - u^2/c^2}}\right),$$

where we have used the relationship $u = dx/dt$ and the substitution rule for integrals (see formula 6.21 on page 110). The derivative of the relativistic momentum P with respect to the velocity u is obtained using the chain and quotient rules analogous to above

$$d\left(\frac{m_0\,u}{\sqrt{1 - u^2/c^2}}\right)\bigg/du = \frac{m_0\sqrt{1 - u^2/c^2} - m_0\,u\left(-2u/c^2\right)\dfrac{1}{2}\dfrac{1}{\sqrt{1 - u^2/c^2}}}{\left(\sqrt{1 - u^2/c^2}\right)^2}$$

$$= \frac{m_0 \left(1 - u^2/c^2\right) + m_0 \, u^2/c^2}{\left(\sqrt{1 - u^2/c^2}\right)^3}$$

$$= \frac{m_0}{\left(\sqrt{1 - u^2/c^2}\right)^3}.$$

Substituting this into the last integral yields

$$E_{kin} = \int_0^v u \, d\left(\frac{m_0 \, u}{\sqrt{1 - u^2/c^2}}\right) = \int_0^v \frac{m_0 \, u}{\left(\sqrt{1 - u^2/c^2}\right)^3} \, du.$$

To calculate this integral, we apply the substitution method for integrals and set $w = u^2/c^2$, then $dw/du = 2u/c^2$ and $du/dw = c^2/2u$. Note that with this substitution the limits of integration also change $(0 \to 0, v \to v^2/c^2)$, so we get

$$E_{kin} = \int_0^v \frac{m_0 \, u}{\left(\sqrt{1 - u^2/c^2}\right)^3} \, du = \int_0^{v^2/c^2} \frac{m_0 \, u}{\left(\sqrt{1 - w}\right)^3} \frac{du}{dw} \, dw$$

$$= \int_0^{v^2/c^2} \frac{m_0 \, u}{\left(\sqrt{1 - w}\right)^3} \frac{c^2}{2u} \, dw = \frac{m_0 \, c^2}{2} \int_0^{v^2/c^2} \frac{1}{\left(\sqrt{1 - w}\right)^3} \, dw.$$

The integrand

$$\frac{1}{\left(\sqrt{1 - w}\right)^3} = (1 - w)^{-3/2}$$

has the antiderivative

$$2 \left(1 - w\right)^{-1/2}.$$

This leads to

$$E_{kin} = \frac{m_0 \, c^2}{2} \int_0^{v^2/c^2} \frac{1}{\left(\sqrt{1 - w}\right)^3} \, dw$$

$$= \frac{m_0 \, c^2}{2} \left[\frac{2}{\sqrt{1 - w}}\right]_0^{v^2/c^2}$$

$$= m_0 \, c^2 \left(\frac{1}{\sqrt{1 - v^2/c^2}} - 1\right)$$

$$= \frac{m_0 \, c^2}{\sqrt{1 - v^2/c^2}} - m_0 \, c^2.$$

This expression for the kinetic energy consists of two terms. The second is independent of the velocity v of the particle and is the rest energy $E_0 = m_0 \, c^2$

introduced above. The **total relativistic energy** (see above) is the sum of kinetic energy and rest energy

$$E = E_{kin} + m_0\, c^2 = \frac{m_0\, c^2}{\sqrt{1 - v^2/c^2}} = m_r\, c^2. \tag{12.18}$$

The total relativistic energy of a particle corresponds to the work, done by a force on the particle. This work increases the energy of the resting particle from $m_0\, c^2$ to $m_r\, c^2$, where m_r is the relativistic mass. The exact expression for the relativistic kinetic energy

$$E_{kin} = \frac{m_0\, c^2}{\sqrt{1 - v^2/c^2}} - m_0\, c^2 \tag{12.19}$$

initially has little resemblance to the classical size

$$E_{kin} = \frac{1}{2}\, mv^2.$$

However, if one applies the approximation used above for $v \ll c$

$$\frac{1}{\sqrt{1 - v^2/c^2}} \approx 1 + \frac{1}{2}\frac{v^2}{c^2}$$

it follows that

$$E_{kin} = \frac{m_0\, c^2}{\sqrt{1 - v^2/c^2}} - m_0\, c^2 \approx m_0\, c^2 \left(1 + \frac{1}{2}\frac{v^2}{c^2}\right) - m_0\, c^2 = \frac{1}{2}\, m_0\, v^2.$$

An important interpretation of Eq. (12.18) is that the work/energy invested in acceleration not only increases speed, but also causes an increase in mass. And the relativistic mass m_r becomes larger the closer the particle's speed approaches the speed of light. In this sense, in relativistic mechanics, energy conservation is synonymous with mass conservation, mass and energy are two aspects of the same thing. This relationship is also called **mass-energy equivalence**.

Example 12.5.

- Even in tiny masses, due to the proportionality factor c^2 between rest mass and rest energy, there are huge amounts of energy. If it were possible to completely convert one gram of dust into energy, then the result would be

$$E = m_0\, c^2 = 0.001\ \text{kg·}9\text{·}10^{16}\ \text{m}^2/\text{s}^2 = 9\text{·}10^{13}\ \text{Wattsek.} = 25,000,000\ \text{kWh},$$

which is approximately the annual consumption of 5000 households.

- Another important example of mass-energy equivalence, even vital for us humans, is the energy radiation of the Sun. Inside the Sun, under high pressure and high temperatures, helium nuclei are produced from hydrogen protons by fusion. Each helium nucleus weighs slightly less than the building blocks from which it is formed. This minimal mass difference is converted into energy according to the formula $E = mc^2$ and radiated.

- And finally, an example from everyday life. When you wind a mechanical clock, you add energy to it. This energy supply only tensions a spring and does not set the clock in motion. This spring tension increases the mass of the clock, i.e., *wound clocks weigh more than unwound ones!* $\square$

Relativistic Energy Theorem

If one multiplies Eq. (12.16) by c^2, one obtains

$$mc^2 = \frac{m_0\,c^2}{\sqrt{1 - v^2/c^2}}$$

and from this by squaring

$$m^2 c^4 = \frac{m_0^2\,c^4}{1 - v^2/c^2}$$
$$\Rightarrow \quad m^2 c^4 \left(1 - v^2/c^2\right) = m_0^2\,c^4$$
$$\Rightarrow \quad m^2 c^4 - m^2 v^2 c^2 = m_0^2\,c^4.$$

Further noting that

$$\begin{aligned}
m^2 c^4 &= E^2 \\
m^2 v^2 &= \gamma^2 m_0^2\,v^2 = P^2 \\
m_0^2\,c^4 &= E_0^2
\end{aligned}$$

the equation

$$E^2 - P^2 c^2 = E_0^2 \tag{12.20}$$

follows, known as the **relativistic energy theorem**. All observers, regardless of the inertial system they are in, always measure the same rest energy E_0 for the same rest mass m_0. Therefore, the left side of the last equation must also be a quantity that must take the same value in all inertial systems. It is said that the difference $E^2 - P^2 c^2$ is *invariant* to a transformation from one inertial system to another. We will deal more intensively with such invariant quantities in the following sections.

If a particle has zero rest mass, it follows from (12.20)

$$E^2 - P^2 c^2 = 0 \Longrightarrow P = E/c, \tag{12.21}$$

i.e., even the massless photons have a momentum P and can thus exert a so-called **radiation pressure** on other particles. The momentum of photons is directly proportional to the energy and thus to the frequency of the light, as Einstein showed in one of his famous works ("light quantum hypothesis"). The radiation pressure of the photons ensures, for example, that asteroids with their own rotation receive a "push" on their orbit around the Sun: The side heated and illuminated by the Sun rotates "back" or "forward" depending on the direction of rotation and radiates the received heat energy back there. According to Newton's third law ("action equals reaction"), the radiation pressure of these photons exerts a thrust on the asteroid, which changes the asteroid's orbit (so-called **Yarkovsky effect**).

12.7. Spacetime Intervals

In this section, we want to start dealing with the question of what spacetime means and how to identify further (geometric) quantities that are observer-independent. A first such quantity, namely the speed of light c in a vacuum, is set as a postulate of the Special Theory of Relativity. Another, namely $E^2 - P^2 c^2$, we derived in section 12.6.

In the last sections, we also showed that observers in different inertial systems perceive time periods and distances differently, i.e., in the Special Theory of Relativity, unlike in Newtonian physics, there are no independent location or time specifications anymore. The so-called **spacetime**, which connects space and time, replaces the separate constructs of space and time. Specifically, we define spacetime as the set of all possible events. Since an event is always described by space *and* time specifications, four quantities (t, x, y, z) are needed to define an event A, where t is the time coordinate and x, y, z are the space coordinates in an inertial system S. In another inertial system S', the same event A is described with (generally different) coordinates (t', x', y', z'). Because of these four required specifications, spacetime is also referred to as four-dimensional or time as the fourth dimension. In a later chapter, we will delve deeper into the structures of spacetime and show, for example, that spacetime can be represented as a set of four-dimensional vectors.

Here we want to determine another invariant (reference system-independent) quantity of spacetime and for this purpose, we do the following thought experiment. A train passes a railway embankment at a constant speed v to the right, see Fig. 12.12.

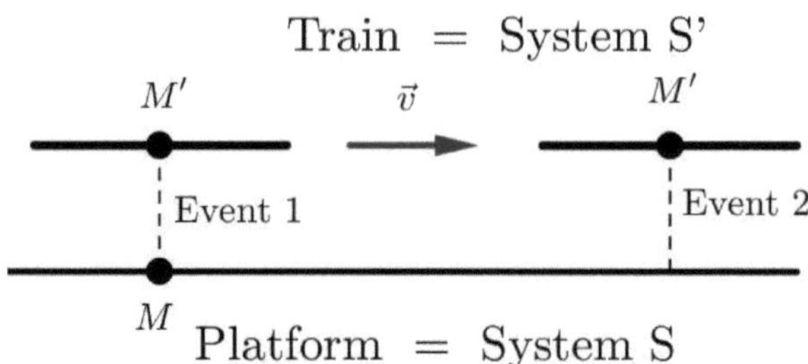

Figure 12.12.: Spacetime Interval

The event 1 is defined by the fact that the observer M' in the middle of the train (reference system S') encounters an observer M on the railway platform (reference system S). The event 2 happens from the perspective of the observer M' after a period of time $\Delta t'$. Since from the perspective of M, the observer M' is moving to the right at speed v and he knows that $\Delta t'$ is the proper time of M', he reads the larger time interval

$$\Delta t = \gamma \Delta t' = \frac{\Delta t'}{\sqrt{1 - v^2/c^2}} \Rightarrow \Delta t' = \Delta t \sqrt{1 - v^2/c^2}$$

on his clock. The last equation is squared

$$\Delta t'^{\,2} = \Delta t^2 \left(1 - v^2/c^2\right) = \Delta t^2 - \Delta t^2 \cdot v^2/c^2.$$

Since the speed $v = \Delta x/\Delta t$, where Δx denotes the spatial distance between event 1 and event 2 from the perspective of M, it follows that $\Delta t = \Delta x/v$. Substituting this gives

$$\Delta t'^{\,2} = \Delta t^2 - \Delta t^2 v^2/c^2 = \Delta t^2 - \Delta x^2/c^2,$$

and from this, after multiplying by $-c^2$, we get

$$-c^2 \Delta t'^{\,2} = \Delta x^2 - c^2 \Delta t^2.$$

If we consider that for the observer M' the two events occur at the same place, so the spatial distance in S' is zero, $\Delta x' = 0$, then the last equation can also be written as

$$\Delta x'^{\,2} - c^2 \Delta t'^{\,2} = \Delta x^2 - c^2 \Delta t^2$$

The expression on both sides of the equation is called the **square of the spacetime interval**, so $\Delta x^2 - c^2 \Delta t^2$ from the perspective of M and $\Delta x'^{\,2} -$

$c^2 \Delta t'^{\,2}$ from the perspective of M'. Both quantities are identical, although M and M' each measure completely different spatial and temporal distances. We have assumed in our thought experiment that S and S' move along the same line. If we generalize the result to arbitrary spatial directions, we must use the square spatial distance in three dimensions between two events, which according to Eq. 2.14 on page 42 is $\Delta x^2 + \Delta y^2 + \Delta z^2$. This means that we obtain the general expression for the square of the spacetime interval, which is identical in all inertial systems, by

$$\Delta x^2 + \Delta y^2 + \Delta z^2 - c^2 \Delta t^2.$$

Note that the expression $c \cdot t$ has the dimension of a length and can therefore be considered as an equal, time-representing fourth dimension alongside the three spatial dimensions x, y, z. However, the minus sign before the quantity $c \cdot t$ shows us that we are not measuring "normal" intervals (normal, in the sense of a (Euclidean) extension from three to four dimensions, would be the square of the distance: $x^2 + y^2 + z^2 + c^2 t^2$. In spacetime, (invariant) distances are therefore measured differently than in space. And since the way distances are measured determines the geometry of a space, the geometry of spacetime is not Euclidean. We want to examine this in a little more detail and take a closer look at the square of the spacetime interval Δs^2

$$\Delta s^2 = \underbrace{\Delta x^2 + \Delta y^2 + \Delta z^2}_{\substack{\text{Square of the} \\ \text{spatial distance}}} - \underbrace{c^2 \Delta t^2}_{\substack{\text{Square of the distance} \\ \text{that light travels in time } \Delta t}}.$$

The symbol Δs^2 is just a *designation*, one should *not* assume that it always represents a (positive) square number. On the contrary, in most interesting cases, the expression $\Delta x^2 + \Delta y^2 + \Delta z^2 - c^2 \Delta t^2$ is negative! If the time and space intervals are infinitesimally small, then we write for the infinitesimal spacetime interval

$$ds^2 = dx^2 + dy^2 + dz^2 - c^2 dt^2. \tag{12.22}$$

We can distinguish three cases:

1. A beam of light is emitted at location 1 (event 1) and received at location 2 (event 2). Then the spatial distance $\sqrt{x^2 + y^2 + z^2}$ is the same as the light path $c \cdot t$, so it holds

$$\Delta s^2 = 0. \tag{12.23}$$

   Two events that can be connected by light rays, are called **light-like** to each other and have the spacetime interval zero! This is a first major difference to Euclidean distance measurement. There, distances only have the value zero when the points coincide.

2. A beam of light is emitted at location 1 (event 1). The event 2 takes place at location 2, *after* the beam of light has passed location 2. The path of the light is therefore greater than the spatial distance between the two events

$$c \cdot t > \sqrt{x^2 + y^2 + z^2} \text{ i.e. } \Delta s^2 < 0. \tag{12.24}$$

Two events, whose spacetime interval is less than zero, are called **time-like** to each other. Since the spatial distance between the two events is less than the light path, event 2 can be influenced by event 1, e.g. by the beam of light, which is received at location 2, triggering a mechanism that initiates event 2. If material particles are to move from event 1 (start at location 1) to event 2 (arrival at location 2), then the spacetime interval between the two events *must be time-like*.

In the rest system of the material particles, the two events happen one after the other at the same location, because from the particles' point of view, location 1 flies past the particle (event 1) and after a period of time location 2 (event 2). In the inertial system of the particles, the spatial distance between the two events is zero $\sqrt{x^2 + y^2 + z^2} = 0$, i.e., it is

$$\Delta s^2 = -c^2 t^2.$$

The square of the spacetime interval becomes maximally negative in the rest system, namely $-c^2 t^2$. Since a clock, which is carried along in the rest system, always measures the proper time t_E, the maximum (negative) square spacetime distance is

$$\Delta s^2 = -c^2 t_E^2. \tag{12.25}$$

This also sounds unusual. When you rest at a location, you are moving (even maximally) through spacetime.

The above expression provides another way to define the proper time:

$$\begin{aligned}
t_E^2 &= -\frac{\Delta s^2}{c^2} = -\frac{x^2 + y^2 + z^2 - c^2 t^2}{c^2} = t^2 - \frac{1}{c^2}\left(x^2 + y^2 + z^2\right) \\
&= t^2 \left(1 - \frac{1}{c^2}\frac{x^2 + y^2 + z^2}{t^2}\right),
\end{aligned}$$

where (t, x, y, z) can be any inertial system. If v is the relative speed between the two reference systems, then

$$v = \frac{\sqrt{x^2 + y^2 + z^2}}{t},$$

substituting this into the last equation yields

$$t_E^2 = t^2 \left(1 - \frac{1}{c^2} \frac{x^2 + y^2 + z^2}{t^2} \right) = t^2 \left(1 - \frac{v^2}{c^2} \right),$$

thus

$$t_E = t \sqrt{1 - \frac{v^2}{c^2}},$$

which is nothing other than the time dilation (12.4).

3. A beam of light is sent from location 1 (event 1). Event 2 takes place at location 2, *before* the beam of light has passed location 2. Thus, the path of the light is smaller than the spatial distance between the two events

$$c \cdot t < \sqrt{x^2 + y^2 + z^2} \ i.e. \ \Delta s^2 > 0. \tag{12.26}$$

Two events, whose spacetime interval is greater than zero, are called **space-like** to each other. Since the spatial distance between the two events is greater than the path of light, event 2 cannot be influenced by event 1, because a signal from event 1 to event 2 would have to move faster than the speed of light.

In the extreme case for an observer the two spatially separated events occur simultaneously, i.e. $c^2 t^2 = 0$, so the spacetime interval

$$\Delta s^2 = x^2 + y^2 + z^2$$

becomes maximally positive and corresponds to the proper length. For other observers, although simultaneity is destroyed, for some event 1 occurs before event 2, for others it is exactly the opposite, but all measure the same spacetime interval.

Since the spacetime interval is an invariant quantity, all observers always agree on whether two events are light-, time- or spacelike to each other. Thus, spacetime itself can be divided into three regions without overlap, which we will illustrate a bit more precisely in Chap. 13.

To conclude this section, we calculate a numerical example for the spacetime intervals.

Example 12.6.
We refer to the events from the twin paradox example (see Sect. 12.4). Event
1 is the start of twin M' from Earth, event 2 is the arrival on the planet. The
distance between Earth and planet is eight light years from Earth's perspective,
the speed of the rocket is 0.8 c. The spatial coordinate system is arranged so
that the x-axis represents the direct connection between Earth and planet, i.e.,
the y and z distances between the two events are zero.

1. From the perspective of twin M on Earth, the temporal distance between
   the two events is

$$t = x/v = 8\,Lj/0.8\,c = 10\,\text{years}.$$

   This calculates the square of the spacetime interval to

$$\Delta S^2 = x^2 - c^2 t^2 = 64\,(\text{Lj})^2 - c^2 \cdot 100\,\text{years} = -36\,(\text{Lj})^2.$$

2. From the perspective of the flying twin M', the spatial distance between
   the two events is zero, $x' = 0$, because the astronaut considers himself
   at rest. The distance $\Delta x'$ Earth - Planet is calculated with the length
   contraction from the spaceship's perspective to

$$\Delta x' = x \cdot \sqrt{1 - \frac{v^2}{c^2}} = 8 \cdot \sqrt{1 - \frac{(0.8\,c)^2}{c^2}} = 8 \cdot 0.6 = 4.8\,\text{lightyears}.$$

   From this the temporal distance t' between both events is derived

$$t' = \frac{\Delta x'}{v} = \frac{4.8 \cdot c}{0.8 \cdot c} = 6\,\text{years}.$$

   In total, twin M' calculates the following spacetime interval

$$\Delta S'^2 = x'^2 - c^2 t'^2 = 0 - c^2 \cdot 36\,\text{years} = -36\,(\text{Lj})^2.$$

Although both twins determine very different spatial and temporal distances,
they agree on the spacetime interval. $\square$

13. The Geometry of Spacetime

In this chapter, we want to investigate the geometric properties of spacetime in more detail. This includes rules on how to transform physical quantities from one inertial system to another, in which units relativistic phenomena can be described most easily, which optical aids are available for illustrating the relationships, etc.

13.1. Lorentz Transformation

In physics, one is interested in formulating the laws in such a way that they are independent of coordinate systems. This is particularly true in Special (and later also in General) Relativity, where each observer carries his own reference system, in which he is at rest. It would be unfortunate if, on the one hand, according to Einstein's postulates, the physical laws are the same in every inertial system, but on the other hand, one would not be able to formulate them in such a way that they take the same form in any arbitrarily chosen inertial system. In section 12.1 we have shown that the laws of Newtonian mechanics can be formulated unchanged if we move from one coordinate system (t, x, y, z) to another (t', x', y', z') and the transition is made with the Galilean transformation (see formula 12.1 on page 227). We want to generalize this invariance of the physical laws to Special Relativity. For this, we need other transformation rules than the Galilean ones, which we have already shown by means of time dilation, length contraction and velocity addition that they do not apply in Special Relativity. However, we expect that the new transformation rules to be found for small relative velocities between two inertial systems ($v \ll c$) will converge to the Galilean ones.

For the derivation, we consider two inertial systems S and S', where S is considered to be at rest and S' is supposed to move with velocity v to the right on the same line (x'-axis equals x-axis). The clocks in both systems were set to zero at the time when the two zero points met. For clarity, the reference systems are drawn separately in Fig. 13.1 and the z-axis is omitted.

© The Author(s), under exclusive license to Springer-Verlag GmbH, DE, part of Springer Nature 2026

M. Ruhrländer, *Ascent to the Einstein Equations*,

https://doi.org/10.1007/978-3-662-72672-3_13

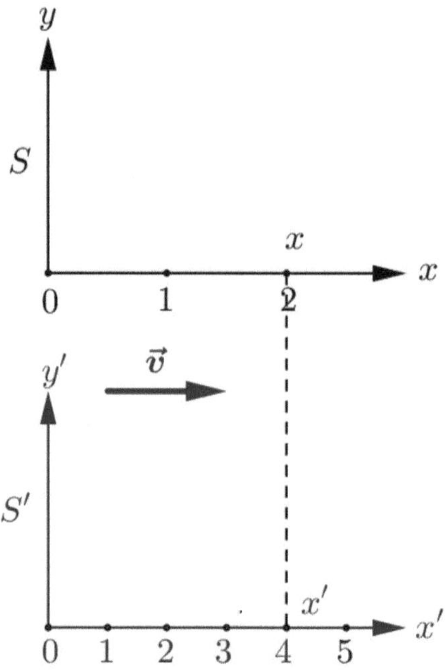

Figure 13.1.: Lorentz transformation I

From the perspective of S, the moving x'-axis is shortened due to the length contraction (in the figure by 50%, e.g. the interval $[0, 2]$ on the x-axis is twice as large as on the x'-axis). This means, that larger x'-values are needed to establish a correspondence between x- and x'-values (dashed line). From the length contraction it follows that

$$x' = \frac{x}{\sqrt{1 - v^2/c^2}}. \tag{13.1}$$

For an observer in the resting system S, a clock at position x' lags behind the clock at the zero point by the value $x' \cdot v/c^2$ due to the relativity of simultaneity (see section 12.4). If we substitute the above expression for x', we obtain for the clock reading at x' the value

$$-\frac{x \cdot v/c^2}{\sqrt{1 - v^2/c^2}}, \tag{13.2}$$

where the minus sign indicates that from the perspective of S, the clock at x' lags behind the clock at the zero point, which shows zero.

We now consider in Fig. 13.2 from the perspective of S an arbitrary point in time t, i.e., the coordinate system S' has moved to the right by the distance $v \cdot t$.

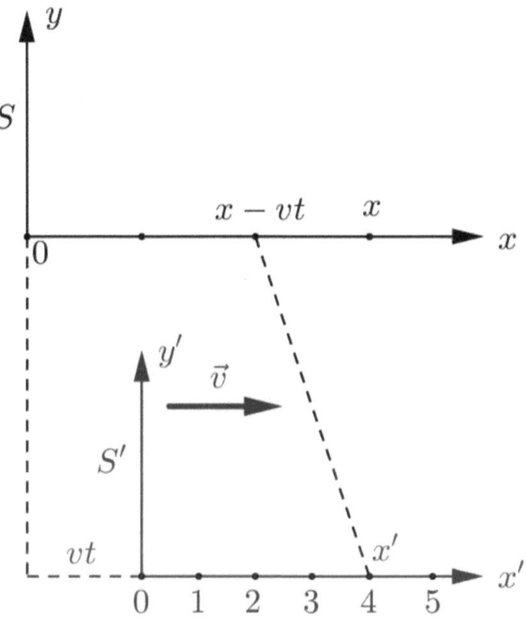

Figure 13.2.: Lorentz transformation II

The point x' of the moving system S' is at time t at the level of the point x in the system S. At time $t = 0$ it was therefore at the level of the point $x - vt$, we substitute this value into (13.1) and obtain

$$x' = \frac{x - vt}{\sqrt{1 - v^2/c^2}}.$$

Similarly, for a clock that is at the level of the point x at time t, it was at the level of the point $x - vt$ at time $t = 0$. We use (13.2) to determine the value of the clock display at time zero:

$$-\frac{(x - vt) \cdot v/c^2}{\sqrt{1 - v^2/c^2}}$$

A clock in the system S', which is moving at speed v, shows from the perspective of S due to time dilation over the period t a value that is $t \cdot \sqrt{1 - v^2/c^2}$ larger than at time zero. Since it was lagging by the above amount from the perspective of S at time zero, S reads the following clock reading t' in the system S' at time t

$$
\begin{aligned}
t' &= -\frac{(x - vt) \cdot v/c^2}{\sqrt{1 - v^2/c^2}} + t \cdot \sqrt{1 - v^2/c^2} \\[2ex]
&= \frac{-\dfrac{x \cdot v}{c^2} + \dfrac{t \cdot v^2}{c^2} + t\left(1 - \dfrac{v^2}{c^2}\right)}{\sqrt{1 - v^2/c^2}} \\[2ex]
&= \frac{t - \dfrac{x \cdot v}{c^2}}{\sqrt{1 - v^2/c^2}}.
\end{aligned}
$$

We have thus found two transformation formulas that allow us to calculate the values of x' and t' from those of x and t. If we also consider that the lengths perpendicular to the direction of motion do not change, i.e., are the same in both inertial systems, we obtain the four so-called **special Lorentz transformations**:

$$
\begin{aligned}
x' &= \frac{x - vt}{\sqrt{1 - v^2/c^2}} \\[2ex]
y' &= y \\[1ex]
z' &= z \\[2ex]
t' &= \frac{t - \dfrac{x \cdot v}{c^2}}{\sqrt{1 - v^2/c^2}}.
\end{aligned}
\tag{13.3}
$$

These four equations were already established at the end of the 19th century (i.e., before Einstein) by the Dutch physicist H. Lorentz in the context of his investigations in the theory of ether. If the relative speed v is small compared to the speed of light, then $\dfrac{v^2}{c^2} \approx 0$ and $\dfrac{x \cdot v}{c^2} \approx 0$, i.e., for small speeds the Lorentz transformations become the Galilei transformations (see formula 12.1 on page 227).

   If the event coordinates (x', y', z', t') of the moving system S' are known, one can convert these into the coordinates of one's own system. If one considers that the sign of the relative speed v changes, if one assumes S' to be stationary and S to be moving, one obtains the **inverse Lorentz transformations**:

$$
\begin{aligned}
x &= \frac{x' + vt'}{\sqrt{1 - v^2/c^2}} \\[2ex]
y &= y' \\[1ex]
z &= z'
\end{aligned}
\tag{13.4}
$$

$$t = \frac{t' + \dfrac{x' \cdot v}{c^2}}{\sqrt{1 - v^2/c^2}}.$$

The Lorentz transformations can also be used for distance intervals. For example, if

$$\Delta x' = x_2' - x_1', \Delta x = x_2 - x_1, \Delta t = t_2 - t_1,$$

then it follows

$$\begin{aligned}
\Delta x' &= x_2' - x_1' \\
&= \frac{x_2 - vt_2}{\sqrt{1 - v^2/c^2}} - \frac{x_1 - vt_1}{\sqrt{1 - v^2/c^2}} \\
&= \frac{(x_2 - x_1) - v\,(t_2 - t_1)}{\sqrt{1 - v^2/c^2}} \\
&= \frac{\Delta x - v \cdot \Delta t}{\sqrt{1 - v^2/c^2}}.
\end{aligned}$$

In the same way, one calculates $\Delta t'$, so that one obtains the **Lorentz transformations for intervals**:

$$\begin{aligned}
\Delta x' &= \frac{\Delta x - v \cdot \Delta t}{\sqrt{1 - v^2/c^2}} \\
\Delta y' &= \Delta y \\
\Delta z' &= \Delta z \\
\Delta t' &= \frac{\Delta t - \dfrac{\Delta x \cdot v}{c^2}}{\sqrt{1 - v^2/c^2}}
\end{aligned} \tag{13.5}$$

From the Lorentz transformations, some of the results derived directly from the Einstein postulates in the last sections immediately follow:

1. If two events occur at the same place ($\Delta x = 0$) with a time difference Δt_E (= proper time), then from the last equation you get

$$\Delta t' = \frac{\Delta t_E}{\sqrt{1 - v^2/c^2}},$$

   hence the time dilation (see formula 12.4 on page 234).

2. If two events occur simultaneously ($\Delta t = 0$), then from the fourth and first equation of (13.5)

$$\Delta t' = \frac{-\dfrac{\Delta x \cdot v}{c^2}}{\sqrt{1 - v^2/c^2}} = -\frac{\Delta x' \cdot v}{c^2},$$

hence the formula for the relativity of simultaneity (see section 12.4).

3. A rod of length L'_E (= proper length) rests in the system S', the coordinates of the rod ends are x'_1 and x'_2, i.e.

$$L'_E = x'_2 - x'_1.$$

If an observer in the system S wants to measure the length of the rod, he must measure the beginning and end of the rod *simultaneously* (for him), otherwise the measurement result would be distorted by the relative speed between the two systems. This follows from (13.3)

$$
\begin{aligned}
L'_E \;&=\; x'_2 - x'_1 \\[2mm]
&=\; \frac{x_2 - v \cdot t_2}{\sqrt{1 - v^2/c^2}} - \frac{x_1 - v \cdot t_1}{\sqrt{1 - v^2/c^2}} \\[2mm]
&\underset{t_2 = t_1}{=}\; \frac{(x_2 - x_1)}{\sqrt{1 - v^2/c^2}} \\[2mm]
&=\; \frac{L}{\sqrt{1 - v^2/c^2}},
\end{aligned}
$$

hence the length contraction (see formula 12.5 on page 235).

4. We calculate the squares of the spacetime intervals using the Lorentz transformations. It applies

$$
\begin{aligned}
x' - c \cdot t' \;&=\; \frac{x - v \cdot t}{\sqrt{1 - v^2/c^2}} - \frac{c\left(t - \dfrac{x \cdot v}{c^2}\right)}{\sqrt{1 - v^2/c^2}} \\[3mm]
&=\; \frac{1}{\sqrt{1 - v^2/c^2}}\,(x - v \cdot t - c \cdot t + x \cdot v/c) \\[3mm]
&=\; \frac{1}{\sqrt{1 - v^2/c^2}}\,(x\,(1 + v/c) - c \cdot t\,(1 + v/c)) \\[3mm]
&=\; \frac{1 + v/c}{\sqrt{1 - v^2/c^2}}\,(x - c \cdot t)\,.
\end{aligned}
$$

Similarly, it follows

$$
x' + c \cdot t' = \frac{1 - v/c}{\sqrt{1 - v^2/c^2}}\,(x + c \cdot t)\,.
$$

Multiplying both equations, we get

$$
x'^2 - \left(c \cdot t'\right)^2 \;=\; \left(x' - c \cdot t'\right)\left(x' + c \cdot t'\right)
$$

$$
\begin{aligned}
&= \left[\frac{1 + v/c}{\sqrt{1 - v^2/c^2}} \, (x - c \cdot t) \right] \left[\frac{1 - v/c}{\sqrt{1 - v^2/c^2}} \, (x + c \cdot t) \right] \\
&= \left[(x - c \cdot t)\,(x + c \cdot t) \right] \left[\frac{1 + v/c}{\sqrt{1 - v^2/c^2}} \, \frac{1 - v/c}{\sqrt{1 - v^2/c^2}} \right] \\
&= \left[x^2 - (c \cdot t)^2 \right] \left[\frac{(1 + v/c)\,(1 - v/c)}{1 - v^2/c^2} \right] \\
&= x^2 - (c \cdot t)^2 \,,
\end{aligned}
$$

thus the invariance of the spacetime interval.

13.2. Natural Units and General Lorentz Transformations

In the previous sections, we have repeatedly pointed out that the Special Theory of Relativity only significantly differs from Newtonian mechanics and other non-relativistic physical theories when the relative velocities between reference systems are large. For example, the γ factor only significantly differs from 1 when $v > 0.1\,c$. Furthermore, in the Special Theory of Relativity, the speed of light in a vacuum is constant, i.e., all observers in all inertial systems measure the same value for c, so c is a universal constant. We take advantage of these two properties of the Special Theory of Relativity and define a new unit system, which we mainly want to use from now on. So far, we have represented physical quantities in the so-called International System of Units (SI). In the SI system, length is measured in meters m, time in seconds s, and mass in kilograms kg (see also Tab. 30.1). In the SI system, the dimension of velocity v is equal to distance L divided by time T ($\dim v = \mathsf{L}/\mathsf{T}$).

We now want to define the unit for time also in meters, for which we use the constancy of the speed of light: A time meter should be the time that light needs to traverse a distance of one meter (we are more accustomed to the light year, namely the distance that light travels in one year). If we choose this length unit for time, we work instead of the SI units with **natural units**. Let's see what value the speed of light takes in these units:

$$
c = \frac{1\,\mathsf{m}}{\text{time that light needs to cross 1 meter}} = \frac{1\,\mathsf{m}}{1\,\mathsf{m}} = 1
$$

The value of the speed of light is therefore equal to 1 and moreover, c is dimensionless! Of course, you can also go back, i.e., if results are available in natural units, you use the relationships

$$
3 \cdot 10^8 \, \mathsf{m/s} = 1 \Rightarrow 1\,\mathsf{s} = 3 \cdot 10^8 \, \mathsf{m} \Rightarrow 1\,\mathsf{m} = \frac{1}{3 \cdot 10^8} \mathsf{s},
$$

to obtain SI units. The SI system contains a lot of derived units, such as Joule and Newton, see Tab. 30.2. When using natural units, the dimensions of many SI units simplify considerably, e.g. for Joule in SI units

$$\mathsf{J} = \mathsf{N} \cdot \mathsf{m} = \mathsf{kg} \cdot \mathsf{m}^2/\mathsf{s}^2$$

(N = Newton) and in natural units

$$\mathsf{J} = \mathsf{kg} \cdot \mathsf{m}^2/\mathsf{m}^2 = \mathsf{kg}.$$

Let's look at how the essential quantities of Special Relativity can be represented in natural units.

Example 13.1.

- Let's start with Einstein's famous equation: $E = mc^2$ degenerates to $E = m$. The recognition value of the last equation is not very high, which is why we will also present the results in SI units at some points in this book. On the other hand, the equation $E = m$ most clearly expresses the equivalence between energy and mass, in natural units energy is equal to mass!

- Next, we want to compare the Lorentz transformations in SI units with those in natural units:

SI	Nat
$x' = \dfrac{x - v \cdot t}{\sqrt{1 - v^2/c^2}}$	$x' = \dfrac{x - v \cdot t}{\sqrt{1 - v^2}} = \gamma(x - vt)$
$y' = y$	$y' = y$
$z' = z$	$z' = z$
$t' = \dfrac{t - \dfrac{v \cdot x}{c^2}}{\sqrt{1 - v^2/c^2}}$	$t' = \dfrac{t - v \cdot x}{\sqrt{1 - v^2}} = \gamma(t - vx)$

In natural units, there is complete symmetry between time t and space x, the Lorentz transformations have not only simplified, but also unified.

- As a final example, we present the spacetime interval in natural coordinates:

$$\Delta s^2 = x^2 - t^2 \quad \square$$

General Lorentz Transformation

So far, in deriving the Lorentz transformation, we have always assumed for simplicity that the observer in system S' is moving parallel to the x-axis with the velocity

$$\vec{v} \rightarrow \begin{pmatrix} v \\ 0 \\ 0 \end{pmatrix}.$$

We now drop this assumption and derive the Lorentz transformations for arbitrary relative speeds in natural units. We assume that observer A in system S and observer B in system S' were at the same place at time $t = t' = 0$

$$x = y = z = x' = y' = z' = 0$$

that is, the coordinate origins of S and S' coincide. B is moving relative to A with the velocity

$$\vec{v} = \begin{pmatrix} v_x \\ v_y \\ v_z \end{pmatrix}.$$

The spatial components of the event E in system S are

$$\vec{r} \rightarrow \begin{pmatrix} x \\ y \\ z \end{pmatrix}.$$

We will use the fact that a three-dimensional vector $\vec{r}$ can be divided into a vector $\vec{r}_\parallel$ parallel to $\vec{v}$ and a vector $\vec{r}_\perp$ perpendicular to $\vec{v}$.

Remark 13.1. **MT: Division of a Vector**
If $\vec{a}$ and $\vec{b}$ are arbitrary vectors, the vector $\vec{b}$ can be divided into a vector $\vec{b}_{a\perp}$, which is perpendicular to the vector $\vec{a}$, and a vector $\vec{b}_{a\parallel}$, which is parallel to $\vec{a}$. From Fig. 13.3, it is already apparent that

$$\vec{b} = \vec{b}_{a\perp} + \vec{b}_{a\parallel} \tag{13.6}$$

holds.

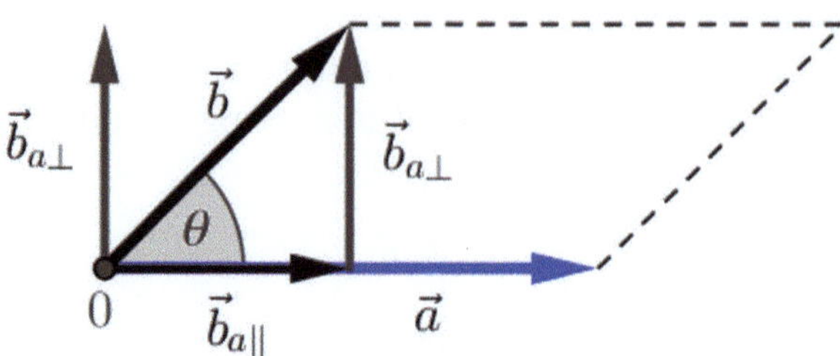

Figure 13.3.: Division of a vector

We want to calculate the vectors $\vec{b}_{a\parallel}$ and $\vec{b}_{a\perp}$ concretely. For the length of $\vec{b}_{a\parallel}$, it holds

$$b_{a\parallel} = b \cdot \cos\theta.$$

The scalar product of the vectors $\vec{a}$ and $\vec{b}$ was defined by

$$\vec{a} \cdot \vec{b} = ab \cos\theta,$$

so it follows

$$b_{a\parallel} = b \cdot \cos\theta = \frac{ab \cdot \cos\theta}{a} = \frac{\vec{a} \cdot \vec{b}}{a}.$$

Since the vector $\vec{b}_{a\parallel}$ should point in the direction of the vector $\vec{a}$, we have to multiply $b_{a\parallel}$ with the unit vector in the direction of $\vec{a}$

$$\vec{e}_a = \frac{\vec{a}}{a}$$

and obtain

$$\vec{b}_{a\parallel} = b_{a\parallel}\vec{e}_a = \frac{\vec{a} \cdot \vec{b}}{a}\frac{\vec{a}}{a} = \frac{\vec{a} \cdot \vec{b}}{a^2}\vec{a}. \tag{13.7}$$

We define

$$\vec{b}_{a\perp} = \vec{b} - \vec{b}_{a\parallel}, \tag{13.8}$$

then it immediately follows

$$\vec{b}_{a\parallel} + \vec{b}_{a\perp} = \vec{b}.$$

To show that the vector $\vec{b}_{a\perp}$ is perpendicular to the vector $\vec{a}$, we calculate the scalar product of the two vectors

$$\vec{b}_{a\perp} \cdot \vec{a} = \left(\vec{b} - \vec{b}_{a\parallel}\right) \cdot \vec{a} = \vec{b} \cdot \vec{a} - \vec{b}_{a\parallel} \cdot \vec{a} = \vec{a} \cdot \vec{b} - \frac{\vec{a} \cdot \vec{b}}{a^2}\vec{a} \cdot \vec{a} = \vec{a} \cdot \vec{b} - \frac{\vec{a} \cdot \vec{b}}{a^2}a^2 = 0.$$

Here we have used that the scalar product is commutative ($\vec{a} \cdot \vec{b} = \vec{b} \cdot \vec{a}$) and that $\vec{a} \cdot \vec{a} = a^2$. The scalar product is zero, so the two vectors are perpendicular to each other. $\square$

With (13.7) and (13.8) it follows

$$\vec{r}_\parallel = \frac{\vec{v} \cdot \vec{r}}{v^2} \vec{v}$$

$$\vec{r}_\perp = \vec{r} - \vec{r}_\parallel = \vec{r} - \frac{\vec{v} \cdot \vec{r}}{v^2} \vec{v}.$$

By definition of the scalar product (see Eq. 4.3 on page 55) it is

$$\begin{aligned}
\vec{v} \cdot \vec{r} &= \vec{v} \cdot \left(\vec{r}_\parallel + \vec{r}_\perp \right) \\
&= vr_\parallel \cos \angle \left(\vec{v}, \vec{r}_\parallel \right) + vr_\perp \cos \angle \left(\vec{v}, \vec{r}_\perp \right) \\
&= vr_\parallel \cos \left(0° \right) + vr_\perp \cos \left(90° \right) \\
&= vr_\parallel.
\end{aligned}$$

The following Lorentz transformation applies to the system S'

$$\begin{aligned}
t' &= \gamma \left(t - vr_\parallel \right) = \gamma \left(t - \vec{v} \cdot \vec{r} \right) \\
\vec{r}\,'_\parallel &= \gamma \left(\vec{r}_\parallel - \vec{v}t \right) \\
\vec{r}\,'_\perp &= \vec{r}_\perp.
\end{aligned}$$

The last equation follows from the fact that lengths perpendicular to the relative speed are not changed. We now substitute the expressions for $\vec{r}_\parallel$ and $\vec{r}_\perp$ and obtain

$$\begin{aligned}
\vec{r}\,' &= \vec{r}\,'_\parallel + \vec{r}\,'_\perp \\
&= \gamma \left(\vec{r}_\parallel - \vec{v}t \right) + \vec{r}_\perp \\
&= \gamma \left(\frac{\vec{v} \cdot \vec{r}}{v^2} \vec{v} - \vec{v}t \right) + \vec{r} - \frac{\vec{v} \cdot \vec{r}}{v^2} \vec{v} \\
&= \vec{r} + (\gamma - 1) \frac{\vec{v} \cdot \vec{r}}{v^2} \vec{v} - \gamma \vec{v}t.
\end{aligned}$$

Overall, we thus obtain the **general Lorentz transformation**

$$t' = \gamma \left(t - \vec{v} \cdot \vec{r} \right) \tag{13.9}$$

$$\vec{r}\,' = \vec{r} + (\gamma - 1) \frac{\vec{v} \cdot \vec{r}}{v^2} \vec{v} - \gamma \vec{v}t.$$

In the further course of our considerations, however, we mostly resort to the simple form of the Lorentz transformation (i.e., relative motion parallel to the x-axis), unless explicitly stated otherwise.

13.3. Spacetime Diagrams (STD)

In this section, we introduce the so-called **Minkowski diagrams**, which are excellent for graphically representing the phenomena of Special Relativity. Using these diagrams, almost all of the previous results can be derived quantitatively with geometric considerations. However, we do not do this consistently, but only with selected examples. Our main focus is on gaining a deeper understanding of the structure of spacetime. It should be noted at this point that in addition to the Minkowski diagrams, there are other ways of representing relativistic phenomena, in particular the (not so widespread) approach by Epstein [8] is very well suited to provide an overview of the essential facts with simple geometric means. But back to the Minkowski diagrams. Hermann Minkowski (1854 - 1908) was Einstein's mathematics professor and, shortly after Einstein published his Special Theory of Relativity, made the phenomenon of spacetime more understandable with his diagrams and other structuring aids. In his honor, the spacetime of the Special Theory of Relativity is also called **Minkowski space**.

We can identify each observer with the inertial system that is his rest system, i.e., coordinate system and observer are interchangeable. To represent spacetime, we use a spacetime diagram (see section 2.1), which maps time t (in natural coordinates) on the vertical axis. For simplicity, we only represent space one-dimensionally - usually on the x-axis - which is completely sufficient for illustrating most phenomena of the Special Theory of Relativity. Each point in a spacetime diagram („STD") represents an event. The angle between the two coordinate axes can fundamentally be chosen arbitrarily, but for reasons of convenience, we want to assign a stationary observer a perpendicular axis cross and the coordinates (t, x) (y and z are of course also part of it, but - as said - are not represented in a spacetime diagram). Inertial systems are characterized by the fact that the relative speed between them remains constant over time. Equal distances are covered in equal times. Therefore, inertial movements in spacetime diagrams are represented by straight lines (see also section 2.1).

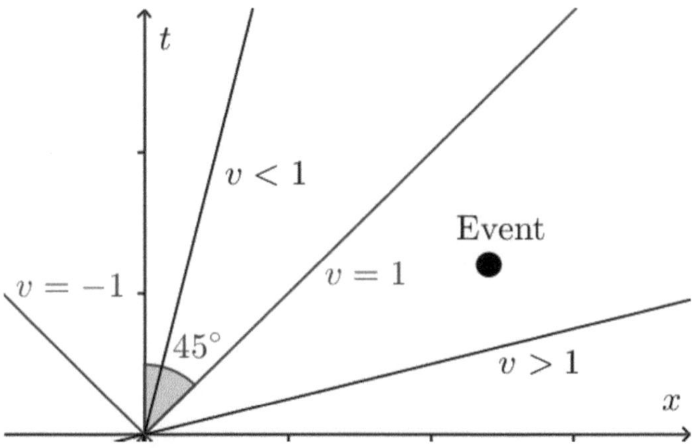

Figure 13.4.: Spacetime diagram I

Fig. 13.4 shows, in addition to a point that is supposed to represent an event, three worldlines of particles (see section 2.1), which are determined by a relationship $x = x(t)$ or $t = t(x)$. The slope of a worldline is related to the speed of the particles:

$$\text{Slope} = \frac{dt}{dx} = \frac{1}{\dfrac{dx}{dt}} = \frac{1}{v}$$

Here it was used that the slope of a worldline is equal to the first derivative of the function $t = t(x)$ with respect to x. Since time and distance are measured in natural units by the same units and the speed of light takes the value 1, light rays (worldlines of photons) are represented by straight lines that have a slope of ± 1, i.e., from the perspective of the rest system, they enclose an angle of 45° with the t-axis. The bisecting line to the right has a slope of 1 due to $v = 1$, the bisecting line to the left has a slope of -1 due to $v = -1$. worldlines of particles whose speed is less than the speed of light ($v < 1$) have a slope greater than 1, and events that lie between the x-axis and the bisecting line would only be reachable with superluminal speed ($v > 1$). Therefore, we can divide the space-time diagram into three regions.

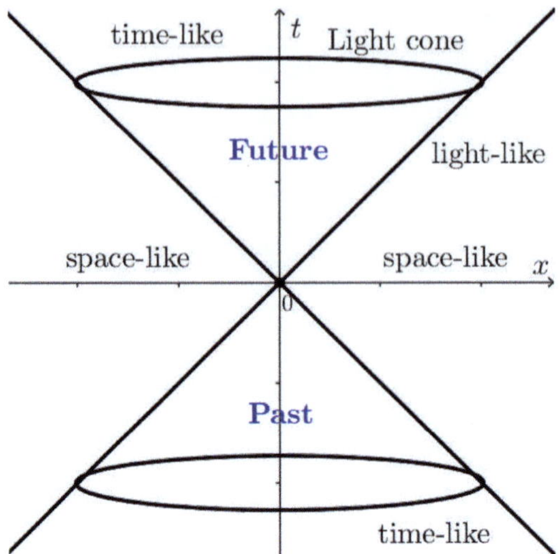

Figure 13.5.: Division of spacetime

1. Events that are connected to the origin by light rays lie on the bisectors. We have called these events *light-like*.

2. Events whose worldline lies within the areas marked by the bisectors, which are also called **light cones** (in Fig. 13.5 indicated by adding another spatial dimension using ellipses), can be reached from the origin at a speed less than the speed of light, so they are *time-like*. The upper half of the light cone is called the **future**, due to $t > 0$, the lower half is called the **past** of the event at the origin.

3. Events that lie outside the light cones. These cannot be reached by signals from the origin, they are *space-like*.

We now consider another observer with coordinates (t', x'), who is supposed to move at a speed $v < 1$ in the positive x-direction relative to the stationary observer, and ask ourselves how and where the axes for t' and x' are to be drawn in Fig. 13.4. We again assume that the two observers met at the time $t = t' = 0$. The t'-axis is the location of all events for which $x' = 0$ applies.

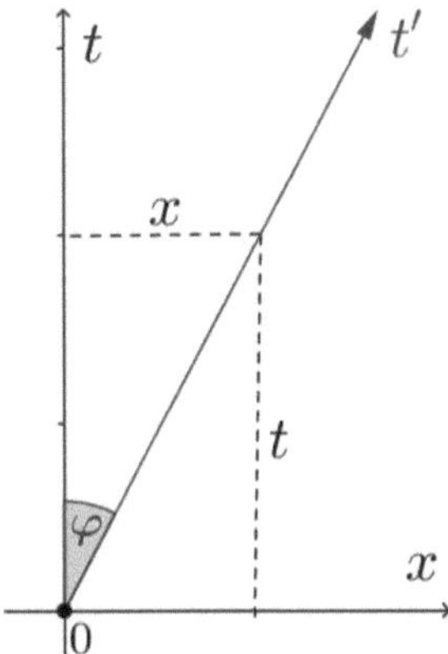

Figure 13.6.: Spacetime diagram II

In Fig. 13.6 the angle φ between the t- and the t'-axis is given by:

$$\tan \varphi = \frac{x}{t} = v$$

Since the relative speed v is known, the t'-axis can thus be easily constructed.

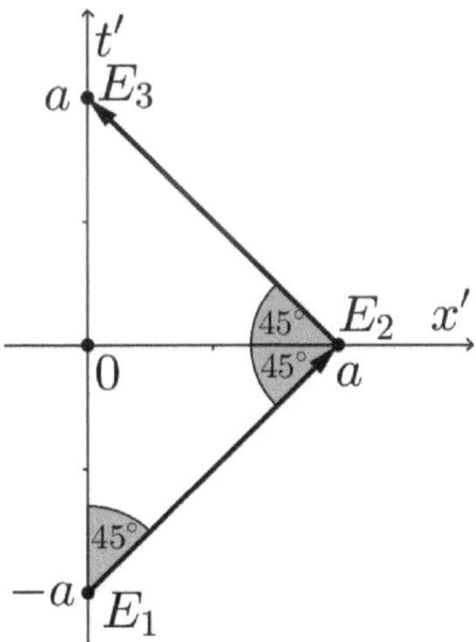

Figure 13.7.: Spacetime diagram III

To construct the x'-axis, we use the fact that light propagates at an angle of $45°$ to the right or left. If we start a light beam at the space-time point/event $E_1\,(t', x') = (-a, 0)$ to the right in Fig. 13.7, the light beam reaches the x'-axis at the space-time point $E_2 = (0, a)$. There it is reflected by a mirror without delay and reaches the t'-axis at $t' = a$, i.e., at the space-time point $E_3 = (a, 0)$.

Note that the angle between the two light beams is $45° + 45° = 90°$, the light beams are perpendicular to each other in Fig. 13.7. We can therefore define the x'-axis as the location of all events that reflect light in the described manner. To interpret Fig. 13.7: The light does *not* first run to the right and then to the left "up", but directly from the location $x' = 0$ to the location $x' = a$ and then back! The apparent upward movement is caused by the time the light takes to cover the two distances.

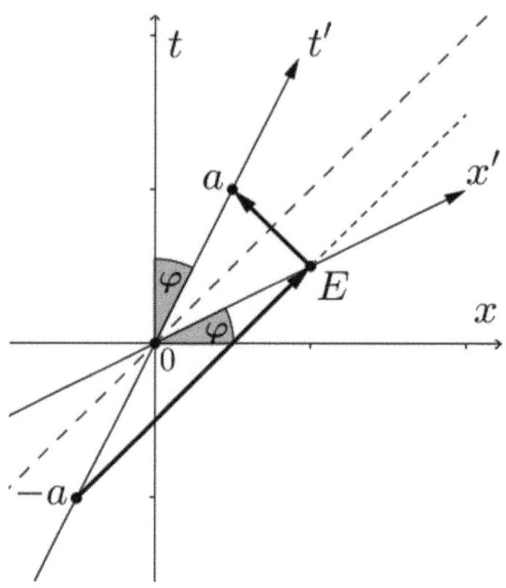

Figure 13.8.: Spacetime diagram IV

With Fig. 13.8 we construct the x'-axis, by performing the following steps:

1. The position of the t'-axis is constructed using the method described above. We then choose two arbitrary points $-a$ and a on the t'-axis.

2. At the event point $-a$ we start a light beam (thick arrow and dotted line, slope 45°) to the right.

3. We construct the point E by dropping the perpendicular from the point a to the light beam, so that the distances $\overline{-aE}$ and $\overline{Ea}$ form a right angle with each other. We now interpret the event E in such a way that we imagine that a mirror is placed there, which reflects a light beam coming from $-a$ to a (thick arrow to the left). The point E thus has, as shown in Fig. 13.7, in the (t', x') coordinate system the coordinates $E(t', x') = (0, a)$ and therefore lies on the x'-axis.

4. Since the origin of the (t', x') coordinate system coincides with the origin of the (t, x) coordinate system, the point 0 also lies on the x'-axis. The

x'-axis is thus the line, which passes through the points 0 and E. Since the distances $\overline{0E}$ and $\overline{0a}$ are of the same length, the triangle $0aE$ is isosceles and the bisector (dashed black line) divides the distance $\overline{aE}$ in half. Therefore, the angle between x- and x'-axis is equal to the angle φ between t- and t'-axis.

The x'-axis does not coincide with the x-axis! Since on the x-axis simultaneous events ($t = 0$) for the first observer and on the x'-axis simultaneous events for the second observer are located and since both axes are not parallel to each other, it can be inferred from the figure that the two observers cannot agree on the simultaneity of events.

When comparing the last two diagrams, it is found that light rays always move on a 45° line, while the t- and t'-axis have different slopes. This illustrates once again that light rays for each observer have the speed $c = 1$, i.e., the slope one.

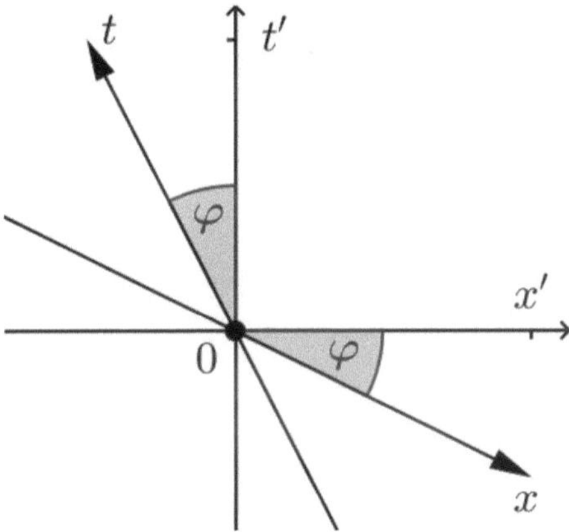

Figure 13.9.: Spacetime diagram V

Fig. 13.9 represents the same situation as above, only that the observer (x', t') is considered to be at rest while the observer (x, t) is moving at the speed $-v$ (i.e. to the left). Here too, the angle φ is calculated by $\tan \varphi = |v|$, where $|v|$ is the magnitude of the relative speed between the two inertial systems.

Now that we know where the axes of the (x', t') coordinate system are located, the next step is to define the units on these axes. For this, we use the invariance of the space-time interval

$$\Delta S^2 = x^2 - t^2 = x'^2 - t'^2.$$

The curve $x^2 - t^2 = 1$ represents a hyperbola in the coordinate system (x, t), see 6.23 on page 115. This describes all space-like events (since $\Delta S^2 > 0$, see (12.26)), whose space-time interval is equal to 1. Similarly, the curve $x^2 - t^2 = -1$ describes all time-like events, whose space-time interval is equal to -1.

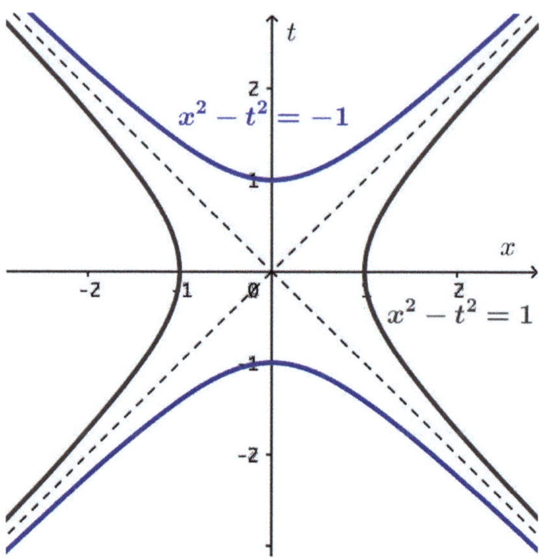

Figure 13.10.: Gauge hyperbolas

Fig. 13.10 shows that the four hyperbola branches asymptotically approach the bisectors (dashed). Since these hyperbolas are used to define the units on the x'- and t'-axis (to gauge), they are also called **gauge hyperbolas**. To determine the length unit on the t'-axis, we note that for all events on the t'-axis $x' = 0$ applies and that every point on the t'-axis has a time-like distance to the origin of the coordinate system (x, t). This results in

$$-1 = x'^2 - t'^2 = -t'^2 = x^2 - t^2,$$

i.e., the intersection of the t'-axis with the gauge hyperbola $x^2 - t^2 = -1$ provides the space-time point $(0, 1)$ in the coordinate system (x', t'). Similarly, the space-time point $(1, 0)$ in the coordinate system (x', t') is obtained as the intersection of the hyperbola $x^2 - t^2 = 1$ with the x'-axis.

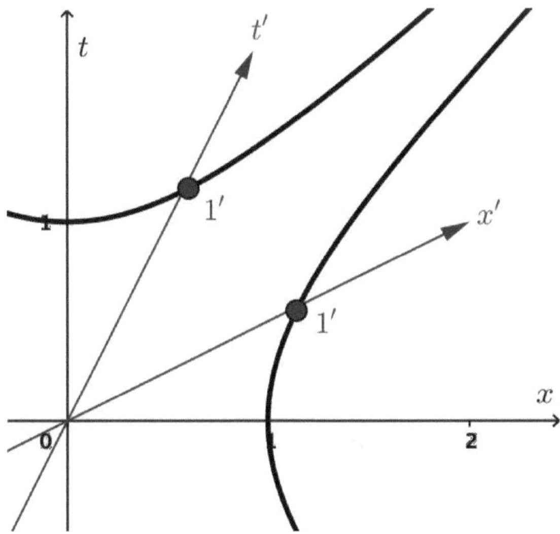

Figure 13.11.: Spacetime diagram VI

The space-time points $(x', t') = (0, 1)$ and $(x', t') = (1, 0)$ are in Fig. 13.11 each marked with $1'$. These points appear "further away" from the origin than the unit points on the t- or x-axis. This again shows that we are not in Euclidean geometry, distances are measured by the spacetime interval $(-t^2 + x^2)$ (and then the points are equally far away) and not by the Euclidean distance $t^2 + x^2$!

The line equations of the axes of the moving observer can also be determined directly from the Lorentz transformations

$$x' = \frac{x - v \cdot t}{\sqrt{1 - v^2}}$$

$$t' = \frac{t - v \cdot x}{\sqrt{1 - v^2}}$$

If you set $x' = 0$ in the first equation and $t' = 0$ in the second, you get

$$0 = \frac{x - v \cdot t}{\sqrt{1 - v^2}} \Rightarrow t = \frac{1}{v} x \text{ line equation for the } t'\text{-axis}$$

$$0 = \frac{t - v \cdot x}{\sqrt{1 - v^2}} \Rightarrow t = v \cdot x \text{ line equation for the } x'\text{-axis.}$$

As an application example of the Minkowski diagrams, we now consider the length contraction again. For this, we look at a rod of length 1 between the points $(0, 0)$ and $(1, 0)$ in the rest system in Fig. 13.12.

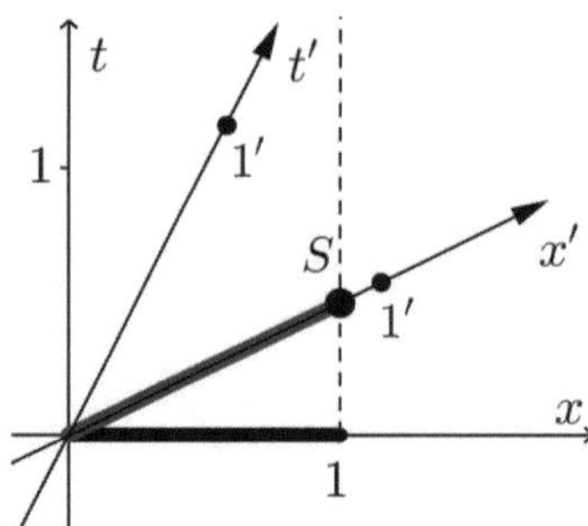

Figure 13.12.: Length contraction I

The worldlines of the rod, the *t*-axis for the front end of the rod and the dashed line for the rear end of the rod, run vertically in the rest system. The moving observer must, in order to determine the length of the rod, measure both ends of the rod *simultaneously*, i.e., the beginning and end of the rod must lie on the x'-axis (more generally: on a parallel to the x'-axis) for him. The moving observer checks where the worldlines of the rod ends intersect his x'-axis. This is the case at 0 and at S, from his point of view the distance $\overline{0S}$ is shorter than 1. For an observer moving relative to the rest system, distances in the rest system (which lie in the direction of motion) are therefore contracted. Due to the principle of relativity, it also applies that the rest observer sees distances in the system S' as shortened, which Fig. 13.13 illustrates.

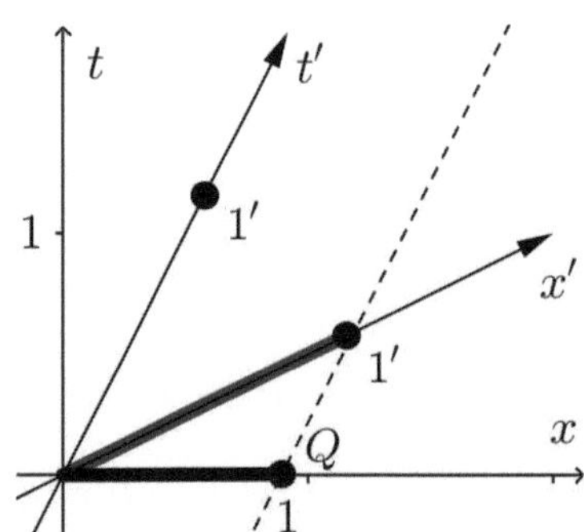

Figure 13.13.: Length contraction II

A rod of length 1 lies on the x'-axis. The worldlines of the rod ends are the t'-axis and the dashed line. If the rest observer simultaneously measures the rod ends on his x-axis, he gets the distance $\overline{0Q}$, which from his point of view is shorter than 1.

Time dilation can also be graphically explained, as Fig. 13.14 shows.

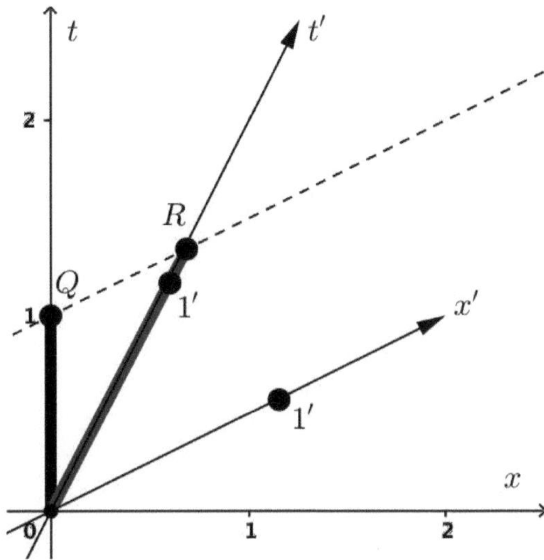

Figure 13.14.: Time dilation diagram

For the stationary observer, the distance $\overline{0Q}$ (thick black bar) represents a unit of time. The moving observer draws a parallel (dashed) through Q to his x'-axis and finds that it intersects his time axis t' at point R. For him, the events Q and R are simultaneous, as they are on lie parallel to the x'-axis. On his time axis, he reads a larger time value for R. Therefore, for the moving observer, the time intervals in the rest system are extended („dilated").

14. Vector Calculus in Special Relativity Theory

Spacetime can be understood as a (four-dimensional) vector space. In extension of vector calculus for Euclidean space, the necessary basics, which are predominantly mathematical or technical in nature, are introduced. These sections are essential for further understanding, even if the presented definitions and abbreviating notations, which build on the results of vector calculus in the Euclidean plane and further develop them, appear difficult at first glance. However, they help enormously when we delve further into the subject of arbitrary coordinate systems. In the first sections of this chapter, some results are derived using very similar means as we have also used for the Euclidean plane. Thus, these sections offer a good opportunity for repetition and further practice of the sometimes quite abstract vector calculus. After the initially technically oriented sections, we will then pick up some of the results of Special Relativity Theory achieved so far and show that they fit well into the new vector notations.

14.1. Definition of Spacetime Vectors

In this section, we extend vector calculus to four-dimensional space-time and first define the displacement vector $\Delta \vec{x}$, which points from a space-time point with the coordinates (t_1, x_1, y_1, z_1) to a second one with the coordinates (t_2, x_2, y_2, z_2) and whose components are the coordinate differences

$$\Delta \vec{x} \underset{S}{\rightarrow} (t_2 - t_1, x_2 - x_1, y_2 - y_1, z_2 - z_1) = (\Delta t, \Delta x, \Delta y, \Delta z).$$

Just like in the two-dimensional plane, the arrow above a symbol denotes a (four-dimensional) vector, the arrow after the vector $\Delta \vec{x}$ means something like "has the components", where we list the inertial system S, from which the coordinates originate, below the arrow, if it is necessary for a better understanding, otherwise we leave this hint out. The symbol $\underset{S}{\rightarrow}$ thus emphasizes the difference between a vector, which - as in two and three dimensions - is a geometric object, i.e. independent of the inertial system, and its components,

© The Author(s), under exclusive license to Springer-Verlag GmbH, DE, part of Springer Nature 2026

M. Ruhrländer, *Ascent to the Einstein Equations*,

https://doi.org/10.1007/978-3-662-72672-3_14

which depend on the inertial system. If S' is another inertial system, one can express *the same* vector $\Delta \vec{x}$ also by its components in S'

$$\Delta \vec{x} \underset{S'}{\rightarrow} \left(\Delta t', \Delta x', \Delta y', \Delta z' \right).$$

If we want to represent any coordinate, we use small Greek letters, e.g. α, β, μ, ν as indices. α, β, ν, μ can each take the values t, x, y, z in the Cartesian coordinate system. We want to transfer the abbreviating notations of *Einstein's summation convention* and the *free indices* introduced in the chapter on tensor calculus in the Euclidean plane to space-time.

Einstein's summation convention: Whenever an expression contains an index that is superscripted in one variable and subscripted in another, a summation is performed over the values that the index can take. For example, the expression $A_\alpha B^\alpha$ is an abbreviating notation for the sum

$$A_\alpha \, B^\alpha = A_t \, B^t + A_x \, B^x + A_y \, B^y + A_z \, B^z,$$

if the index α can take the values t, x, y, z.

If α is a free index, then the expression $\Delta \alpha$ can take *any* value from $\Delta t, \Delta x, \Delta y, \Delta z$. We denote the coordinates in an inertial system S' as $\alpha', \beta', \mu', \nu'$, which in the Cartesian coordinate system can each take the values t', x', y', z'. Another shorthand notation is to use the free index α to represent *all possible values simultaneously*. For this, we write the quantity indexed with α in curly brackets, for example

$$\{\Delta \alpha\} = (\Delta t, \Delta x, \Delta y, \Delta z)$$

or

$$\{\alpha\} = (t, x, y, z)$$

or

$$\{A^\alpha\} = \left(A^t, A^x, A^y, A^z \right).$$

If we want to know how the coordinates in S' can be calculated from those in S, we use the Lorentz transformations, which in natural units can be represented as follows:

$$
\begin{aligned}
\Delta t' &= \frac{\Delta t - v \cdot \Delta x}{\sqrt{1 - v^2}} = \frac{\Delta t}{\sqrt{1 - v^2}} - \frac{v \cdot \Delta x}{\sqrt{1 - v^2}} \\
\Delta x' &= \frac{\Delta x - v \cdot \Delta t}{\sqrt{1 - v^2}} = \frac{\Delta x}{\sqrt{1 - v^2}} - \frac{v \cdot \Delta t}{\sqrt{1 - v^2}} \\
\Delta y' &= \Delta y \\
\Delta z' &= \Delta z
\end{aligned}
$$

Since the expressions

$$\gamma = \frac{1}{\sqrt{1 - v^2}} \text{ and } v\gamma = \frac{v}{\sqrt{1 - v^2}}$$

are numbers that depend not on the displacement vector $\Delta\vec{x}$, but only on the relative speed v, the Lorentz transformations represent *linear* transformations, i.e., there are 16 numbers, which we denote with

$$L^{\alpha'}_{\alpha} \quad \text{with } \alpha = t, x, y, z \text{ and } \alpha' = t', x', y', z',$$

so that applies:

$$
\begin{aligned}
\Delta t' &= L^{t'}_{t} \Delta t + L^{t'}_{x} \Delta x + L^{t'}_{y} \Delta y + L^{t'}_{z} \Delta z \\
\Delta x' &= L^{x'}_{t} \Delta t + L^{x'}_{x} \Delta x + L^{x'}_{y} \Delta y + L^{x'}_{z} \Delta z \\
\Delta y' &= L^{y'}_{t} \Delta t + L^{y'}_{x} \Delta x + L^{y'}_{y} \Delta y + L^{y'}_{z} \Delta z \\
\Delta z' &= L^{z'}_{t} \Delta t + L^{z'}_{x} \Delta x + L^{z'}_{y} \Delta y + L^{z'}_{z} \Delta z
\end{aligned}
$$

These four equations can be combined into one using the free index notation

$$\Delta\alpha' = L^{\alpha'}_{\alpha} \Delta\alpha. \tag{14.1}$$

By comparing the previous four equations with the Lorentz transformations, we obtain the values of the $L^{\alpha'}_{\alpha}$:

$$
\begin{array}{ll}
L^{t'}_{t} = \gamma & L^{t'}_{x} = -v\gamma \\
L^{x'}_{t} = -v\gamma & L^{x'}_{x} = \gamma \\
L^{y'}_{y} = 1 & L^{z'}_{z} = 1
\end{array}
$$

All other values of $L^{\alpha'}_{\alpha}$ are equal to zero.

We now define a **general vector** $\vec{A}$ (also called a **four-vector** or **spacetime vector**) as a collection of four numbers

$$A^{t}, A^{x}, A^{y}, A^{z},$$

which can be understood as the coordinates of a point in spacetime. In relativity theory, we write vectors and their components with capital letters. Just as in two- and three-dimensional Euclidean space, where we identified points in space with position vectors, we consider the four coordinates of any spacetime point in the inertial system S as components of a vector

$$\vec{A} \underset{S}{\rightarrow} \left(A^{t}, A^{x}, A^{y}, A^{z} \right).$$

So we write the components of a vector, separated by commas, in brackets and not in column form as in earlier chapters. With the free index $\alpha = t, x, y, z$ and the above-introduced abbreviating notation

$$\{A^\alpha\} = \left(A^t, A^x, A^y, A^z\right)$$

one can also briefly write

$$\vec{A} \underset{S}{\rightarrow} \{A^\alpha\}.$$

We require that the components of the vector transform in the same way as the coordinates, if we want to calculate these in another reference system S', i.e.

$$A^{\alpha'} = L^{\alpha'}_\alpha A^\alpha, \tag{14.2}$$

where Einstein's summation convention must be observed. If the components of a vector are known in an inertial system, they are uniquely determined for every other inertial system by the transformation rule.

Just like in two and three dimensions, two vectors $\vec{A}$ and $\vec{B}$ can be added

$$\vec{A} + \vec{B} \rightarrow \left(A^t + B^t, A^x + B^x, A^y + B^y, A^z + B^z\right)$$

and multiplied by a number c

$$c\vec{A} \rightarrow \left(cA^t, cA^x, cA^y, cA^z\right).$$

We explicitly check the addition rule, i.e., we check whether the sum of two vectors obeys the transformation rule. Let $A^{\alpha'}$ and $B^{\alpha'}$ be the components of $\vec{A}$ and $\vec{B}$ in the system S', then it follows

$$
\begin{aligned}
A^{\alpha'} + B^{\alpha'} &= L^{\alpha'}_\alpha A^\alpha + L^{\alpha'}_\alpha B^\alpha \\
&= \left(L^{\alpha'}_t A^t + L^{\alpha'}_x A^x + L^{\alpha'}_y A^y + L^{\alpha'}_z A^z\right) \\
&\quad + \left(L^{\alpha'}_t B^t + L^{\alpha'}_x B^x + L^{\alpha'}_y B^y + L^{\alpha'}_z B^z\right) \\
&= L^{\alpha'}_t \left(A^t + B^t\right) + L^{\alpha'}_x \left(A^x + B^x\right) \\
&\quad + L^{\alpha'}_y \left(A^y + B^y\right) + L^{\alpha'}_z \left(A^z + B^z\right) \\
&= L^{\alpha'}_\alpha \left(A^\alpha + B^\alpha\right),
\end{aligned}
$$

so the components of $\vec{A} + \vec{B}$ transform as required.

Example 14.1.
To practice the new terminology and notation, let's work through a specific

example. The inertial system S' may move with speed $v = 0.8$ in the positive x-direction relative to S. Then it follows

$$\gamma = \frac{1}{\sqrt{1 - v^2}} = \frac{1}{\sqrt{1 - 0.64}} = \frac{1}{\sqrt{0.36}} = \frac{1}{0.6} = 1.\bar{6}$$

and

$$-\gamma v = -1.67 \cdot 0.8 = -1.\bar{3}.$$

The vector $\vec{A}$ has the components $\vec{A} \underset{S}{\to} (5, 4, 2, 1)$ in system S. Its components in S' are calculated with (14.2) to

$$
\begin{aligned}
A^{t'} &= L^{t'}_{t} A^t + L^{t'}_{x} A^x + L^{t'}_{y} A^y + L^{t'}_{z} A^z \\
&= 1.\bar{6} \cdot 5 + (-1.\bar{3}) \cdot 4 + 0 \cdot 2 + 0 \cdot 1 = 3 \\
A^{x'} &= L^{x'}_{t} A^t + L^{x'}_{x} A^x + L^{x'}_{y} A^y + L^{x'}_{z} A^z \\
&= -1.33 \cdot 5 + 1.67 \cdot 4 + 0 \cdot 2 + 0 \cdot 1 = 0 \\
A^{y'} &= L^{y'}_{t} A^t + L^{y'}_{x} A^x + L^{y'}_{y} A^y + L^{y'}_{z} A^z = 0 \cdot 5 + 0 \cdot 4 + 1 \cdot 2 + 0 \cdot 1 = 2 \\
A^{z'} &= L^{z'}_{t} A^t + L^{z'}_{x} A^x + L^{z'}_{y} A^y + L^{z'}_{z} A^z = 0 \cdot 5 + 0 \cdot 4 + 0 \cdot 2 + 1 \cdot 1 = 1,
\end{aligned}
$$

so

$$\vec{A} \underset{S'}{\to} (3, 0, 2, 1),$$

and we recognize that the third and fourth components do not differ, which is of course due to the fact that the relative motion takes place in the x-direction and lengths perpendicular to the x-axis ($y = 2$, $z = 1$) do not change. $\square$

14.2. Vector Algebra

As in the chapter on tensor calculus in the Euclidean plane, the shorthand notation can be used to use matrix calculus for an alternative representation of the transformation rules. Since the calculation rules for the required (4×4) matrices can be largely taken over one to one from the two-dimensional case, we limit ourselves to a few remarks.

Remark 14.1. **MT: Matrix calculus in four dimensions**

A (4×4) **matrix** $M = (M^{\mu}_{\nu})$ is a square number scheme, consisting of 16 numbers, which are arranged in four rows and four columns. The upper index μ denotes the row and the lower ν the column. The row and column indices run from t to z:

$$
M = (M^{\mu}_{\nu}) = \begin{pmatrix}
M^t_t & M^t_x & M^t_y & M^t_z \\
M^x_t & M^x_x & M^x_y & M^x_z \\
M^y_t & M^y_x & M^y_y & M^y_z \\
M^z_t & M^z_x & M^z_y & M^z_z
\end{pmatrix}
$$

We now define some of the possible calculation operations that can be performed with matrices.

1. A matrix $M = (M^\mu{}_\nu)$ can be multiplied with a vector

$$\vec{A} \to \left(A^t, A^x, A^y, A^z\right),$$

the result is again a vector

$$M \cdot \vec{A} = \vec{B} \to \left(B^t, B^x, B^y, B^z\right),$$

whose components B^μ are calculated as follows:

$$B^\mu = M^\mu{}_\nu A^\nu$$

with Einstein's summation convention.

2. A matrix $M = (M^\mu{}_\nu)$ can be multiplied with another matrix $N = (N^\mu{}_\nu)$, the result is again a matrix

$$M \cdot N = O = (O^\mu{}_\nu),$$

whose matrix elements are calculated as follows

$$O^\mu{}_\nu = M^\mu{}_\alpha N^\alpha{}_\nu.$$

This equation makes sense, since on the left as well as on the right side the same free indices μ, ν appear and also stand in the correct place, the index α is the summation index.

3. If the matrix M has an inverse $M^{-1} = \left(M^{-1}\right)^\alpha{}_\beta$, then in index notation

$$M^\alpha{}_\gamma \left(M^{-1}\right)^\gamma{}_\beta = \delta^\alpha{}_\beta. \ \Box \tag{14.3}$$

We have already encountered some examples of matrices.

Example 14.2.

- The 16 numbers in Eq. (14.1), which we have called $L^{\alpha'}{}_\alpha$, can be considered as a matrix:

$$L = \left(L^{\alpha'}{}_\alpha\right) = \begin{pmatrix} \gamma & -\gamma v & 0 & 0 \\ -\gamma v & \gamma & 0 & 0 \\ 0 & 0 & 1 & 0 \\ 0 & 0 & 0 & 1 \end{pmatrix} \tag{14.4}$$

This matrix L is called **Lorentz transformation matrix**. With the help of the matrix, the transformations for the components of a vector can now also be compactly expressed by a vector equation:

$$\begin{pmatrix} A^{t'} \\ A^{x'} \\ A^{y'} \\ A^{z'} \end{pmatrix} = L \cdot \begin{pmatrix} A^{t} \\ A^{x} \\ A^{y} \\ A^{z} \end{pmatrix} \tag{14.5}$$

- The general Lorentz transformation $\Lambda^{\mu}{}_{\nu}$ (13.9) has the matrix representation

$$\begin{pmatrix} \gamma & -\gamma v^{x} & -\gamma v^{y} & -\gamma v^{z} \\ -\gamma v^{x} & 1 + (\gamma - 1)\dfrac{(v^{x})^{2}}{v^{2}} & (\gamma - 1)\dfrac{v^{x} v^{y}}{v^{2}} & (\gamma - 1)\dfrac{v^{x} v^{z}}{v^{2}} \\ -\gamma v^{y} & (\gamma - 1)\dfrac{v^{x} v^{y}}{v^{2}} & 1 + (\gamma - 1)\dfrac{(v^{y})^{2}}{v^{2}} & (\gamma - 1)\dfrac{v^{y} v^{z}}{v^{2}} \\ -\gamma v^{z} & (\gamma - 1)\dfrac{v^{x} v^{z}}{v^{2}} & (\gamma - 1)\dfrac{v^{y} v^{z}}{v^{2}} & 1 + (\gamma - 1)\dfrac{(v^{z})^{2}}{v^{2}} \end{pmatrix} . \tag{14.6}$$

- The Kronecker delta introduced above can also be written as a matrix:

$$E = \left(\delta_{\alpha}^{\beta} \right) = \begin{cases} 1 & if\ \alpha = \beta \\ 0 & if\ \alpha \neq \beta \end{cases} = \begin{pmatrix} 1 & 0 & 0 & 0 \\ 0 & 1 & 0 & 0 \\ 0 & 0 & 1 & 0 \\ 0 & 0 & 0 & 1 \end{pmatrix}$$

The matrix E is called **identity matrix**. $\square$

We now want to show that the Lorentz transformation matrix is nothing more than the Jacobian matrix of the coordinate transformation $(t, x, y, z) \rightarrow (t', x', y', z')$. Thus, the Lorentz transformations have the same status as the general coordinate transformations, which we have extensively presented in the chapter on vector calculation in the Euclidean plane. We now start from the Lorentz transformation

$$\begin{aligned} t' &= \frac{t - v \cdot x}{\sqrt{1 - v^{2}}} \\ x' &= \frac{x - v \cdot t}{\sqrt{1 - v^{2}}} \\ y' &= y \\ z' &= z \end{aligned}$$

and calculate the Jacobian matrix of the partial derivatives

$$
\begin{pmatrix}
\partial t'/\partial t & \partial t'/\partial x & \partial t'/\partial y & \partial t'/\partial z \\
\partial x'/\partial t & \partial x'/\partial x & \partial x'/\partial y & \partial x'/\partial z \\
\partial y'/\partial t & \partial y'/\partial x & \partial y'/\partial y & \partial y'/\partial z \\
\partial z'/\partial t & \partial z'/\partial x & \partial z'/\partial y & \partial z'/\partial z
\end{pmatrix}
=
\begin{pmatrix}
\gamma & -v\gamma & 0 & 0 \\
-v\gamma & \gamma & 0 & 0 \\
0 & 0 & 1 & 0 \\
0 & 0 & 0 & 1
\end{pmatrix}
= \left(L^{\alpha'}{}_{\alpha} \right).
$$

The Jacobian matrix thus coincides with the Lorentz matrix. Note that in the Jacobian matrix the primed coordinates are derived after the unprimed ones, hence there is also in the expression for the Lorentz transformation $L^{\alpha'}_{\alpha}$ the primed upper index α'.

In Special Relativity, the Lorentz matrices have exactly the same role as the Jacobian matrices of the general coordinate transformations in Euclidean space. The Lorentz matrix has more rows and columns, but on the other hand is not dependent on the points of spacetime, but always takes the same value. This simplifies - as we will see - especially the calculation of the (covariant) derivatives of vectors or tensors in SR quite considerably.

We have derived the inverse Lorentz transformations in Eq. (13.4). They are

$$
\begin{aligned}
t &= \frac{t' + vx'}{\sqrt{1 - v^2}}. \\
x &= \frac{x' + vt'}{\sqrt{1 - v^2}} \\
y &= y' \\
z &= z.'
\end{aligned}
$$

These differ from the Lorentz transformations formally in that the speed v is replaced by $-v$. So for clarity, let's write the Lorentz transformation matrix as $L^{\alpha'}_{\alpha}(v)$ and for the matrix of the inverse Lorentz transformation $L^{\alpha}_{\alpha'}(-v)$, so it follows, taking into account

$$
\gamma^2 = \frac{1}{1 - v^2}
$$

for the product of the two matrices

$$
L^{\alpha'}{}_{\alpha}(v)\, L^{\beta}{}_{\alpha'}(-v) =
\begin{pmatrix}
\gamma & -\gamma v & 0 & 0 \\
-\gamma v & \gamma & 0 & 0 \\
0 & 0 & 1 & 0 \\
0 & 0 & 0 & 1
\end{pmatrix}
\begin{pmatrix}
\gamma & \gamma v & 0 & 0 \\
\gamma v & \gamma & 0 & 0 \\
0 & 0 & 1 & 0 \\
0 & 0 & 0 & 1
\end{pmatrix}
$$

$$
= \begin{pmatrix} \gamma^2 - \gamma^2 v^2 & 0 & 0 & 0 \\ 0 & -\gamma^2 v^2 + \gamma^2 & 0 & 0 \\ 0 & 0 & 1 & 0 \\ 0 & 0 & 0 & 1 \end{pmatrix}
$$

$$
= \begin{pmatrix} \gamma^2 \left(1 - v^2\right) & 0 & 0 & 0 \\ 0 & \gamma^2 \left(1 - v^2\right) & 0 & 0 \\ 0 & 0 & 1 & 0 \\ 0 & 0 & 0 & 1 \end{pmatrix}
$$

$$
= \begin{pmatrix} 1 & 0 & 0 & 0 \\ 0 & 1 & 0 & 0 \\ 0 & 0 & 1 & 0 \\ 0 & 0 & 0 & 1 \end{pmatrix} = E = \left(\delta^\beta{}_\alpha\right).
$$

The matrix $L^\beta{}_{\alpha'}\left(-v\right)$ is thus the inverse matrix to the Lorentz matrix $L^{\alpha'}{}_\alpha\left(v\right)$. Of course, this relationship also follows from the fact that $L^\beta{}_{\alpha'}\left(-v\right)$ is the Jacobian matrix of the inverse mapping of the Lorentz transformation. It applies

$$
\begin{pmatrix} \partial t/\partial t' & \partial t/\partial x' & \partial t/\partial y' & \partial t/\partial z' \\ \partial x/\partial t' & \partial x/\partial x' & \partial x/\partial y' & \partial x/\partial z' \\ \partial y/\partial t' & \partial y/\partial x' & \partial y/\partial y' & \partial y/\partial z' \\ \partial z/\partial t' & \partial z/\partial x' & \partial z/\partial y' & \partial z/\partial z' \end{pmatrix} = \begin{pmatrix} \gamma & v\gamma & 0 & 0 \\ v\gamma & \gamma & 0 & 0 \\ 0 & 0 & 1 & 0 \\ 0 & 0 & 0 & 1 \end{pmatrix} = \left(L^\beta{}_{\alpha'}\left(-v\right)\right)
$$

and the lower dashed index in the inverse Lorentz matrix can be explained with the derivative after the dashed coordinates. In Eq. 8.29 on page 160 it was shown that the two Jacobian matrices are inverse to each other.

Let's look again at the transformation rule of the components of a vector, which in Eq. (14.5) were formulated using the Lorentz matrix,

$$
\vec{A}' = L\left(v\right) \cdot \vec{A},
$$

and multiply both sides of this equation (from the left) with the inverse matrix $L\left(-v\right)$, then we get

$$
L\left(-v\right) \cdot \vec{A}' = \underbrace{L\left(-v\right) \cdot L\left(v\right)}_{=E} \cdot \vec{A} = E \cdot \vec{A} = \vec{A},
$$

so the inverse transformation formula for the vector components, which in index notation looks like this:

$$
A^\alpha = L^\alpha{}_{\alpha'}\left(-v\right) A^{\alpha'} \tag{14.7}
$$

14.3. The Four-Velocity, the Four-Momentum

In three dimensions, we have defined the instantaneous velocity of a particle as the slope of the tangent to the worldline $\vec{x}(t)$ of the particle at a point (see section 2.1 on page 16. We transfer this definition to spacetime.

For a particle moving at constant velocity v along the positive x-axis, the spatial coordinates in its rest system S' are zero (the particle is at rest at the origin of the system S') and its worldline is the t'-axis. We *define* the **four-velocity** $\vec{U}$ of such a particle by

$$\vec{U} \underset{S'}{\rightarrow} \left(U^{t'}, U^{x'}, U^{y'}, U^{z'} \right) = (1, 0, 0, 0).$$

The four-velocity of a particle moving at constant velocity is thus a vector that is tangent to its worldline and in the rest system of the particle has the components $(1, 0, 0, 0)$. To get the components of $\vec{U}$ in a system S at rest relative to S', we apply the formula (14.7) in matrix form

$$U^\alpha \;=\; L^\alpha{}_{\alpha'}(-v)\, U^{\alpha'} \Longleftrightarrow$$

$$\begin{pmatrix} U^t \\ U^x \\ U^y \\ U^z \end{pmatrix} = \begin{pmatrix} \gamma & \gamma v & 0 & 0 \\ \gamma v & \gamma & 0 & 0 \\ 0 & 0 & 1 & 0 \\ 0 & 0 & 0 & 1 \end{pmatrix} \begin{pmatrix} 1 \\ 0 \\ 0 \\ 0 \end{pmatrix} = \begin{pmatrix} \gamma \\ \gamma v \\ 0 \\ 0 \end{pmatrix} = \begin{pmatrix} \dfrac{1}{\sqrt{1-v^2}} \\ \dfrac{v}{\sqrt{1-v^2}} \\ 0 \\ 0 \end{pmatrix}.$$

For small velocities $v \ll 1$ it holds

$$\frac{v}{\sqrt{1-v^2}} \approx v,$$

i.e., the *spatial components* of $\vec{U}$ in S are equal to $\vec{v} = (v, 0, 0)$, which corresponds to the velocity of the particle as perceived by S. Therefore, the vector $\vec{U}$ is called **four-velocity**. If the particle does not move along the x-axis, but with arbitrary constant velocity $\vec{v} = (v^x, v^y, v^z)$, the components of the four-velocity in system S take the general form

$$\vec{U} \underset{S}{\rightarrow} \left(U^t, U^x, U^y, U^z \right) = (\gamma, \gamma v^x, \gamma v^y, \gamma v^z), \tag{14.8}$$

whose derivation is given in section 14.4.

The spacetime interval Δs^2 between the origin and the spacetime point $(1, 0, 0, 0)$ in system S' is calculated as

$$\Delta s^2 = -t'^2 + x'^2 + y'^2 + z'^2 = -1$$

and is independent of the reference system. The square of the proper time in S' is according to Eq. (12.25)

$$t_E^2 = -\Delta s^2 = 1,$$

thus corresponding to the negative spacetime interval. The proper time thus provides a way to measure the lengths of straight worldlines in spacetime independently of the observer. If the worldline of a particle is not straight, but curved, which means that the particle is not moving at a constant speed, but is being accelerated, the same results can be obtained by the following considerations. To this end, we first explain the concept of **worldline in spacetime.**

Worldlines in Spacetime

In any arbitrary inertial system S we define the (usually curved) worldline of a particle in spacetime as a *timelike* four-vector

$$\vec{X}\left(t\right) \underset{S}{\rightarrow} \left(t, x\left(t\right), y\left(t\right), z\left(t\right)\right),$$

whose spatial components can depend on the coordinate time t in any way. It is not assumed that the particle must move along a straight line. For simplicity and transparency Fig. 14.1 again assumes that the particle only moves in the x-direction, i.e., the y- and z-components are assumed to be constant $(y\left(t\right) = z\left(t\right) = const)$ and are not shown in Fig. 14.1.

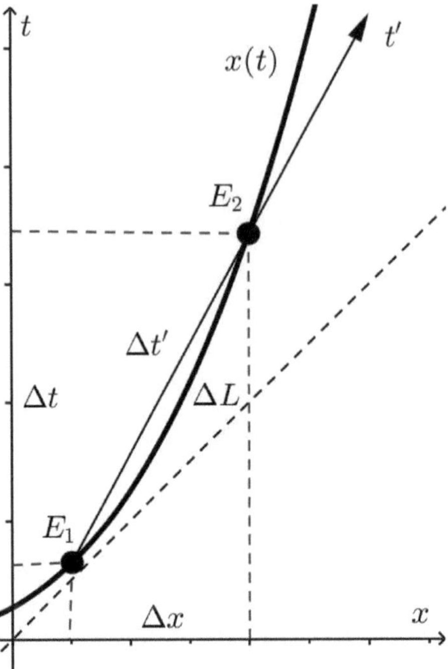

Figure 14.1.: Curved worldline

We divide the curved worldline into many segments ΔL, which are delimited by two spacetime points (in the graphic E_1 and E_2) and introduce moving reference systems $S' = (t', x')$ in such a way (in the graphic only the t'-axis is drawn), that the segments ΔL lie (approximately) parallel to the time axis t', i.e. $\Delta L \approx \Delta t'$. The particle under consideration then rests momentarily in the respective system S', which is therefore also called **instantaneous rest system**. For the moving particle, the measured time between the spacetime points E_1 and E_2 corresponds to the proper time $\Delta t'$ in the system S'. These statements are of course only approximate and become correct when the quantities Δx, Δt and ΔL become infinitesimally small. If the clock moves along the entire worldline, it goes from one instantaneous rest system to the next, so there are infinitely many rest systems. The moving clock sums all the times it spends in the respective rest system. The time indication on the clock can therefore be considered as a measure for the length of the worldline. The time difference of an arbitrarily moving clock gives the length of its worldline between two events. Since the time displayed on the moving clock is composed of a sum of proper times and the proper times of a system are invariant, the length of a curved worldline is also observer-independent. We want to make these rather qualitative statements a bit more precise in the following.

To do this, we first consider the curve of a worldline

$$\vec{r}(t) = \begin{pmatrix} x(t) \\ y(t) \\ z(t) \end{pmatrix}$$

in *three-dimensional Euclidean* space. In Fig. 14.2, for simplicity, it is assumed that $z(t) = 0$ and the z-axis is not depicted.

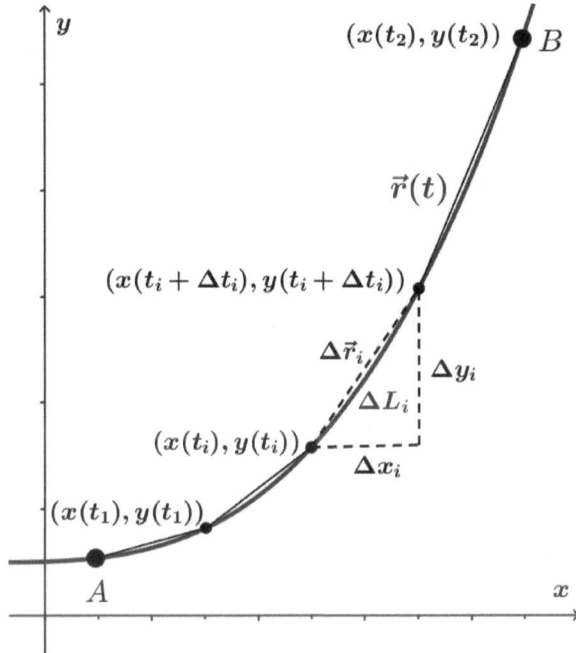

Figure 14.2.: Arc length of a curve

If you want to calculate the length of this curve between two points $A \to (x(t_1), y(t_1))$ and $B \to (x(t_2), y(t_2))$, you divide the curve piece into small segments ΔL_i. Each segment is approximated by the length of the dashed chord $\Delta \vec{r}_i$. This length is obtained with the Pythagorean theorem and with

$$\Delta x_i = x(t_i + \Delta t_i) - x(t_i) \quad \text{und} \quad \Delta y_i = y(t_i + \Delta t_i) - y(t_i)$$

to

$$|\Delta \vec{r}_i| = \sqrt{(\Delta x_i)^2 + (\Delta y_i)^2} = \sqrt{[x(t_i + \Delta t_i) - x(t_i)]^2 + [y(t_i + \Delta t_i) - y(t_i)]^2}.$$

We now use the differential again (see Eq. 4.8 on page 59) to estimate the differences under the root sign, i.e., we assume that Δt_i is small. Then it holds

$$x(t_i + \Delta t_i) - x(t_i) \approx \dot{x}(t_i)\,\Delta t_i$$

$$y\left(t_i + \Delta t_i\right) - y\left(t_i\right) \; \approx \; \dot{y}\left(t_i\right)\Delta t_i$$

and thus

$$\left|\Delta \vec{r}_i\right| \approx \sqrt{\left[\dot{x}\left(t_i\right)\Delta t_i\right]^2 + \left[\dot{y}\left(t_i\right)\Delta t_i\right]^2} = \sqrt{\dot{x}\left(t_i\right)^2 + \dot{y}\left(t_i\right)^2} \cdot \Delta t_i = \left|\dot{\vec{r}}\left(t_i\right)\right|\Delta t_i,$$

where $\dot{x}\left(t\right)$ again denotes the derivative of $x\left(t\right)$ with respect to time. If we sum up all the chord pieces, we get

$$L_n = \sum_{i=1}^{n}\left|\Delta \vec{r}_i\right| \approx \sum_{i=1}^{n}\left|\dot{\vec{r}}\left(t_i\right)\right|\Delta t_i.$$

By refining the decomposition with $n \to \infty$, the graph of $\vec{r}\left(t\right)$ is approximated arbitrarily closely by the polyline L_n. This limit formation turns the sum into an integral and we obtain for the length L of the curve between A and B

$$L = \lim_{n \to \infty} L_n = \int_{t_1}^{t_2}\left|\dot{\vec{r}}\left(t\right)\right|dt.$$

The quantity L is called the **arc length** of the curve between A and B . A very similar derivation leads us to the definition of the length of a curve in spacetime.

In spacetime, as above, we have to "make do" with two dimensions in Fig. 14.3. We divide the worldline $\vec{X}\left(t\right)$ of the particle (bold line) between two events $A \to \left(x\left(t_1\right), t_1\right)$ and $B \to \left(x\left(t_2\right), t_2\right)$ into (infinitesimal) small curve pieces ΔL_i. Each piece is approximated by the proper time $\Delta t_{i'}$ of the momentary rest system between t_i and $t_i + \Delta t_i$. This proper time is calculated due to

$$\left(\Delta t_{i'}\right)^2 = -\Delta \vec{X}_i{}^2 = -\left(-\Delta t_i^2 + \Delta x_i^2\right),$$

see equation (12.25) to

$$\Delta t_{i'} = \sqrt{-\Delta \vec{X}_i{}^2} = \sqrt{-\left(-\Delta t_i^2 + \Delta x_i^2\right)}.$$

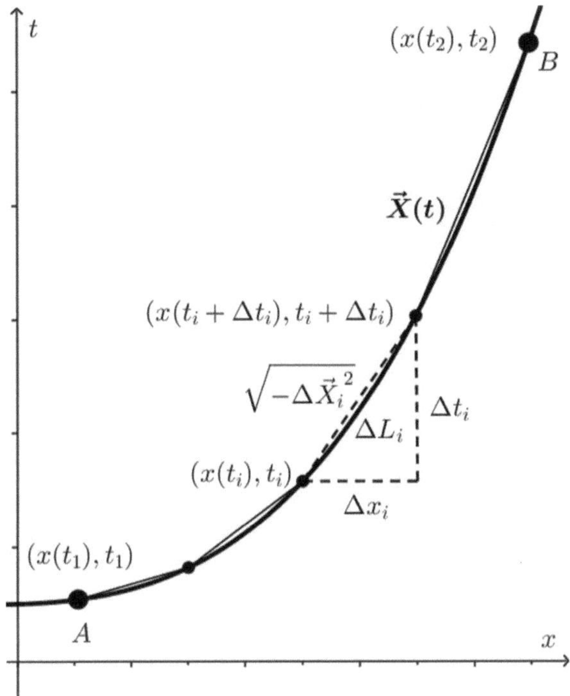

Figure 14.3.: Worldline of a curve in spacetime

The length of all chord pieces is thus

$$L_n = \sum_{i=1}^{n} \Delta t_{i'} = \sum_{i=1}^{n} \sqrt{\Delta t_i^2 - \Delta x_i^2} = \sum_{i=1}^{n} \sqrt{1 - \frac{\Delta x_i^2}{\Delta t_i^2}} \cdot \Delta t_i.$$

By refining the subdivision of the curve with $n \to \infty$, the graph of $\vec{X}(t)$ is approximated arbitrarily closely by the polyline L_n. This limit formation turns the sum into an integral and the expression $\Delta x_i / \Delta t_i$ converges to the derivative $\dot{x}(t) = v(t)$. We then obtain for the length L of the curve between A and B

$$L = \int_{t_1}^{t_2} \sqrt{1 - v^2(t)}\, dt.$$

If S'' is another reference system and $v''(t)$ is the respective relative speed between S'' and the momentary rest systems S', the following follows analogously

$$L = \int_{t_1''}^{t_2''} \sqrt{1 - v''^2(t'')}\, dt'',$$

i.e., the value of L does not depend on the reference system, but on the speed of the particle in the respective reference system. If the worldline of a particle is curved, forces act on this particle. Or in other words: According to the postulates of SR, a force-free particle moves along a straight line. Let's assume that the particle moves force-free from point A to point B. Then the worldline of the particle is a straight line. The length of this straight line corresponds in all other inertial systems S to what has just been said

$$L_{force-free} = \int_{t_1}^{t_2} \sqrt{1 - v^2}\, dt,$$

where now v is the *constant* relative speed between S and the rest system of the particle. Since the length is independent of the reference system, we can also calculate it in the rest system of the particle. There, $v = 0$ and it results

$$L_{force-free} = \int_{t_E(A)}^{t_E(B)} \sqrt{1 - 0^2}\, dt_E = \int_{t_E(A)}^{t_E(B)} dt_E = t_E(B) - t_E(A) = \Delta_{AB}\, t_E.$$

If we now consider the length of any other (curvilinear) worldline of the particle between A and B in any inertial system S, we find that for us the particle does not move at a constant speed, but that at least in a partial interval the speed $v\,(t)$ is different from zero, i.e., that in this partial interval

$$\sqrt{1 - v^2\,(t)} < 1$$

applies. This follows

$$L_{non-force-free} = \int_{t_1}^{t_2} \sqrt{1 - v^2\,(t)}\, dt < L_{force-free}.$$

In other words, the following mnemonic applies.

The worldline of a force-free particle between two events A and B is the worldline with the *greatest* proper time among all worldlines between these two events.

We will revisit and generalize this result in the context of General Relativity Theory.

Four-Momentum

The **four-momentum** of a particle $\vec{P}$ of rest mass m is defined in analogy to Newtonian mechanics in any reference system S by

$$\vec{P} = m\vec{U} \underset{S}{\rightarrow} \left(mU^t, mU^x, mU^y, mU^z\right) \underset{(14.8)}{=} \left(m\gamma, m\gamma v^x, m\gamma v^y, m\gamma v^z\right) \quad (14.9)$$

Note that the temporal component of the four-momentum $m\gamma$ corresponds to the relativistic energy, see Eq. (12.18), and the three spatial components correspond to those of the relativistic momentum

$$\left(m\gamma v^x, m\gamma v^y, m\gamma v^z\right) = \left(P^x, P^y, P^z\right),$$

see Eq. (12.17). This is the reason why the components of the four-momentum are often written as

$$\vec{P} \underset{S}{\rightarrow} \left(E, P^x, P^y, P^y\right)$$

In the rest system S' of the particle, it is

$$\vec{P} = m\vec{U} \underset{S'}{\rightarrow} m\left(1, 0, 0, 0\right) = \left(m, 0, 0, 0\right).$$

A particle has no spatial momentum in its rest system, and his energy is $E = m$, which in SI units is again the Einstein equation $E = mc^2$.

In analogy to the spacetime interval, we define the **square of a vector**

$$\vec{A} \underset{S}{\rightarrow} \left(A^t, A^x, A^y, A^z\right)$$

by

$$\vec{A}^{\,2} = -\left(A^t\right)^2 + \left(A^x\right)^2 + \left(A^y\right)^2 + \left(A^z\right)^2. \quad (14.10)$$

Like the spacetime interval, the square of a vector is a reference system-independent number, a so-called **Lorentz scalar**. Just like the square of the spacetime interval, the square of a vector can be positive, then $\vec{A}$ is called **space-like**. If the square is equal to zero, then $\vec{A}$ is called **light-like**. And if the square is negative, then $\vec{A}$ is called **time-like**. It is also the case that from $\vec{A}^{\,2} = 0$ it *cannot* generally be concluded that all components of $\vec{A}$ are zero. Since the four-velocity vector of a particle is equal to the vector $\vec{U} \rightarrow (1, 0, 0, 0)$ in its rest system, it therefore applies in all reference systems

$$\vec{U}^{\,2} = -1.$$

14.4. Relativistic Dynamics

We want to approach the four-velocity again from a different side and define it as in three dimensions as "derivative of the path with respect to time". For this, we assume that a particle with mass undergoes a small displacement

$$d\vec{x} \underset{S}{\rightarrow} (dt, dx, dy, dz)$$

along its worldline. The magnitude of the displacement is equal to equation (14.10)

$$ds^2 = -dt^2 + dx^2 + dy^2 + dz^2,$$

which, according to formula (12.22), corresponds to the infinitesimal spacetime interval:

$$ds^2 = d\vec{x}^2 = -dt_E^2$$

Since the worldline of the particle is time-like, ds^2 is negative and thus $dt_E^2 = -ds^2$ is a positive number. We can therefore take the square root and get

$$dt_E = \sqrt{-ds^2} = \sqrt{-d\vec{x}^2}.$$

We now consider the vector $d\vec{x}/dt_E$, which, since dt_E is a number, is a multiple of the vector $d\vec{x}$, so parallel to $d\vec{x}$ and thus also tangential to the worldline of the particle. It applies

$$\frac{d\vec{x}}{dt_E} \cdot \frac{d\vec{x}}{dt_E} = \frac{d\vec{x}^2}{dt_E^2} = \frac{-dt_E^2}{dt_E^2} = -1,$$

therefore $d\vec{x}/dt_E$ is a time-like unit vector, which is tangential to the worldline. In a momentary rest system S', $dt = dt_E$, so

$$d\vec{x} \underset{S'}{\rightarrow} (dt, 0, 0, 0) = (dt_E, 0, 0, 0),$$

from which

$$d\vec{x}/dt_E \underset{S'}{\rightarrow} (1, 0, 0, 0)$$

follows, which exactly corresponds to the definition of the four-vector $\vec{U}$ in the rest system. So there it applies

$$\vec{U} = d\vec{x}/dt_E. \tag{14.11}$$

The "natural" definition of velocity by $d\vec{x}/dt$ would have led to a quantity that is *not* a four-vector, since in Lorentz transformations not only the numerator

$d\vec{x}$, but also the denominator dt would have to be transformed, whereas the proper time dt_E is an invariant quantity!

The components of $d\vec{x}/dt_E$ in any inertial system S are calculated due to

$$dt_E = dt\,\sqrt{1 - v^2} = \frac{dt}{\gamma}$$

with the chain rule to

$$\frac{d\vec{x}}{dt_E} = \frac{d\vec{x}}{dt}\frac{dt}{dt_E} = \frac{1}{\sqrt{1 - v^2}}\frac{d\vec{x}}{dt} \xrightarrow{S} \gamma\left(\frac{dt}{dt}, \frac{dx}{dt}, \frac{dy}{dt}, \frac{dz}{dt}\right) = \left(\gamma, \gamma v^x, \gamma v^y, \gamma v^z\right),$$

which, according to (14.8), corresponds exactly to the definition of $\vec{U}$ in S.

We now consider the quantity

$$\vec{A} = \frac{d\vec{U}}{dt_E} = \frac{d^2\vec{x}}{dt_E^2},$$

which is a four-vector for the same reasons as above. The vector $\vec{A}$ is called **four-acceleration** . We choose an instantaneous rest system S', in which the particle is accelerated from rest. In S', the proper time coincides with the coordinate time: $dt_E = dt$, hence follows

$$\vec{A} = \frac{d^2\vec{x}}{dt_E^2} \xrightarrow{S'} \left(\frac{d^2t}{dt^2}, \frac{d^2x}{dt^2}, \frac{d^2y}{dt^2}, \frac{d^2z}{dt^2}\right) = (0, a^x, a^y, a^z),$$

since due to $dt/dt = 1$ it follows that $d^2t/dt^2 = 0$. In the instantaneous rest system, the four-acceleration thus only has a spatial component, which is equal to the acceleration $\vec{a}$ of the particle. The square of $\vec{A}$ is the same in all inertial systems

$$\vec{A}^2 = (a^x)^2 + (a^y)^2 + (a^z)^2 = |\vec{a}|^2$$

and positive.

With the four-acceleration, we can write down the **relativistic generalization of Newton's equation of motion**

$$m\,\vec{A} = m\,\frac{d^2\vec{x}}{dt_E^2} = \vec{F}, \tag{14.12}$$

where the four-vector $\vec{F}$ on the right side is called **Minkowski force**. Since $\vec{A}$ is a four-vector and m is the rest mass, i.e., a Lorentz scalar, the equation is Lorentz invariant. We require that the equation reduces to the Newtonian limit for small velocities $v \ll 1$. In the instantaneous rest system S' ($v = 0$), the left side yields

$$m\,\vec{A} \underset{S'}{\rightarrow} (0, ma^x, ma^y, ma^y),$$

i.e., for the Minkowski force in S' it follows

$$\vec{F} \underset{S'}{\rightarrow} (0, F^x, F^y, F^z) = (0, \vec{F}_N),$$

where $\vec{F}_N \rightarrow (F^x, F^y, F^z)$ denotes the *three-dimensional* force vector of Newtonian mechanics. For the components of the Minkowski force in the inertial system S, it then follows

$$F^\alpha = L^\alpha{}_{\alpha'}(-v)\,F^{\alpha'} = \begin{pmatrix} \gamma & \gamma v & 0 & 0 \\ \gamma v & \gamma & 0 & 0 \\ 0 & 0 & 1 & 0 \\ 0 & 0 & 0 & 1 \end{pmatrix} \begin{pmatrix} 0 \\ F^x \\ F^y \\ F^z \end{pmatrix} = \begin{pmatrix} \gamma v F^x \\ \gamma F^x \\ F^y \\ F^z \end{pmatrix}.$$

We want to use the new tools of vector calculation for some derive the results achieved in earlier chapters. For the four-momentum

$$\vec{P} \rightarrow (E, P^x, P^y, P^z)$$

it holds on the one hand

$$\vec{P}^2 = m^2 \vec{U}^2 = -m^2$$

and on the other hand

$$\vec{P}^2 = -E^2 + (P^x)^2 + (P^y)^2 + (P^z)^2,$$

from which

$$E^2 = m^2 + (P^x)^2 + (P^y)^2 + (P^z)^2$$

follows, which exactly corresponds to the *relativistic energy theorem* (see formula 12.20 on page 258) when considering that the rest energy $E_0 = m$.

   Light particles (photons) have no mass and move at the speed of light, so they are light-like, i.e., for a light particle

$$d\vec{x}^2 = 0$$

and thus also $dt_E = 0$, which with Eq. (14.11) means that no four-speed can be defined for photons. So there is no inertial system in which photons rest. Of course, vectors can be defined that are tangential to a beam of light, which is a straight line, but no one is found whose square is not zero. The four-momentum of a photon is due to

$$\vec{P}^2_{photon} = -m^2 = 0$$

light-like, but usually not a unit vector. Its components consist of the energy and the (spatial) momentum of the particle. A photon with energy E, moving in system S in x-direction, has the four-momentum components

$$\vec{P}_{photon} \underset{S}{\rightarrow} (E, P^x, 0, 0),$$

from which

$$\vec{P}^2_{photon} = -E^2 + (P^x)^2 = 0$$

follows and thus we again get (see also Eq. 12.21 on page 259) the result that the spatial momentum of a photon corresponds to its energy.

According to Einstein (see Sect. 29.3), a photon has the energy

$$E = h \cdot f,$$

where f denotes the frequency of the photon, and h is the so-called **Planck constant**

$$h = 6,626 \cdot 10^{-34} \, \text{Js}^{-1} \quad \text{in SI units,}$$

see Tab. 30.3. Assuming, in a reference system S a photon has the frequency f and moves in x-direction. Then the particle in a reference system S', which moves with speed v relative to S also in x-direction, with the Lorentz transformation and because $E = hf = P^x$ has the energy

$$
\begin{aligned}
E' &= P^{t'} = \gamma P^t - v\gamma P^x = E/\sqrt{1-v^2} - P^x v/\sqrt{1-v^2} \\
&= hf/\sqrt{1-v^2} - hfv/\sqrt{1-v^2} = E \frac{1-v}{\sqrt{1-v^2}}.
\end{aligned}
$$

So, because $E' = hf'$ for the frequency f' in S'

$$\frac{f'}{f} = \frac{E'}{E} = \frac{1-v}{\sqrt{1-v^2}} = \frac{\sqrt{(1-v)^2}}{\sqrt{(1-v)(1+v)}} = \sqrt{\frac{(1-v)}{(1+v)}}.$$

This is again (compare Eq. 12.12 on page 242) the Doppler effect formula for photons.

15. Tensor Calculus in Special Relativity Theory

In the last chapter we have found that physical laws in Special Relativity Theory have the same form in every reference system if they are expressed as four-vector equations. The physical laws of spacetime thus behave like the classical laws of Newtonian physics: They are invariant under Galilean transformations when formulated in (three-dimensional) vector form, cf. Sect. 12.1. The four-vectors, which we will often simply call "vectors" from now on, are geometric objects that have different components in different coordinate systems, but remain unchanged in spacetime. We have also seen in some examples that some special calculations with vectors always have the same result in any reference system, e.g., the square of a vector is an invariant quantity. In this chapter, we want to extend the concept of vector calculation in spacetime to **tensor calculation**. The reason for this abstraction is that some physical relationships in the Special Theory of Relativity (e.g., the unification of electric and magnetic fields into the electromagnetic field) can no longer be represented as vector quantities, but require complicated expressions (i.e., tensors). Of course, we demand from tensor equations that represent physical laws that they take the same form in every reference system, which we have already shown in detail for the two-dimensional Euclidean space.

15.1. $(0, N)$-Tensoren

In this section, we repeat and generalize some statements about tensors that we have derived for the Euclidean plane. The terms and calculations often have a great similarity with those for the Euclidean plane, so we limit ourselves to a brief presentation of what is already known and only deal with the new constructs in more detail.

A $(0, N)$-tensor is a mapping that takes N four-vectors as arguments and returns values in the real numbers and is linear, i.e., *multilinear*, in *each* of its arguments. In the theory of relativity, the number N can take the values $0, 1, 2, 3, 4$. As a continuous example in this section, we want to consider a $(0, 3)$-tensor $\mathbf{T}$. This tensor $\mathbf{T}$ thus maps three vectors to a real number, its

$4 \cdot 4 \cdot 4 = 64$ components have *three lower* indices in the inertial system S, i.e.,

$$\mathbf{T} \underset{S}{\rightarrow} T_{\alpha\beta\gamma},$$

where each index α, β, γ can take the values t, x, y, z in the Cartesian coordinate system. As in the two-dimensional case, the tensor components must fulfill the following transformation formula when changing to the inertial system S':

$$T_{\alpha'\beta'\gamma'} = \frac{\partial\alpha}{\partial\alpha'} \frac{\partial\beta}{\partial\beta'} \frac{\partial\gamma}{\partial\gamma'} T_{\alpha\beta\gamma}$$

with a *triple* sum on the right side! Since the last chapter showed that in SR the Jacobian matrices $(\partial\alpha/\partial\alpha')$ are equal to the Lorentz matrices $L^{\alpha}_{\ \alpha'}(-v)$, the transformation rule can also be written as

$$T_{\alpha'\beta'\gamma'} = L^{\alpha}_{\ \alpha'}(-v)\, L^{\beta}_{\ \beta'}(-v)\, L^{\gamma}_{\ \gamma'}(-v)\, T_{\alpha\beta\gamma} \tag{15.1}$$

For three arbitrary vectors $\vec{A}, \vec{B}, \vec{C}$, the function value of $\mathbf{T}$ is defined by

$$\mathbf{T}\left(\vec{A}, \vec{B}, \vec{C}\right) = A^{\alpha} B^{\beta} C^{\gamma}\, T_{\alpha\beta\gamma}$$

and we have to show that this expression is a Lorentz scalar, i.e., it always delivers the same number regardless of the chosen inertial system. It applies with (14.2) and (15.1)

$$
\begin{aligned}
& A^{\alpha'} B^{\beta'} C^{\gamma'} T_{\alpha'\beta'\gamma'} \\
=\ & L^{\alpha'}_{\ \alpha}(v)\, A^{\alpha} L^{\beta'}_{\ \beta}(v)\, B^{\beta} L^{\gamma'}_{\ \gamma}(v)\, C^{\gamma} L^{\mu}_{\ \alpha'}(-v)\, L^{\nu}_{\ \beta'}(-v)\, L^{\rho}_{\ \gamma'}(-v)\, T_{\mu\nu\rho} \\
=\ & \underbrace{L^{\alpha'}_{\ \alpha}(v)\, L^{\mu}_{\ \alpha'}(-v)}_{=\delta^{\mu}_{\alpha}}\, \underbrace{L^{\beta'}_{\ \beta}(v)\, L^{\nu}_{\ \beta'}(-v)}_{=\delta^{\nu}_{\beta}}\, \underbrace{L^{\gamma'}_{\ \gamma}(v)\, L^{\rho}_{\ \gamma'}(-v)}_{=\delta^{\rho}_{\gamma}}\, A^{\alpha} B^{\beta} C^{\gamma}\, T_{\mu\nu\rho} \\
=\ & \delta^{\mu}_{\alpha} \cdot \delta^{\nu}_{\beta} \cdot \delta^{\rho}_{\gamma}\, A^{\alpha} B^{\beta} C^{\gamma}\, T_{\mu\nu\rho} \\
=\ & A^{\alpha} B^{\beta} C^{\gamma}\, T_{\alpha\beta\gamma},
\end{aligned}
$$

since the Lorentz matrices are inverses of each other. Thus, we have demonstrated the independence of $\mathbf{T}(\vec{A}, \vec{B}, \vec{C})$ from the inertial system.

Since a real-valued function $\phi(t, z, y, x)$, which is also called a **scalar field** dependent on spacetime points, does not have any vector as an argument, such functions are *defined* as $(0,0)$-tensors.

In the following sections, we will examine the $(0,1)$- and $(0,2)$-tensors that primarily occur in the theory of special relativity in more detail.

15.2. One-Forms

A $(0,1)$-tensor is called a **one-form**, also known as a **covector** or **covariant vector**. We define a one-form $\tilde{p}$ as a linear mapping that maps a vector to a real number, and which can be represented by four numbers, i.e., its components, when choosing an inertial system S

$$\tilde{p} \underset{S}{\to} (p_t, p_x, p_y, p_z) = \{p_\alpha\},$$

where the components should transform according to

$$p_{\alpha'} = \frac{\partial \alpha}{\partial \alpha'} p_\alpha = L^\alpha{}_{\alpha'}(-v) p_\alpha \tag{15.2}$$

when changing the inertial system $S \to S'$. If $\vec{A}$ is any vector, then $\tilde{p}\left(\vec{A}\right)$ is defined by

$$\tilde{p}\left(\vec{A}\right) = A^t p_t + A^x p_x + A^y p_y + A^z p_z = A^\alpha p_\alpha \tag{15.3}$$

and as above, it is shown that the expression on the right side is a Lorentz scalar.

In Remark 4.8 on page 66, we defined the constructs of partial derivative and gradient for a function $\phi(x, y, z)$ that depends on the three spatial coordinates x, y, z and takes values in the real numbers. We want to transfer these definitions to spacetime, and we will see that the gradient of a scalar function is naturally a one-form. So let ϕ be a function that depends in a reference system S on the four spacetime coordinates $\phi = \phi(t, x, y, z)$ and takes values in the real numbers. We define the partial derivative $\partial\phi/\partial t$ of ϕ with respect to the variable t at the point (t_0, x_0, y_0, z_0) as a derivative, which arises when we consider the three spatial variables as constants, as in Eq. 4.14 on page 66 by

$$\frac{\partial \phi(t_0, x_0, y_0, z_0)}{\partial t} = \lim_{\Delta t \to 0} \frac{\phi(t_0 + \Delta t, x_0, y_0, z_0) - \phi(t_0, x_0, y_0, z_0)}{\Delta t}.$$

Similarly, the partial derivatives of ϕ with respect to x, y and z, namely $\partial\phi/\partial x, \partial\phi/\partial y$ and $\partial\phi/\partial z$ are defined. We want to consider the four numbers

$$\frac{\partial \phi(t_0, x_0, y_0, z_0)}{\partial t}, \frac{\partial \phi(t_0, x_0, y_0, z_0)}{\partial x}, \frac{\partial \phi(t_0, x_0, y_0, z_0)}{\partial y}, \frac{\partial \phi(t_0, x_0, y_0, z_0)}{\partial z}$$

as components of a one-form $\tilde{d}\phi$, so

$$\tilde{d}\phi \underset{S}{\to} (d_t\,\phi, d_x\,\phi, d_y\,\phi, d_z\,\phi) = \left(\frac{\partial\phi}{\partial t}, \frac{\partial\phi}{\partial x}, \frac{\partial\phi}{\partial y}, \frac{\partial\phi}{\partial z}\right), \tag{15.4}$$

where for simplicity the argument (t_0, x_0, y_0, z_0) was omitted from the function ϕ. However, one must not forget in the sequence that the one-form $\tilde{d}\phi$ is always considered at a *fixed* point in space (t_0, x_0, y_0, z_0). The one-form $\tilde{d}\phi$ is called the **gradient** of ϕ. If one applies the gradient $\tilde{d}\phi$ to a vector $\vec{A} \underset{S}{\to} \left(A^t, A^x, A^y, A^z\right)$, one obtains

$$\tilde{d}\phi(\vec{A}) = d_\alpha \phi \, A^\alpha = \frac{\partial \phi}{\partial t} A^t + \frac{\partial \phi}{\partial x} A^x + \frac{\partial \phi}{\partial y} A^y + \frac{\partial \phi}{\partial z} A^z,$$

from which the linearity of $\tilde{d}\phi$ immediately follows. Because if

$$\vec{A} \to \left(A^t, A^x, A^y, A^z\right), \vec{B} \to \left(B^t, B^x, B^y, B^z\right)$$

are two vectors and c_1 and c_2 are two numbers, then

$$\begin{aligned}
\tilde{d}\phi\left(c_1\vec{A} + c_2\vec{B}\right) &= \frac{\partial \phi}{\partial t}\left(c_1 A^t + c_2 B^t\right) + \frac{\partial \phi}{\partial x}\left(c_1 A^x + c_2 B^x\right) \\
&+ \frac{\partial \phi}{\partial y}\left(c_1 A^y + c_2 B^y\right) + \frac{\partial \phi}{\partial z}\left(c_1 A^z + c_2 B^z\right) \\
&= c_1\left(\frac{\partial \phi}{\partial t} A^t + \frac{\partial \phi}{\partial x} A^x + \frac{\partial \phi}{\partial y} A^y + \frac{\partial \phi}{\partial z} A^z\right) \\
&+ c_2\left(\frac{\partial \phi}{\partial t} B^t + \frac{\partial \phi}{\partial x} B^x + \frac{\partial \phi}{\partial y} B^y + \frac{\partial \phi}{\partial z} B^z\right) \\
&= c_1\tilde{d}\phi\left(\vec{A}\right) + c_2\tilde{d}\phi\left(\vec{B}\right).
\end{aligned}$$

We are still checking whether $\tilde{d}\phi$ also has the desired transformation behavior, i.e., whether in a system S' with the spacetime coordinates (t', x', y', z') also

$$\tilde{d}\phi \underset{S'}{\to} \left(d_{t'}\,\phi, d_{x'}\,\phi, d_{y'}\,\phi, d_{z'}\,\phi\right) = \left(\frac{\partial \phi}{\partial t'}, \frac{\partial \phi}{\partial x'}, \frac{\partial \phi}{\partial y'}, \frac{\partial \phi}{\partial z'}\right)$$

applies. If one writes out the compressed transformation formula

$$d_{\alpha'}\,\phi = L^\alpha{}_{\alpha'}\left(-v\right) d_\alpha\,\phi$$

one obtains the four components

$$\begin{aligned}
d_{t'}\,\phi &= \gamma \frac{\partial \phi}{\partial t} + \gamma v \frac{\partial \phi}{\partial x} \\
d_{x'}\,\phi &= \gamma v \frac{\partial \phi}{\partial t} + \gamma \frac{\partial \phi}{\partial x} \\
d_{y'}\,\phi &= \frac{\partial \phi}{\partial y}
\end{aligned} \tag{15.5}$$

$$d_{z'}\,\phi \;=\; \frac{\partial\phi}{\partial z}.$$

The spacetime coordinates (t, x, y, z) result from (t', x', y', z') with the inverse Lorentz transformation, i.e., it applies

$$t \;=\; \gamma t' + \gamma v x' \Rightarrow \frac{\partial t}{\partial t'} = \gamma,\, \frac{\partial t}{\partial x'} = \gamma v$$

$$x \;=\; \gamma v t' + \gamma x' \Rightarrow \frac{\partial x}{\partial t'} = \gamma v,\, \frac{\partial x}{\partial x'} = \gamma$$

$$y \;=\; y' \Rightarrow \frac{\partial y}{\partial y'} = 1$$

$$z \;=\; z' \Rightarrow \frac{\partial z}{\partial z'} = 1.$$

We now replace in Eq. (15.5) the quantities γ and γv by the partial derivatives of the inverse transformation and obtain

$$d_{t'}\,\phi \;=\; \gamma\frac{\partial\phi}{\partial t} + \gamma v\frac{\partial\phi}{\partial x} = \frac{\partial\phi}{\partial t}\frac{\partial t}{\partial t'} + \frac{\partial\phi}{\partial x}\frac{\partial x}{\partial t'}$$

$$d_{x'}\,\phi \;=\; \gamma v\frac{\partial\phi}{\partial t} + \gamma\frac{\partial\phi}{\partial x} = \frac{\partial\phi}{\partial t}\frac{\partial t}{\partial x'} + \frac{\partial\phi}{\partial x}\frac{\partial x}{\partial x'}$$

$$d_{y'}\,\phi \;=\; \frac{\partial\phi}{\partial y} = \frac{\partial\phi}{\partial y}\frac{\partial y}{\partial y'}$$

$$d_{z'}\,\phi \;=\; \frac{\partial\phi}{\partial z} = \frac{\partial\phi}{\partial z}\frac{\partial z}{\partial z'}.$$

Since the other elements of the Jacobian matrix $\left(\partial x^\beta/\partial x^{\alpha'}\right)$ are equal to zero, it also applies

$$d_{t'}\,\phi \;=\; \frac{\partial\phi}{\partial t}\frac{\partial t}{\partial t'} + \frac{\partial\phi}{\partial x}\frac{\partial x}{\partial t'} + \frac{\partial\phi}{\partial y}\frac{\partial y}{\partial t'} + \frac{\partial\phi}{\partial z}\frac{\partial z}{\partial t'}$$

$$d_{x'}\,\phi \;=\; \frac{\partial\phi}{\partial t}\frac{\partial t}{\partial x'} + \frac{\partial\phi}{\partial x}\frac{\partial x}{\partial x'} + \frac{\partial\phi}{\partial y}\frac{\partial y}{\partial x'} + \frac{\partial\phi}{\partial z}\frac{\partial z}{\partial x'}$$

$$d_{y'}\,\phi \;=\; \frac{\partial\phi}{\partial t}\frac{\partial t}{\partial y'} + \frac{\partial\phi}{\partial x}\frac{\partial x}{\partial y'} + \frac{\partial\phi}{\partial y}\frac{\partial y}{\partial y'} + \frac{\partial\phi}{\partial z}\frac{\partial z}{\partial y'} \qquad (15.6)$$

$$d_{z'}\,\phi \;=\; \frac{\partial\phi}{\partial t}\frac{\partial t}{\partial z'} + \frac{\partial\phi}{\partial x}\frac{\partial x}{\partial z'} + \frac{\partial\phi}{\partial y}\frac{\partial y}{\partial z'} + \frac{\partial\phi}{\partial z}\frac{\partial z}{\partial z'}.$$

If we again use the index notation and observe Einstein's summation convention, the last four equations can be compressed as

$$d_{\alpha'}\,\phi = \frac{\partial\phi}{\partial\alpha}\frac{\partial\alpha}{\partial\alpha'},$$

which once again demonstrates the advantage of this notation. To further process the last equation, we need a generalized *chain rule*.

Remark 15.1. **MT: Generalized Chain Rule**

Let ϕ be a real-valued function that depends on a spacetime point (t, x, y, z), $\phi = \phi(t, x, y, z)$. The coordinates of the spacetime point may themselves depend on other variables

$$
\begin{aligned}
t &= t(t', x', y', z') \\
x &= x(t', x', y', z') \\
y &= y(t', x', y', z') \\
z &= z(t', x', y', z')
\end{aligned}
$$

Then ϕ also indirectly depends on the variables t', x', y', z' as well. Analogous to Eq. 4.24 on page 71 the partial derivatives of ϕ with respect to the variables t', x', y', z' follow this chain rule

$$
\begin{aligned}
\frac{\partial \phi}{\partial t'} &= \frac{\partial \phi}{\partial t}\frac{\partial t}{\partial t'} + \frac{\partial \phi}{\partial x}\frac{\partial x}{\partial t'} + \frac{\partial \phi}{\partial y}\frac{\partial y}{\partial t'} + \frac{\partial \phi}{\partial z}\frac{\partial z}{\partial t'} \\
\frac{\partial \phi}{\partial x'} &= \frac{\partial \phi}{\partial t}\frac{\partial t}{\partial x'} + \frac{\partial \phi}{\partial x}\frac{\partial x}{\partial x'} + \frac{\partial \phi}{\partial y}\frac{\partial y}{\partial x'} + \frac{\partial \phi}{\partial z}\frac{\partial z}{\partial x'} \\
\frac{\partial \phi}{\partial y'} &= \frac{\partial \phi}{\partial t}\frac{\partial t}{\partial y'} + \frac{\partial \phi}{\partial x}\frac{\partial x}{\partial y'} + \frac{\partial \phi}{\partial y}\frac{\partial y}{\partial y'} + \frac{\partial \phi}{\partial z}\frac{\partial z}{\partial y'} \\
\frac{\partial \phi}{\partial z'} &= \frac{\partial \phi}{\partial t}\frac{\partial t}{\partial z'} + \frac{\partial \phi}{\partial x}\frac{\partial x}{\partial z'} + \frac{\partial \phi}{\partial y}\frac{\partial y}{\partial z'} + \frac{\partial \phi}{\partial z}\frac{\partial z}{\partial z'},
\end{aligned}
\tag{15.7}
$$

which can be compressed in index notation as

$$
\frac{\partial \phi}{\partial \alpha'} = \frac{\partial \phi}{\partial \alpha}\frac{\partial \alpha}{\partial \alpha'}. \qquad \Box
\tag{15.8}
$$

With (15.7) we can rewrite the right side of (15.6) and obtain

$$
\begin{aligned}
d_{t'}\,\phi &= \frac{\partial \phi}{\partial t'} \\
d_{x'}\,\phi &= \frac{\partial \phi}{\partial x'} \\
d_{y'}\,\phi &= \frac{\partial \phi}{\partial y'} \\
d_{z'}\,\phi &= \frac{\partial \phi}{\partial z'}
\end{aligned}
\tag{15.9}
$$

or in index notation

$$
d_{\alpha'}\,\phi = \frac{\partial \phi}{\partial \alpha'}
$$

and thus the desired result

$$
\tilde{d}\phi \underset{S'}{\to} \left(d_{t'}\,\phi, d_{x'}\,\phi, d_{y'}\,\phi, d_{z'}\,\phi\right) = \left(\frac{\partial \phi}{\partial t'}, \frac{\partial \phi}{\partial x'}, \frac{\partial \phi}{\partial y'}, \frac{\partial \phi}{\partial z'}\right).
$$

If we consider the event (t, x, y, z) as a spacetime point on the worldline of a particle at proper time t_E, we get

$$t = t\left(t_E\right), x = x\left(t_E\right), y = y\left(t_E\right), z = z\left(t_E\right).$$

Then the function ϕ is also implicitly a function of the proper time t_E

$$\phi\left(t_E\right) = \phi\left(t\left(t_E\right), x\left(t_E\right), y\left(t_E\right), z\left(t_E\right)\right).$$

Now we know from Eq. (14.11) that the four-velocity in an instantaneous inertial system S has the components

$$\vec{U} \underset{S}{\to} \left(U^t, U^x, U^y, U^z\right) = \left(\frac{dt}{dt_E}, \frac{dx}{dt_E}, \frac{dy}{dt_E}, \frac{dz}{dt_E}\right)$$

With the chain rule 4.23 on page 70, the derivative of ϕ with respect to t_E follows

$$
\begin{aligned}
\frac{d\phi}{dt_E} &= \frac{\partial\phi}{\partial t}\frac{dt}{dt_E} + \frac{\partial\phi}{\partial x}\frac{dx}{dt_E} + \frac{\partial\phi}{\partial y}\frac{dy}{dt_E} + \frac{\partial\phi}{\partial z}\frac{dz}{dt_E} \\
&= \frac{\partial\phi}{\partial t}U^t + \frac{\partial\phi}{\partial x}U^x + \frac{\partial\phi}{\partial y}U^y + \frac{\partial\phi}{\partial z}U^z.
\end{aligned}
\tag{15.10}
$$

If we read these equations from back to front, we get a rule for calculating the rate of change of ϕ from the components of the four-velocity along a curve, with respect to which $\vec{U}$ is tangential. For the components of the gradient of a scalar function ϕ, we also write briefly

$$\tilde{d}\phi \to \frac{\partial\phi}{\partial\alpha} = \partial_\alpha\phi = \phi_{,\alpha}.
\tag{15.11}$$

15.3. $(0, 2)$-**Tensors**

A $(0, 2)$-tensor $\mathbf{T}$ is a linear mapping that maps two vectors to a real number and can be represented by 16 numbers, i.e., its components, when choosing an inertial system S

$$\mathbf{T} \underset{S}{\to} T_{\alpha\beta},$$

where the components should transform according to a change of inertial system $S \to S'$ as

$$T_{\alpha'\beta'} = \frac{\partial\alpha}{\partial\alpha'}\frac{\partial\beta}{\partial\beta'}T_{\alpha\beta} = L^\alpha{}_{\alpha'}\left(-v\right)L^\beta{}_{\beta'}\left(-v\right)T_{\alpha\beta}.
\tag{15.12}$$

If $\vec{A}, \vec{B}$ are arbitrary vectors, then $\mathbf{T}\left(\vec{A}, \vec{B}\right)$ is defined by

$$\mathbf{T}\left(\vec{A}, \vec{B}\right) = A^\alpha B^\beta T_{\alpha\beta} \tag{15.13}$$

and it can be shown as above that the expression on the right side is a Lorentz scalar.

An important $(0,2)$-tensor is the **metric tensor g**, which for two arbitrary vectors $\vec{A}, \vec{B}$ is defined by

$$\mathbf{g}\left(\vec{A}, \vec{B}\right) = -A^t B^t + A^x B^x + A^y B^y + A^z B^z.$$

The metric tensor has the components in the inertial system S

$$\mathbf{g} \underset{S}{\to} g_{\alpha\beta} = \begin{cases} -1 & \alpha = \beta = t \\ 1 & \alpha = \beta \neq t \\ 0 & \alpha \neq \beta \end{cases}.$$

The components of $\mathbf{g}$ are usually called $g_{\alpha\beta} = \eta_{\alpha\beta}$, and they are in matrix notation

$$(g_{\alpha\beta}) = (\eta_{\alpha\beta}) = \begin{pmatrix} -1 & 0 & 0 & 0 \\ 0 & 1 & 0 & 0 \\ 0 & 0 & 1 & 0 \\ 0 & 0 & 0 & 1 \end{pmatrix}.$$

We first show that $\mathbf{g}$ has tensor properties. $\mathbf{g}$ is a bilinear mapping, because for the first variable

$$\begin{aligned} \mathbf{g}\left(c_1 \vec{A}_1 + c_2 \vec{A}_2, \vec{B}\right) &= -\left(c_1 A_1^t + c_2 A_2^t\right) B^t + \left(c_1 A_1^x + c_2 A_2^x\right) B^x \\ &\quad + \left(c_1 A_1^y + c_2 A_2^y\right) B^y + \left(c_1 A_1^z + c_2 A_2^z\right) B^z \\ &= -c_1 A_1^t B^t - c_2 A_2^t B^t + c_1 A_1^x B^x + c_2 A_2^x B^x \\ &\quad + c_1 A_1^y B^y + c_2 A_2^y B^y + c_1 A_1^z B^z + c_2 A_2^z B^z \\ &= c_1 \mathbf{g}\left(\vec{A}_1, \vec{B}\right) + c_2 \mathbf{g}\left(\vec{A}_2, \vec{B}\right). \end{aligned}$$

Similarly, the linearity in the second variable of $\mathbf{g}$ follows. We calculate the components of $\mathbf{g}$ in an inertial system S' with the transformation formula (9.11) for matrices

$$\begin{aligned} \left(\eta_{\alpha'\beta'}\right) &= L^\alpha{}_{\alpha'}(-v)\,\eta_{\alpha\beta}\,L^\beta{}_{\beta'}(-v) \\ &= \begin{pmatrix} \gamma & \gamma v & 0 & 0 \\ \gamma v & \gamma & 0 & 0 \\ 0 & 0 & 1 & 0 \\ 0 & 0 & 0 & 1 \end{pmatrix}^T \begin{pmatrix} -1 & 0 & 0 & 0 \\ 0 & 1 & 0 & 0 \\ 0 & 0 & 1 & 0 \\ 0 & 0 & 0 & 1 \end{pmatrix} \begin{pmatrix} \gamma & \gamma v & 0 & 0 \\ \gamma v & \gamma & 0 & 0 \\ 0 & 0 & 1 & 0 \\ 0 & 0 & 0 & 1 \end{pmatrix} \end{aligned}$$

$$
= \begin{pmatrix} \gamma & \gamma v & 0 & 0 \\ \gamma v & \gamma & 0 & 0 \\ 0 & 0 & 1 & 0 \\ 0 & 0 & 0 & 1 \end{pmatrix} \begin{pmatrix} -\gamma & -\gamma v & 0 & 0 \\ \gamma v & \gamma & 0 & 0 \\ 0 & 0 & 1 & 0 \\ 0 & 0 & 0 & 1 \end{pmatrix}
$$

$$
= \begin{pmatrix} -\gamma^2 + \gamma^2 v^2 & -\gamma^2 v + \gamma^2 v & 0 & 0 \\ -\gamma^2 v + \gamma^2 v & -\gamma^2 v^2 + \gamma^2 & 0 & 0 \\ 0 & 0 & 1 & 0 \\ 0 & 0 & 0 & 1 \end{pmatrix}
$$

$$
= \begin{pmatrix} -\gamma^2(1 - v^2) & 0 & 0 & 0 \\ 0 & \gamma^2(1 - v^2) & 0 & 0 \\ 0 & 0 & 1 & 0 \\ 0 & 0 & 0 & 1 \end{pmatrix}
$$

$$
= \begin{pmatrix} -1 & 0 & 0 & 0 \\ 0 & 1 & 0 & 0 \\ 0 & 0 & 1 & 0 \\ 0 & 0 & 0 & 1 \end{pmatrix} = (\eta_{\alpha\beta}) , \tag{15.14}
$$

since $\gamma^2 \left(1 - v^2\right) = 1$. The components of the metric tensor are therefore the same and constant in all inertial systems. This also implies the coordinate system independence of $\mathbf{g}(\vec{A}, \vec{B})$. Furthermore, from the defining equation of $\mathbf{g}$, it follows that

$$
\mathbf{g}\left(\vec{A}, \vec{B}\right) = \mathbf{g}\left(\vec{B}, \vec{A}\right)
$$

i.e., the metric tensor is symmetric. If we set $\vec{B} = \vec{A}$, we obtain the (invariant) square of $\vec{A}$ with the metric tensor,

$$
\vec{A}^2 = -A^t A^t + A^x A^x + A^y A^y + A^z A^z = \eta_{\alpha\beta} A^\alpha A^\beta = \mathbf{g}\left(\vec{A}, \vec{A}\right) ,
$$

i.e., the metric tensor measures the (square) length of vectors.

15.4. Correspondence of Vectors and One-Forms

As in two dimensions, the metric tensor can be used to define a corresponding one-form $\tilde{v}$ to a fixed vector $\vec{V}$:

$$
\tilde{v} \underset{S}{\rightarrow} v_\alpha = \eta_{\alpha\beta} V^\beta \tag{15.15}
$$

We check whether the components fulfill the transformation rules for one-forms. To do this, let's choose another inertial system S', then with (15.12)

$$
v_{\alpha'} = \eta_{\alpha'\beta'} V^{\beta'} = \left(L^\alpha{}_{\alpha'}(-v) L^\beta{}_{\beta'}(-v) \eta_{\alpha\beta}\right) \left(L^{\beta'}{}_\gamma(v) V^\gamma\right)
$$

$$
\begin{aligned}
&= \left(L^{\alpha}{}_{\alpha'}\left(-v\right)\eta_{\alpha\beta}\,V^{\gamma}\right)\underbrace{\left(L^{\beta'}{}_{\gamma}\left(v\right)L^{\beta}{}_{\beta'}\left(-v\right)\right)}_{=\delta^{\beta}_{\gamma}} \\
&= L^{\alpha}{}_{\alpha'}\left(-v\right)\eta_{\alpha\beta}\,V^{\gamma}\,\delta^{\beta}_{\gamma} \\
&= L^{\alpha}{}_{\alpha'}\left(-v\right)\left(\eta_{\alpha\beta}\,V^{\beta}\right),
\end{aligned}
$$

follows, i.e., $\tilde{v}$ is a one-form. The following relationship holds between the components of $\vec{V}$ and those of $\tilde{v}$

$$
\begin{pmatrix} v_t \\ v_x \\ v_y \\ v_z \end{pmatrix} = \begin{pmatrix} -1 & 0 & 0 & 0 \\ 0 & 1 & 0 & 0 \\ 0 & 0 & 1 & 0 \\ 0 & 0 & 0 & 1 \end{pmatrix} \begin{pmatrix} V^t \\ V^x \\ V^y \\ V^z \end{pmatrix} = \begin{pmatrix} -V^t \\ V^x \\ V^y \\ V^z \end{pmatrix}, \tag{15.16}
$$

i.e., the components of the one-form and the normal vector coincide in the space components, the time component differs by a minus sign.

We now ask the reverse question: Can we find a corresponding vector $\vec{P}$ for any one-form $\tilde{p}$? To answer this question, we first need to determine the inverse matrix $\left(\eta^{\alpha\beta}\right)$ to $\left(\eta_{\alpha\beta}\right)$. It holds

$$
\left(\eta_{\alpha\beta}\right)\left(\eta_{\alpha\beta}\right) = \begin{pmatrix} -1 & 0 & 0 & 0 \\ 0 & 1 & 0 & 0 \\ 0 & 0 & 1 & 0 \\ 0 & 0 & 0 & 1 \end{pmatrix} \begin{pmatrix} -1 & 0 & 0 & 0 \\ 0 & 1 & 0 & 0 \\ 0 & 0 & 1 & 0 \\ 0 & 0 & 0 & 1 \end{pmatrix} = \begin{pmatrix} 1 & 0 & 0 & 0 \\ 0 & 1 & 0 & 0 \\ 0 & 0 & 1 & 0 \\ 0 & 0 & 0 & 1 \end{pmatrix},
$$

i.e., the metric tensor is inverse to itself:

$$
\left(\eta^{\alpha\beta}\right) = \left(\eta_{\alpha\beta}\right)
$$

If p_α are the components of the one-form $\tilde{p}$, the same applies to $\vec{P}$:

$$
\vec{P} \rightarrow P^{\alpha} = \eta^{\alpha\beta}p_{\beta} \tag{15.17}
$$

15.5. (N, M)-Tensors

Due to the found one-to-one correspondence between one-forms and vectors, we can now also consider a vector $\vec{V}$ as a tensor, which applied to any one-form $\tilde{p}$ yields a real number. The rule for this is

$$
\vec{V}(\tilde{p}) = \tilde{p}(V) = p_{\alpha}V^{\alpha}. \tag{15.18}
$$

For this reason, a vector is also called a $(1,0)$-tensor. More generally, a $(M,0)$-tensor is a multilinear mapping that maps M one-forms $(M = 1,2,3,4)$ to the real numbers. The components of $(M, 0)$-tensors have M *upper* indices. The statements of the last sections about $(0, N)$-tensors apply analogously to the $(M, 0)$-tensors. As a continuous example for this section, we consider the properties of a $(2, 0)$-tensor

$$\mathbf{T} \to T^{ij},$$

whose components transform according to a coordinate system change as

$$T^{\alpha'\beta'} = \frac{\partial \alpha'}{\partial \alpha} \frac{\partial \beta'}{\partial \beta} T^{\alpha\beta} = L^{\alpha'}_{\alpha}(v) L^{\beta'}_{\beta}(v) T^{\alpha\beta}.$$

If $\tilde{p} \to p_\alpha$ and $\tilde{q} \to q_\beta$ are two one-forms, then

$$\mathbf{T}(\tilde{p}, \tilde{q}) = p_\alpha\, q_\beta\, T^{\alpha\beta}.$$

A special $(2, 0)$-tensor is the **metric tensor for one-forms**, which we want to denote with

$$\mathbf{G} \to \eta^{\alpha\beta},$$

i.e.

$$\mathbf{G}(\tilde{p}, \tilde{q}) = \eta^{\alpha\beta} p_\alpha\, q_\beta.$$

Finally, a (M, N)-tensor is a multilinear mapping that maps M one-forms and N vectors to the real numbers, where M, N can again be the numbers one to four. The components of a (M, N)-tensor thus have M *upper* and N *lower* indices. As a continuous example, a $(1, 1)$-tensor should serve us in this section. A $(1, 1)$-tensor $\mathbf{T}$ requires a one-form $\tilde{p}$ and a vector $\vec{V}$ to produce the number $\mathbf{T}(\tilde{p}, \vec{V})$. The components of

$$\mathbf{T} \to T^{\alpha}_{\beta}$$

have a superscript (from the one-form) and a subscript (from the vector) index and transform during a change of coordinate system according to

$$T^{\alpha'}_{\beta'} = \frac{\partial \alpha'}{\partial \alpha} \frac{\partial \beta}{\partial \beta'} T^{\alpha}_{\beta} = L^{\alpha'}_{\alpha}(v) L^{\beta}_{\beta'}(-v) T^{\alpha}_{\beta}.$$

If $\vec{V}$ and $\tilde{p}$ are a vector and a one-form, then it follows that

$$\mathbf{T}(\tilde{p}, \vec{V}) = p_\alpha V^\beta T^{\alpha}_{\beta}.$$

In generalizing the correspondence between vectors and one-forms, one can use the metric tensor $\mathbf{g}$ to generate a corresponding $(M - 1, N + 1)$-tensor from a (M, N)-tensor. Similarly, one uses the tensor $\mathbf{G}$ to generate a corresponding $(M + 1, N - 1)$-tensor from a (M, N)-tensor. It is customary to name these new tensors with the same name and distinguish them only by the index position of their components. For example, the corresponding tensor of a $(0,2)$-tensor $\mathbf{T}$ is a $(1,1)$-tensor, which is also denoted by $\mathbf{T}$. The components of the corresponding tensor $\mathbf{T}$ are derived from the components $T_{\alpha\beta}$ of the $(0,2)$-tensor and from the components $\eta^{\alpha\beta}$ of $\mathbf{G}$ by

$$\mathbf{T} \to T^{\alpha}_{\ \beta} = \eta^{\alpha\gamma}\, T_{\gamma\beta}.$$

That is said, one "pulls the first index α of $T_{\alpha\beta}$ upwards with the inverse matrix $\eta^{\alpha\gamma}$". With exactly the same steps, one can also pull the second index of the components $T_{\alpha\beta}$ upwards and obtains

$$T^{\ \beta}_{\alpha} = g^{\beta\gamma}\, T_{\alpha\gamma}.$$

If we look at the corresponding tensor of a $(2,0)$-tensor $\mathbf{T}$ as another example, we get just as well

$$T^{\alpha}_{\ \beta} = T^{\alpha\gamma}\, \eta_{\gamma\beta}.$$

In this case, one says, "one pulls the second index β of $T^{\alpha\beta}$ downwards with the matrix $\eta_{\gamma\beta}$".

If we choose the metric $\mathbf{g}$ with its components $\eta_{\alpha\beta}$ as a $(0,2)$-tensor, we obtain

$$\eta^{\alpha}_{\ \beta} = \eta^{\alpha\gamma}\, \eta_{\gamma\beta} = \delta^{\alpha}_{\ \beta},$$

since the two matrices $(\eta_{\alpha\beta})$ and $(\eta^{\alpha\beta})$ are inverse to each other. If we also pull up the other index, we get

$$\eta^{\alpha\beta} = \eta^{\beta\gamma}\, \delta^{\alpha}_{\ \gamma} = \eta^{\beta\alpha} = \eta^{\alpha\beta},$$

i.e., as in two dimensions, we obtain the components of $\mathbf{G}$ by pulling up the indices of the components of $\mathbf{g}$ twice. The metric is the only tensor with this property.

15.6. (Covariant) Derivatives of Tensors

In Eq. (15.4), we showed that in an inertial system S, the gradient, i.e., the derivative of a real-valued function (scalar field) $\phi(t, x, y, z)$, at a spacetime point (t_0, x_0, y_0, z_0) is a one-form:

$$\tilde{d}\phi \to \left(\frac{\partial \phi}{\partial t}, \frac{\partial \phi}{\partial x}, \frac{\partial \phi}{\partial y}, \frac{\partial \phi}{\partial z} \right)$$

Since a real-valued function can be considered as a $(0,0)$-tensor, we thus obtain a one-form, i.e., a $(0,1)$-tensor by deriving a function. As in two dimensions, the (normal) gradient of a scalar field is a tensor, i.e., the covariant derivative of a scalar field coincides with the partial derivative. We want to show that in special relativity, the components of the covariant derivative of an *arbitrary* tensor coincide with the partial derivatives of its components. To this end, we first consider a vector field

$$\vec{V} \rightarrow V^\alpha = V^\alpha\,(t, x, y, z)$$

and show that the partial derivatives of V^α are the components of a $(1,1)$ tensor. We first note that the Lorentz matrices are constant, so their derivatives with respect to the coordinates t, x, y, z vanish, i.e., it holds

$$\frac{\partial}{\partial\beta'}\left(\frac{\partial\alpha}{\partial\alpha'}\right) = \frac{\partial}{\partial\beta'}\left(L^\alpha_{\ \alpha'}\,(-v)\right) = \frac{\partial}{\partial\beta'}\begin{pmatrix} \gamma & \gamma v & 0 & 0 \\ \gamma v & \gamma & 0 & 0 \\ 0 & 0 & 1 & 0 \\ 0 & 0 & 0 & 1 \end{pmatrix} = \begin{pmatrix} 0 & 0 & 0 & 0 \\ 0 & 0 & 0 & 0 \\ 0 & 0 & 0 & 0 \\ 0 & 0 & 0 & 0 \end{pmatrix}.$$

$$(15.19)$$

We show that the partial derivatives of V^α transform like the components of a $(1,1)$-tensor, proceeding analogously to section 9.4. It holds with (15.19) and the product and chain rule

$$\begin{aligned}
\frac{\partial V^\alpha}{\partial\beta} &= \frac{\partial}{\partial\beta}\left(\frac{\partial\alpha}{\partial\alpha'}V^{\alpha'}\right) = \frac{\partial}{\partial\beta}\left(\frac{\partial\alpha}{\partial\alpha'}\right)V^{\alpha'} + \left(\frac{\partial\alpha}{\partial\alpha'}\right)\frac{\partial V^{\alpha'}}{\partial\beta} \\[2mm]
&\underset{chain\,rule}{=} \frac{\partial\beta'}{\partial\beta}\,\underbrace{\frac{\partial}{\partial\beta'}\left(\frac{\partial\alpha}{\partial\alpha'}\right)}_{=0}V^{\alpha'} + \frac{\partial\alpha}{\partial\alpha'}\frac{\partial\beta'}{\partial\beta}\frac{\partial V^{\alpha'}}{\partial\beta'} \\[2mm]
&= \left(\frac{\partial\alpha}{\partial\alpha'}\frac{\partial\beta'}{\partial\beta}\right)\frac{\partial V^{\alpha'}}{\partial\beta'} = \left(L^\alpha_{\ \alpha'}\,(-v)\,L^{\beta'}_{\ \beta}\,(v)\right)\frac{\partial V^{\alpha'}}{\partial\beta'}.
\end{aligned}$$

So, the ordinary derivative of a vector is also a tensor. In Eq. 9.24 on page 200, we defined the Christoffel symbols using coordinate transformations. This definition can be transferred one-to-one to spacetime, and we again obtain with (15.19)

$$\Gamma^\alpha_{\alpha'\beta} = \frac{\partial\beta'}{\partial\beta}\,\underbrace{\frac{\partial}{\partial\beta'}\left(\frac{\partial\alpha}{\partial\alpha'}\right)}_{=0} = 0,$$

i.e., in special relativity, all Christoffel symbols are zero and thus the following mnemonic applies:

The components of the covariant derivative of a tensor in special relativity coincide with the partial derivatives of its components.

For example, for a $(1,1)$-tensor $\mathbf{T}$, one calculates for *each* component

$$T^\alpha_\beta = T^\alpha_\beta\,(t, x, y, z)$$

the gradient and obtains the 16 one-forms

$$\tilde{d}T^\alpha_\beta \underset{S}{\rightarrow} \left(\frac{\partial T^\alpha_\beta}{\partial t}, \frac{\partial T^\alpha_\beta}{\partial x}, \frac{\partial T^\alpha_\beta}{\partial y}, \frac{\partial T^\alpha_\beta}{\partial z} \right).$$

With the compact notation as in (15.11) we can also write the components of $\tilde{d}T^\alpha{}_\beta$ as

$$\left(\frac{\partial T^\alpha_\beta}{\partial t}, \frac{\partial T^\alpha_\beta}{\partial x}, \frac{\partial T^\alpha_\beta}{\partial y}, \frac{\partial T^\alpha_\beta}{\partial z} \right) = T^\alpha{}_{\beta,\gamma}.$$

We now *define* the **(covariant) derivative** of the $(1,1)$-tensor $\mathbf{T}$ as the $(1,2)$-tensor $\nabla\mathbf{T}$, whose components are the gradients of the tensor components of $\mathbf{T}$:

$$\nabla\mathbf{T} \underset{S}{\rightarrow} T^\alpha{}_{\beta,\gamma} \tag{15.20}$$

16. Energy-Momentum Tensors in Special Relativity

In this chapter, we examine two important energy-momentum tensors, the energy-momentum tensor of **incoherent matter** (of **dust**) and that of an **ideal fluid**. By fluid, we mean a continuous collection of particles that occupy a certain volume but have no specific shape. Fluids thus include liquids and gases. The continuum should be so large that the dynamics of individual particles do not matter, but only "average values", such as the number of particles in a unit volume, energy and momentum density, pressure, temperature, etc., are used for description. These properties of a certain collection of particles, which we want to call an **element**, should be observable in an arbitrarily small volume, so we can simplify the assumption that they can vary from spacetime point to spacetime point.

The energy-momentum tensor of the electromagnetic field, which is cited in many publications (see e.g. [36]) as a "parade example" of tensor systematics in special relativity, we do not consider here, as we will not deal with electromagnetic phenomena in the description of general relativity.

16.1. Incoherent Matter, Dust

We start with the simplest form of a fluid/matter field, namely dust, which we define as a collection of non-interacting particles that are all momentarily at rest in a certain inertial system S, i.e., the particles do not move relative to each other. If we want to answer the question of how many particles are in a volume of size 1 (unit volume), we simply count the particles and divide by the subvolume in which they are located. If there are N particles in a rectangular volume V with side lengths $\Delta x, \Delta y, \Delta z$, i.e., $V = \Delta x \Delta y \Delta z$, we define the **particle density** n in the volume by

$$n = \frac{N}{V} = \frac{N}{\Delta x \Delta y \Delta z}.$$

The particle density can vary from spacetime point to spacetime point if we choose the volume correspondingly small. In the current rest system S frame of the particles the particle density does not depend on time t, but in another

M. Ruhrländer, *Ascent to the Einstein Equations*, https://doi.org/10.1007/978-3-662-72672-3_16

reference system S', it does depend on time t' due to the Lorentz transformation. What does the particle density look like in an inertial system S'? From the perspective of the system S', all particles are moving at the same speed v. If one counts the particles from S', one obtains the same number as in the system S, but from the perspective of S', the volume in which the particles are located changes due to the Lorentz contraction.

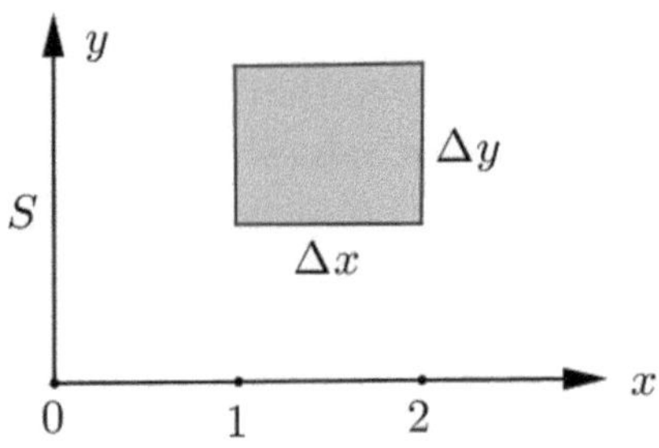

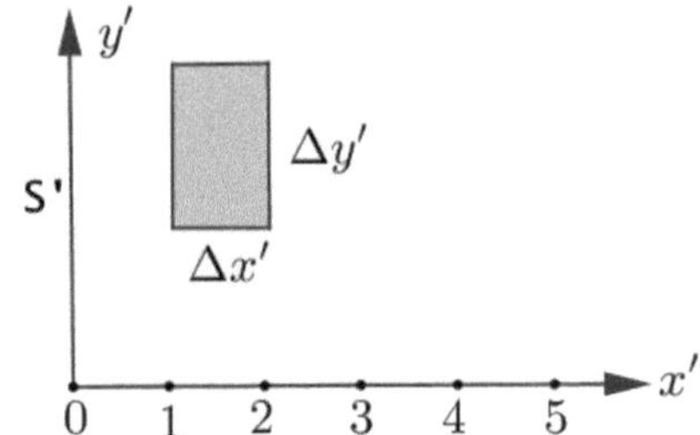

Figure 16.1.: Particle Density

In Fig. 16.1, for simplicity, the z-coordinate is omitted and it is again assumed that the particles are moving in the positive x-direction. Thus, the side length $\Delta x'$ in S' is reduced by the Lorentz contraction to

$$\Delta x' = \Delta x \sqrt{1 - v^2},$$

while the sides perpendicular to the velocity v are the same in both systems, i.e.

$$\Delta y' = \Delta y$$
$$\Delta z' = \Delta z.$$

This implies that the particle density n' in the system S', in which the particles have the velocity v, is greater than in the rest system S:

$$n' = \frac{N}{V'} = \frac{N}{\Delta x' \Delta y' \Delta z'} = \frac{N}{\Delta x \sqrt{1 - v^2}\, \Delta y \Delta z} = \frac{1}{\sqrt{1 - v^2}}\, n = \gamma n$$

Next, we want to consider the **flux of particles** through a surface. This is defined by the number of particles that cross a unit area on this surface in a unit of time. The flux depends on the inertial system, as area and time are system-dependent quantities. Another dependency is the orientation of the surface. If the surface is parallel to the direction of motion of the particles, no single particle crosses the surface, if the surface is perpendicular to the direction of motion of the particles, the flux of particles through the surface is maximal. In the rest system of the particles, the flux is zero, the particles do not move there. We again consider an inertial system S', in which all particles may move with velocity v in the x'-direction. The surface A' is initially oriented so that it is perpendicular to the x'-axis.

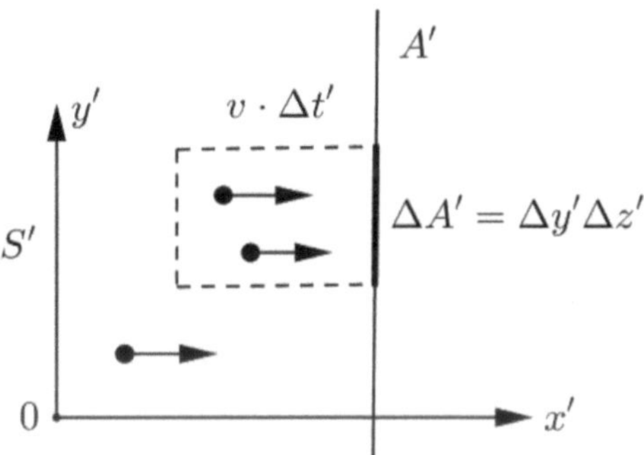

Figure 16.2.: Particle flux perpendicular to the surface

In Fig. 16.2, three particles (thick dots) are drawn that move in the x'-direction. The rectangular volume V', which is bounded by the dashed line, contains exactly the particles that cross the partial surface $\Delta A' = \Delta y' \Delta z'$ in the time $\Delta t'$. This volume is therefore calculated to be

$$V' = v \Delta t' \Delta A' = v \Delta t' \Delta y' \Delta z'$$

and contains

$$N = n' \cdot V' = \frac{1}{\sqrt{1 - v^2}}\, n \cdot v \Delta t' \Delta A'$$

particles. The particle flux $F^{x'}$ in the x'-direction is obtained from this by dividing by $\Delta t'$ and $\Delta A'$:

$$F^{x'} = \frac{n \cdot v}{\sqrt{1 - v^2}} = \gamma n \cdot v$$

We now generalize the situation and assume that the particles are moving with a velocity $\vec{v}$, which also has a component in the y'-direction

$$\vec{v} \underset{S'}{\rightarrow} (v^{x'}, v^{y'}, 0).$$

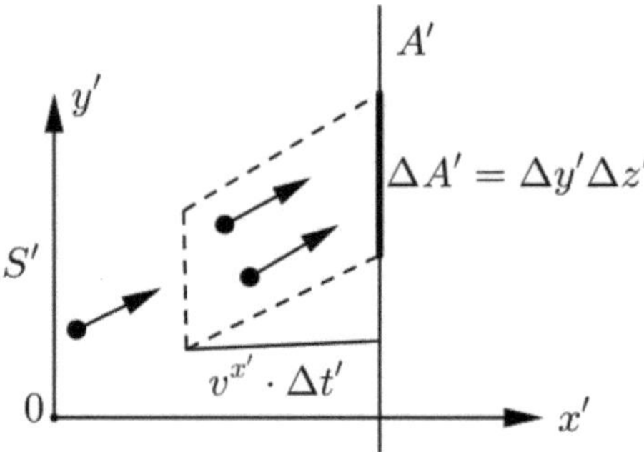

Figure 16.3.: Particle flow oblique to surface

In Fig. 16.3 all particles that cross the surface $\Delta A'$ in time $\Delta t'$ are located in the dashed volume („**parallelepiped**"). The volume of the parallelepiped is calculated from the base area times the height H, so $\Delta A' \cdot H$. However, the height is equal to $H = v^{x'} \Delta t'$, so

$$V' = v^{x'} \Delta t' \Delta A' = v^{x'} \Delta t' \Delta y' \Delta z',$$

and for the flow through A' in x' direction applies

$$F^{x'} = \frac{n \cdot v^{x'}}{\sqrt{1 - v^2}} = \gamma n \cdot v^{x'}.$$

Generally, we *define* the **particle flow four-vector** $\vec{N}$ by

$$\vec{N} = n\vec{U},$$

where $\vec{U}$ is the four-velocity of the particles. If the particles in an inertial system S' have the velocity $\vec{v} \rightarrow (v^{x'}, v^{y'}, v^{z'})$, then with Eq. (14.8) we get

$$\vec{U} \underset{S'}{\rightarrow} (\gamma, \gamma v^{x'}, \gamma v^{y'}, \gamma v^{z'})$$

and thus

$$\vec{N} \underset{S'}{\rightarrow} (\gamma n, \gamma v^{x'} n, \gamma v^{y'} n, \gamma v^{z'} n).$$

In every inertial system, the time component of $\vec{N}$ is the particle density, and the spatial components are the particle flows through the surface portions perpendicular to the spatial coordinate axes. In classical physics, where there is no Lorentz contraction, the particle density is a frame-independent quantity, whereas the particle flow due to the velocity components occurring therein is frame-dependent. The combination of these two physical quantities into a frame-invariant particle flow four-vector is thus very similar to the combination of energy and momentum into the four-momentum vector. From the last equation follows similarly to the rest mass

$$
\begin{aligned}
\vec{N}^2 &= -\gamma^2 n^2 + \gamma^2 \left(v^{x'}\right)^2 n^2 + \gamma^2 \left(v^{y'}\right)^2 n^2 + \gamma^2 \left(v^{z'}\right)^2 n^2 \\
&= n^2\gamma^2\left(-1 + \left(v^{x'}\right)^2 + \left(v^{y'}\right)^2 + \left(v^{z'}\right)^2\right) \\
&= n^2\gamma^2(-1 + v^2) \\
&= -n^2\gamma^2(1 - v^2) \\
&= -n^2
\end{aligned}
$$

and from this

$$
n = \sqrt{-\vec{N}^2},
$$

i.e., the **rest particle density** n is a Lorentz scalar, just like the rest mass m of a particle is a Lorentz scalar, see the relativistic energy theorem 12.20 on page 258.

In the current rest system, the energy of a particle is equal to the rest mass m, and the number of particles per unit volume is equal to n, i.e., the energy per unit volume, the so-called **energy density** ρ, is for dust in the rest system equal to

$$
\rho = nm.
$$

ρ is like n and m a scalar. In an inertial system S', in which the particles move with velocity v, the particle density is equal to γn and the energy of a particle according to Eq. 12.18 on page 257 is equal to γm, from which the energy density in S'

$$
\rho' = \gamma n \cdot \gamma m = \frac{\rho}{1 - v^2} = \rho\gamma^2
$$

results. Note that to transform from ρ to ρ' twice the factor γ is required, therefore ρ and ρ' cannot be the components of a vector. To find out which object could have the energy density ρ' as a component, let's first look at two arbitrary vectors $\vec{A}$ and $\vec{B}$ and define a mapping $\mathbf{T}$, whose components consist of the products of the components of the two vectors

$$\mathbf{T} \to T^{\alpha\beta} = A^{\alpha}B^{\beta},$$

and which applied to two one-forms $\tilde{p}$ and $\tilde{q}$ yields the number

$$\mathbf{T}\left(\tilde{p}, \tilde{q}\right) = A^{\alpha}B^{\beta}p_{\alpha}\,q_{\beta}.$$

The mapping $\mathbf{T}$ is bilinear by construction, and the components transform with (14.2) according to

$$T^{\alpha'\beta'} = A^{\alpha'}B^{\beta'} = \left(L^{\alpha'}_{\alpha}\left(v\right)A^{\alpha}\right)\left(L^{\beta'}_{\beta}\left(v\right)B^{\beta}\right) = L^{\alpha'}_{\alpha}\left(v\right)L^{\beta'}_{\beta}\left(v\right)T^{\alpha\beta}.$$

So $\mathbf{T}$ is a $(2,0)$-tensor, which is also called the **tensor product** of the vectors $\vec{A}$ and $\vec{B}$ and is denoted by

$$\mathbf{T} = \vec{A} \otimes \vec{B},$$

where the multiplication symbol $\otimes$ is intended to highlight the difference from the multiplication of numbers.

Looking at the two vectors $\vec{N}$ and with Eq. (14.9) the four-momentum

$$\vec{P} \underset{S'}{\to} \left(P^{t'}, P^{x'}, P^{y'}, P^{z'}\right) = \left(m\gamma, m\gamma v^{x}, m\gamma v^{y}, m\gamma v^{z}\right)$$

one sees that the energy density in the system S' is the product of the t'-components of the two vectors

$$\rho' = \gamma n \cdot \gamma m.$$

We *define* the **energy-momentum tensor $\mathbf{T}$** for dust by the tensor product

$$\mathbf{T} = \vec{P} \otimes \vec{N}.$$

Since $\vec{N} = n\vec{U}$ and $\vec{P} = m\vec{U}$, it follows

$$\mathbf{T} = \vec{P} \otimes \vec{N} = nm\vec{U} \otimes \vec{U} = \rho\vec{U} \otimes \vec{U}.$$

In any arbitrary inertial system S', in which the particles move with velocity v, it holds

$$\vec{U} \underset{S'}{\to} \left(\gamma, \gamma v^{x'}, \gamma v^{y'}, \gamma v^{z'}\right),$$

and thus for the components of $\mathbf{T}$

$$\mathbf{T} \to \rho U^{\alpha'}U^{\beta'}$$

and in matrix notation

$$
T^{\alpha'\beta'} = \begin{pmatrix}
\rho\gamma^2 & \rho\gamma^2 v^{x'} & \rho\gamma^2 v^{y'} & \rho\gamma^2 v^{z'} \\
\rho\gamma^2 v^{x'} & \rho\gamma^2 \left(v^{x'}\right)^2 & \rho\gamma^2 v^{x'} v^{y'} & \rho\gamma^2 v^{x'} v^{z'} \\
\rho\gamma^2 v^{y'} & \rho\gamma^2 v^{y'} v^{x'} & \rho\gamma^2 \left(v^{y'}\right)^2 & \rho\gamma^2 v^{y'} v^{z'} \\
\rho\gamma^2 v^{z'} & \rho\gamma^2 v^{z'} v^{x'} & \rho\gamma^2 v^{z'} v^{y'} & \rho\gamma^2 \left(v^{z'}\right)^2
\end{pmatrix}
$$

$$
= \rho' \begin{pmatrix}
1 & v^{x'} & v^{y'} & v^{z'} \\
v^{x'} & \left(v^{x'}\right)^2 & v^{x'} v^{y'} & v^{x'} v^{z'} \\
v^{y'} & v^{y'} v^{x'} & \left(v^{y'}\right)^2 & v^{y'} v^{z'} \\
v^{z'} & v^{z'} v^{x'} & v^{z'} v^{y'} & \left(v^{z'}\right)^2
\end{pmatrix}.
$$

Note that

$$
T^{\alpha'\beta'} = T^{\beta'\alpha'}
$$

applies, i.e., $\mathbf{T}$ is a symmetric tensor. The components $T^{\alpha'\beta'}$ can be interpreted as flows of different densities:

- $T^{t't'} = \rho'$ is the energy density, which can be seen as energy flow through a surface with $t' = const.$

- $T^{t'i'} = \rho' v^{i'}$ is the energy flow through a surface with $i' = const.$

- $T^{i't'} = \rho' v^{i'}$ is the i'-th momentum density, i.e. the i'-th momentum flow through a surface with $t' = const.$

- $T^{i'j'} = \rho' v^{i'} v^{j'}$ is the i'-th momentum flow through a surface with $j' = const.$

In the current rest system S of the particles, it applies

$$
\vec{U} \underset{S}{\to} (1,0,0,0), \quad \vec{v} = \vec{0}, \quad \gamma = 1
$$

and thus in S

$$
T^{\alpha\beta} = \begin{pmatrix}
\rho & 0 & 0 & 0 \\
0 & 0 & 0 & 0 \\
0 & 0 & 0 & 0 \\
0 & 0 & 0 & 0
\end{pmatrix}.
$$

In the current rest system, the particle density n and the energy m, thus the energy density ρ, are constant over time, so

$$
\frac{\partial \rho}{\partial t} = \frac{\partial T^{tt}}{\partial t} = 0.
$$

Since in the rest system all other components of **T** are equal to zero, it follows
for all α

$$
\begin{aligned}
T^{\alpha\beta}_{\ ,\beta} &= \frac{\partial T^{\alpha\beta}}{\partial\beta} \\
&= \frac{\partial T^{\alpha t}}{\partial t} + \frac{\partial T^{\alpha x}}{\partial x} + \frac{\partial T^{\alpha y}}{\partial y} + \frac{\partial T^{\alpha z}}{\partial z} \\
&= 0 + 0 + 0 + 0 = 0.
\end{aligned}
$$

The components $T^{\alpha\beta}_{\ ,\beta}$ are obtained by contraction (summation over β) from
the components of the derivative of **T** (15.20) and are thus the components of
a vector, which is also called **four-divergence**. The components $T^{\alpha\beta}_{\ ,\beta}$ of the
four-divergence transform from the rest system S into any inertial system S'
by

$$
T^{\alpha'\beta'}_{\ ,\beta'} = L^{\alpha'}_{\alpha}(v)\,\underbrace{L^{\beta'}_{\beta}(v)\,L^{\gamma}_{\beta'}(-v)}_{=\delta^{\gamma}_{\beta}}\,T^{\alpha\beta}_{\ ,\gamma} = L^{\alpha'}_{\alpha}(v)\,\delta^{\gamma}_{\beta}\,T^{\alpha\beta}_{\ ,\gamma} = L^{\alpha'}_{\alpha}(v)\,T^{\alpha\beta}_{\ ,\beta} \quad (16.1)
$$

i.e., it also applies in any arbitrary inertial system S'

$$
T^{\alpha'\beta'}_{\ ,\beta'} = 0,
$$

i.e., the divergence of the energy-momentum tensor for dust is equal to zero.
It is also said:

| The energy-momentum tensor is divergence-free. |

We want to interpret this result in more detail physically. If we write out the
equation for $\alpha' = t'$, we obtain with $\rho' = \rho\gamma^2$

$$
T^{t'\beta'}_{\ ,\beta'} = \frac{\partial\rho'}{\partial t'} + \frac{\partial(\rho' v^{x'})}{\partial x'} + \frac{\partial(\rho' v^{y'})}{\partial y'} + \frac{\partial(\rho' v^{z'})}{\partial z'} = 0. \quad (16.2)
$$

This equation states that the rate of increase of energy in the volume $(\partial\rho'/\partial t')$
is equal to the net inflow of energy into the volume. Let's clarify this again
with a graphical representation.

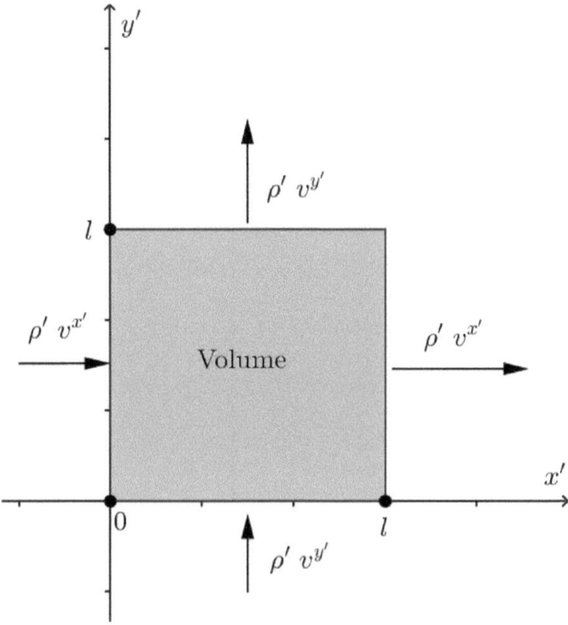

Figure 16.4.: Energy flow

In Fig. 16.4, a cube with side length l is shown, with the z'-axis suppressed again. Energy can flow in and out through each side of the cube. The rate of energy flowing into the cube in the x'-direction through the area of size l^2 at $x' = 0$ is equal to

$$l^2 T^{t'x'} \left(x' = 0\right) = l^2 \rho' \left(x' = 0\right) v^{x'} \left(x' = 0\right),$$

and the outflow at $x' = l$ is given by

$$l^2 T^{t'x'} \left(x' = l\right).$$

The net inflow to the cube in the x'-direction is therefore

$$l^2 \left(T^{t'x'} \left(x' = 0\right) - T^{t'x'} \left(x' = l\right)\right).$$

Similarly for the other two directions

$$l^2 \left(T^{t'y'} \left(y' = 0\right) - T^{t'y'} \left(y' = l\right)\right)$$

and

$$l^2 \left(T^{t'z'} \left(z' = 0\right) - T^{t'z'} \left(z' = l\right)\right).$$

In total, the net inflow is the sum of the last three expressions. This net inflow is now equal to the increase in energy in the cube (conservation of energy!), i.e., equal to the temporal change of energy density times volume, thus equal to

$$\frac{\partial\left(T^{t't'}\cdot l^3\right)}{dt'} = l^3\,\frac{\partial T^{t't'}}{dt'} = l^3\,\frac{\partial\rho'}{dt'}.$$

We thus obtain

$$
\begin{aligned}
l^3\,\frac{\partial\rho'}{dt'} &= l^2\left(T^{t'x'}\left(x'=0\right) - T^{t'x'}\left(x'=l\right)\right) \\
&+ l^2\left(T^{t'y'}\left(y'=0\right) - T^{t'y'}\left(y'=l\right)\right) \\
&+ l^2\left(T^{t'z'}\left(z'=0\right) - T^{t'z'}\left(z'=l\right)\right).
\end{aligned}
$$

Dividing by l^3 and rearranging yields

$$
\begin{aligned}
\frac{\partial\rho'}{dt'} &= -\frac{T^{t'x'}\left(x'=l\right) - T^{t'x'}\left(x'=0\right)}{l} \\
&- \frac{T^{t'y'}\left(y'=l\right) - T^{t'y'}\left(y'=0\right)}{l} \\
&- \frac{T^{t'z'}\left(z'=l\right) - T^{t'z'}\left(z'=0\right)}{l}.
\end{aligned}
$$

If we now perform the limit $l \to 0$, the difference quotients on the right-hand side become partial derivatives and it follows

$$
\begin{aligned}
\frac{\partial\rho'}{dt'} &= -\left(\frac{\partial T^{t'x'}}{\partial x'} + \frac{\partial T^{t'y'}}{\partial y'} + \frac{\partial T^{t'z'}}{\partial z'}\right) \\
&= -\left(\frac{\partial\left(\rho'v^{x'}\right)}{\partial x'} + \frac{\partial\left(\rho'v^{y'}\right)}{\partial y'} + \frac{\partial\left(\rho'v^{z'}\right)}{\partial z'}\right).
\end{aligned}
$$

The bracket expression on the right side is a (three-dimensional) *divergence* (see remark 6.6 on page 123) and one writes

$$\frac{\partial\left(\rho'v^{x'}\right)}{\partial x'} + \frac{\partial\left(\rho'v^{y'}\right)}{\partial y'} + \frac{\partial\left(\rho'v^{z'}\right)}{\partial z'} = div\left(\rho'\vec{v}\right),$$

i.e., one can also write Eq. (16.2) as

$$\frac{\partial\rho'}{\partial t'} + div\left(\rho'\vec{v}\right) = 0. \tag{16.3}$$

It is also called **continuity equation** (see e.g. [10]) in classical hydrodynamics. From the continuity equation thus follows the *conservation of energy* for dust or also vice versa, as it is stated in most textbooks: The conservation of energy in dust inevitably leads to the continuity equation.

We still want to examine the other components of the four-divergence of **T**. For $\alpha' = x'$ one obtains from the divergence-free nature of **T**

$$T^{x'\beta'}{}_{,\beta'} = \frac{\partial\left(\rho' v^{x'}\right)}{\partial t'} + \frac{\partial\left(\rho'\left(v^{x'}\right)^2\right)}{\partial x'} + \frac{\partial\left(\rho' v^{x'} v^{y'}\right)}{\partial y'} + \frac{\partial\left(\rho' v^{x'} v^{z'}\right)}{\partial z'} = 0.$$

The partial derivatives are calculated with the product rule to

$$\frac{\partial\left(\rho' v^{x'}\right)}{\partial t'} = \frac{v^{x'}\partial\left(\rho'\right)}{\partial t'} + \frac{\rho'\partial\left(v^{x'}\right)}{\partial t'}$$

$$\frac{\partial\left(\rho'\left(v^{x'}\right)^2\right)}{\partial x'} = \frac{v^{x'}\partial\left(\rho' v^{x'}\right)}{\partial x'} + \frac{\rho' v^{x'}\partial\left(v^{x'}\right)}{\partial x'}$$

$$\frac{\partial\left(\rho' v^{x'} v^{y'}\right)}{\partial y'} = \frac{v^{x'}\partial\left(\rho' v^{y'}\right)}{\partial y'} + \frac{\rho' v^{y'}\partial\left(v^{x'}\right)}{\partial y'}$$

$$\frac{\partial\left(\rho' v^{x'} v^{z'}\right)}{\partial z'} = \frac{v^{x'}\partial\left(\rho' v^{z'}\right)}{\partial z'} + \frac{\rho' v^{z'}\partial\left(v^{x'}\right)}{\partial z'}.$$

One thus obtains

$$\begin{aligned}
T^{x'\beta'}{}_{,\beta'} &= \frac{v^{x'}\partial\left(\rho'\right)}{\partial t'} + \frac{\rho'\partial\left(v^{x'}\right)}{\partial t'} + \frac{v^{x'}\partial\left(\rho' v^{x'}\right)}{\partial x'} + \frac{\rho' v^{x'}\partial\left(v^{x'}\right)}{\partial x'} \\
&\quad + \frac{v^{x'}\partial\left(\rho' v^{y'}\right)}{\partial y'} + \frac{\rho' v^{y'}\partial\left(v^{x'}\right)}{\partial y'} + \frac{v^{x'}\partial\left(\rho' v^{z'}\right)}{\partial z'} + \frac{\rho' v^{z'}\partial\left(v^{x'}\right)}{\partial z'} \\
&= v^{x'}\left(\frac{\partial\left(\rho'\right)}{\partial t'} + \frac{\partial\left(\rho' v^{x'}\right)}{\partial x'} + \frac{\partial\left(\rho' v^{y'}\right)}{\partial y'} + \frac{\partial\left(\rho' v^{z'}\right)}{\partial z'}\right) \\
&\quad + \rho'\left(\frac{\partial\left(v^{x'}\right)}{\partial t'} + \frac{v^{x'}\partial\left(v^{x'}\right)}{\partial x'} + \frac{v^{y'}\partial\left(v^{x'}\right)}{\partial y'} + \frac{v^{z'}\partial\left(v^{x'}\right)}{\partial z'}\right).
\end{aligned}$$

The first bracket in the last equation is due to the continuity equation (16.2) equals zero, and we obtain

$$
\begin{aligned}
T^{x'\beta'}{}_{,\beta'} &= \rho' \left(\frac{\partial\left(v^{x'}\right)}{\partial t'} + \frac{v^{x'}\partial\left(v^{x'}\right)}{\partial x'} + \frac{v^{y'}\partial\left(v^{x'}\right)}{\partial y'} + \frac{v^{z'}\partial\left(v^{x'}\right)}{\partial z'} \right) \\[2mm]
&= \rho' \left(\frac{\partial\left(v^{x'}\right)}{\partial t'} + \left(\frac{\partial\left(v^{x'}\right)}{\partial x'}, \frac{\partial\left(v^{x'}\right)}{\partial y'}, \frac{\partial\left(v^{x'}\right)}{\partial z'} \right) \cdot \vec{v} \right) \\[2mm]
&= \rho' \left(\frac{\partial\left(v^{x'}\right)}{\partial t'} + grad\left(v^{x'}\right) \cdot \vec{v} \right) = 0.
\end{aligned}
$$

Similarly, for $\alpha = y', z'$

$$
\begin{aligned}
T^{y'\beta'}{}_{,\beta'} &= \rho' \left(\frac{\partial\left(v^{y'}\right)}{\partial t'} + grad\left(v^{y'}\right) \cdot \vec{v} \right) = 0 \\[2mm]
T^{z'\beta'}{}_{,\beta'} &= \rho' \left(\frac{\partial\left(v^{z'}\right)}{\partial t'} + grad\left(v^{z'}\right) \cdot \vec{v} \right) = 0.
\end{aligned}
$$

These three equations are combined into a three-dimensional vector equation

$$
\rho' \left(\frac{\partial \vec{v}}{\partial t'} + (grad\,\vec{v}) \cdot \vec{v} \right) = \vec{0}, \tag{16.4}
$$

where the expression $(grad\,\vec{v})$ denotes the tensor

$$
(grad\,\vec{v}) =
\begin{pmatrix}
\dfrac{\partial\left(v^{x'}\right)}{\partial x'} & \dfrac{\partial\left(v^{x'}\right)}{\partial y'} & \dfrac{\partial\left(v^{x'}\right)}{\partial z'} \\[4mm]
\dfrac{\partial\left(v^{y'}\right)}{\partial x'} & \dfrac{\partial\left(v^{y'}\right)}{\partial y'} & \dfrac{\partial\left(v^{y'}\right)}{\partial z'} \\[4mm]
\dfrac{\partial\left(v^{z'}\right)}{\partial x'} & \dfrac{\partial\left(v^{z'}\right)}{\partial y'} & \dfrac{\partial\left(v^{z'}\right)}{\partial z'}
\end{pmatrix}
$$

This gives the **Euler's equation** of hydrodynamics (see [10]) for the case that no external forces and no pressure are present. The Euler equation provides, analogous to energy conservation, the conservation of momentum when traversing the unit volume.

> The divergence-free nature of the energy-momentum tensor for dust is equivalent to the conservation of energy and momentum.

In our model of incoherent matter, we also assume that no particles are destroyed or created, i.e., the number of particles remains constant. This conservation law can be derived in the same way as the above, by comparing the rate of change of the number of particles in a fluid element to the inflow and outflow of particles through the boundary surfaces, from which we obtain for the particle flow vector $\vec{N}$ the conservation law for the number of particles

$$\frac{\partial N^t}{\partial t} = -\frac{\partial N^x}{\partial x} - \frac{\partial N^y}{\partial y} - \frac{\partial N^z}{\partial z}$$

or

$$N^\alpha{}_{,\alpha} = (nU^\alpha)_{,\alpha} = 0, \tag{16.5}$$

i.e., the divergence-free nature of the particle flow vector, is obtained.

16.2. Ideal Fluids

We now want to extend the dust model by two aspects so that we come to the **general fluids**. On the one hand, the particles can move randomly relative to each other, and on the other hand, there can be various forces between the particles that contribute to the total energy of the particle quantity. For each individual element (which is a small collection of particles) there is also an instantaneous inertial system (inertial system that depends on space and time), in which the relative speed v is momentarily equal to zero, but unlike dust, two elements usually no longer have a common instantaneous inertial system, since they can move relative to each other. All physical quantities associated with the fluid elements, such as four-velocity, particle and energy density, flux vector, temperature, as well as pressure, tension and strain, are defined as *values in the instantaneous inertial systems*. In general fluids, surface tensions, frictions, heat exchange, etc. can occur between the elements. The **ideal fluids** on the other hand are defined by the fact that no heat exchange takes place and no forces appear parallel to the surface of the elements. In other words, they have constant temperature, the particles can be moved without force and the total force acts perpendicular to the surface. The **pressure** p in a fluid element is defined as the quotient of force F by the area A of the element

$$p = \frac{F}{A}.$$

We consider in the instantaneous inertial system S an arbitrary cuboid volume element $\Delta V = \Delta x \Delta y \Delta z$.

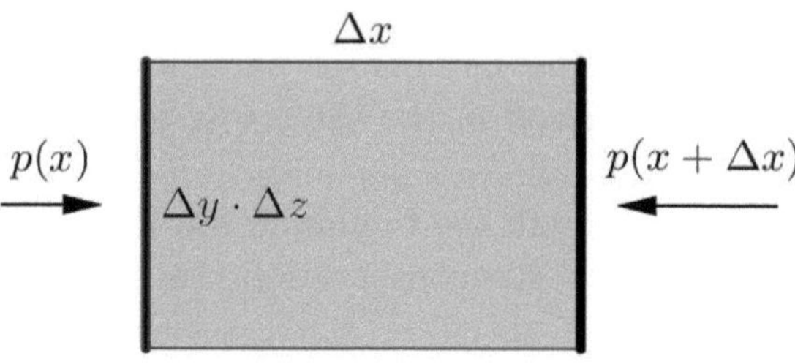

Figure 16.5.: Pressure difference

In Fig. 16.5 the z-axis is again suppressed. The pressure $p(x)$ may act on the left surface element $\Delta y \Delta z$. If the pressure changes in the x-direction, the pressure $p(x + \Delta x)$ acts on the opposite side. The pressure difference in the x-direction is therefore

$$p(x) - p(x + \Delta x).$$

For small Δx we can estimate the difference by the differential (see formula 4.16 on page 67) and get

$$p(x) - p(x + \Delta x) \approx -\frac{\partial p}{\partial x} \Delta x.$$

The resulting force in the x-direction is then

$$F^x = (p(x) - p(x + \Delta x)) \Delta A \approx -\frac{\partial p}{\partial x} \Delta x \Delta y \Delta z = -\frac{\partial p}{\partial x} \Delta V.$$

Similarly, one obtains for the other force components

$$F^y \approx -\frac{\partial p}{\partial y} \Delta V$$

$$F^z \approx -\frac{\partial p}{\partial z} \Delta V.$$

With the limit transition $\Delta V \to dV$ this results in the vector equation

$$\vec{F} = -grad(p) \cdot dV = -(F^x, F^y, F^z) \cdot dV.$$

Due to the free mobility of the fluid particles, the total force in the rest system must be zero (the own weight of the fluid element is neglected), i.e., it holds

$$grad\,(p) = \vec{0},$$

which in turn means that the pressure in the fluid element is constant. Therefore, the same pressure acts on each surface element dA of the surrounding walls in the rest system of the element.

With this, we are able to set up the energy-momentum tensor for ideal fluids in the instantaneous rest system. First, as with dust, that $T^{tt} = \rho$ is the energy density. Since in the instantaneous rest system the velocity v is equal to zero and no other (heat) energy can flow, it holds

$$T^{ti} = T^{it} = 0, \quad i = x, y, z.$$

Since no forces act parallel to the element surfaces and the momentum flow is also zero due to $v = 0$, $T^{ij} = 0$ applies for $i \neq j$. The only forces per unit area acting on the fluid element are perpendicular to the surfaces and, according to the above consideration, are equal to the pressure p in every direction. This means:

$$T^{xx} = T^{yy} = T^{zz} = p$$

In summary, for the components of the energy-momentum tensor $\mathbf{T}$ of an ideal fluid in the rest system S

$$\mathbf{T} \underset{S}{\rightarrow} T^{\alpha\beta} = \begin{pmatrix} \rho & 0 & 0 & 0 \\ 0 & p & 0 & 0 \\ 0 & 0 & p & 0 \\ 0 & 0 & 0 & p \end{pmatrix},$$

which means that the energy-momentum tensor for an ideal fluid without pressure ($p = 0$) is equal to that of incoherent matter. For an arbitrary inertial system S', which moves relative to S with the velocity $\vec{v} \rightarrow (v^x, v^y, v^z)$, this results in

$$\vec{U} \underset{S'}{\rightarrow} (U^{t'}, U^{x'}, U^{y'}, U^{z'}) = (\gamma, \gamma v^x, \gamma v^y, \gamma v^z)$$

and

$$\eta^{\alpha'\beta'} = \begin{pmatrix} -1 & 0 & 0 & 0 \\ 0 & 1 & 0 & 0 \\ 0 & 0 & 1 & 0 \\ 0 & 0 & 0 & 1 \end{pmatrix}$$

for the components of $\mathbf{T}$

$$\mathbf{T} \underset{S'}{\rightarrow} T^{\alpha'\beta'} = (\rho + p)U^{\alpha'}U^{\beta'} + p\eta^{\alpha'\beta'}, \tag{16.6}$$

which in matrix notation leads to

$$T^{\alpha'\beta'} \;=\; (\rho+p)U^{\alpha'}U^{\beta'} + p\eta^{\alpha'\beta'} \tag{16.7}$$

$$= (\rho+p)\gamma^2 \begin{pmatrix} 1 & v^x & v^y & v^z \\ v^x & (v^x)^2 & v^x v^y & v^x v^z \\ v^y & v^y v^x & (v^y)^2 & v^y v^z \\ v^z & v^z v^x & v^z v^y & (v^z)^2 \end{pmatrix} + \begin{pmatrix} -p & 0 & 0 & 0 \\ 0 & p & 0 & 0 \\ 0 & 0 & p & 0 \\ 0 & 0 & 0 & p \end{pmatrix}.$$

First, we check whether this formula for the case of the rest system, i.e. $v^x = v^y = v^z = 0$ and $\gamma = 1$, again delivers the result from above:

$$(\rho+p)\begin{pmatrix} 1 & 0 & 0 & 0 \\ 0 & 0 & 0 & 0 \\ 0 & 0 & 0 & 0 \\ 0 & 0 & 0 & 0 \end{pmatrix} + \begin{pmatrix} -p & 0 & 0 & 0 \\ 0 & p & 0 & 0 \\ 0 & 0 & p & 0 \\ 0 & 0 & 0 & p \end{pmatrix} = \begin{pmatrix} \rho & 0 & 0 & 0 \\ 0 & p & 0 & 0 \\ 0 & 0 & p & 0 \\ 0 & 0 & 0 & p \end{pmatrix},$$

so we get the tensor components of the rest system. However, this does not yet prove that the components of $\mathbf{T}$ in S' can be calculated as in (16.7). This is shown by applying the transformation rule for $(0,2)$-tensors, taking into account that the system S' is moving at the speed of $-v$ from the point of view of the rest system and therefore the inverse Lorentz transformation must be applied. For simplicity, we assume that the particles move only in the positive x-direction, so $\vec{v} = (v^x, 0, 0)$, i.e. $v = v^x$, $v^y = 0$, $v^z = 0$. The resulting Lorentz transformation then has the matrix representation

$$L^{\alpha'}_{\alpha}(-v) = \begin{pmatrix} \gamma & \gamma v^x & 0 & 0 \\ \gamma v^x & \gamma & 0 & 0 \\ 0 & 0 & 1 & 0 \\ 0 & 0 & 0 & 1 \end{pmatrix},$$

and the components $T^{\alpha'\beta'}$ result from (16.7) to

$$\left(T^{\alpha'\beta'}\right) = (\rho+p)\gamma^2 \begin{pmatrix} 1 & v^x & 0 & 0 \\ v^x & (v^x)^2 & 0 & 0 \\ 0 & 0 & 0 & 0 \\ 0 & 0 & 0 & 0 \end{pmatrix} + \begin{pmatrix} -p & 0 & 0 & 0 \\ 0 & p & 0 & 0 \\ 0 & 0 & p & 0 \\ 0 & 0 & 0 & p \end{pmatrix}$$

$$= \begin{pmatrix} (\rho+p)\gamma^2 - p & v^x(\rho+p)\gamma^2 & 0 & 0 \\ v^x(\rho+p)\gamma^2 & (v^x)^2(\rho+p)\gamma^2 + p & 0 & 0 \\ 0 & 0 & p & 0 \\ 0 & 0 & 0 & p \end{pmatrix}. \tag{16.8}$$

The transformation formula can be represented in matrix notation as follows:

$$\left(T^{\alpha'\beta'}\right) \;=\; \left(L^{\alpha'}_{\alpha}(-v)\, L^{\beta'}_{\beta}(-v)\, T^{\alpha\beta}\right)$$

$$= \begin{pmatrix} \gamma & \gamma v^x & 0 & 0 \\ \gamma v^x & \gamma & 0 & 0 \\ 0 & 0 & 1 & 0 \\ 0 & 0 & 0 & 1 \end{pmatrix} \begin{pmatrix} \rho & 0 & 0 & 0 \\ 0 & p & 0 & 0 \\ 0 & 0 & p & 0 \\ 0 & 0 & 0 & p \end{pmatrix} \begin{pmatrix} \gamma & \gamma v^x & 0 & 0 \\ \gamma v^x & \gamma & 0 & 0 \\ 0 & 0 & 1 & 0 \\ 0 & 0 & 0 & 1 \end{pmatrix},$$

i.e., two matrix multiplications have to be performed. Multiplication of the two left matrices results in

$$\begin{pmatrix} \gamma & \gamma v^x & 0 & 0 \\ \gamma v^x & \gamma & 0 & 0 \\ 0 & 0 & 1 & 0 \\ 0 & 0 & 0 & 1 \end{pmatrix} \begin{pmatrix} \rho & 0 & 0 & 0 \\ 0 & p & 0 & 0 \\ 0 & 0 & p & 0 \\ 0 & 0 & 0 & p \end{pmatrix} = \begin{pmatrix} \gamma\rho & \gamma v^x p & 0 & 0 \\ \gamma v^x \rho & \gamma p & 0 & 0 \\ 0 & 0 & p & 0 \\ 0 & 0 & 0 & p \end{pmatrix}$$

and

$$\begin{pmatrix} \gamma\rho & \gamma v^x p & 0 & 0 \\ \gamma v^x \rho & \gamma p & 0 & 0 \\ 0 & 0 & p & 0 \\ 0 & 0 & 0 & p \end{pmatrix} \begin{pmatrix} \gamma & \gamma v^x & 0 & 0 \\ \gamma v^x & \gamma & 0 & 0 \\ 0 & 0 & 1 & 0 \\ 0 & 0 & 0 & 1 \end{pmatrix} =$$

$$\begin{pmatrix} \gamma^2\rho + \gamma^2 (v^x)^2 p & \gamma^2 v^x \rho + \gamma^2 v^x p & 0 & 0 \\ \gamma^2 v^x \rho + \gamma^2 v^x p & \gamma^2 (v^x)^2 \rho + \gamma^2 p & 0 & 0 \\ 0 & 0 & p & 0 \\ 0 & 0 & 0 & p \end{pmatrix}.$$

The matrix elements $T^{t't'}, T^{t'x'}, T^{x'x'}$ can be transformed with $\gamma^2 = 1/\left(1 - (v^x)^2\right)$ to

$$\begin{aligned}
T^{t't'} &= \gamma^2\rho + \gamma^2 (v^x)^2 p \\
&= \gamma^2\rho + \gamma^2 (v^x)^2 p - \gamma^2 p + \gamma^2 p \\
&= \gamma^2\rho + \gamma^2 p \left((v^x)^2 - 1\right) + \gamma^2 p \\
&= \gamma^2\rho + \gamma^2 p - \gamma^2 p \left(1 - (v^x)^2\right) \\
&= \gamma^2 (\rho + p) - p,
\end{aligned}$$

$$\begin{aligned}
T^{t'x'} &= \gamma^2 v^x \rho + \gamma^2 v^x p \\
&= \gamma^2 v^x (\rho + p),
\end{aligned}$$

$$\begin{aligned}
T^{x'x'} &= \gamma^2 (v^x)^2 \rho + \gamma^2 p \\
&= \gamma^2 (v^x)^2 \rho + \gamma^2 p - \gamma^2 (v^x)^2 p + \gamma^2 (v^x)^2 p \\
&= \gamma^2 (v^x)^2 \rho + \gamma^2 p \left(1 - (v^x)^2\right) + \gamma^2 (v^x)^2 p
\end{aligned}$$

$$= \gamma^2 \left(v^x\right)^2 \left(\rho + p\right) + p.$$

That is, we obtain the same matrix as in (16.8). For the complete derivation of the formula (16.7) one uses instead of $\left(L^{\alpha'}_{\alpha}\left(-v\right)\right)$ the (inverse) general Lorentz transformation matrix (see 14.6 on page 293)

$$\Lambda^{\mu}_{\nu} = \begin{pmatrix} \gamma & -\gamma v^x & -\gamma v^y & -\gamma v^z \\ -\gamma v^x & 1 + (\gamma - 1)\dfrac{(v^x)^2}{v^2} & (\gamma - 1)\dfrac{v^x v^y}{v^2} & (\gamma - 1)\dfrac{v^x v^z}{v^2} \\ -\gamma v^y & (\gamma - 1)\dfrac{v^x v^y}{v^2} & 1 + (\gamma - 1)\dfrac{(v^y)^2}{v^2} & (\gamma - 1)\dfrac{v^y v^z}{v^2} \\ -\gamma v^z & (\gamma - 1)\dfrac{v^x v^z}{v^2} & (\gamma - 1)\dfrac{v^y v^z}{v^2} & 1 + (\gamma - 1)\dfrac{(v^z)^2}{v^2} \end{pmatrix}.$$

However, this requires considerably more computational effort, which we want to save ourselves at this point. We can summarize the **Energy-Momentum tensor for ideal fluids** as

$$\mathbf{T} = (\rho + p)\vec{U} \otimes \vec{U} + p\mathbf{G}, \tag{16.9}$$

where $\mathbf{G}$ again denotes the metric tensor for one-forms

$$\mathbf{G} \to \eta^{\alpha\beta}.$$

The energy-momentum tensor for dust is obtained in the case $p = 0$. That means, an ideal fluid can only be without pressure if its particles do not make any random movements. Or in other words: pressure arises from the random velocities of the particles. Even a very thin gas exerts a certain pressure, e.g. on the surfaces of the container in which it moves.

Just as in the case of incoherent matter, the energy-momentum tensor for an ideal fluid is symmetric and divergence-free, as can be calculated with identical derivation as above. Therefore, it applies

$$T^{\alpha\beta}_{\ ,\beta} = \left[(\rho + p)U^{\alpha}U^{\beta} + p\eta^{\alpha\beta}\right]_{,\beta} = 0. \tag{16.10}$$

Symmetry and divergence-freeness are outstanding features of energy-momentum tensors. They also apply to general fluids or to the (not considered here) electromagnetic field.

Non-relativistic limit case

When the velocities are non-relativistic, the rest energy of matter exceeds the kinetic energy contributions by many orders of magnitude. In particular, the velocities of the fluid particles are low, so the pressure compared to the particle density can be neglected. We therefore assume that

$$\begin{aligned}
\gamma &\approx 1, \\
\rho + p &\approx \rho, \\
p &\approx 0, \\
v^x &\ll 1
\end{aligned}$$

applies. Then for the non-relativistic limit case

$$\left(T^{\alpha\beta}\right) \approx \begin{pmatrix} \rho & 0 & 0 & 0 \\ 0 & 0 & 0 & 0 \\ 0 & 0 & 0 & 0 \\ 0 & 0 & 0 & 0 \end{pmatrix}. \tag{16.11}$$

17. Literature References and Further Information on Part III

With Special Relativity Theory, it is similar to mechanics: There are a multitude of books on it, especially popular science presentations. In this category two books are recommended here. One is by Einstein himself: The Special and General Theory [7], which first appeared in 1916 and has become the standard work of generally understandable introductions over the years. The book comes with very few formulas (e.g., the Lorentz transformations are derived with simple means) and describes in everyday language the physical basics of both theories of relativity.

The second popular science book is by Lewis C. Epstein: Relativity visualizad [8], in which the basic ideas of Special Relativity Theory with many thought experiments, instructive drawings and vivid images are conveyed. Anyone who wants to have a pictorial representation of the physical principles of relativity theory in addition to the formulaic one, is well served with this book.

The third part of this book is divided into two. First, the Special Relativity Theory is developed directly from the Einstein's postulates with the simplest means. This idea is also pursued by the book "Special Relativity for beginners" by J. Freund [12], which has been acted as a model for the formulation of the first chapters in this part. The treatise by Freund is about on the level of this book. There four-vectors are also introduced and the transformations of the electromagnetic field up to the Lorentz invariance of the Maxwell equations are presented in very detailed form. Thus, Freund goes in some parts beyond what is treated in this book. In a similar direction Max Born's book "Einstein's Theory of Relativity" [3], which first appeared in 1920, is also recommended. In his description of the Special Theory of Relativity, Born uses only simple mathematics (e.g., without calculus) and focuses primarily on phenomena in the electromagnetic field. At the end of his book, he also provides an easy-to-understand outlook on the General Theory of Relativity.

The presentation of vector and tensor calculus in the field of Special Relativity in this third part follows the description in the book by Schutz [29]. There, the energy-momentum tensor is also derived in a comprehensible way. The other literature recommendations for deepening the understanding of vector

M. Ruhrländer, *Ascent to the Einstein Equations*,
https://doi.org/10.1007/978-3-662-72672-3_17

and tensor calculus have already been listed at the end of part II.

Part IV.

Fundamentals of General Relativity

Einstein's General Theory of Relativity (GR), which he presented at the end of 1915, is a theory of gravity. The name already suggests that the GR is a generalization of the Special Theory of Relativity, which Einstein published in 1905. An immediate question arises. Why is a new theory of gravity necessary at all, when Newton's gravitation has proven for centuries that it is capable of explaining most of the observed phenomena related to gravity? What keeps the planets/moons in their orbits? Why are these orbits elliptical, and how can I calculate them? Why are there tides on Earth, and how can they be explained? Why and when is there a solar eclipse? Why does an apple fall from a tree? For all these questions and many more, Newton's gravitation provides satisfactory answers.

A second obvious question is: Why is a generalization of the Special Theory of Relativity a theory of gravity? Why not a theory of electromagnetism or elementary particles or a completely new theory? What is special about gravity that a generalization of the SR, which is a theory of space and time, necessarily leads to it?

Here, in this introductory section, we primarily want to try to answer the first question. The simple reason for the necessity of a new theory of gravity is that Newton's theory is not compatible with the Special Theory of Relativity. So if we assume that the SR is correct (and we do), then Newton's model of gravity must be discarded. To better understand this, let's look again at Newton's law of gravitation (see equation 6.4 on page 88), which states that the force exerted by a mass M on a second mass m is given by

$$\vec{F} = -\frac{GMm}{r^2} \cdot \vec{e}_r,$$

where the vector $\vec{e}_r$ is the unit vector in the direction $M \to m$ and r denotes the distance between the centers of mass. If we assume that the mass M changes over time, i.e., is a function of the time variable t, the equation becomes

$$\vec{F}(t) = -\frac{GM(t)\,m}{r^2} \cdot \vec{e}_r.$$

This means that the force experienced by the mass m at time t depends on the mass $M(t)$ at the same time t. Changes to the mass M at time t at the location of mass M would therefore be *simultaneously* noticeable at the location of mass m. However, this would require an *immediate* („**instantaneous**") transmission of force, contradicting the Special Theory of Relativity, which states that nothing can move faster than the speed of light. Newton's law of gravitation is therefore *not compatible* with the Special Theory of Relativity.

We can also demonstrate the incompatibility of the two theories with the following, more formal considerations. We define the vector field

$$\vec{g} = \frac{\vec{F}}{m} = -\frac{GM}{r^2} \cdot \vec{e}_r,$$

i.e., $\vec{g}$ is a kind of gravitational field strength, a gravitational field per unit mass. The vector field $\vec{g}$ depends on r, but not on t. Such a (three-dimensional) vector field is however not compatible with the Special Theory of Relativity, as it cannot be Lorentz-invariant. For this to happen, as we saw in part III, $\vec{g}$ would have to be a four-vector.

However, it should be clear that regardless of the fact that Newton's theory is "wrong", any new theory of gravity must "contain" Newton's theory, i.e., the new theory must include Newton's law of gravity as a special case.

This part IV of the book is structured similarly to part I and III. First, we want to make simple physical considerations that introduce the problems of the General Theory of Relativity and make it clear that we must take new physical and mathematical paths to arrive at a quantitative description of the phenomena surrounding gravity. After that, we will deal with the mathematical foundations that are necessary to establish the physical laws of the General Theory of Relativity. In doing so, we will further generalize and expand the vector and tensor calculus already practiced in the first parts, to finally arrive at the Einstein equations. Based on these basic, very general and abstractly formulated laws we then want to investigate phenomena in the near-Earth environment (i.e., in our solar system) and make comparisons with the corresponding results from Newton's theory of gravitation. We will find that Einstein's theory eliminates some of the weaknesses of the predecessor theory and thus provides a satisfactory explanation of actually measured physical phenomena.

Because we have already laid a good foundation for the necessary extensions of tensor calculus in the earlier chapters, these are, in my view, easier to handle than the concrete conclusions from the abstract Einstein equations, which we will derive in the last chapters. Every (even so simple) solution of these equations requires a tremendous amount of computational effort if one wants to calculate concrete phenomena, such as the motion of a particle in a gravitational field, in detail. But here too, the principle applies that we proceed step by step and document the derivations and transformations in detail.

18. Gravitation and Spacetime Model

We start with some considerations that show that the space-time model of the Special Theory of Relativity must be extended so that the physical phenomena associated with gravitation can be explained. We first revisit the equivalence between inertial and gravitational mass and show that clocks slow down in the Newtonian gravitational field.

18.1. Principle of Equivalence

The equality of gravitational and inertial mass, described in section 6.5 as the principle of equivalence, leads to the fact that bodies - regardless of their mass - experience exactly the same acceleration due to gravity near the Earth. If m_g denotes the gravitational mass and m_i the inertial mass of a body, then according to the second law of Newton and Galileo's law of falling bodies

$$F = m_i a = -m_g g,$$

where a is the acceleration and g is the acceleration due to gravity directed towards the center of the Earth. So if m_i and m_g are equal, we get the relationship valid for every body

$$a = -g,$$

i.e., the acceleration due to gravity a is the same for all bodies, assuming that air resistance and other possible forces are negligible. Now imagine that we are falling from a high tower together with a heavy ball. The ball would stay by our side and we would get the impression that no forces are acting on it. If we give the ball a push, it would move away from us in a straight line at a constant speed (even downwards!). Of course, both the ball and we would be falling towards the Earth due to the common acceleration due to gravity, but the relative movements between us would be constant. The accelerations due to gravity acting on us and the ball cancel each other out. We can therefore reformulate the principle of equivalence:

M. Ruhrländer, *Ascent to the Einstein Equations*,
https://doi.org/10.1007/978-3-662-72672-3_18

> *Principle of Equivalence*: In the near-Earth gravitational field, all objects appear to a freely falling observer as if there were no gravitational field. That is, for a freely falling observer, the laws of physics are as they would be "far out" in the universe without gravitational influence by any masses.

Of course, even without air friction, we would feel that we are in free fall. But this is only because our body is accustomed to the gravitational field of the Earth. From a purely physical point of view, free fall is equivalent to a state without gravity, assuming a homogeneous gravitational field. Here, **homogeneous** means that the field is constant in time and space and always points in the same direction.

The argument that freely falling observers are free of gravity can also be reversed. We use a thought experiment, similar to one invented by Einstein. We assume that a rocket is moving in an empty space without gravity with a constant acceleration g. To an observer inside the rocket, it appears as if there is a gravitational field inside the rocket. If he lets objects fall, they all fall to the floor of the rocket with the same acceleration, regardless of their internal composition, from his point of view. From the outside, the objects do not fall to the floor, but stay where they are released. However, the floor moves towards the objects due to the constant acceleration in such a way that the rocket occupant has the impression that the objects are falling to the floor. If he holds an object in his hand, he feels a weight that exactly matches the weight the object would have on Earth. Assuming that he can only perceive what happens inside the rocket, he cannot distinguish his state from that in which the rocket is standing on the Earth's surface. In this sense, homogeneous gravitational fields are equivalent to reference systems that are uniformly accelerated relative to an inertial system.

18.2. Gravitational Redshift

We want to use the newly formulated principle of equivalence to show that gravity causes a redshift of light. For this, we assume that an observer is located on the ground and sends a light beam with the frequency f_d up to a second observer on a tower of height h. The observer on the tower measures the frequency f_u of the light reaching him. We further assume that there is a third observer who falls from the tower when the light flash is triggered, thus becoming a freely falling observer. He perceives the light beam as if there were no gravity and he himself were at rest. For him, the frequency of the emitted light does not change, he measures the same frequency f_d as the observer on the ground. This also applies to the moment when the light has reached the top of the tower. In this short period of time, the third observer has fallen, i.e., relative to him, the second observer on the tower has moved away from the light source. According to the Doppler effect, the observer on the tower measures a smaller frequency of light, for him a redshift occurs. To determine how large this is, we calculate the speed of the free-falling observer when the light reaches the top of the tower. Since the falling speed of the third observer is small compared to the speed of light, we use the classical Doppler effect formula 12.11 on page 242 for simplicity in the calculation. To reach the top of the tower, the light needs time h (note that we continue to work with natural units, i.e., the speed of light is $c = 1$). In this time, the falling observer has fallen with the Earth's acceleration g, i.e., his final speed is $v = gh$. Thus, with the classical Doppler formula (12.11), we get

$$f_u = (1 - v)\, f_d = (1 - gh)\, f_d. \tag{18.1}$$

The difference between the two frequencies is very small, for example, if the tower is $100\,\mathrm{m}$ high, then the speed $v = gh$ is only 1.1×10^{-14}. Nevertheless, the effect was accurately detected in many experiments in the 1960s, see [39]. Nowadays, it is routinely taken into account, for example, when correcting time differences in the transmission of GPS signals.

As already shown in section 12.3, the redshift is both a property of the observer and of the light itself. Light has different frequencies depending on the height in the Earth's gravitational field. If we measure the frequency at a certain height above the Earth's surface and compare it with that on the Earth's surface, we see a redshift, if we measure the frequency of the light in free fall, we see no redshift.

The gravitational redshift also has effects on time itself: gravity slows down time. Let's imagine there are two identical clocks that "tick" at the same frequency f_d as the light emitted from the ground in the above thought experiment. We bring one clock to the top of the tower, the other stays on the ground. We send light with the frequency f_d from the ground to the top of the tower for about 10^{20} ticks. Since visible light oscillates at about 10^{15} per second, this would mean about one day. Now the clock on the top of the tower receives the light from the ground redshifted, the frequency is about one tick in 10^{14} less than its own. This means, the clock below appears to go slower from the observer above! Within one day, the clock above has ticked about 10^6 times more often than the clock below. If you then bring both clocks back together, you will notice this time difference. Although it is only a nanosecond ($1\,\mathrm{ns} = 10^{-9}\,\mathrm{s}$), it is measurable and has been proven in many experiments. The design of the clocks does not matter for the gravitational slowing down of time: *All* clocks run faster when they are "higher" in the gravitational field. And physically, time is anyway only what can be measured with clocks.

18.3. Light Deflection at the Sun in the Newtonian Gravitational Field

We want to show with the equivalence principle that light changes its direction when it passes the Sun. The falling acceleration in the Newtonian gravitational field of the Sun is according to Eq. 6.30 on page 125

$$a = \frac{GM_S}{d^2}$$

for a particle that is at a distance d from the Sun, where M_S denotes the mass of the Sun, see Tab. 30.3.

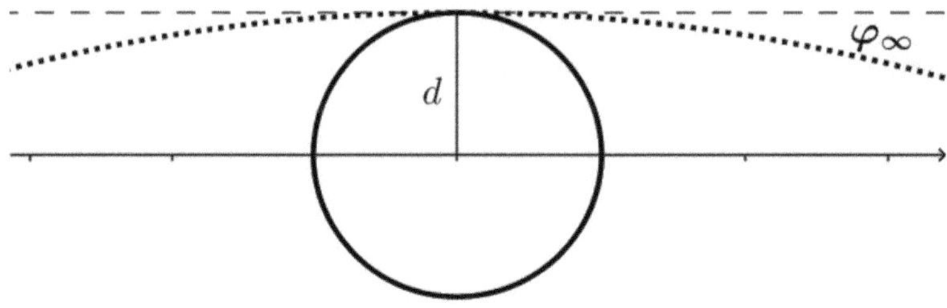

Figure 18.1.: Light deflection at the Sun according to Newton

In Fig. 18.1 a light beam (dotted line) is drawn, which passes exactly on the surface of the Sun, i.e., the distance d is the radius of the Sun. The dashed line shows how the light beam would run if it were not deflected by the sun's gravitational field. The angle φ_∞ indicates the size of the deflection. We want to determine this deflection angle approximately. To do this, we conduct a similar thought experiment as in the last section. We consider a light beam that passes the Sun. For an observer falling freely in the sun's gravitational field, the light beam moves in a straight line. However, since the freely falling observer is moving towards the gravitational center, the light must continuously change its direction so that the observer can perceive it as a straight line. To estimate the size of this effect, we imagine that the observer is at a distance d, i.e., where the light beam comes closest to the Sun, at rest relative to the Sun. At the moment when the light beam passes him, he falls with the above acceleration towards the interior of the Sun. The light moves at the speed of light c and experiences the greatest deflection in a time of the order of magnitude $t \approx d/c$, i.e., the time it takes for the light to be significantly away from the Sun again. During this time, the observer has reached a speed perpendicular to the light beam of

$$v = a \cdot t \approx \frac{GM_S\, d}{d^2}\frac{}{c} = \frac{GM_S}{cd}.$$

According to the equivalence principle, the light must have received the same speed perpendicular to its original direction. Since the speed of light in the original direction has changed only slightly, we calculate the deflection angle by simple geometry. The ratio v/c roughly indicates the slope of the dotted line, i.e.

$$\tan \varphi_\infty \approx \frac{v}{c}.$$

Since the deflection angle φ_∞ is small, we can estimate the tangent with the differential

$$\tan \varphi_\infty \approx \underbrace{\tan 0}_{=0} + \varphi_\infty \tan'(0) = \varphi_\infty(1 + \tan^2(0)) = \varphi_\infty.$$

Here we have used that the derivative of $\tan x$ is equal to $1 + \tan^2 x$. We get

$$\varphi_\infty \approx \frac{v}{c} \approx \frac{GM_S}{c^2 d}.$$

The total deviation is twice as large, as both the incoming and the outgoing light are deflected in the same way, i.e., the prediction from Newton's theory of gravitation for the total deflection of light at the Sun is

$$\frac{2GM_S}{c^2 d}.$$

We will see in section 23.7 that the general theory of relativity makes a prediction about the deflection of light by the Sun that is twice as large as Newton's. This prediction by Einstein was first experimentally tested in 1919 during a solar eclipse and has since been confirmed in numerous other experiments. The Newtonian prediction for the deflection of light at the Sun does not agree with the experiments. It is, along with other phenomena (such as the perihelion rotation of Mercury), an indication that an extension of the current theory of gravitation is needed, especially for objects with high speeds or large masses.

18.4. Gravitation and Curvature

In the special theory of relativity, it is assumed that there are inertial systems that fill the entire spacetime. Each event can be described by a single inertial system, whose coordinate points are at rest relative to the origin and whose clocks are synchronized with the clock at the origin at every point in space. In this section, we will show that in a non-homogeneous gravitational field, it is not possible to construct a **global inertial system** in which all clocks run the same. In this sense, gravitational fields are not compatible with the Special Theory of Relativity. However, there is the possibility to define so-called **local inertial systems** in "small" spacetime regions (small enough so that the inhomogeneities of the gravitational forces are not measurable).

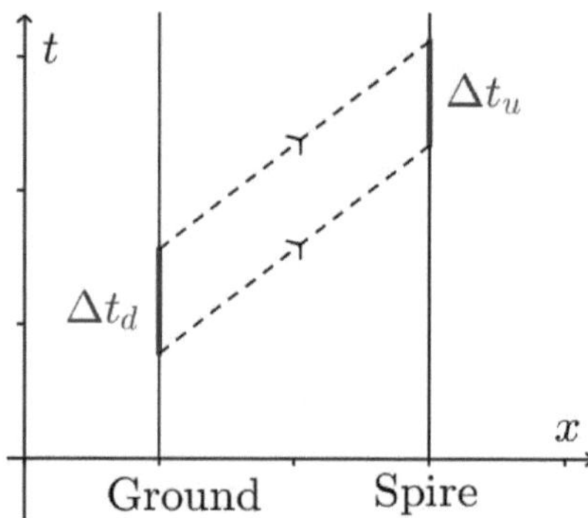

Figure 18.2.: The Earth is not an inertial system

First, we show that it is not possible, for example, to consider a laboratory on the Earth as a rest system in the sense of the Special Theory of Relativity. We again use the thought experiment from above and imagine that shortly after one another two light signals are sent from the ground to the top of the tower. In Fig. 18.2 a Minkowski diagram is drawn, which represents the ground and the top of the tower as resting, so their worldlines run vertically. In the time interval Δt_d the two light signals are sent from the ground. The dashed lines are meant to suggest that it is fundamentally possible that light rays in the gravitational field of the Earth *do not* move in a straight line, as we have assumed in the graphic for simplicity. But however the worldlines of light run in a gravitational field, the effect must be the same for both light rays, since we assume that the gravitational field is temporally and spatially homogeneous, i.e., constant. This means that the worldlines of the two light rays run congruently and we conclude from the assumed Minkowski geometry, that $\Delta t_d = \Delta t_u$. But we know from the redshift due to the gravitational field, that $\Delta t_u > \Delta t_d$. Thus, the assumed Minkowski geometry is wrong and the laboratory system on Earth is not an inertial system.

However, this does not mean that there are no inertial systems in gravitational fields at all. An important property of inertial systems is that a particle moves at a constant speed as long as no force is exerted. The gravitational force differs from all other forces in that all bodies with the same initial speed follow the same worldline. Therefore, a reference system that falls freely in a gravitational field is a candidate for an inertial system, since - as explained in the last section - in such a reference system unaccelerated particles maintain a uniform speed. We have shown in section 18.2 that no gravitational redshift occurs in a freely falling reference system. One could therefore assume that a freely falling reference system in the Earth's gravitational field represents a (global) inertial system. But this is also not the case, since freely falling reference systems on different sides of the Earth fall in different directions. So it is still not possible to construct a global inertial system.

The highest we can aspire to are so-called *local* inertial systems, which we will now deal with. We consider a freely falling reference system in the Earth's gravitational field. On the left side of Fig. 18.3 four particles T_1, T_2, T_3, T_4 are drawn, which at the time $t = 0$ in a freely falling reference system (dashed box) move towards the Earth. The four particles are connected to each other for purely optical reasons, but move independently of each other.

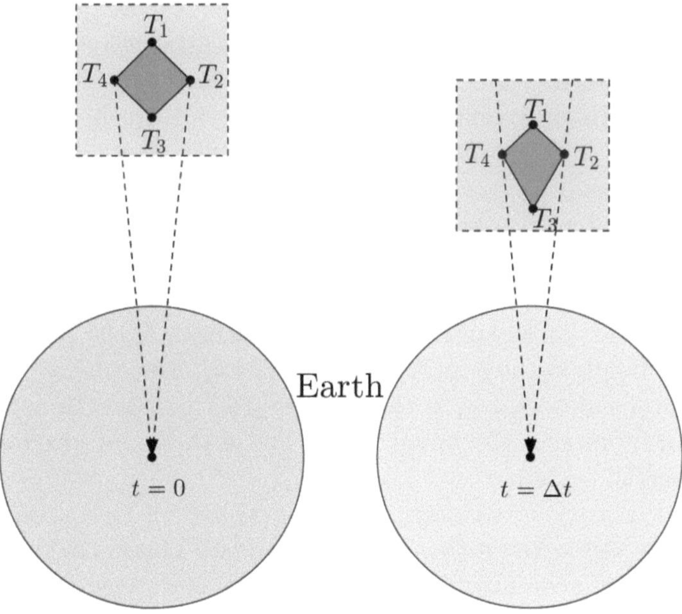

Figure 18.3.: The gravitational field of the Earth is not homogeneous

At the time $t = 0$ the four particles form the corners of a (imaginary) square.

According to Newton's law of gravitation

$$\vec{F} = -\frac{Gm_1 m_2}{r^2}\,\vec{e}_r$$

the gravitational force is directed towards the center of the Earth, i.e., particles T_1 and T_3 fall straight down, the particles T_2 and T_4 along the dashed arrows. After a time span Δt, the reference system has reached the position on the right side of the figure. We observe that the positions of the particles relative to each other have changed to a *kite quadrilateral*. On the one hand, particles T_2 and T_4 have fallen towards each other due to the movement along the dashed arrows, they have approached each other due to the horizontal component of the gravitational force. On the other hand, the distance between T_1 and T_3 has increased. This is due to the fact that in the Newtonian gravitational field the force is constant in time, but not in space. Particles that are closer to the center of gravity (T_3), experience a stronger gravitational acceleration than particles that are further away (T_1). Particles that fall freely in the Earth's gravitational field thus experience forces that compress them horizontally and pull them apart vertically. These forces are called **tidal forces** and are ultimately responsible for the low and high tide on Earth. The inhomogeneities (different directions, spatially different accelerations) in the gravitational field ensure that particles falling freely with the same initial velocity do *not* move along parallel trajectories, i.e., they ensure that there can be no global inertial system. If the inhomogeneities can be neglected, a freely falling reference system is an inertial system. Every gravitational field can be assumed to be homogeneous in a "small" spacetime region, so that local inertial systems can be constructed. Local inertial systems are similar to the momentary reference systems of fluids, here the reference system is only inertial in a small region and for a short time. How small the spacetime region must be chosen for physical investigations depends on the one hand on the strength of the inhomogeneities and on the other hand on the experimental situation. Since every inhomogeneity is fundamentally detectable, a reference system can mathematically only be inertial in an infinitesimally small region and a theory of gravitation must include local inertial systems.

In the Special Theory of Relativity, the worldlines of two particles always remain parallel to each other if they started parallel. This is also an important property of Euclidean geometry ("parallel axiom"). The spacetime geometry of the Special Theory of Relativity is not Euclidean, because the metric $\mathbf{g} \to \eta_{\alpha\beta}$ of SR is different from the Euclidean metric $\delta_{\alpha\beta}$, as we have shown in part III. Nevertheless, the parallel axiom also applies in the spacetime geometry of SR. We have seen that in an inhomogeneous gravitational field the worldlines of two particles do not remain parallel, even if they started parallel. We want to briefly call spaces that do not fulfill the parallel axiom **curved spaces** (otherwise **flat spaces**), even if this terminology does not exactly correspond to the precise mathematical definition.

For example, the Earth's surface is curved. If two people at the equator start parallel to each other heading north and always go straight ahead, they meet at the North Pole. Or more generally: Locally straight lines on a spherical surface form so-called great circles when attached to each other, and great circles have common intersection points. Nevertheless, one can claim that the Earth's surface is locally flat. It does not lead to major distortions (i.e., distortions of distances) when printing a city map on a flat piece of paper, while the attempt to "flatten" the entire Earth fails. The spherical surface is therefore locally flat, but the locally straight lines, which are called **geodesics** (great circles on a spherical surface are thus geodesics), generally do not remain parallel.

Einstein's great advance in the development of the General Theory of Relativity was the realization that the gravitational theory he wanted to establish should be based on a geometry that could describe a curved spacetime. He identified the worldlines of freely falling particles with the geodesics of this curved spacetime. The worldlines are straight lines in the local inertial systems, but globally they do not remain parallel.

In the following chapters, we will follow this thought and look for a theory of gravitation that assumes a curved spacetime to explain the effects of gravitation on the worldlines of particles. For this, we need to know more about (the mathematics of) spacetime and curvature. The simplest way to learn this is to adopt some of the terms and concepts about curvilinear coordinate systems introduced in Chap. 8 from the Euclidean plane and then transfer them to the four-dimensional spacetime, which we initially want to demonstrate with some examples.

18.5. General Coordinate Systems

In this section, we deal with arbitrary non-inertial systems in the spacetime of SR. The laws of Special Relativity formulated as tensor equations apply only in inertial systems, but that does not mean that other reference systems are not permissible.

Accelerated Reference Systems in SR

In classical mechanics, for example, rotating reference systems are used to describe gyroscopic motions, although Newton's laws initially only apply in inertial systems. As an example of an accelerated reference system, we consider a reference system K' (coordinates $\alpha', \beta', \mu', \nu'$, which can each take the values

t', x', y', z'), which rotates uniformly around the z-axis compared to an inertial system S (coordinates $\alpha, \beta, \mu, \nu = t, x, y, z$). A simple form of transformation between the two coordinate systems is given by

$$
\begin{aligned}
t &= t' \\
x &= x' \cos(\omega t') - y' \sin(\omega t') \\
y &= x' \sin(\omega t') + y' \cos(\omega t') \\
z &= z',
\end{aligned}
$$

where ω denotes the (constant) angular velocity (cf. the derivation of the formula 8.14 on page 150). We calculate the infinitesimal spacetime interval (12.22) in inertial system S

$$
ds^2 = \eta_{\alpha\beta}\, d\alpha d\beta = -dt^2 + dx^2 + dy^2 + dz^2. \tag{18.2}
$$

To express the right side in α' coordinates, we note that the coordinates t, x, y, z are functions of the coordinates t', x', y', z', so

$$
\begin{aligned}
t &= t(t', x', y', z') \\
x &= x(t', x', y', z') \\
y &= y(t', x', y', z') \\
z &= z(t', x', y', z')
\end{aligned}
$$

and calculate with Eq. 4.8 on page 59 the differentials

$$
\begin{aligned}
dt &= dt' \\
dx &= \frac{\partial x}{\partial t'}\, dt' + \frac{\partial x}{\partial x'}\, dx' + \frac{\partial x}{\partial y'}\, dy' + \frac{\partial x}{\partial z'}\, dz' \\
&= \omega\left(-x' \sin(\omega t') - y' \cos(\omega t')\right) dt' + \cos(\omega t)\, dx' - \sin(\omega t)\, dy' \\
dy &= \frac{\partial y}{\partial t'}\, dt' + \frac{\partial y}{\partial x'}\, dx' + \frac{\partial y}{\partial y'}\, dy' + \frac{\partial y}{\partial z'}\, dz' \\
&= \omega\left(x' \cos(\omega t') - y' \sin(\omega t')\right) dt' + \sin(\omega t)\, dx' + \cos(\omega t)\, dy' \\
dz &= dz'.
\end{aligned}
$$

Squaring and adding results after some simple calculations

$$
\begin{aligned}
ds^2 &= \left(\omega^2(x'^2 + y'^2) - 1\right) dt'^2 + dx'^2 + dy'^2 + dz'^2 \\
&\quad - 2\omega y'\, dt' dx' + 2\omega x'\, dt' dy' \\
&= g_{\mu'\nu'}\, d\mu' d\nu'.
\end{aligned}
$$

So, we have defined the components of a quantity $g_{\mu'\nu'}$ („metric") by the differentials

$$d\alpha = \frac{\partial \alpha}{\partial \mu'}\, d\mu'$$

and by the spacetime interval

$$ds^2 = \eta_{\alpha\beta}\, d\alpha d\beta = \eta_{\alpha\beta}\, \frac{\partial \alpha}{\partial \mu'}\frac{\partial \beta}{\partial \nu'}\, d\mu'\, d\nu'$$

by

$$g_{\mu'\nu'} = \eta_{\alpha\beta}\, \frac{\partial \alpha}{\partial \mu'}\frac{\partial \beta}{\partial \nu'}. \tag{18.3}$$

We will show later that these are the components of a tensor, although the way of transformation clearly indicates this. In the accelerated reference system, the spacetime interval thus has a more complicated shape, the quantities $g_{\mu'\nu'}$ are symmetric $(g_{\mu'\nu'} = g_{\nu'\mu'})$, but time- and location-dependent, and in contrast to

$$(\eta_{\mu\nu}) = \begin{pmatrix} -1 & 0 & 0 & 0 \\ 0 & 1 & 0 & 0 \\ 0 & 0 & 1 & 0 \\ 0 & 0 & 0 & 1 \end{pmatrix}$$

is

$$(g_{\mu'\nu'}) = \begin{pmatrix} \omega^2\left(x'^2 + y'^2\right) - 1 & -\omega y' & \omega x' & 0 \\ -\omega y' & 1 & 0 & 0 \\ \omega x' & 0 & 1 & 0 \\ 0 & 0 & 0 & 1 \end{pmatrix}$$

not a diagonal matrix. We calculate with (18.3) again explicitly the component $g_{t't'}$ of the metric

$$g_{t't'} = \eta_{\alpha\beta}\, \frac{\partial \alpha}{\partial t'}\frac{\partial \beta}{\partial t'} = -\frac{\partial t}{\partial t'}\frac{\partial t}{\partial t'} + \frac{\partial x}{\partial t'}\frac{\partial x}{\partial t'} + \frac{\partial y}{\partial t'}\frac{\partial y}{\partial t'} + \frac{\partial z}{\partial t'}\frac{\partial z}{\partial t'},$$

since only the diagonal elements of $\eta_{\alpha\beta}$ are different from zero. If we insert the transformations, we get

$$
\begin{aligned}
g_{t't'} &= -1 + \left(\frac{\partial\left(x'\cos\left(\omega t'\right) - y'\sin\left(\omega t'\right)\right)}{\partial t'}\right)^2 \\
&\quad + \left(\frac{\partial\left(x'\sin\left(\omega t'\right) + y'\cos\left(\omega t'\right)\right)}{\partial t'}\right)^2 \\
&= -1 + \left(-x'\omega\sin\left(\omega t'\right) - y'\omega\cos\left(\omega t'\right)\right)^2 \\
&\quad + \left(x'\omega\cos\left(\omega t'\right) - y'\omega\sin\left(\omega t'\right)\right)^2
\end{aligned}
$$

$$
\begin{aligned}
&= -1 + \omega^2 \left(x'^2 \sin^2 \left(\omega t' \right) + 2x'y' \sin \left(\omega t' \right) \cos \left(\omega t' \right) + y'^2 \cos^2 \left(\omega t' \right) \right) \\
&+ \;\; \omega^2 \left(x'^2 \cos^2 \left(\omega t' \right) - 2x'y' \sin \left(\omega t' \right) \cos \left(\omega t' \right) + y'^2 \sin^2 \left(\omega t' \right) \right) \\
&= -1 + \omega^2 \left(x'^2 + y'^2 \right),
\end{aligned}
$$

so again the above result.

In Eq. 2.11 on page 40 we calculated the centripetal acceleration of rotating particles in the plane. With formula (2.13) and

$$
r' = \sqrt{x'^2 + y'^2}
$$

the magnitude of the centripetal acceleration results in

$$
a = \frac{v^2}{r'} = \frac{(\omega r')^2}{r'} = \omega^2 r' = \omega^2 \sqrt{x'^2 + y'^2}.
$$

The **centrifugal acceleration** is equal in magnitude to the centripetal acceleration, but it acts outwards, i.e.

$$
\vec{a}_{centrifugal} = \omega^2 \begin{pmatrix} x' \\ y' \end{pmatrix}.
$$

If you set

$$
\phi = -\frac{\omega^2}{2} \left(x'^2 + y'^2 \right),
$$

then for the **centrifugal force** $\vec{Z}$ we get

$$
\vec{Z} = -m \nabla \phi = -m \begin{pmatrix} \partial \phi / \partial x' \\ \partial \phi / \partial y' \end{pmatrix} = -m \begin{pmatrix} -\omega^2 x' \\ -\omega^2 y' \end{pmatrix} = m \vec{a}_{centrifugal}.
$$

So ϕ can be considered as *centrifugal potential* (see definition 4.18 on page 68) and it can be noted that $g_{t't'}$ is related to ϕ:

$$
g_{t't'} = - \left(1 + 2\phi \right) \tag{18.4}
$$

The centrifugal potential thus appears in the metric. We will see later that the first derivatives of the metric determine the forces in the relativistic equations of motion. Therefore, it can already be suspected that the gravitational fields are described by the $g_{\mu'\nu'}$. And since the metric also determines the spacetime geometry, gravity and spacetime geometry are related.

The representation of the spacetime interval

$$
ds^2 = g_{\mu'\nu'} \, d\mu' d\nu'
$$

by the metric does not change its meaning, but only the way of calculation. In particular, for the display of a clock in the co-moving system K (proper time) because

$$dt = dt_E, \quad dx' = dy' = dz' = 0$$

and because $dt_E^2 = -ds^2$ according to Eq. (12.25) it follows

$$dt_E = \sqrt{-ds^2} = \sqrt{-g_{t't'}}\, dt' = \sqrt{1 + 2\phi}\, dt'.$$

Conversely, due to time dilation (see Eq. (12.4))

$$dt_E = \sqrt{1 - v^2}\, dt',$$

which, however, due to

$$2\phi = -\omega^2(x'^2 + y'^2) = -\omega^2 r'^2 = -v^2$$

matches.

In Chap. 8 we have shown that the components of the metric tensor in polar coordinates in the Euclidean plane depend on the coordinates. This also applies to the (natural) extension of polar coordinates to four-dimensional Euclidean **cylindrical coordinates**, which are defined by

$$
\begin{aligned}
t &= t' \\
x &= r \cos\varphi \\
y &= r \sin\varphi \\
z &= z'.
\end{aligned}
$$

For the differentials follows

$$
\begin{aligned}
dt &= dt' \\
dx &= \frac{\partial x}{\partial t'}\, dt' + \frac{\partial x}{\partial r}\, dr + \frac{\partial x}{\partial \varphi}\, d\varphi + \frac{\partial x}{\partial z'}\, dz' = \cos\varphi\, dr - r \sin\varphi\, d\varphi \\
dy &= \frac{\partial y}{\partial t'}\, dt' + \frac{\partial y}{\partial r}\, dr + \frac{\partial y}{\partial \varphi}\, d\varphi + \frac{\partial y}{\partial z'}\, dz' = \sin\varphi\, dr + r \cos\varphi\, d\varphi \\
dz &= dz'.
\end{aligned}
$$

And thus

$$
\begin{aligned}
ds^2 &= -dt^2 + dx^2 + dy^2 + dz^2 \\
&= -dt'^2 + \cos^2\varphi\, dr^2 - 2r \cos\varphi \sin\varphi\, dr\, d\varphi + r^2 \sin^2\varphi\, d\varphi^2 \\
&\quad + \sin^2\varphi\, dr^2 + 2r \cos\varphi \sin\varphi\, dr\, d\varphi + r^2 \cos^2\varphi\, d\varphi^2 + dz'^2
\end{aligned}
$$

$$
\begin{aligned}
&= -dt'^2 + \left(\sin^2\varphi + \cos^2\varphi\right)dr^2 + r^2\left(\sin^2\varphi + \cos^2\varphi\right)d\varphi^2 + dz'^2 \\
&= -dt'^2 + dr^2 + r^2 d\varphi^2 + dz'^2 \\
&= g_{\mu'\nu'}\, d\mu'\, d\nu'
\end{aligned}
$$

with $\mu', \nu' = t', r, \varphi, z'$, i.e., the components of the metric for cylindrical coordinates are also coordinate-dependent. The coordinate dependence of the metric tensor can therefore be based on the acceleration of the considered reference system *or* on the use of a non-Cartesian coordinate system.

In summary, we can state that accelerated reference systems in SR are permissible, however, the spacetime interval does not have the simple form as in (18.2), it rather applies

$$
ds^2 = g_{\mu'\nu'}(\{\alpha'\})\, d\mu' d\nu', \tag{18.5}
$$

where the $g_{\mu'\nu'}$ are functions of the spacetime points and thus of the coordinate

$$
\{\alpha'\} = (t', x', y', z').
$$

The $g_{\mu'\nu'}$ are determined in such a way that the spacetime interval ds^2 remains unchanged in the transformation $\alpha \to \alpha'$.

Extended Equivalence Principle

We want to return to the equivalence principle, i.e., to the equality of gravitational and inertial mass (see sections 6.5 and 18.1). If gravitational and inertial mass are equal, gravitational forces are simultaneously inertial forces. This means that homogeneous gravitational fields can be eliminated by transitioning to an accelerated coordinate system. As a simple example, we consider the gravitational field assumed to be homogeneous at the Earth's surface, in which the Newtonian equation of motion for a point mass

$$
m_i \ddot{\vec{r}} = m_g \vec{g} \tag{18.6}
$$

applies, where m_i is the inertial mass and m_g is the gravitational mass of the particle and

$$
\vec{g} \to (0, 0 - g)
$$

denotes the constant Earth's acceleration. We consider, for approximation, a system at rest on the Earth's surface as an inertial system and the following transformation to an accelerated reference system K'

$$
\begin{aligned}
t &= t' \\
x &= x'
\end{aligned}
$$

$$
\begin{aligned}
y &= y' \\
z &= z' - \frac{1}{2} g t'^2.
\end{aligned}
$$

The origin of K' ($\vec{r}' = 0$) moves from the perspective of the resting system such that $\vec{r} = \vec{g} t^2/2$ holds. If we insert the transformation into the equation of motion (18.6), we obtain

$$
m_i \ddot{\vec{r}}' = m_i \begin{pmatrix} \ddot{x}' \\ \ddot{y}' \\ \ddot{z}' \end{pmatrix} = m_i \begin{pmatrix} \ddot{x} \\ \ddot{y} \\ \ddot{z} + g \end{pmatrix} = m_i \begin{pmatrix} \ddot{x} \\ \ddot{y} \\ \ddot{z} \end{pmatrix} - m_i \begin{pmatrix} 0 \\ 0 \\ -g \end{pmatrix}.
$$

For the first term on the right side, according to Newton's equation of motion (18.6)

$$
m_i \begin{pmatrix} \ddot{x} \\ \ddot{y} \\ \ddot{z} \end{pmatrix} = m_g \begin{pmatrix} 0 \\ 0 \\ -g \end{pmatrix},
$$

so it follows

$$
m_i \ddot{\vec{r}}' = m_g \begin{pmatrix} 0 \\ 0 \\ -g \end{pmatrix} - m_i \begin{pmatrix} 0 \\ 0 \\ -g \end{pmatrix} = (m_g - m_i) \begin{pmatrix} 0 \\ 0 \\ -g \end{pmatrix} = \begin{pmatrix} 0 \\ 0 \\ 0 \end{pmatrix},
$$

since $m_i = m_g$. The equality of inertial and gravitational mass thus allows us to choose a reference system („free falling") in which the gravitational forces disappear. In section 18.4 we had already shown that in inhomogeneous gravitational fields only local inertial systems exist. With these considerations, we can formulate the principle of equivalence more precisely:

> *Principle of Equivalence*: In local inertial systems, *all* processes proceed as if there were no gravitational field. In local inertial systems, the laws of Special Relativity apply.

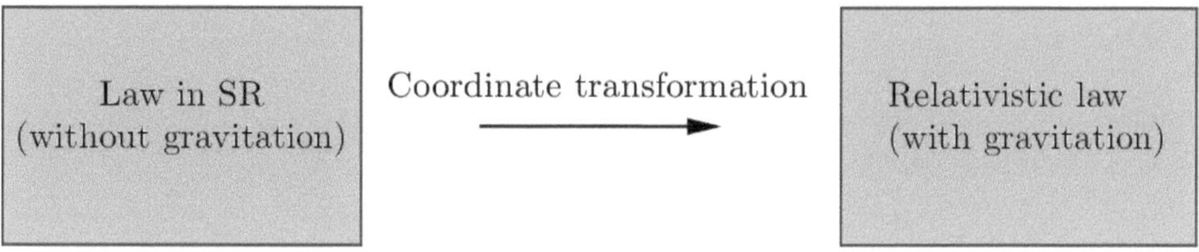

Figure 18.4.: From SR by coordinate transformation to GR

The principle of equivalence allows the formulation of relativistic laws with gravity. To do this, one starts from known laws in Special Relativity and transforms them by a general coordinate transformation into another reference system, see Fig. 18.4. The coordinate transformation contains the relative acceleration of the two reference systems. The law on the right side contains the gravitational field corresponding to the accelerations.
If we read the above example backwards, then

$$m\ddot{\vec{r}}' = 0$$

is the law in the (free falling) inertial system, and if we apply the coordinate transformations, we obtain

$$m\ddot{\vec{r}} = m\vec{g},$$

i.e., the equation of motion in the gravitational field. In the general case of an inhomogeneous gravitational field, such a coordinate transformation is, however, not globally possible as in this simple example, but only locally.

19. The mathematical foundations of Curved Spacetime

In this chapter, we want to take a look at the mathematical concepts needed to describe four-dimensional curved spacetime. The required mathematical aids were essentially first researched by Bernhard Riemann ($1826 - 1866$) in the middle of the $19th$ century and were thus available to Einstein at the beginning of the $20th$ century. The theory developed by Riemann (the so-called **differential geometry**) is partly very abstract and "unwieldy", so we choose a greatly simplified representation here, which omits many (mathematical) details and only briefly describes the essential constructs. The first sections of this chapter follow the initial explanations in the book by Jaenich [17], but are stripped down to our needs.

19.1. Manifolds

In the last chapter, we worked out that in the presence of gravitational fields there can be no global, but only local inertial systems in spacetime. So spacetime looks locally like a Minkowski space, what it looks like globally depends on the physical situation, i.e., on the existing gravitational fields at the different spacetime points. This means that the measurements of times and distances can be different from spacetime point to spacetime point, which usually leads to the need to calculate with different coordinate systems at different spacetime points. And the requirement to have to use different coordinate systems leads to the fact that spacetime itself is a curved space. The suitable mathematical object for describing spacetime is a so-called **differentiable manifold**. A (four-dimensional) manifold is essentially a continuous collection of points (a "space") that locally looks like the (four-dimensional) Euclidean space.

Remark 19.1. **MT: The four-dimensional Euclidean space**
We generalize the concept of a Euclidean space to four dimensions. This means we consider the set of all points P, which can be uniquely described in the Cartesian coordinate system K (represented by four perpendicular axes) by four real numbers t, x, y, z, where we will occasionally also use the abbreviating

© The Author(s), under exclusive license to Springer-Verlag GmbH, DE, part of Springer Nature 2026
M. Ruhrländer, *Ascent to the Einstein Equations*,
https://doi.org/10.1007/978-3-662-72672-3_19

index notation $\{\alpha\} = (t, x, y, z)$ again. We again call the four numbers the coordinates of the point and write as in the earlier chapters

$$P \underset{K}{\rightarrow} \{\alpha\} = (t, x, y, z).$$

Later we will talk about the four-dimensional (curved) spacetime, then as in the Special Theory of Relativity the coordinate t denotes time and x, y, z are supposed to represent the three-dimensional Euclidean space. At the moment, however, we consider the four coordinates as arbitrarily selectable real numbers. Of course, we cannot vividly imagine four perpendicular coordinate axes, for this we do not have adequate sensory perception possibilities. So when we talk about something four-dimensional, we have to think purely formally and abstractly. If we have, for example, two different points $P_1 \rightarrow (t_1, x_1, y_1, z_1)$ and $P_2 \rightarrow (t_2, x_2, y_2, z_2)$, then the **Euclidean distance** d of the two points is defined by

$$d\,(P_1, P_2) = \sqrt{(t_2 - t_1)^2 + (x_2 - x_1)^2 + (y_2 - y_1)^2 + (z_2 - z_1)^2},$$

and as in the Euclidean plane or in three-dimensional space, it can be shown that this distance does not change when switching to another coordinate system. With the Euclidean distance, the Euclidean geometry is defined, and we denote the four-dimensional Euclidean space with $\mathbb{R}^4$. On this space, a vector calculation can also be introduced, in such a way that each vector $\vec{A}$ can be represented as a linear combination the four Cartesian basis vectors e_t, e_x, e_y, e_z with

$$e_t \rightarrow \begin{pmatrix} 1 \\ 0 \\ 0 \\ 0 \end{pmatrix}, e_x \rightarrow \begin{pmatrix} 0 \\ 1 \\ 0 \\ 0 \end{pmatrix}, e_y \rightarrow \begin{pmatrix} 0 \\ 0 \\ 1 \\ 0 \end{pmatrix}, e_z \rightarrow \begin{pmatrix} 0 \\ 0 \\ 0 \\ 1 \end{pmatrix}$$

can be written as

$$\vec{A} = A^t e_t + A^x e_x + A^y e_y + A^z e_z = A^\alpha e_\alpha,$$

where the four real numbers A^t, A^x, A^y, A^z are again called the **components** of $\vec{A}$. Most vector calculation rules transfer one-to-one from the two- and three-dimensional case to the four-dimensional, so we do not want to go into further detail here. $\square$

What does it mean for a manifold to look locally like four-dimensional Euclidean space? Well, it means the following. One can divide a manifold into subareas and for each subarea U there is a unique assignment that assigns four real numbers, i.e., a point (t, x, y, z) from $\mathbb{R}^4$, to each point P from the subarea U. In other words and a bit more precisely: For each subarea U of the manifold there is a mapping

$$h : U \to \mathbb{R}^4,$$

which assigns uniquely four real numbers, which we want to call the **coordinates** of P, to each point P:

$$h\left(P\right) = \left(h^1\left(P\right), h^2\left(P\right), h^3\left(P\right), h^4\left(P\right)\right) = \left(t\left(P\right), x\left(P\right), y\left(P\right), z\left(P\right)\right)$$

Here we have represented the component functions of $h = \left(h^1, h^2, h^3, h^4\right)$ by the coordinate(functions) t, x, y, z of P, which depend on the point P. For simplification reasons, we usually leave out the P and write

$$h\left(P\right) = (t, x, y, z) = \{\alpha\} = \alpha,$$

but we are aware that this is just a shorthand for the penultimate expression, and hope that the use of (t, x, y, z) sometimes as coordinates of a fixed point and sometimes as coordinate functions of h does not lead to any confusion. The requirement for the function h to be unique means that the inverse mapping h^{-1} exists, i.e., that

$$h^{-1}(t, x, y, z) = P.$$

We want to call the subarea U together with the mapping h a **local coordinate system**. The unique identification of points by coordinates does not sound different than it applies to every point of a flat space. The crucial point here, however, is the fact that the chosen coordinate system only has to be valid in a *subarea U*, which includes the point P. For every other point of the manifold, there should also be such local coordinate systems, so that every point can be uniquely described by four coordinates, which generally come from different coordinate systems. A local coordinate system is also vividly called a **map** and the set of all maps an **atlas**. These terminologies stem from the fact that - as we will see further below - the surface of a sphere (Earth's surface) is an example of a (curved) manifold. It is further required that in areas where the various local coordinate systems overlap, the so-called **map change** is reversibly unique and differentiable. Fig. 19.1 explains this property.

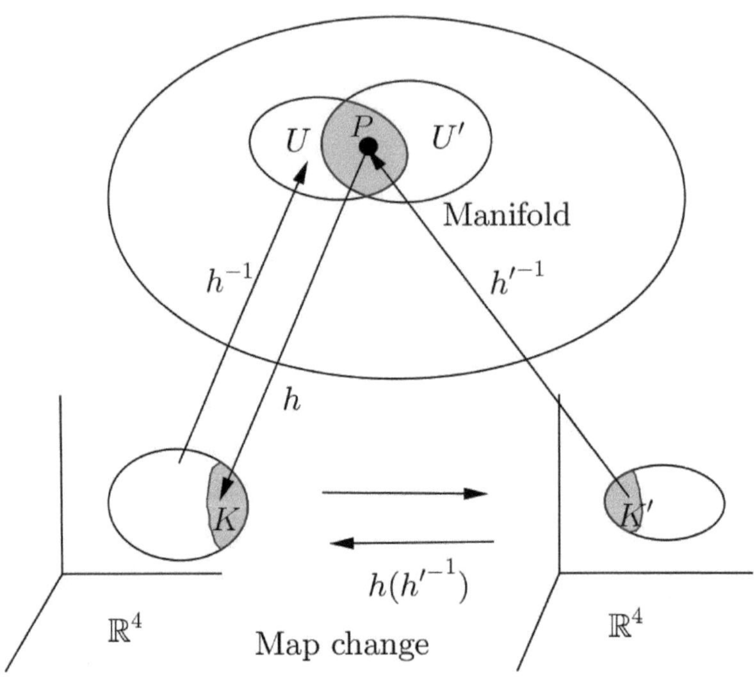

Figure 19.1.: Map change in manifolds

For the point P there are two local coordinate systems (U, h) and (U', h'), whose overlap area is the gray filled part of the manifold. The mapping h maps the subarea U to the left subarea of $\mathbb{R}^4$, the mapping h' (not shown in the graphic) the subarea U' to the right subarea of $\mathbb{R}^4$. These two subareas of $\mathbb{R}^4$ are also called **parameter areas** of (U, h) and (U', h') respectively. The gray marked areas in the left (K) and right (K') lower area are each the images of the overlap area on the manifold under the mappings h and h'. In Fig. 19.1 the map change from K' to K is explicitly given: First the inverse mapping h'^{-1} is applied to a point $h'(P) = (t', x', y', z')$ from the gray area K' and thus ends up at the point P in the gray area of the manifold. The mapping h is then applied to this point of the manifold and one ends up in the gray area K. That is, by the "detour" over the manifold one has by the composite function $h\left(h'^{-1}\right)$ (always to be read as: first h'^{-1} and then h) a mapping from K' to K, thus defining a map change in $\mathbb{R}^4$. We write this map change as in the formula 8.15 on page 152 as coordinate transformation, i.e. briefly as

$$\left[h\left(h'^{-1}\right)\right]\left(t',x',y',z'\right) = \begin{pmatrix} t \\ x \\ y \\ z \end{pmatrix} = \begin{pmatrix} t\left(t',x',y',z'\right) \\ x\left(t',x',y',z'\right) \\ y\left(t',x',y',z'\right) \\ z\left(t',x',y',z'\right) \end{pmatrix}.$$

This transformation should be a reversibly unique, differentiable mapping, i.e., for each point P of the overlap area the Jacobian matrix

$$\left(\frac{\partial\alpha}{\partial\alpha'}\right)_{h(P)} = \begin{pmatrix} \dfrac{\partial t}{\partial t'} & \dfrac{\partial t}{\partial x'} & \dfrac{\partial t}{\partial y'} & \dfrac{\partial t}{\partial z'} \\[2mm] \dfrac{\partial x}{\partial t'} & \dfrac{\partial x}{\partial x'} & \dfrac{\partial x}{\partial y'} & \dfrac{\partial x}{\partial z'} \\[2mm] \dfrac{\partial y}{\partial t'} & \dfrac{\partial y}{\partial x'} & \dfrac{\partial y}{\partial y'} & \dfrac{\partial y}{\partial z'} \\[2mm] \dfrac{\partial z}{\partial t'} & \dfrac{\partial z}{\partial x'} & \dfrac{\partial z}{\partial y'} & \dfrac{\partial z}{\partial z'} \end{pmatrix}_{h(P)}$$

should exist and be invertible. The same should also apply to the inverse mapping. The subscript $h\left(P\right)$ is intended to indicate that we have to calculate the Jacobian matrix at the point $h\left(P\right)$ in $\mathbb{R}^4$. However, we will usually omit this hint in the future for the sake of simpler notation.

We want to consider some examples of manifolds.

Example 19.1.
The simplest are the Euclidean plane and also the Minkowski space. For both spaces, there exists a single (global) coordinate system for the identification of points, namely for the Euclidean plane

$$(U,h) = \left(\mathbb{R}^2, id\right)$$

and analogously for the Minkowski space

$$(U,h) = (\text{Minkowski space}, id),$$

where the function id is supposed to be the **identity**, i.e.

$$id\left(x,y\right) = \left(x,y\right)$$

in the Euclidean plane and

$$id\left(t,x,y,z\right) = \left(t,x,y,z\right)$$

in the Minkowski space. A manifold that can be described by a single coordinate system, we call **flat**. The flat manifolds are not of interest to us in general relativity, since we already know from the physically based considerations of the last sections that we need a curved spacetime to describe gravitational phenomena. $\square$

The "prime example" of a curved manifold, albeit only two-dimensional, is the surface of a sphere, which we want to represent in detail in the following. One way to describe the points on a sphere's surface with coordinates is to use longitude and latitude. To do this, we first introduce a new coordinate system for the three-dimensional Euclidean space.

Remark 19.2. **MT: Spherical coordinates**
Spherical coordinates offer the possibility to uniquely determine each point $P = (x, y, z)$ in three-dimensional space (except the origin) by two angles and the distance of the point from the origin.

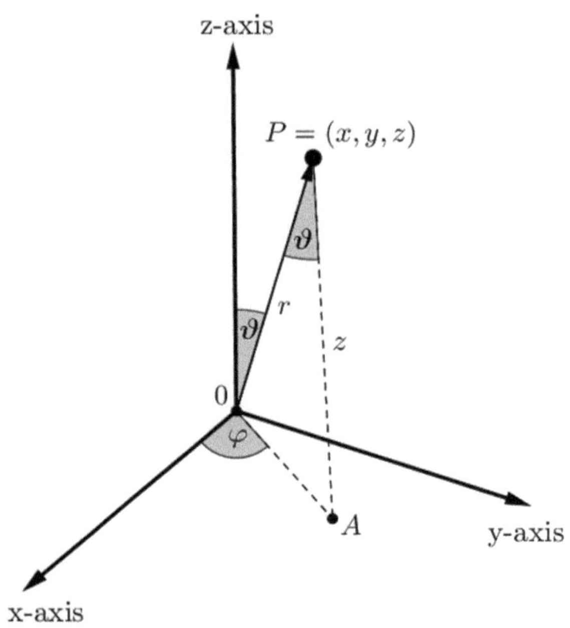

Figure 19.2.: Spherical coordinates

In 19.2, the point $P = (x, y, z)$, the distance r from P to the origin, and the angles ϑ and φ can be seen. The angle ϑ is the angle between the z-axis and the radius vector r and runs from 0 (North Pole) to π (South Pole). On a sphere's surface with radius r, the (x, y)-plane would be the equatorial plane and the angle ϑ the geographical latitude. The angle φ (geographical longitude) is between the x-axis and the foot point A of P in the (x, y)-plane (foot of the vertical dashed line) and runs from 0 to 2π. We now want to express the coordinates x, y, z by r, ϑ, φ. It can be immediately seen from the graphic that

$$z = r\cos\vartheta$$

holds. Also from the graphic, it is evident that the distance in the (x, y)-plane from the origin to the foot point A of P (lower dashed line) is equal to $r\sin\vartheta$. This distance is used to determine the coordinates x and y of the foot point A using polar coordinates in the (x, y)-plane, where the distance $r\sin\vartheta$ (instead of just r) must be taken, so it follows

$$\begin{aligned} x &= (r\sin\vartheta)\cos\varphi \\ y &= (r\sin\vartheta)\sin\varphi. \,\Box \end{aligned}$$

Example 19.2.
The surface of a sphere with the center at the origin and radius r is defined as the set of all points (x, y, z) that satisfy the equation

$$x^2 + y^2 + z^2 = r^2.$$

In the following, for simplicity, we consider the surface the **unit sphere**, i.e., all points in $\mathbb{R}^3$, which satisfy

$$x^2 + y^2 + z^2 = 1$$

with

$$\begin{aligned} x &= \sin\vartheta\cos\varphi \\ y &= \sin\vartheta\sin\varphi \\ z &= \cos\vartheta, \end{aligned} \tag{19.1}$$

and we want to represent these as a manifold with two local coordinate systems. The three equations (19.1) already provide us with a mapping that maps the angles ϑ, φ to the points x, y, z of the sphere surface. We call this mapping h^{-1}:

$$h^{-1}(\vartheta, \varphi) = (x(\vartheta, \varphi), y(\vartheta, \varphi), z(\vartheta, \varphi)) = (\sin\vartheta\cos\varphi, \sin\vartheta\sin\varphi, \cos\vartheta)$$

It must be unique, which means that for every point (x, y, z) on the sphere surface there must be exactly one pair (ϑ, φ) with

$$h^{-1}(\vartheta, \varphi) = (x, y, z).$$

This requirement is not met for $\vartheta = 0$, because no matter how I choose the angle φ, I always get

$$h^{-1}(0, \varphi) = (0, 0, 1),$$

i.e., the "North Pole". The same applies for $\vartheta = \pi$:

$$h^{-1}(\pi, \varphi) = (0, 0, -1),$$

i.e., for the South Pole. The function h^{-1} is therefore only unique, if the angles ϑ, φ are chosen such that

$$0 < \vartheta < \pi, 0 \le \varphi < 2\pi$$

applies. By the mapping h^{-1}, only points on the sphere surface are reached, because it holds:

$$
\begin{aligned}
x^2 + y^2 + z^2 &= \sin^2 \vartheta \cos^2 \varphi + \sin^2 \vartheta \sin^2 \varphi + \cos^2 \vartheta \\
&= \sin^2 \vartheta \left(\underbrace{\cos^2 \varphi + \sin^2 \varphi}_{=1} \right) + \cos^2 \vartheta \\
&= \left(\underbrace{\sin^2 \vartheta + \cos^2 \vartheta}_{=1} \right) = 1
\end{aligned}
$$

To get to a chart (U, h) for the sphere surface, we need the inverse mapping of the function h^{-1}, because $h = \left(h^{-1}\right)^{-1}$. To calculate this concretely, we look again at

$$
h^{-1}(\vartheta, \varphi) = \begin{pmatrix} \sin \vartheta \cos \varphi \\ \sin \vartheta \sin \varphi \\ \cos \vartheta \end{pmatrix} = \begin{pmatrix} x \\ y \\ z \end{pmatrix}
$$

and have to express the two angles ϑ, φ by x, y, z. Since $z = \cos \vartheta$ and ϑ comes from the open interval $(0, \pi)$, where the cosine strictly decreases, i.e., is invertible, it holds

$$\vartheta = \arccos z.$$

We obtain the angle φ as in the polar coordinates by the arc function (see Remark 8.6 on page 158), so

$$
\varphi = arc(x, y) = \begin{cases} \arccos \left(\dfrac{x}{\sin \vartheta} \right) & y \ge 0 \\[3mm] -\arccos \left(\dfrac{x}{\sin \vartheta} \right) & y < 0 \end{cases}.
$$

Now with the penultimate equation and because $\sin x = \sqrt{1 - \cos^2 x}$

$$\sin \vartheta = \sin(\arccos z) = \sqrt{1 - \cos^2(\arccos z)} = \sqrt{1 - z^2} = \sqrt{x^2 + y^2},$$

since $x^2 + y^2 + z^2 = 1$. Thus, the angle φ can be expressed as follows:

$$\varphi = arc(x, y) = \begin{cases} \arccos\left(\dfrac{x}{\sqrt{x^2 + y^2}}\right) & y \geq 0 \\[2em] -\arccos\left(\dfrac{x}{\sqrt{x^2 + y^2}}\right) & y < 0 \end{cases},$$

and the wanted map (U, h) is

$U \;=\;$ All points on the sphere surface excluding the North and South Pole

$$h \;=\; (\vartheta, \varphi) = \begin{pmatrix} \arccos(z), & \begin{cases} \arccos\left(\dfrac{x}{\sqrt{x^2 + y^2}}\right) & y \geq 0 \\[2em] -\arccos\left(\dfrac{x}{\sqrt{x^2 + y^2}}\right) & y < 0 \end{cases} \end{pmatrix}.$$

We have restricted the domain of h so that the North and South Poles are not mapped. Therefore, at least one more coordinate system is needed to cover the missing points unambiguously. This second coordinate system (U', h') can now be defined, for example, by considering the point $(1, 0, 0)$ as the new "North Pole" and its **antipodal** point $(-1, 0, 0)$ as the new "South Pole", and again using the spherical coordinates. We rotate our original sphere by $\alpha = 90°$ around the y-axis and obtain with the rotation matrix around the y-axis (for the two-dimensional case see formula 8.13 on page 149)

$$\begin{pmatrix} \cos\alpha & 0 & \sin\alpha \\ 0 & 1 & 0 \\ -\sin\alpha & 0 & \cos\alpha \end{pmatrix} = \begin{pmatrix} \cos 90° & 0 & \sin 90° \\ 0 & 1 & 0 \\ -\sin 90° & 0 & \cos 90° \end{pmatrix} = \begin{pmatrix} 0 & 0 & 1 \\ 0 & 1 & 0 \\ -1 & 0 & 0 \end{pmatrix}$$

for $0 < \vartheta' < \pi, 0 \leq \varphi' < 2\pi$ the mapping

$$h'^{-1}(\vartheta', \varphi') = \begin{pmatrix} 0 & 0 & 1 \\ 0 & 1 & 0 \\ -1 & 0 & 0 \end{pmatrix} \begin{pmatrix} \sin\vartheta' \cos\varphi' \\ \sin\vartheta' \sin\varphi' \\ \cos\vartheta' \end{pmatrix} = \begin{pmatrix} \cos\vartheta' \\ \sin\vartheta' \sin\varphi' \\ -\sin\vartheta' \cos\varphi' \end{pmatrix}.$$

The "old" North Pole is now uniquely reached by choosing $\vartheta' = 90°, \varphi' = 180°$, because

$$h'^{-1}(90°, 180°) = \begin{pmatrix} \cos 90° \\ \sin 90° \sin 180° \\ -\sin 90° \cos 180° \end{pmatrix} = \begin{pmatrix} 0 \\ 0 \\ 1 \end{pmatrix}.$$

For the South Pole, the following applies analogously

$$h'^{-1}\left(90°, 0°\right) = \begin{pmatrix} \cos 90° \\ \sin 90° \sin 0° \\ -\sin 90° \cos 0° \end{pmatrix} = \begin{pmatrix} 0 \\ 0 \\ -1 \end{pmatrix}.$$

The second map (U', h') is thus defined analogously to above by

$$U' = \text{All points on the sphere surface excluding the points } (1,0,0) \text{ and } (-1,0,0)$$

$$h' = (\vartheta', \varphi') = \begin{pmatrix} \arccos(x), & \begin{cases} \arccos\left(\dfrac{y}{\sqrt{y^2 + z^2}}\right) & z \geq 0 \\[4mm] -\arccos\left(\dfrac{y}{\sqrt{y^2 + z^2}}\right) & z < 0 \end{cases} \end{pmatrix}.$$

So, we have achieved that the two subareas U and U' together cover the entire sphere surface. The two subareas overlap in all points except the four points

$$(0,0,1), (0,0,-1), (1,0,0), (-1,0,0).$$

The coordinates we have chosen are not the only ones that exist for the sphere surface, there are even infinitely many. However, it is important to note that you always need *at least two local coordinate systems* to cover all points of the sphere surface. And this is a first confirmation that the sphere surface is not flat, but curved. $\square$

19.2. Tangent Space, Tangential Vectors, Tensors

We want to formulate physical laws on a manifold, which do not change their form even when the local coordinate system is changed. For this, we need objects like vectors or more generally tensors with the corresponding transformation behavior, as we have also derived it in two-dimensional Euclidean space for curvilinear coordinate systems or in the special theory of relativity.

First, we want to adjust the notations of the last section to those in the earlier chapters and thereby discard some ballast. If P is a point in the manifold and (U, h) and (U', h') are two local coordinate systems around P, we write instead

$$h(P) = \{\alpha\} = (t, x, y, z) \text{ und } h'(P) = \{\alpha\}' = (t', x', y', z')$$

from now on

$$P \underset{h}{\to} (t, x, y, z) \text{ und } P \underset{h'}{\to} (t', x', y', z'),$$

i.e., we assume that there are always suitable subareas U and U' on which the mappings h and h' are defined without explicitly listing them. If it is irrelevant in which local coordinate system we are, we write again shortly

$$P \to (t, x, y, z).$$

We define a **tangential vector** $\vec{A}$ at P as an object that can be represented in any local coordinate system around P by four real numbers

$$\vec{A}(P) \underset{h}{\to} \{A^\alpha(P)\} = \left(A^t(P), A^x(P), A^y(P), A^z(P)\right)$$

or

$$\vec{A}(P) \underset{h'}{\to} \left\{A^{\alpha'}(P)\right\} = \left(A^{t'}(P), A^{x'}(P), A^{y'}(P), A^{z'}(P)\right).$$

As in the two-dimensional Euclidean space or in the Special Theory of Relativity, the numbers $A^t(P), A^x(P), A^y(P), A^z(P)$ are also called the *components* of $\vec{A}$ in h. Also with the tangential vectors and later with the more general tensors, we want to omit the argument (P) for simplicity in the future, so it must always be inferred from the context whether we mean a fixed vector $\vec{A}$ or a vector field $\vec{A}(P)$, when we speak of $\vec{A}$. We require from the components that they transform by the following transformation

$$A^\alpha = \left(\frac{\partial \alpha}{\partial \alpha'}\right) A^{\alpha'} \tag{19.2}$$

bzw.

$$A^{\alpha'} = \left(\frac{\partial \alpha'}{\partial \beta}\right) A^\beta, \tag{19.3}$$

where again Einstein's summation convention must be observed. The set of all tangential vectors at P is called the **tangent space** at P and is written as $T_P M$. The tangent space at P forms a so-called **abstract vector space** and is the place where the vectorial physical laws of General Relativity are defined. It is particularly important to note that the tangent space is *always only defined for a specific point P* of the manifold, another point Q of the manifold has a different tangent space, since the transformation properties of the (different) tangential vectors depend on the Jacobian matrices of the coordinate transformations at the location P or Q, and these are generally different. In particular, one cannot add or scalar multiply tangential vectors from different tangent spaces or perform other operations, which are allowed with vectors that come from a single vector space. One can imagine the different tangent spaces as different vector spaces "attached" to each point of the manifold.

The flat tangent space is used in a small vicinity of a point P as an approximation for the manifold. Therefore, a manifold is also called **locally flat**. So if we consider a small part of the Earth's surface around a certain point P as flat, we are moving (at least purely mathematically) in the tangential plane T_PM and no longer on the Earth's surface! And of course, the local flatness of a manifold, i.e. the existence of the tangent space, is a mathematical means to represent the local inertial system in spacetime.

From now on, we will briefly call tangential vectors vectors or spacetime vectors. When we talk about vectors in the following, it always means that we are moving in a (fixed) tangent space T_PM of a point of a manifold. If the components of a vector are known in a local coordinate system, they are uniquely determined by the transformation rule for any other local coordinate system.

We want to define **one-forms** in a similar way. A one-form $\tilde{p}$ is (as in Sec. 9.1) defined as an object that can be identified in any local coordinate system h by four numbers p_t, p_x, p_y, p_z, which are referred to as components of the one-form

$$\tilde{p} \underset{h}{\rightarrow} \{p_\alpha\} = (p_t, p_x, p_y, p_z)\,.$$

The components of the one-form should transform inversely to the components of a vector, i.e., if h and h' are two local coordinate systems around P, then it should hold

$$p_{\alpha'} = \left(\frac{\partial \beta}{\partial \alpha'}\right)_{h(P)} p_\beta. \tag{19.4}$$

Note that one-forms are also defined at a fixed point P of the manifold, since the Jacobian matrix at the point $h(P)$ serves as a transformation matrix. The set of one-forms also forms an (abstract) vector space, since one-forms can be added like vectors and multiplied by a number and the result each time transforms like a one-form. As in Sec. 9.1 about the one-forms in the Euclidean plane, a one-form can also be considered as a linear mapping that maps a vector to a real number, and in the local coordinate system h

$$\tilde{p}(\vec{A}) = p_\alpha\, A^\alpha = p_t\, A^t + p_x\, A^x + p_y\, A^y + p_z\, A^z$$

applies. The set of all these mappings is called the **dual vector space** to T_PM. This is the place where the one-forms "live". However, in the following, when introducing tensors of higher order, we will not list their living areas so precisely, but will briefly refer to these as "tangential spaces" and focus on the transformation properties of their components. We still have to show that the definition of a one-form as a linear mapping is unique, i.e., that the expression $\tilde{p}\left(\vec{A}\right)$ always yields the same real number, no matter in which local coordinate system we are. So let h and h' be two local coordinate systems around the point P, then with the transformation formulas (19.3) and (19.3) we get

$$\tilde{p}(\vec{A}) = p_{\alpha'} A^{\alpha'} = \left[\left(\frac{\partial \alpha}{\partial \alpha'}\right) p_\alpha\right]\left[\left(\frac{\partial \alpha'}{\partial \beta}\right) A^\beta\right] = \underbrace{\left(\frac{\partial \alpha}{\partial \alpha'}\right)\left(\frac{\partial \alpha'}{\partial \beta}\right)}_{=(\delta^\alpha_\beta)} p_\alpha A^\beta = p_\alpha A^\alpha,$$

since the two Jacobian matrices are inverse to each other, i.e.

$$\left(\frac{\partial \alpha}{\partial \alpha'}\right)\left(\frac{\partial \alpha'}{\partial \beta}\right) = \begin{pmatrix} 1 & 0 & 0 & 0 \\ 0 & 1 & 0 & 0 \\ 0 & 0 & 1 & 0 \\ 0 & 0 & 0 & 1 \end{pmatrix} = (\delta^\alpha_\beta).$$

In the Euclidean plane and in Minkowski space, the gradient of a scalar field was an important example of a one-form. We want to show that this also applies to manifolds. If $f : M \to \mathbb{R}$ is a function that assigns a real number to each point P from the manifold M, we choose a local coordinate system h and consider the function $f\left(h^{-1}\left(t, x, y, z\right)\right)$.

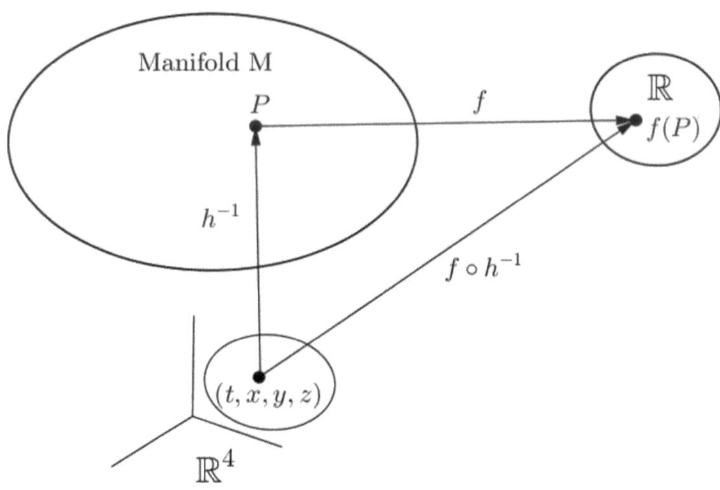

Figure 19.3.: Real-valued function on a manifold

Fig. 19.3 shows a point P in the manifold M and the mapping f, which maps the point P to the real number $f(P)$. Furthermore, the coordinates of

$$P \underset{h}{\rightarrow} (t, x, y, z)$$

as well as the coordinate inversion mapping

$$h^{-1}(t, x, y, z) = P$$

are shown. If you form the composite function

$$f\left(h^{-1}(t, x, y, z)\right),$$

you start from (t, x, y, z) and initially land at P and then it goes from P to $f(P)$, i.e., you get a mapping from (t, x, y, z) to $f(P)$, which we also want to call f again and with which we can calculate as usual, since its domain is a subset of $\mathbb{R}^4$ and its range are the real numbers.

We now consider a scalar field $f(t, x, y, z)$ on a manifold and define the gradient of f by

$$\tilde{d}f(t, x, y, z) \underset{h}{\to} \left(\frac{\partial f(t, x, y, z)}{\partial t}, \frac{\partial f(t, x, y, z)}{\partial x}, \frac{\partial f(t, x, y, z)}{\partial y}, \frac{\partial f(t, x, y, z)}{\partial z}\right).$$

If h' is another coordinate system, then by the chain rule 15.7 on page 314

$$\frac{\partial f}{\partial \alpha'} = \frac{\partial \alpha}{\partial \alpha'}\frac{\partial f}{\partial \alpha}, \tag{19.5}$$

where the arguments (t, x, y, z) were omitted in the derivatives, i.e., even for scalar fields on a manifold, the gradient is a one-form. For $\partial f/\partial \alpha$ we write again briefly

$$\frac{\partial f}{\partial \alpha} = f_{,\alpha}.$$

Higher order tensors are defined in the same way as in the Euclidean plane or in the Special Theory of Relativity by their transformation properties. For example, $\mathbf{T}$ is a $(1, 2)$-tensor if its components $T^{\alpha'}{}_{\beta'\gamma'}$ can be calculated in the transformation to an h'-coordinate system from

$$T^{\alpha'}{}_{\beta'\gamma'} = \frac{\partial \alpha'}{\partial \mu}\frac{\partial \nu}{\partial \beta'}\frac{\partial \rho}{\partial \gamma'}T^{\mu}{}_{\nu\rho}.$$

We will return to the treatment of tensors later. But first, we extend the present manifold by an additional property.

19.3. Riemannian Spaces

So far, we have not introduced a metric on the manifold M, i.e., on each tangent space $T_P M$. In the two-dimensional Euclidean space this was simple, there we defined the components of the metric by the scalar product of the (coordinate system dependent) basis vectors, see section 9.3. In our case, however, it is not clear whether the tangent spaces even possess something like a scalar product. We therefore require that there is a *symmetric, non-degenerate* $(0,2)$-tensor

$$\mathbf{g} \underset{h}{\to} g_{\mu\nu},$$

that locally looks like the Minkowski metric. We call such a tensor $\mathbf{g}$ the **metric tensor** (shortly **metric**). What does this mean? Well, quite analogous to tensor calculus in the Euclidean plane or in SR, we call a $(0,2)$-tensor $\mathbf{g}$ *symmetric* if for two arbitrary vectors $\vec{A}$ and $\vec{B}$ it holds that

$$\mathbf{g}(\vec{A},\vec{B}) = \mathbf{g}(\vec{B},\vec{A}).$$

A $(0,2)$-tensor $\mathbf{g}$ is called **non-degenerate** if its components $(g_{\mu\nu})$ form an invertible matrix. The inverse matrix, which we again want to call $(g^{\mu\nu})$, is then also symmetric. Since the metric $\mathbf{g}$ is symmetric, only 10 of the 16 matrix elements of $g_{\mu\nu}$ (those on the main diagonal and to the right of it) are independent, the others are obtained by reflection on the main diagonal, e.g. $g_{xt} = g_{tx}$. If a $(0,2)$-tensor $\mathbf{g}$ has an invertible component matrix in a certain local coordinate system, it also does in all other coordinate systems. For if h' is another local coordinate system around P, then it follows

$$(g_{\mu'\nu'}) = \left(\frac{\partial\mu}{\partial\mu'}\right)\left(\frac{\partial\nu}{\partial\nu'}\right)(g_{\mu\nu}),$$

and due to the invertibility of the two Jacobian matrices, $g_{\mu'\nu'}$ is also invertible as the product of three invertible matrices.

Now to the demand "looks locally like the Minkowski metric". This means that for every point P of the manifold there is a special local coordinate system, which we want to emphasize by the *notation* $\hat{h}$. We write the coordinates of a point in this coordinate system as

$$P \underset{\hat{h}}{\to} \{\hat{\alpha}\} = \left(\hat{t}, \hat{x}, \hat{y}, \hat{z}\right).$$

The components of $\mathbf{g}$ formed with the local coordinate system $\hat{h}$ should be the components of the Minkowski metric, i.e.

$$(g_{\hat{\mu}\hat{\nu}}) = \begin{pmatrix} -1 & 0 & 0 & 0 \\ 0 & 1 & 0 & 0 \\ 0 & 0 & 1 & 0 \\ 0 & 0 & 0 & 1 \end{pmatrix} = (\eta_{\hat{\mu}\hat{\nu}}),$$

where we want to use the symbol $\eta_{\hat\mu\hat\nu}$ again for the components of the Minkowski metric. With this demand on the tensor $\mathbf{g}$, the mathematical means is found, with which one can represent the **local inertial system** justified in the last sections with physical arguments. The last demand also implies that *near the point P* the last equation is "almost" true, which means that for all $\hat\mu, \hat\nu, \hat\rho$ the first derivatives are

$$\frac{\partial g_{\hat\mu\hat\nu}(P)}{\partial\hat\rho} = 0,$$

but the second ones are generally not for some $\hat\mu, \hat\nu, \hat\rho, \hat\sigma$

$$\frac{\partial^2 g_{\hat\mu\hat\nu}(P)}{\partial\hat\sigma\partial\hat\rho} \neq 0.$$

The latter applies to all manifolds that are not flat everywhere. The proof of this conclusion is long and complicated (see e.g. [29]), so we leave it out here. The metric is therefore approximately the Minkowski metric near P and we want to call the local coordinate system $\hat h$ also again *local inertial system*. A differentiable manifold that has a metric, we call **Riemannian space** and do not mind that actually "semi-Riemannian" or "pseudo-Riemannian" would be the exact designation. The mathematical model of spacetime in general relativity is therefore a Riemannian space, with which we want to deal more closely in the next sections.

Tensor Analysis in Riemannian Space

In the local inertial system, the laws of special relativity apply, in particular for the spacetime interval

$$ds^2 = \eta_{\hat\alpha\hat\beta}\, d\hat\alpha\, d\hat\beta.$$

In a neighboring point, the local inertial system will generally be accelerated. There, the spacetime interval will be of the form

$$ds^2 = g_{\mu\nu}\, d\mu\, d\nu$$

(see equation (18.5)). The transition from the local inertial system to any reference system K is done by a (reversible-unique) coordinate transformation

$$\hat\alpha = \hat\alpha\left(t, x, y, z\right).$$

Reversible-unique means that also the inverse transformation

$$\alpha = \alpha\left(\hat t, \hat x, \hat y, \hat z\right)$$

exists and the two Jacobian matrices $\dfrac{\partial\hat{\alpha}}{\partial\alpha}$ and $\dfrac{\partial\alpha}{\partial\hat{\alpha}}$ are invertible. Considering the differentials

$$d\hat{\alpha} = \frac{\partial\hat{\alpha}}{\partial\mu}\,d\mu, \tag{19.6}$$

it follows

$$ds^2 = \eta_{\hat{\alpha}\hat{\beta}}\,d\hat{\alpha}d\hat{\beta} = \eta_{\hat{\alpha}\hat{\beta}}\frac{\partial\hat{\alpha}}{\partial\mu}\frac{\partial\hat{\beta}}{\partial\nu}\,d\mu\,d\nu$$

and from this, that the components of the metric tensor $g_{\mu\nu}$ are given by

$$g_{\mu\nu} = \eta_{\hat{\alpha}\hat{\beta}}\frac{\partial\hat{\alpha}}{\partial\mu}\frac{\partial\hat{\beta}}{\partial\nu}. \tag{19.7}$$

With this, one can concretely demonstrate the transformation property of the metric:

$$
\begin{aligned}
g_{\mu'\nu'} &= \eta_{\hat{\alpha}\hat{\beta}}\frac{\partial\hat{\alpha}}{\partial\mu'}\frac{\partial\hat{\beta}}{\partial\nu'} \\[2mm]
&= \eta_{\hat{\alpha}\hat{\beta}}\left(\frac{\partial\hat{\alpha}}{\partial\sigma}\frac{\partial\sigma}{\partial\mu'}\right)\left(\frac{\partial\hat{\beta}}{\partial\tau}\frac{\partial\tau}{\partial\nu'}\right) \\[2mm]
&= \frac{\partial\sigma}{\partial\mu'}\frac{\partial\tau}{\partial\nu'}\left(\eta_{\hat{\alpha}\hat{\beta}}\frac{\partial\hat{\alpha}}{\partial\sigma}\frac{\partial\hat{\beta}}{\partial\tau}\right) \\[2mm]
&= \frac{\partial\sigma}{\partial\mu'}\frac{\partial\tau}{\partial\nu'}\,g_{\sigma\tau}
\end{aligned}
$$

For the spacetime interval in Riemannian space, we can therefore write

$$ds^2 = g_{\mu\nu}\,(t,x,y,z)\,d\mu\,d\nu. \tag{19.8}$$

The spacetime interval determines the distance ds^2 between the neighboring points

$$\{\mu\} = (t,x,y,z)$$

and

$$\{\mu + d\mu\} = (t+dt,\, x+dx,\, y+dy,\, z+dz)$$

of the coordinate system. Just as in section 12.7, we define the distance between two points as *timelike*, when $ds^2 < 0$, as *lightlike*, when $ds^2 = 0$, and as *spacelike*, when $ds^2 > 0$.

The components $g_{\mu\nu}$ are determined by the coordinate transformation, which in turn depends on the accelerations between K and the local inertial system. For two different spacetime points, the accelerations are generally different. It applies:

- For "apparent" gravitational fields (e.g., rotating reference systems like in section 18.5, there is a (single) coordinate transformation that leads everywhere to the Minkowski metric $\eta_{\hat{\mu}\hat{\nu}}$.

- For "true" gravitational fields, there is no global coordinate transformation that transforms $g_{\mu\nu}(\{\alpha\})$ everywhere into $\eta_{\hat{\mu}\hat{\nu}}$.

We will later see that the second statement is equivalent to a curvature of the Riemannian space. According to the equivalence principle, the gravitational fields can be locally eliminated, i.e., all physical effects appear locally as if a Minkowski space were present. This in turn means that all information about the gravitational fields is contained in the coordinate transformations, which lead at the respective point of spacetime to the local inertial system, and thus in the metric. The metric therefore contains the relativistic description of the gravitational field.

Although the equivalence principle follows the relativistic laws in the gravitational field, the gravitational field itself is not determined by the equivalence principle. The transformation equations and thus the determination of the $g_{\mu\nu}$ have no correspondence in the Special Relativity Theory. The relationship between the gravitational fields $g_{\mu\nu}$ and their sources is determined by Einstein's field equations.

Pulling indices up or down

Just as in Special Relativity Theory or as in the Euclidean plane, the metric tensor can be used to make a corresponding one-form $\tilde{V}$ from a vector $\vec{V}$ (or vice versa). The components of $\tilde{V}$ are calculated from those of $\vec{V}$ by

$$V_\alpha = g_{\alpha\beta}\, V^\beta$$

or vice versa

$$V^\alpha = g^{\alpha\beta} V_\beta,$$

where $g^{\mu\nu}$ is the inverse matrix to $g_{\mu\nu}$, i.e., it holds

$$g^{\tau\nu}\, g_{\nu\sigma} = \delta^\tau_\sigma.$$

The contraction of indices is also a permitted operation for more general tensors, it generates new tensors from existing ones. Let for example $\mathbf{T}$ be a $(1,2)$ tensor, then the items

$$T^\mu{}_{\nu\mu} = \delta^\rho_\mu\, T^\mu{}_{\nu\rho} = g_{\mu\sigma}\, g^{\sigma\rho}\, T^\mu{}_{\nu\rho}$$

are by construction the components of a one-form.

Covariant Derivative of Tensors in Riemannian Space

In this section, we proceed very similarly to how we did for the Euclidean plane. Therefore, some derivations are repeated, which once again represents a good opportunity to consolidate the knowledge acquired so far about tensor calculus.

Unlike a scalar field ϕ, the ordinary derivative of a spacetime vector is generally not a tensor. To show this, we follow the argument in section 8.3 and assume that two local coordinate systems h and h' are given. Then, if the ordinary derivative of a spacetime vector $\vec{V}$ were a $(1,1)$-tensor, the components $V^{\alpha'}_{\ ,\beta'}$ would transform as follows

$$V^{\alpha'}_{\ ,\beta'} = \frac{\partial \alpha'}{\partial \mu}\frac{\partial \nu}{\partial \beta'}\, V^{\mu}_{\ ,\nu}.$$

However, with the transformation formula (19.3) and the product rule it holds

$$V^{\alpha'}_{\ ,\beta'} = \frac{\partial V^{\alpha'}}{\partial \beta'} = \frac{\partial}{\partial \beta'}\left(\frac{\partial \alpha'}{\partial \mu} V^{\mu}\right) = \frac{\partial}{\partial \beta'}\left(\frac{\partial \alpha'}{\partial \mu}\right) V^{\mu} + \left(\frac{\partial \alpha'}{\partial \mu}\right)\frac{\partial V^{\mu}}{\partial \beta'}. \tag{19.9}$$

If you apply the chain rule to both terms, you get

$$\begin{aligned}
V^{\alpha'}_{\ ,\beta'} &= \frac{\partial \nu}{\partial \beta'}\left(\frac{\partial^2 \alpha'}{\partial \nu \partial \mu}\right) V^{\mu} + \left(\frac{\partial \alpha'}{\partial \mu}\right)\frac{\partial \nu}{\partial \beta'}\frac{\partial V^{\mu}}{\partial \nu} \\
&= \frac{\partial \nu}{\partial \beta'}\left(\frac{\partial^2 \alpha'}{\partial \nu \partial \mu}\right) V^{\mu} + \frac{\partial \alpha'}{\partial \mu}\frac{\partial \nu}{\partial \beta'}\, V^{\mu}_{\ ,\nu}.
\end{aligned} \tag{19.10}$$

The second term on the right side in the last equation is exactly what one would expect if a tensor transformation were present. The first term, which is generally not zero, is therefore the cause for the ordinary derivative of a spacetime vector not being a tensor. As we have already seen in the Euclidean plane, this non-tensorial behavior is not due to a space curvature, rather curvilinear coordinates ensure that the transformation matrices $\partial \alpha'/\partial \mu$ are not constant.

To define a tensorial derivative of a vector $\vec{V}$ at a point in spacetime, we use the fact that there is a local inertial system at every point in spacetime. From special relativity, we know that in inertial systems the ordinary derivatives of four-vectors are tensors, i.e., if S and S' are two different *inertial systems*, then for the components of the derivatives of $\vec{V}$ with the corresponding Lorentz transformations follows

$$V^{\alpha'}_{\ ,\beta'} = L^{\alpha'}_{\ \mu}(v)\, L^{\nu}_{\ \beta'}(-v)\, V^{\mu}_{\ ,\nu} = \frac{\partial \alpha'}{\partial \mu}\frac{\partial \nu}{\partial \beta'}\, V^{\mu}_{\ ,\nu}.$$

The reason for this is that the Lorentz transformations are constant, i.e., the first term in the above Eq. (19.9) becomes zero, cf. also (15.19):

$$\frac{\partial}{\partial \beta'}\left(\frac{\partial \alpha'}{\partial \mu}\right) = 0$$

If $\hat{\alpha}$ are the Minkowski coordinates in the local inertial system and for the components of the spacetime vector $\vec{V}$

$$\vec{V} \underset{\hat{\alpha}}{\to} \left\{V^{\hat{\alpha}}\right\} = \left(V^{\hat{t}}, V^{\hat{x}}, V^{\hat{y}}, V^{\hat{z}}\right)$$

and for an arbitrary coordinate system $h \to \{\alpha\}$

$$\vec{V} \underset{\alpha}{\to} \{V^{\alpha}\} = \left(V^{t}, V^{x}, V^{y}, V^{z}\right),$$

we *define* the **covariant derivative** ∇ of a vector $\vec{V}$ by

$$\nabla \vec{V} \underset{\alpha}{\to} \{V^{\alpha}_{;\beta}\}$$

with

$$V^{\alpha}_{;\beta} = \frac{\partial \alpha}{\partial \hat{\mu}}\frac{\partial \hat{\nu}}{\partial \beta} V^{\hat{\mu}}_{,\hat{\nu}}. \tag{19.11}$$

We first show that the covariant derivative is indeed a $(1,1)$ tensor. To do this, note that when changing the coordinate system $\alpha \to \alpha'$ at a fixed point P, the local inertial system remains the same, i.e., the components $V^{\hat{\mu}}_{,\hat{\nu}}$ of the derivative of the vector $\vec{V}$ in the inertial system do not change due to the coordinate change. The components of the covariant derivative in the coordinate system α' are

$$V^{\alpha'}_{;\beta'} = \frac{\partial \alpha'}{\partial \hat{\mu}}\frac{\partial \hat{\nu}}{\partial \beta'} V^{\hat{\mu}}_{,\hat{\nu}}$$

and can be rewritten using the chain rule to

$$\begin{aligned}
V^{\alpha'}_{;\beta'} &= \frac{\partial \alpha'}{\partial \hat{\mu}}\frac{\partial \hat{\nu}}{\partial \beta'} V^{\hat{\mu}}_{,\hat{\nu}} \\
&= \left(\frac{\partial \alpha'}{\partial \alpha}\frac{\partial \alpha}{\partial \hat{\mu}}\right)\left(\frac{\partial \hat{\nu}}{\partial \beta}\frac{\partial \beta}{\partial \beta'}\right) V^{\hat{\mu}}_{,\hat{\nu}} \\
&= \frac{\partial \alpha'}{\partial \alpha}\frac{\partial \beta}{\partial \beta'}\left(\frac{\partial \alpha}{\partial \hat{\mu}}\frac{\partial \hat{\nu}}{\partial \beta} V^{\hat{\mu}}_{,\hat{\nu}}\right) \\
&= \frac{\partial \alpha'}{\partial \alpha}\frac{\partial \beta}{\partial \beta'} V^{\alpha}_{,\beta}.
\end{aligned}$$

The components of the covariant derivative thus show the desired transformation behavior, so the covariant derivative is indeed a $(1,1)$-tensor. We now want to find an explicit expression for the $V^{\alpha}_{;\beta}$ that does *not* require the components of $\vec{V}$ in the local inertial system. To do this, we insert the expression (19.10), which looks like this for the components of $\vec{V}$ in the inertial system:

$$V^{\hat{\mu}}_{,\hat{\nu}} = \frac{\partial\hat{\mu}}{\partial\sigma}\frac{\partial\tau}{\partial\hat{\nu}}V^{\sigma}_{,\tau} + \frac{\partial\tau}{\partial\hat{\nu}}\cdot\frac{\partial^2\hat{\mu}}{\partial\tau\partial\sigma}V^{\sigma},$$

into the formula (19.11) and again obtain with the chain rule and taking into account

$$\frac{\partial\alpha}{\partial\sigma} = \delta^{\alpha}_{\sigma} = \begin{cases} 1 & \alpha = \sigma \\ 0 & \alpha \neq \sigma \end{cases}$$

the relationship

$$\begin{aligned} V^{\alpha}_{;\beta} &= \frac{\partial\alpha}{\partial\hat{\mu}}\frac{\partial\hat{\nu}}{\partial\beta}V^{\hat{\mu}}_{,\hat{\nu}} \\ &= \frac{\partial\alpha}{\partial\hat{\mu}}\frac{\partial\hat{\nu}}{\partial\beta}\left(\frac{\partial\hat{\mu}}{\partial\sigma}\frac{\partial\tau}{\partial\hat{\nu}}V^{\sigma}_{,\tau} + \frac{\partial\tau}{\partial\hat{\nu}}\cdot\frac{\partial^2\hat{\mu}}{\partial\tau\partial\sigma}V^{\sigma}\right) \\ &= \underbrace{\frac{\partial\alpha}{\partial\hat{\mu}}\frac{\partial\hat{\mu}}{\partial\sigma}}_{=\frac{\partial\alpha}{\partial\sigma}}\underbrace{\frac{\partial\hat{\nu}}{\partial\beta}\frac{\partial\tau}{\partial\hat{\nu}}}_{=\frac{\partial\tau}{\partial\beta}}V^{\sigma}_{,\tau} + \frac{\partial\alpha}{\partial\hat{\mu}}\underbrace{\frac{\partial\hat{\nu}}{\partial\beta}\frac{\partial\tau}{\partial\hat{\nu}}}_{=\frac{\partial\tau}{\partial\beta}}\cdot\frac{\partial^2\hat{\mu}}{\partial\tau\partial\sigma}V^{\sigma} \\ &= \frac{\partial\alpha}{\partial\sigma}\frac{\partial\tau}{\partial\beta}V^{\sigma}_{,\tau} + \frac{\partial\alpha}{\partial\hat{\mu}}\frac{\partial\tau}{\partial\beta}\cdot\frac{\partial^2\hat{\mu}}{\partial\tau\partial\sigma}V^{\sigma} \\ &= \delta^{\alpha}_{\sigma}\delta^{\tau}_{\beta}V^{\sigma}_{,\tau} + \frac{\partial\alpha}{\partial\hat{\mu}}\delta^{\tau}_{\beta}\frac{\partial^2\hat{\mu}}{\partial\tau\partial\sigma}V^{\sigma} \\ &= V^{\alpha}_{,\beta} + \frac{\partial\alpha}{\partial\hat{\mu}}\frac{\partial^2\hat{\mu}}{\partial\beta\partial\sigma}V^{\sigma}. \end{aligned}$$

As in the Euclidean plane (see formula 9.24 on page 200) we *define* the **Christoffel symbols** by

$$\Gamma^{\alpha}_{\sigma\beta} = \frac{\partial\alpha}{\partial\hat{\mu}}\frac{\partial^2\hat{\mu}}{\partial\beta\partial\sigma} \tag{19.12}$$

and finally obtain

$$V^{\alpha}_{;\beta} = V^{\alpha}_{,\beta} + \Gamma^{\alpha}_{\sigma\beta}V^{\sigma}. \tag{19.13}$$

Since the gradient of a scalar function according to Eq. (19.5) transforms like a one-form, i.e., is already a tensor, we define the covariant derivative of a scalar function ϕ as the ordinary partial derivative

$$\nabla\phi \rightarrow \phi_{;\beta} = \phi_{,\beta}.$$

To calculate the covariant derivative of a one-form $\tilde{p}$, we use the property that a one-form applied to a vector is a scalar function, i.e., yields independent of the coordinate system a real number

$$\tilde{p}(\vec{V}) = p_\alpha V^\alpha.$$

So, using the product rule, we get

$$(p_\alpha V^\alpha)_{;\beta} = (p_\alpha V^\alpha)_{,\beta} = \frac{\partial}{\partial\beta}(p_\alpha V^\alpha) = \frac{\partial p_\alpha}{\partial\beta}V^\alpha + p_\alpha\frac{\partial V^\alpha}{\partial\beta}.$$

We now replace the partial derivative of V^α with the covariant one, i.e., with (19.13) it follows

$$\frac{\partial V^\alpha}{\partial\beta} = V^\alpha_{;\beta} - V^\gamma\,\Gamma^\alpha_{\gamma\beta}.$$

Inserting this into the above equation and renaming the indices $\alpha \leftrightarrow \gamma$

$$
\begin{aligned}
(p_\alpha V^\alpha)_{;\beta} &= \frac{\partial p_\alpha}{\partial\beta}V^\alpha + p_\alpha V^\alpha_{;\beta} - p_\alpha V^\gamma\,\Gamma^\alpha_{\gamma\beta}.\\
&= \left(\frac{\partial p_\alpha}{\partial\beta} - p_\gamma\,\Gamma^\gamma_{\alpha\beta}\right)V^\alpha + p_\alpha V^\alpha_{;\beta}.
\end{aligned}
$$

All terms in the last equation except the one in the bracket are components of one-forms. So that the bracket multiplied by V^α again yields the components of a one-form, the expressions in the bracket must be the components of a $(0,2)$-tensor. We define this tensor as the covariant derivative of $\tilde{p}$ and write for it

$$\nabla\tilde{p} \rightarrow p_{\alpha;\beta} = \frac{\partial p_\alpha}{\partial x^\beta} - p_\gamma\Gamma^\gamma_{\alpha\beta}. \tag{19.14}$$

So we can write the penultimate equation as

$$(p_\alpha V^\alpha)_{;\beta} = p_{\alpha;\beta}\,V^\alpha + p_\alpha V^\alpha_{;\beta} \tag{19.15}$$

and thus obtain a product rule for the covariant derivative.

To determine the covariant derivative of a $(0,2)$-tensor $\mathbf{T}$, we start from the form

$$T^{\alpha\beta} = A^\alpha B^\beta,$$

which is no restriction, as a $(0,2)$-tensor transforms component-wise like a vector. With the product rule for the covariant derivative it follows

$$T^{\alpha\beta}_{;\gamma} = \left(A^\alpha B^\beta\right)_{;\gamma} = A^\alpha_{;\gamma}\,B^\beta + A^\alpha B^\beta_{;\gamma}.$$

Since the right side is a tensor, the quantity defined by this is also a tensor and we can continue with (19.13)

$$
\begin{aligned}
T^{\alpha\beta}_{\ ;\gamma} &= A^{\alpha}_{\ ;\gamma} B^{\beta} + A^{\alpha} B^{\beta}_{\ ;\gamma} \\
&= A^{\alpha}_{\ ,\gamma} B^{\beta} + \Gamma^{\alpha}_{\sigma\gamma} A^{\sigma} B^{\beta} + A^{\alpha} B^{\beta}_{\ ,\gamma} + \Gamma^{\beta}_{\sigma\gamma} A^{\alpha} B^{\sigma} \\
&= A^{\alpha}_{\ ,\gamma} B^{\beta} + A^{\alpha} B^{\beta}_{\ ,\gamma} + \Gamma^{\alpha}_{\sigma\gamma} A^{\sigma} B^{\beta} + \Gamma^{\beta}_{\sigma\gamma} A^{\alpha} B^{\sigma} \\
&= T^{\alpha\beta}_{\ ,\gamma} + \Gamma^{\alpha}_{\sigma\gamma} T^{\sigma\beta} + \Gamma^{\beta}_{\sigma\gamma} T^{\alpha\sigma}.
\end{aligned}
$$

In the same way, one obtains

$$
T_{\alpha\beta;\gamma} = T_{\alpha\beta,\gamma} - \Gamma^{\sigma}_{\alpha\gamma} T_{\sigma\beta} - \Gamma^{\sigma}_{\beta\gamma} T_{\alpha\sigma}
$$

and

$$
T^{\alpha}_{\ \beta;\gamma} = T^{\alpha}_{\ \beta,\gamma} + \Gamma^{\alpha}_{\sigma\gamma} T^{\sigma}_{\ \beta} - \Gamma^{\sigma}_{\beta\gamma} T^{\alpha}_{\ \sigma}
$$

and analogously the covariant derivatives of higher tensors. The following rules apply for the covariant derivative:

- The covariant mapping of a scalar field, i.e., a $(0,0)$-tensor, is a $(0,1)$-tensor (i.e., a one-form), the covariant mapping of a vector, i.e., a $(1,0)$-tensor, is a $(1,1)$-tensor, and the covariant derivative of a one-form, i.e., a $(0,1)$-tensor, is a $(0,2)$ tensor. In general, the covariant derivative of a (M,N)-tensor is a $(M,N+1)$-tensor.

- In the local inertial system the covariant derivative reduces to the partial derivative due to
$$
g_{\mu\nu} = \eta_{\mu\nu}.
$$

- The usual rules of differentiation apply, in particular the product rule.

Christoffel Symbols by Metric

In the following, we use the relations (19.7)

$$
g_{\mu\nu} = \eta_{\hat{\alpha}\hat{\beta}} \frac{\partial \hat{\alpha}}{\partial \mu} \frac{\partial \hat{\beta}}{\partial \nu}
$$

and (19.12)

$$
\Gamma^{\alpha}_{\sigma\beta} = \frac{\partial \alpha}{\partial \hat{\mu}} \frac{\partial^2 \hat{\mu}}{\partial \beta \partial \sigma},
$$

to express the Christoffel symbols by the ordinary first derivatives of the metric tensor. For this, we consider the following combination of first derivatives of the metric and note that the derivative of $\eta_{\hat{\alpha}\hat{\beta}}$ is zero. We then obtain by repeatedly applying the product rule

$$
\begin{aligned}
g_{\mu\nu,\lambda} + g_{\lambda\nu,\mu} - g_{\mu\lambda,\nu} &= \frac{\partial}{\partial\lambda}\left(\eta_{\hat{\alpha}\hat{\beta}}\frac{\partial\hat{\alpha}}{\partial\mu}\frac{\partial\hat{\beta}}{\partial\nu}\right) + \frac{\partial}{\partial\mu}\left(\eta_{\hat{\alpha}\hat{\beta}}\frac{\partial\hat{\alpha}}{\partial\lambda}\frac{\partial\hat{\beta}}{\partial\nu}\right) \\
&\quad - \frac{\partial}{\partial\nu}\left(\eta_{\hat{\alpha}\hat{\beta}}\frac{\partial\hat{\alpha}}{\partial\mu}\frac{\partial\hat{\beta}}{\partial\lambda}\right) \\
&= \eta_{\hat{\alpha}\hat{\beta}}\frac{\partial^2\hat{\alpha}}{\partial\lambda\partial\mu}\frac{\partial\hat{\beta}}{\partial\nu} + \eta_{\hat{\alpha}\hat{\beta}}\frac{\partial\hat{\alpha}}{\partial\mu}\frac{\partial^2\hat{\beta}}{\partial\lambda\partial\nu} \\
&\quad + \eta_{\hat{\alpha}\hat{\beta}}\frac{\partial^2\hat{\alpha}}{\partial\mu\partial\lambda}\frac{\partial\hat{\beta}}{\partial\nu} + \eta_{\hat{\alpha}\hat{\beta}}\frac{\partial\hat{\alpha}}{\partial\lambda}\frac{\partial^2\hat{\beta}}{\partial\mu\partial\nu} \\
&\quad - \eta_{\hat{\alpha}\hat{\beta}}\frac{\partial^2\hat{\alpha}}{\partial\nu\partial\mu}\frac{\partial\hat{\beta}}{\partial\lambda} - \eta_{\hat{\alpha}\hat{\beta}}\frac{\partial\hat{\alpha}}{\partial\mu}\frac{\partial^2\hat{\beta}}{\partial\nu\partial\lambda}.
\end{aligned}
$$

Since the terms are symmetric in $\hat{\alpha}$ and $\hat{\beta}$ and the order of the second derivatives can be swapped according to Schwarz's theorem (9.27), the second and sixth as well as the fourth and fifth terms cancel out. Since the first and the third term are equal, the total result is

$$
g_{\mu\nu,\lambda} + g_{\lambda\nu,\mu} - g_{\mu\lambda,\nu} = 2\eta_{\hat{\alpha}\hat{\beta}}\frac{\partial^2\hat{\alpha}}{\partial\lambda\partial\mu}\frac{\partial\hat{\beta}}{\partial\nu}.
$$

On the other hand, from (19.12)

$$
\begin{aligned}
g_{\nu\sigma}\,\Gamma^{\sigma}_{\mu\lambda} &= \left(\eta_{\hat{\alpha}\hat{\beta}}\frac{\partial\hat{\alpha}}{\partial\nu}\frac{\partial\hat{\beta}}{\partial\sigma}\right)\left(\frac{\partial\sigma}{\partial\hat{\mu}}\frac{\partial^2\hat{\mu}}{\partial\lambda\partial\mu}\right) \\
&= \eta_{\hat{\alpha}\hat{\beta}}\frac{\partial\hat{\alpha}}{\partial\nu}\underbrace{\left(\frac{\partial\hat{\beta}}{\partial\sigma}\frac{\partial\sigma}{\partial\hat{\mu}}\right)}_{=\delta^{\hat{\beta}}_{\hat{\mu}}}\frac{\partial^2\hat{\mu}}{\partial\lambda\partial\mu} \\
&= \eta_{\hat{\alpha}\hat{\beta}}\frac{\partial\hat{\alpha}}{\partial\nu}\frac{\partial^2\hat{\beta}}{\partial\mu\partial\lambda}.
\end{aligned}
$$

So, by comparing the last two results, we get

$$
g_{\nu\sigma}\,\Gamma^{\sigma}_{\mu\lambda} = \frac{1}{2}\left(g_{\mu\nu,\lambda} + g_{\lambda\nu,\mu} - g_{\mu\lambda,\nu}\right).
$$

Let $g^{\mu\nu}$ be the matrix inverse to $g_{\mu\nu}$, i.e.

$$
g^{\tau\nu}g_{\nu\sigma} = \delta^{\tau}_{\sigma}.
$$

Multiplying the penultimate equation by $g^{\tau\nu}$, we get

$$\Gamma^{\tau}_{\mu\lambda} = \frac{g^{\tau\nu}}{2}\left(g_{\mu\nu,\lambda} + g_{\lambda\nu,\mu} - g_{\mu\lambda,\nu}\right), \tag{19.16}$$

from which, due to the symmetry of $g_{\mu\nu}$, the symmetry in the lower two indices of the Christoffel symbols follows

$$\Gamma^{\tau}_{\mu\lambda} = \Gamma^{\tau}_{\lambda\mu}. \tag{19.17}$$

We also calculate the covariant derivative of the metric tensor

$$\begin{aligned}
g_{\alpha\beta;\gamma} &= g_{\alpha\beta,\gamma} - \Gamma^{\sigma}_{\alpha\gamma}\,g_{\sigma\beta} - \Gamma^{\sigma}_{\beta\gamma}\,g_{\alpha\sigma} \\
&= g_{\alpha\beta,\gamma} - \frac{g^{\sigma\nu}}{2}\left(g_{\alpha\nu,\gamma} + g_{\gamma\nu,\alpha} - g_{\alpha\gamma,\nu}\right)g_{\sigma\beta} \\
&\quad - \frac{g^{\sigma\nu}}{2}\left(g_{\beta\nu,\gamma} + g_{\gamma\nu,\beta} - g_{\beta\gamma,\nu}\right)g_{\alpha\sigma} \\
&= g_{\alpha\beta,\gamma} - \frac{1}{2}\left(g_{\alpha\beta,\gamma} + g_{\gamma\beta,\alpha} - g_{\alpha\gamma,\beta}\right) - \frac{1}{2}\left(g_{\beta\alpha,\gamma} + g_{\gamma\alpha,\beta} - g_{\beta\gamma,\alpha}\right) \\
&= 0,
\end{aligned}$$

since all terms cancel out. Since in the local inertial system the covariant derivative is equal to the partial derivative, there it holds

$$g_{\hat{\mu}\hat{\nu};\hat{\gamma}} = g_{\hat{\mu}\hat{\nu},\hat{\gamma}} = 0, \tag{19.18}$$

from which, with (19.16), it follows that the Christoffel symbols in the local inertial system (more generally: in Minkowski space) are equal to zero, which we have already shown in part III by other means.

20. Motion in the Gravitational Field, Geodesic Equation

20.1. General Geodesic Equation

As the first physical example of application using the principle of equivalence and tensor calculus, we want to derive the equation of motion for a particle in a gravitational field. In the local inertial system, the laws of Special Relativity apply, i.e., for a force-free timelike particle with mass $m > 0$ and worldline $\vec{x}(t_E)$, the relativistic Newtonian equation of motion 14.12 on page 305 follows:

$$m\,\vec{A} = m\,\frac{d^2\vec{x}}{dt_E^2} = \vec{F}$$

with $\vec{F} = 0$ and the Minkowski coordinates

$$\vec{x}(t_E) \to \{\hat{\alpha}\,(t_E)\} = \left(\hat{t}\,(t_E)\,,\hat{x}\,(t_E)\,,\hat{y}\,(t_E)\,,\hat{z}\,(t_E)\right)$$

of the local inertial system, the equation of motion

$$\frac{d^2\hat{\alpha}}{dt_E^2} = 0, \tag{20.1}$$

where $dt_E^2 = -ds^2$ again denotes the proper time and $dt_E \neq 0$ should apply. If we integrate the last equation twice, we get

$$\hat{\alpha}(t_E) = a^{\hat{\alpha}}t_E + b^{\hat{\alpha}}$$

with suitable integration constants $a^{\hat{\alpha}}$ and $b^{\hat{\alpha}}$. This equation describes a straight line for each coordinate, i.e., a force-free particle in an inertial system moves in a straight line. The same can also be shown for light, (lightlike) photons move in inertial systems also on straight lines, i.e., also for photons applies

$$\frac{d^2\hat{\alpha}(\lambda)}{d\lambda^2} = 0$$

for a parameter λ, which, however, because $dt_E = 0$ cannot be the proper time. We now set for an arbitrary local coordinate system h the transformation (19.6)

$$d\hat{\alpha} = \frac{\partial\hat{\alpha}}{\partial\mu}\,d\mu$$

into the above equation and obtain with the chain and product rule

$$
\begin{aligned}
0 &= \frac{d^2\hat{\alpha}}{dt_E^2} = \frac{d}{dt_E}\left(\frac{d\hat{\alpha}}{dt_E}\right) = \frac{d}{dt_E}\left(\frac{\partial\hat{\alpha}}{\partial\mu}\frac{d\mu}{dt_E}\right)\\[2mm]
&= \frac{\partial\hat{\alpha}}{\partial\mu}\frac{d^2\mu}{dt_E^2} + \frac{d}{dt_E}\left(\frac{\partial\hat{\alpha}}{\partial\mu}\right)\frac{d\mu}{dt_E}\\[2mm]
&= \frac{\partial\hat{\alpha}}{\partial\mu}\frac{d^2\mu}{dt_E^2} + \frac{\partial^2\hat{\alpha}}{\partial\mu\partial\nu}\frac{d\mu}{dt_E}\frac{d\nu}{dt_E}.
\end{aligned}
$$

We multiply the equation with the Jacobian matrix inverse to $(\partial\hat{\alpha}/\partial\mu)$, namely with $(\partial\gamma/\partial\hat{\alpha})$, note that

$$
\left(\frac{\partial\gamma}{\partial\hat{\alpha}}\right)\left(\frac{\partial\hat{\alpha}}{\partial\mu}\right) = \delta^\gamma_\mu
$$

applies, and obtain

$$
\begin{aligned}
0 &= \frac{\partial\gamma}{\partial\hat{\alpha}}\frac{\partial\hat{\alpha}}{\partial\mu}\frac{d^2\mu}{dt_E^2} + \frac{\partial\gamma}{\partial\hat{\alpha}}\frac{\partial^2\hat{\alpha}}{\partial\mu\partial\nu}\frac{d\mu}{dt_E}\frac{d\nu}{dt_E}\\[2mm]
&= \delta^\gamma_\mu\frac{d^2\mu}{dt_E^2} + \frac{\partial\gamma}{\partial\hat{\alpha}}\frac{\partial^2\hat{\alpha}}{\partial\mu\partial\nu}\frac{d\mu}{dt_E}\frac{d\nu}{dt_E}\\[2mm]
&= \frac{d^2\gamma}{dt_E^2} + \frac{\partial\gamma}{\partial\hat{\alpha}}\frac{\partial^2\hat{\alpha}}{\partial\mu\partial\nu}\frac{d\mu}{dt_E}\frac{d\nu}{dt_E}.
\end{aligned}
$$

According to (19.12) it applies

$$
\frac{\partial\gamma}{\partial\hat{\alpha}}\frac{\partial^2\hat{\alpha}}{\partial\mu\partial\nu} = \Gamma^\gamma_{\mu\nu},
$$

and thus we obtain the equation of motion (**geodesic equation**) in the gravitational field

$$
\frac{d^2\gamma}{dt_E^2} = -\Gamma^\gamma_{\mu\nu}\frac{d\mu}{dt_E}\frac{d\nu}{dt_E}. \tag{20.2}
$$

The

$$
\{\mu(t_E)\} = (t(t_E), x(t_E), y(t_E), z(t_E))
$$

describe the paths of particles in the coordinate system K with the metric $g_{\mu\nu}\{\alpha\}$, i.e. in a reference system with a gravitational field. If you multiply the right side of (20.2) by the mass m, there are the gravitational forces. So there is a great similarity with the relativistic Newtonian equation of motion. However, one must note that the μ are usually curvilinear coordinates. From (20.2)

$$
ds^2 = g_{\mu\nu}\, d\mu\, d\nu
$$

and

$$dt_E^2 = -ds^2$$

it follows

$$g_{\mu\nu} \frac{d\mu}{dt_E} \frac{d\nu}{dt_E} = \frac{ds^2}{dt_E^2} = -\frac{dt_E^2}{dt_E^2} = -1,$$

so that with the speed $d\mu/dt_E$ only three of the four components are independent of each other. This restriction reminds of the four-velocity

$$\vec{U} \to (\gamma, \gamma v_x, \gamma v_y, \gamma v_z)$$

(see Eq. (14.8)) in the Special Theory of Relativity, which can also be expressed by the three components of the relative speed $\vec{v}$.

Similarly as above, the equation of motion for a light particle can be derived:

$$\frac{d^2\gamma}{d\lambda^2} = -\Gamma^\gamma{}_{\mu\nu} \frac{d\mu}{d\lambda} \frac{d\nu}{d\lambda},$$

where here because of $dt_E = 0$ the side condition

$$g_{\mu\nu} \frac{d\mu}{d\lambda} \frac{d\nu}{d\lambda} = ds^2 = -dt_E^2 = 0$$

has to be considered.

If the worldline of a particle fulfills Eq. (20.2), it is said that the motion of the particle takes place along a **geodesic**. Geodesics can be characterized in two ways:

1. A geodesic is the worldline of a particle with an "extreme" length (see also Sect. 14.3) between two points A and B.

2. A geodesic is the straightest possible (i.e. the least curved) connection between two points.

In this section, we want to prove property (1.), the proof of (2.) takes place in a later chapter, when we have introduced the concept of curvature. We have already derived in Sect. 14.3 that force-free particles in Minkowski space move on a straight line, which is the connection with the greatest proper time between two points. In Minkowski space (and in Euclidean space), straight lines are therefore geodesics. To determine the geodesics in Riemannian space, we take advantage of the fact that the distance between two points is measured using

$$ds^2 = g_{\mu\nu} \, d\mu \, d\nu,$$

and subsequently consider, for simplicity's sake, only *timelike worldlines* of particles between two points, where timelike curves are defined by all adjacent points on the curve having a timelike distance. So if two points A and B are given in Riemannian space and $\{\mu(\lambda)\}$ is a timelike curve with

$$A \to \{\mu(\lambda_A)\}$$

and

$$B \to \{\mu(\lambda_B)\},$$

then it follows as in Sect. 14.3 with

$$dt_E^2 = -ds^2$$

for the proper time between A and B

$$
\begin{aligned}
t_E(B) - t_E(A) &= \int \sqrt{-ds^2} = \int_{\lambda_A}^{\lambda_B} \sqrt{-g_{\mu\nu}\{\alpha(\lambda)\}\, d\mu\, d\nu} \\
&= \int_{\lambda_A}^{\lambda_B} \sqrt{-g_{\mu\nu}\{\alpha(\lambda)\} \frac{d\mu}{d\lambda} \frac{d\nu}{d\lambda}}\, d\lambda = \\
&= \int_{\lambda_A}^{\lambda_B} \sqrt{-g_{\mu\nu}\{\alpha(\lambda)\}\, \dot{\mu}\dot{\nu}}\, d\lambda,
\end{aligned}
\tag{20.3}
$$

where we have set $\dot{\mu} = d\mu/d\lambda$ and once again emphasized that the metric depends on the coordinates along the worldline. We want to show that this proper time becomes extremal when the $\mu(\lambda)$ satisfy the equations of motion (20.2). To do this, we apply a **principle of variation** which we do not want to justify strictly mathematically here. Principles of variation are used almost everywhere in physics. They express a state of static or dynamic equilibrium: if one "shakes" the system under consideration, the work done (virtual) is zero. The system is in this sense in an extremal state. The physically real state is distinguished from all other conceivable states by the fact that it is stable against small displacements of the orbits. So if our sought-after curve is the one that makes the proper time extremal, a small variation of the curve should not cause a variation in length.

Remark 20.1. **MT: Principle of Variation**
To prepare for the following, we discuss a special one-dimensional case. A typical task of the calculus of variations in physics is to find, among all possible curves $y(t)$, the one that makes the expression

$$S = \int_{t_A}^{t_B} L(y(t), \dot{y}(t))\, dt$$

extremal, i.e., minimal or maximal. Here, L (called **Lagrange function** after the Italian mathematician Joseph-Louis Lagrange (1736 - 1813)) is a given real-valued function that depends on y and its derivative $\dot{y}$. A and B are two arbitrary, but fixed boundary points. We are looking for those functions $y(t)$ that take the given values $A = y(t_A)$ and $B = y(t_B)$ at the boundary points and make S extremal. For this, the following variation approach is chosen for a $y(t)$ and considered

$$S(\varepsilon) = \int_{t_A}^{t_B} L(y(t,\varepsilon),\dot{y}(t,\varepsilon))\, dt,$$

where $y(t,\varepsilon)$ is defined by

$$y(t,\varepsilon) = y(t) + \varepsilon\eta(t).$$

Here, one can imagine ε as a "small" number and $\eta(t)$ is a function that vanishes at the boundary points

$$\eta(t_A) = \eta(t_B) = 0.$$

This makes $y(t,\varepsilon)$ a so-called **admissible** function. Then, the **variation** δS of S is formed by

$$
\begin{aligned}
\delta S &= \frac{dS}{d\varepsilon}\,d\varepsilon = \frac{d}{d\varepsilon}\left(\int_{t_A}^{t_B} dt\,\{L(y(t,\varepsilon),\dot{y}(t,\varepsilon))\}\right)d\varepsilon \\
&= \left(\int_{t_A}^{t_B} dt\,\frac{d}{d\varepsilon}\{L(y(t,\varepsilon),\dot{y}(t,\varepsilon))\}\right)d\varepsilon \\
&= \left(\int_{t_A}^{t_B} dt\,\left\{\frac{\partial L}{\partial y}\frac{dy}{d\varepsilon} + \frac{\partial L}{\partial \dot{y}}\frac{d\dot{y}}{d\varepsilon}\right\}\right)d\varepsilon,
\end{aligned}
\tag{20.4}
$$

where again the differential and the chain rule were used. The second term in the last equation can be rewritten using Schwartz's theorem as

$$\frac{\partial L}{\partial \dot{y}}\frac{d\dot{y}}{d\varepsilon} = \frac{\partial L}{\partial \dot{y}}\frac{d}{d\varepsilon}\left(\frac{dy}{dt}\right) = \frac{\partial L}{\partial \dot{y}}\frac{d}{dt}\left(\frac{dy}{d\varepsilon}\right)$$

and transformed using partial integration (see Eq. 6.22 on page 111)

$$\int_{t_A}^{t_B} dt\,\frac{\partial L}{\partial \dot{y}}\frac{d}{dt}\left(\frac{dy}{d\varepsilon}\right) = \left[\frac{\partial L}{\partial \dot{y}}\frac{dy}{d\varepsilon}\right]_{t_A}^{t_B} - \int_{t_A}^{t_B} dt\,\frac{dy}{d\varepsilon}\frac{d}{dt}\left(\frac{\partial L}{\partial \dot{y}}\right).$$

Because

$$\frac{dy}{d\varepsilon} = \eta(t) \quad \text{und} \quad \eta(t_A) = \eta(t_B) = 0$$

the term in the square brackets is omitted and we get overall

$$\delta S = \left(\int_{t_A}^{t_B} dt \left\{ \frac{\partial L}{\partial y} - \frac{d}{dt} \left(\frac{\partial L}{\partial \dot{y}} \right) \right\} \frac{dy}{d\varepsilon} \right) d\varepsilon.$$

Since $S(\varepsilon)$ is to become extremal, $\delta S = 0$ must hold and that for arbitrary variations $y(t, \varepsilon)$, i.e., the expression in the curly brackets in the integrand in the last equation must vanish:

$$\frac{d}{dt} \left(\frac{\partial L}{\partial \dot{y}} \right) - \frac{\partial L}{\partial y} = 0 \qquad (20.5)$$

This equation is called **Euler's differential equation** of the calculus of variations. $\square$

We now want to use this differential equation to find the geodesic curve. We set

$$L(\alpha, \dot{\alpha}) = \sqrt{-g_{\mu\nu} \{\alpha(\lambda)\} \dot{\mu}\dot{\nu}},$$

and the four (for $\alpha = t, x, y, z$) Euler differential equations are given by

$$\frac{d}{d\lambda} \left(\frac{\partial L}{\partial \dot{\alpha}} \right) - \frac{\partial L}{\partial \alpha} = 0.$$

We calculate the individual terms. It holds with the chain and product rule

$$\frac{\partial L}{\partial \dot{\alpha}} = \frac{\partial}{\partial \dot{\alpha}} \left(\sqrt{-g_{\mu\nu} \dot{\mu}\dot{\nu}} \right) = \frac{1}{2 \sqrt{-g_{\mu\nu} \dot{\mu}\dot{\nu}}} \left(-g_{\mu\nu} \delta^{\mu}_{\alpha} \dot{\nu} - g_{\mu\nu} \dot{\mu}\delta^{\nu}_{\alpha} \right),$$

since

$$\frac{\partial \dot{\mu}}{\partial \dot{\alpha}} = \begin{cases} 1 & \mu = \alpha \\ 0 & \mu \neq \alpha \end{cases} = \delta^{\mu}_{\alpha},$$

and since the metric depends not on $\dot{\alpha}$, but only on α. If one further notes that the root in the denominator is equal to L, then it follows

$$\frac{\partial L}{\partial \dot{\alpha}} = \frac{1}{2L} \left(-g_{\alpha\nu} \dot{\nu} - g_{\mu\alpha} \dot{\mu} \right) = -\frac{1}{2L} \left(g_{\alpha\nu} \frac{d\nu}{d\lambda} + g_{\mu\alpha} \frac{d\mu}{d\lambda} \right). \qquad (20.6)$$

On the other hand, it follows from Eq. (20.3):

$$t_E(B) - t_E(A) = \int_{\lambda_A}^{\lambda_B} \sqrt{-g_{\mu\nu} (\alpha(\lambda)) \dot{\mu}\dot{\nu}} \, d\lambda$$

with the fundamental theorem of calculus

$$\frac{dt_E}{d\lambda} = \sqrt{-g_{\mu\nu} (\alpha(\lambda)) \dot{\mu}\dot{\nu}} = L \Rightarrow \frac{d\lambda}{dt_E} = \frac{1}{L},$$

which we substitute into Eq. (20.6) for L :

$$\frac{\partial L}{\partial \dot{\alpha}} = -\frac{1}{2L}\left(g_{\alpha\nu}\frac{d\nu}{d\lambda} + g_{\mu\alpha}\frac{d\mu}{d\lambda}\right) = -\frac{1}{2}\left(g_{\alpha\nu}\frac{d\nu}{d\lambda}\frac{d\lambda}{dt_E} + g_{\mu\alpha}\frac{d\mu}{d\lambda}\frac{d\lambda}{dt_E}\right).$$

$$= -\frac{1}{2}\left(g_{\alpha\nu}\frac{d\nu}{dt_E} + g_{\mu\alpha}\frac{d\mu}{dt_E}\right)$$

If one further notes that the $g_{\alpha\mu}$ implicitly depend on λ, then one obtains with the chain and product rule from the last equation

$$\frac{d}{d\lambda}\frac{\partial L}{\partial\dot{\alpha}} = \frac{d}{d\lambda}\left(-\frac{1}{2}\left(g_{\alpha\nu}\frac{d\nu}{dt_E} + g_{\mu\alpha}\frac{d\mu}{dt_E}\right)\right)$$

$$= -\frac{1}{2}\left(\frac{dg_{\alpha\nu}}{d\lambda}\frac{d\nu}{dt_E} + g_{\alpha\nu}\frac{d}{d\lambda}\frac{d\nu}{dt_E} + \frac{dg_{\mu\alpha}}{d\lambda}\frac{d\mu}{dt_E} + g_{\mu\alpha}\frac{d}{d\lambda}\frac{d\mu}{dt_E}\right)$$

$$= -\frac{1}{2}\left(\frac{dg_{\alpha\nu}}{d\mu}\frac{d\mu}{d\lambda}\frac{d\nu}{dt_E} + g_{\alpha\nu}\frac{d}{d\lambda}\frac{d\nu}{dt_E} + \frac{dg_{\mu\alpha}}{d\nu}\frac{d\nu}{d\lambda}\frac{d\mu}{dt_E} + g_{\mu\alpha}\frac{d}{d\lambda}\frac{d\mu}{dt_E}\right),$$

and with that we have calculated the first term of the Euler differential equation. For the second, taking into account that the derivatives $\dot{\mu}$ do not depend on α, we analogously get

$$\frac{\partial L}{\partial\alpha} = -\frac{1}{2L}\frac{\partial}{\partial\alpha}\left(g_{\mu\nu}\,\dot{\mu}\dot{\nu}\right) = -\frac{1}{2L}\frac{\partial g_{\mu\nu}}{\partial\alpha}\frac{d\mu}{d\lambda}\frac{d\nu}{d\lambda}$$

$$= -\frac{1}{2L}g_{\mu\nu,\alpha}\frac{d\mu}{d\lambda}\frac{d\nu}{d\lambda} = -\frac{1}{2}g_{\mu\nu,\alpha}\frac{d\mu}{d\lambda}\frac{d\lambda}{dt_E}\frac{d\nu}{d\lambda}$$

$$= -\frac{1}{2}g_{\mu\nu,\alpha}\frac{d\mu}{dt_E}\frac{d\nu}{d\lambda},$$

where in the penultimate equation we again used $L = dt_E/d\lambda$. The Euler differential equation thus looks overall like this

$$0 = -\frac{1}{2}\left(g_{\alpha\nu,\mu}\frac{d\mu}{d\lambda}\frac{d\nu}{dt_E} + g_{\alpha\nu}\frac{d}{d\lambda}\frac{d\nu}{dt_E} + g_{\mu\alpha,\nu}\frac{d\nu}{d\lambda}\frac{d\mu}{dt_E} + g_{\mu\alpha}\frac{d}{d\lambda}\frac{d\mu}{dt_E}\right)$$

$$- \left(-\frac{1}{2}g_{\mu\nu,\alpha}\frac{d\mu}{dt_E}\frac{d\nu}{d\lambda}\right).$$

We multiply this equation by $\dfrac{d\lambda}{dt_E}$ and again with the chain rule we get

$$0 = -\frac{1}{2}\left(g_{\alpha\nu,\mu}\frac{d\mu}{d\lambda}\frac{d\nu}{dt_E} + g_{\alpha\nu}\frac{d}{d\lambda}\frac{d\nu}{dt_E} + g_{\mu\alpha,\nu}\frac{d\nu}{d\lambda}\frac{d\mu}{dt_E} + g_{\mu\alpha}\frac{d}{d\lambda}\frac{d\mu}{dt_E}\right)$$

$$- \left(-\frac{1}{2}g_{\mu\nu,\alpha}\frac{d\mu}{dt_E}\frac{d\nu}{d\lambda}\right)$$

$$\begin{aligned}
= & -\frac{1}{2}\left(g_{\alpha\nu,\mu}\frac{d\mu}{d\lambda}\frac{d\lambda}{dt_E}\frac{d\nu}{dt_E} + g_{\alpha\nu}\frac{d\lambda}{dt_E}\frac{d}{d\lambda}\frac{d\nu}{dt_E} + g_{\mu\alpha,\nu}\frac{d\nu}{d\lambda}\frac{d\lambda}{dt_E}\frac{d\mu}{dt_E}\right) \\
& -\frac{1}{2}\left(g_{\mu\alpha}\frac{d\lambda}{dt_E}\frac{d}{d\lambda}\frac{d\mu}{dt_E} - g_{\mu\nu,\alpha}\frac{d\mu}{dt_E}\frac{d\nu}{d\lambda}\frac{d\lambda}{dt_E}\right) \\
= & -\frac{1}{2}\left(g_{\alpha\nu,\mu}\frac{d\mu}{dt_E}\frac{d\nu}{dt_E} + g_{\alpha\nu}\frac{d}{dt_E}\frac{d\nu}{dt_E} + g_{\mu\alpha,\nu}\frac{d\nu}{dt_E}\frac{d\mu}{dt_E}\right) \\
& -\frac{1}{2}\left(g_{\mu\alpha}\frac{d}{dt_E}\frac{d\mu}{dt_E} - g_{\mu\nu,\alpha}\frac{d\mu}{dt_E}\frac{d\nu}{dt_E}\right) \\
= & -\frac{1}{2}\left(g_{\alpha\nu,\mu}\frac{d\mu}{dt_E}\frac{d\nu}{dt_E} + g_{\mu\alpha,\nu}\frac{d\nu}{dt_E}\frac{d\mu}{dt_E} + 2g_{\mu\alpha}\frac{d^2\mu}{dt_E^2}\right) \\
& +\frac{1}{2}g_{\mu\nu,\alpha}\frac{d\mu}{dt_E}\frac{d\nu}{dt_E},
\end{aligned}$$

where the last equation results from the symmetry of $g_{\mu\nu}$ and the fact that one can freely choose the summation index, i.e.

$$g_{\alpha\nu}\frac{d}{dt_E}\frac{d\nu}{dt_E} = g_{\mu\alpha}\frac{d}{dt_E}\frac{d\mu}{dt_E}.$$

We thus obtain from the Euler equation

$$0 = -g_{\mu\alpha}\frac{d^2\mu}{dt_E^2} - \frac{1}{2}\left(g_{\alpha\nu,\mu} + g_{\mu\alpha,\nu} - g_{\mu\nu,\alpha}\right)\frac{d\mu}{dt_E}\frac{d\nu}{dt_E}.$$

Multiplying the equation by $g^{\alpha\tau}$, we obtain with Eq. (19.16)

$$\begin{aligned}
\frac{d^2\tau}{dt_E^2} &= g^{\alpha\tau}g_{\mu\alpha}\frac{d^2\mu}{dt_E^2} \\
&= -\frac{1}{2}g^{\alpha\tau}\left(g_{\alpha\nu,\mu} + g_{\mu\alpha,\nu} - g_{\mu\nu,\alpha}\right)\frac{d\mu}{dt_E}\frac{d\nu}{dt_E} \\
&= -\Gamma^{\tau}{}_{\mu\nu}\frac{d\mu}{dt_E}\frac{d\nu}{dt_E},
\end{aligned}$$

thus again the geodesic equation (20.2). This shows that a geodesic is a curve with extremal proper time.

20.2. Geodesics in the Euclidean Plane

We consider some examples of geodesics, first the Euclidean plane with Cartesian coordinates.

Example 20.1.
The line element there is
$$ds^2 = dx^2 + dy^2$$
and the metric
$$g_{\mu\nu} = \begin{pmatrix} 1 & 0 \\ 0 & 1 \end{pmatrix}.$$
All Christoffel symbols are zero, and the geodesic equation reads with the parameter s along the curve
$$\frac{d^2x}{ds^2} = \frac{d^2y}{ds^2} = 0,$$
from which follows by twice integration
$$\begin{aligned} x &= a \cdot s + b \\ y &= c \cdot s + d \end{aligned}$$
with the constants a, b, c, d. If one solves the first equation for s and substitutes this into the second one, one obtains
$$y = \frac{c}{a} x + d - \frac{b}{a} = mx + n,$$
thus a straight line. $\square$

Example 20.2.
The second example considers polar coordinates in the plane. With the Eq. (9.18), (9.16) and (8.36) applies
$$ds^2 = dr^2 + r^2 d\varphi^2$$
as well as
$$(g_{\mu\nu}) = \begin{pmatrix} 1 & 0 \\ 0 & r^2 \end{pmatrix}$$
and
$$\begin{aligned} \Gamma^\varphi_{r\varphi} &= \Gamma^\varphi_{\varphi r} = \frac{1}{r} \\ \Gamma^r_{\varphi\varphi} &= -r, \end{aligned}$$
all other Christoffel symbols vanish. Since the polar coordinates "live" in the Euclidean plane, the geodesics in polar coordinates should also be straight lines. The most general form of a straight line in the plane is
$$ax + by = c$$

with constants a, b, c. If we substitute polar coordinates for $x = r \cos \varphi$ and $y = r \sin \varphi$, we get

$$a\,r \cos \varphi + b\,r \sin \varphi = c. \tag{20.7}$$

So we expect a result of this form for the geodesic equation in polar coordinates. We set $\mu = r, \varphi$ and $\nu = r, \varphi$ and obtain the following two equations from the general geodesic equation

$$\frac{d^2 r}{ds^2} + \Gamma^r_{\mu\nu} \frac{d\mu}{ds}\frac{d\nu}{ds} = 0$$

$$\frac{d^2 \varphi}{ds^2} + \Gamma^\varphi_{\mu\nu} \frac{d\mu}{ds}\frac{d\nu}{ds} = 0. \tag{20.8}$$

The first equation yields due to

$$\Gamma^r_{\mu\nu} = \begin{cases} -r & \mu = \varphi, \nu = \varphi \\ 0 & sonst \end{cases}$$

the relation

$$\frac{d^2 r}{ds^2} + \Gamma^r_{\varphi\varphi} \frac{d\varphi}{ds}\frac{d\varphi}{ds} = \frac{d^2 r}{ds^2} - r \left(\frac{d\varphi}{ds}\right)^2 = 0.$$

Similarly, due to

$$\Gamma^\varphi_{\mu\nu} = \begin{cases} 1/r & \mu = r, \nu = \varphi \\ 1/r & \mu = \varphi, \nu = r \\ 0 & sonst \end{cases}$$

for the second equation of (20.8)

$$\begin{aligned}
0 &= \frac{d^2 \varphi}{ds^2} + \Gamma^\varphi_{\mu\nu} \frac{d\mu}{ds}\frac{d\nu}{ds} = \frac{d^2 \varphi}{ds^2} + \Gamma^\varphi_{r\varphi} \frac{dr}{ds}\frac{d\varphi}{ds} + \Gamma^\varphi_{\varphi r} \frac{dr}{ds}\frac{d\varphi}{ds} \\
&= \frac{d^2 \varphi}{ds^2} + \frac{2}{r} \frac{dr}{ds}\frac{d\varphi}{ds}.
\end{aligned}$$

In summary, we thus obtain the two equations

$$\frac{d^2 r}{ds^2} - r \left(\frac{d\varphi}{ds}\right)^2 = 0$$

$$\frac{d^2 \varphi}{ds^2} + \frac{2}{r} \frac{dr}{ds}\frac{d\varphi}{ds} = 0.$$

If

$$\varphi' = \frac{d\varphi}{ds} = 0,$$

then φ is constant and it also follows $d^2\varphi/ds^2 = 0$, i.e., the second equation is fulfilled. If we substitute $d\varphi/ds = 0$ into the first equation, we get

$$\frac{d^2 r}{ds^2} = 0 \Rightarrow r = a \cdot s + b.$$

This is a straight line through the origin. If $\varphi' \neq 0$, we divide the second equation by $\varphi' = d\varphi/ds$ and get

$$0 = \frac{ds}{d\varphi}\left\{ \frac{d^2\varphi}{ds^2} + \frac{2}{r}\frac{dr}{ds}\frac{d\varphi}{ds} \right\} = \frac{1}{\varphi'}\frac{d\varphi'}{ds} + \frac{2}{r}\frac{dr}{ds}.$$

We integrate this equation, then it follows

$$\begin{aligned} 0 &= \int \left\{ \frac{1}{\varphi'}\frac{d\varphi'}{ds} + \frac{2}{r}\frac{dr}{ds} \right\} ds = \int \frac{1}{\varphi'} d\varphi' + \int \frac{2}{r} dr \\ &= \ln\left|\varphi'\right| + 2\ln r = \ln\left|\varphi'\right| + \ln r^2 = \ln\left(r^2\left|\varphi'\right|\right). \end{aligned}$$

Exponentiating gives

$$r^2\left|\varphi'\right| = e^{\ln\left(r^2|\varphi'|\right)} = e^0 = 1,$$

so

$$r^2\varphi' = \pm 1 = h = const. \Rightarrow \varphi' = \frac{d\varphi}{ds} = \frac{h}{r^2}.$$

We divide the line element

$$ds^2 = dr^2 + r^2 d\varphi^2$$

by ds^2 and get

$$1 = \left(\frac{dr}{ds}\right)^2 + r^2\left(\frac{d\varphi}{ds}\right)^2 = \left(\frac{dr}{ds}\right)^2 + r^2\left(\frac{h}{r^2}\right)^2 = \left(\frac{dr}{ds}\right)^2 + \frac{h^2}{r^2},$$

from which

$$\frac{dr}{ds} = \pm\sqrt{1 - \frac{h^2}{r^2}}$$

follows. We now form

$$\frac{d\varphi}{dr} = \frac{\dfrac{d\varphi}{ds}}{\dfrac{dr}{ds}} = \pm\frac{\dfrac{h}{r^2}}{\sqrt{1 - \dfrac{h^2}{r^2}}} = \pm\frac{h}{r\sqrt{r^2 - h^2}}$$

and look for a function φ, whose derivative with respect to r corresponds to the right-hand side of the equation. If we differentiate the function

$$\varphi(r) = \pm \arccos\left(\frac{h}{r}\right),$$

where the arccosine is the inverse function of the cosine, using the formula for the derivative of the inverse function (4.21) and the chain rule, we get

$$\frac{d}{dr}\varphi(r) = \pm\frac{d}{dr}\arccos\left(\frac{h}{r}\right) = \pm\frac{1}{\cos'\left(\arccos\left(\frac{h}{r}\right)\right)}\cdot\frac{-h}{r^2}$$

$$= \pm\frac{1}{-\sin\left(\arccos\left(\frac{h}{r}\right)\right)}\cdot\frac{-h}{r^2}.$$

For the sine term we use the formula

$$\sin x = \sqrt{1 - \cos^2 x}$$

and get

$$\frac{d}{dr}\varphi(r) = \pm\frac{1}{\sqrt{1 - \cos^2\left(\arccos\left(\frac{h}{r}\right)\right)}}\cdot\frac{h}{r^2}$$

$$= \pm\frac{1}{\sqrt{1 - \left(\frac{h}{r}\right)^2}}\cdot\frac{h}{r^2} = \pm\frac{h}{r\sqrt{r^2 - h^2}},$$

i.e., up to a constant, which we call φ_0, φ is the sought function. It holds

$$\varphi - \varphi_0 = \pm\arccos\left(\frac{h}{r}\right) \Rightarrow \frac{h}{r} = \cos(\varphi - \varphi_0).$$

If we note that with an addition theorem of the angle functions

$$\cos(\varphi - \varphi_0) = \cos\varphi\cos\varphi_0 + \sin\varphi\sin\varphi_0$$

holds (see section 29.1), we finally get the expected line equation (20.7)

$$h = \cos\varphi_0\, r\cos\varphi + \sin\varphi_0\, r\sin\varphi.\ \square$$

20.3. Geodesics on the Surface of a Sphere

The third example deals with the surface of a sphere, i.e., we want to determine the geodesic lines on a spherical surface. We consider a sphere with a fixed radius a and want to measure the points on this sphere by longitude and latitude, for which we again want to use the line element

$$ds^2 = dx^2 + dy^2 + dz^2.$$

First, we calculate the differentials dx, dy and dz in spherical coordinates (see Remark 19.2) by:

$$
\begin{aligned}
dx &= \frac{\partial x}{\partial \vartheta} d\vartheta + \frac{\partial x}{\partial \varphi} d\varphi = a \cos\vartheta \cos\varphi \, d\vartheta - a \sin\vartheta \sin\varphi \, d\varphi \\
dy &= \frac{\partial y}{\partial \vartheta} d\vartheta + \frac{\partial y}{\partial \varphi} d\varphi = a \cos\vartheta \sin\varphi \, d\vartheta + a \sin\vartheta \cos\varphi \, d\varphi \\
dz &= \frac{\partial z}{\partial \vartheta} d\vartheta + \frac{\partial z}{\partial \varphi} d\varphi = -a \sin\vartheta \, d\vartheta
\end{aligned}
$$

From this, by squaring and combining, we get

$$ds^2 = dx^2 + dy^2 + dz^2 = a^2 \, d\vartheta^2 + a^2 \sin^2\vartheta \, d\varphi^2. \tag{20.9}$$

The metric then follows as

$$(g_{\mu\nu}) = \begin{pmatrix} a^2 & 0 \\ 0 & a^2 \sin^2\vartheta \end{pmatrix}.$$

To determine the Christoffel symbols, we need the inverse metric. To calculate this, we use the formula (8.19) valid for (2×2)-matrices

$$A = \begin{pmatrix} a & b \\ c & d \end{pmatrix}$$

for the inverse:

$$A^{-1} = \frac{1}{\det A} \begin{pmatrix} d & -b \\ -c & a \end{pmatrix}$$

Since

$$\det A = a^2 a^2 \sin^2\vartheta - 0 = a^4 \sin^2\vartheta,$$

it follows that

$$(g_{\mu\nu})^{-1} = g^{\mu\nu} = \frac{1}{a^4 \sin^2\vartheta} \begin{pmatrix} a^2 \sin^2\vartheta & 0 \\ 0 & a^2 \end{pmatrix} = \begin{pmatrix} \dfrac{1}{a^2} & 0 \\ 0 & \dfrac{1}{a^2 \sin^2\vartheta} \end{pmatrix}.$$

We calculate the Christoffel symbols with the formula (19.16)

$$\Gamma^{\tau}_{\mu\lambda} = \frac{g^{\nu\tau}}{2}\left(g_{\mu\nu,\lambda} + g_{\lambda\nu,\mu} - g_{\mu\lambda,\nu}\right)$$

and note that, since a is a fixed number, the only non-vanishing derivative of the metric is

$$g_{\varphi\varphi,\vartheta} = \frac{\partial}{\partial\vartheta}\left(a^2\sin^2\vartheta\right) = 2a^2\sin\vartheta\cos\vartheta.$$

From this we get for the Christoffel symbols, very similar to the polar coordinates,

$$\Gamma^{\vartheta}_{\vartheta\vartheta} = \frac{1}{2}g^{\vartheta\vartheta}\left(g_{\vartheta\vartheta,\vartheta} + g_{\vartheta\vartheta,\vartheta} - g_{\vartheta\vartheta,\vartheta}\right) + \frac{1}{2}g^{\varphi\vartheta}\left(g_{\varphi\vartheta,\vartheta} + g_{\varphi\vartheta,\vartheta} - g_{\vartheta\vartheta,\varphi}\right) = 0$$

$$\Gamma^{\varphi}_{\vartheta\vartheta} = \frac{1}{2}g^{\vartheta\varphi}\left(g_{\vartheta\vartheta,\vartheta} + g_{\vartheta\vartheta,\vartheta} - g_{\vartheta\vartheta,\vartheta}\right) + \frac{1}{2}g^{\varphi\varphi}\left(g_{\varphi\vartheta,\vartheta} + g_{\varphi\vartheta,\vartheta} - g_{\vartheta\vartheta,\varphi}\right) = 0$$

$$\Gamma^{\vartheta}_{\vartheta\varphi} = \frac{1}{2}g^{\vartheta\vartheta}\left(g_{\vartheta\vartheta,\varphi} + g_{\vartheta\varphi,\vartheta} - g_{\vartheta\varphi,\vartheta}\right) + \frac{1}{2}g^{\varphi\vartheta}\left(g_{\varphi\vartheta,\varphi} + g_{\varphi\varphi,\vartheta} - g_{\vartheta\varphi,\varphi}\right) = 0$$

$$\Gamma^{\varphi}_{\vartheta\varphi} = \frac{1}{2}g^{\vartheta\varphi}\left(g_{\vartheta\vartheta,\varphi} + g_{\vartheta\varphi,\vartheta} - g_{\vartheta\varphi,\vartheta}\right) + \frac{1}{2}g^{\varphi\varphi}\left(g_{\varphi\vartheta,\varphi} + g_{\varphi\varphi,\vartheta} - g_{\vartheta\varphi,\varphi}\right)$$

$$= \frac{1}{2}\cdot\frac{1}{a^2\sin^2\vartheta}\,2a^2\sin\vartheta\cos\vartheta = \frac{\cos\vartheta}{\sin\vartheta} = \cot\vartheta = \Gamma^{\varphi}_{\varphi\vartheta}$$

$$\Gamma^{\vartheta}_{\varphi\vartheta} = \frac{1}{2}g^{\vartheta\vartheta}\left(g_{\vartheta\varphi,\vartheta} + g_{\vartheta\vartheta,\varphi} - g_{\varphi\vartheta,\vartheta}\right) + \frac{1}{2}g^{\varphi\vartheta}\left(g_{\varphi\varphi,\vartheta} + g_{\varphi\vartheta,\varphi} - g_{\varphi\vartheta,\varphi}\right) = 0$$

$$\Gamma^{\vartheta}_{\varphi\varphi} = \frac{1}{2}g^{\vartheta\vartheta}\left(g_{\vartheta\varphi,\varphi} + g_{\vartheta\varphi,\varphi} - g_{\varphi\varphi,\vartheta}\right) + \frac{1}{2}g^{\varphi\vartheta}\left(g_{\varphi\varphi,\varphi} + g_{\varphi\varphi,\varphi} - g_{\varphi\varphi,\varphi}\right)$$

$$= \frac{1}{2}\cdot\frac{1}{a^2}\left(-2a^2\sin\vartheta\cos\vartheta\right) = -\sin\vartheta\cos\vartheta$$

$$\Gamma^{\varphi}_{\varphi\varphi} = \frac{1}{2}g^{\vartheta\varphi}\left(g_{\vartheta\varphi,\varphi} + g_{\vartheta\varphi,\varphi} - g_{\varphi\varphi,\vartheta}\right) + \frac{1}{2}g^{\varphi\varphi}\left(g_{\varphi\varphi,\varphi} + g_{\varphi\varphi,\varphi} - g_{\varphi\varphi,\varphi}\right) = 0.$$

Therefore, all Christoffel symbols are zero, except for

$$\Gamma^{\varphi}_{\vartheta\varphi} = \Gamma^{\varphi}_{\varphi\vartheta} = \cot\vartheta \qquad\qquad (20.10)$$

$$\Gamma^{\vartheta}_{\varphi\varphi} = -\sin\vartheta\cos\vartheta.$$

We set $\mu,\nu = \vartheta,\varphi$ and obtain from the geodesic equation the two equations

$$\frac{d^2\vartheta}{ds^2} + \Gamma^{\vartheta}_{\mu\nu}\frac{d\mu}{ds}\frac{d\nu}{ds} = \frac{d^2\vartheta}{ds^2} + \Gamma^{\vartheta}_{\varphi\varphi}\frac{d\varphi}{ds}\frac{d\varphi}{ds} = \frac{d^2\vartheta}{ds^2} - \sin\vartheta\cos\vartheta\left(\frac{d\varphi}{ds}\right)^2 = 0$$

$$\frac{d^2\varphi}{ds^2} + \Gamma^{\varphi}_{\mu\nu}\frac{d\mu}{ds}\frac{d\nu}{ds} = \frac{d^2\varphi}{ds^2} + 2\Gamma^{\varphi}_{\vartheta\varphi}\frac{d\vartheta}{ds}\frac{d\varphi}{ds} = \frac{d^2\varphi}{ds^2} + 2\cot\vartheta\frac{d\vartheta}{ds}\frac{d\varphi}{ds} = 0.$$

Similar to the previous example, we again distinguish two cases. If

$$\varphi' = \frac{d\varphi}{ds} = 0,$$

thus φ is constant and the second equation is fulfilled. It follows from the first equation

$$\frac{d^2\vartheta}{ds^2} = 0 \Rightarrow \vartheta = a \cdot s + b.$$

This equation describes all curves on the sphere that have a fixed longitude, i.e., all (half) circles of longitude on the surface of the sphere that pass through the North and South poles. On Earth, these curves are also called **meridians**.

If in the second case $\varphi' \neq 0$, we divide the second equation by φ' and obtain

$$0 = \frac{1}{\varphi'}\left\{\frac{d^2\varphi}{ds^2} + 2\cot\vartheta\,\frac{d\vartheta}{ds}\frac{d\varphi}{ds}\right\} = \frac{1}{\varphi'}\frac{d\varphi'}{ds} + 2\cot\vartheta\,\frac{d\vartheta}{ds}.$$

We want to integrate this equation, for which we need an antiderivative of

$$\cot\vartheta = \frac{\cos\vartheta}{\sin\vartheta}.$$

For example,

$$\ln\left(\sin\vartheta\right)$$

is one, which can be seen when differentiating with the chain rule

$$\left(\ln\left(\sin\vartheta\right)\right)' = \ln'(\sin\vartheta)\cdot(\sin\vartheta)' = \frac{1}{\sin\vartheta}\cdot\cos\vartheta = \cot\vartheta.$$

So it follows

$$\begin{aligned}
0 &= \int\left\{\frac{1}{\varphi'}\frac{d\varphi'}{ds} + 2\cot\vartheta\,\frac{d\vartheta}{ds}\right\}ds = \int\frac{1}{\varphi'}\,d\varphi' + \int 2\cot\vartheta\,d\vartheta \\
&= \ln\left|\varphi'\right| + 2\ln\left(\left|\sin\vartheta\right|\right) = \ln\left|\varphi'\right| + \ln\left(\sin^2\vartheta\right) = \ln\left(\sin^2\vartheta\left|\varphi'\right|\right).
\end{aligned}$$

Exponentiating gives

$$e^{\ln\left(\sin^2\vartheta|\varphi'|\right)} = \sin^2\vartheta\left|\varphi'\right| = e^0 = 1,$$

so

$$\sin^2\vartheta\varphi' = \pm 1 = h = const. \Rightarrow \varphi' = \frac{d\varphi}{ds} = \frac{h}{\sin^2\vartheta}.$$

We divide the line element

$$ds^2 = a^2 d\vartheta^2 + a^2\sin^2\vartheta\,d\varphi^2$$

by ds^2 and obtain

$$1 = a^2\left(\frac{d\vartheta}{ds}\right)^2 + a^2\sin^2\vartheta\left(\frac{d\varphi}{ds}\right)^2$$

$$= a^2 \left(\frac{d\vartheta}{ds}\right)^2 + a^2 \sin^2\vartheta \left(\frac{h}{\sin^2\vartheta}\right)^2 = a^2\left(\frac{d\vartheta}{ds}\right)^2 + \frac{a^2 h^2}{\sin^2\vartheta},$$

from which

$$\frac{d\vartheta}{ds} = \pm\sqrt{\frac{1}{a^2}\left(1 - \frac{a^2 h^2}{\sin^2\vartheta}\right)} = \pm\sqrt{\frac{1}{a^2} - \frac{h^2}{\sin^2\vartheta}}$$

follows. We now form

$$\frac{d\varphi}{d\vartheta} = \frac{\dfrac{d\varphi}{ds}}{\dfrac{d\vartheta}{ds}} = \pm\frac{\dfrac{h}{\sin^2\vartheta}}{\sqrt{\dfrac{1}{a^2} - \dfrac{h^2}{\sin^2\vartheta}}}.$$

Here too, we immediately give an antiderivative and save ourselves the (somewhat complicated) direct integration. The function

$$\varphi(\vartheta) = \pm\arccos\left(\frac{h\cot\vartheta}{\sqrt{\dfrac{1}{a^2} - h^2}}\right)$$

is an antiderivative. We prove this by forming the derivative. As above, we use the fact that the derivative of the inverse cosine is

$$(\arccos x)' = \frac{-1}{\sqrt{1 - x^2}}$$

and that

$$\frac{1}{\sin^2\vartheta} = \frac{\sin^2\vartheta + \cos^2\vartheta}{\sin^2\vartheta} = 1 + \frac{\cos^2\vartheta}{\sin^2\vartheta} = 1 + \cot^2\vartheta$$

as well as

$$(\cot\vartheta)' = \left(\frac{\cos\vartheta}{\sin\vartheta}\right)' = \frac{-\sin^2\vartheta - \cos^2\vartheta}{\sin^2\vartheta} = \frac{-1}{\sin^2\vartheta}$$

holds. It then follows again with the chain rule

$$\varphi'(\vartheta) = \pm\arccos'\left(\frac{h\cot\vartheta}{\sqrt{\dfrac{1}{a^2} - h^2}}\right)\left(\frac{h}{\sqrt{\dfrac{1}{a^2} - h^2}}\right)(\cot\vartheta)'$$

$$= \pm\frac{-1}{\sqrt{1 - \left(\dfrac{h\cot\vartheta}{\sqrt{\dfrac{1}{a^2} - h^2}}\right)^2}}\left(\frac{h}{\sqrt{\dfrac{1}{a^2} - h^2}}\right)\frac{-1}{\sin^2\vartheta}$$

$$= \pm \frac{\dfrac{h}{\sin^2 \vartheta}}{\sqrt{1 - \dfrac{h^2 \cot^2 \vartheta}{\dfrac{1}{a^2} - h^2}} \sqrt{\dfrac{1}{a^2} - h^2}}$$

$$= \pm \frac{\dfrac{h}{\sin^2 \vartheta}}{\sqrt{\dfrac{1}{a^2} - h^2 - h^2 \cot^2 \vartheta}}$$

$$= \pm \frac{\dfrac{h}{\sin^2 \vartheta}}{\sqrt{\dfrac{1}{a^2} - h^2(1 + \cot^2 \vartheta)}}$$

$$= \pm \frac{\dfrac{h}{\sin^2 \vartheta}}{\sqrt{\dfrac{1}{a^2} - \dfrac{h^2}{\sin^2 \vartheta}}},$$

i.e., up to a constant, which we call φ_0, φ is the function we are looking for. It holds with

$$-b = \frac{h}{\sqrt{\dfrac{1}{a^2} - h^2}}$$

$$\varphi - \varphi_0 = \arccos\left(-b \cot \vartheta\right) \Rightarrow \cos(\varphi - \varphi_0) + b \cot \vartheta = 0.$$

If we again consider that with the addition theorem of the angle functions

$$\cos(\varphi - \varphi_0) = \cos \varphi \cos \varphi_0 + \sin \varphi \sin \varphi_0$$

holds, we get

$$\cos \varphi_0 \cos \varphi + \sin \varphi_0 \sin \varphi + b \cot \vartheta = 0.$$

Multiplication with $\sin \vartheta$ finally yields

$$\cos \varphi_0 \sin \vartheta \cos \varphi + \sin \varphi_0 \sin \vartheta \sin \varphi + b \cos \vartheta = 0.$$

If we substitute for the spherical coordinates x, y, z, we get

$$\frac{\cos \varphi_0}{a} x + \frac{\sin \varphi_0}{a} y + \frac{b}{a} z = 0.$$

This equation describes a plane through the origin when x, y, z are chosen arbitrarily. Since we have placed the center of the sphere at the origin and

x, y, z represent points on the sphere's surface, we get all points that lie on curves that result from the intersections between the planes through the center and the sphere's surface. These curves are called **great circles**. And since the meridians also lie on great circles, all geodesics on a sphere's surface are great circles.

21. Curvature in Riemannian Space

So far, we have made intensive use of the existence of local inertial systems in the definitions and derivations of tensors and derivatives, without explicitly addressing the aspect of curvature, which we also need to introduce precisely. Since we can only perceive curvature with our limited sensory organs when it manifests in one-dimensional (e.g., curves) or two-dimensional (e.g., surfaces) objects (we cannot imagine a three-dimensional space as curved, let alone four-dimensional spacetime), we want to explain some of the terminologies and concepts using simple examples before we then define them exactly in four-dimensional spacetime.

In curvature, two different types, called **extrinsic** and **intrinsic**, must be distinguished. We consider a cylinder. Since the cylinder is round in one direction, we perceive its surface as curved from our three-dimensional perspective. This view is referred to as *extrinsic* curvature, as curvature from the outside, viewed from higher dimensions. On the other hand, a cylinder can be made from a flat rectangular piece of paper that is rolled up and glued at the edges. If you draw parallel lines on the flat piece of paper or calculate the distance between two points, the parallels remain parallel after rolling up on the cylinder and the distances remain the same. All Euclidean laws (e.g., that the sum of the angles in a triangle is 180°) apply on the cylinder surface just as they do in the Euclidean plane. In this respect, the cylinder is flat and it is said to be *intrinsically* flat. The intrinsic geometry only considers the relationships between points on curves, which are completely within the considered surface. When we describe the surface of the cylinder as curved, we compare it with straight lines or planes in three-dimensional space. We want to deal exclusively with the intrinsic properties of spacetime and assume that all worldlines of particles lie in spacetime. Whether there is a higher-dimensional space in which the four-dimensional spacetime is embedded does not interest us here, apart from the fact that we also have no physically motivated reason to gain access to this, which actually exists in the mathematical theory.

The cylinder is intrinsically flat, the surface of a sphere is not. If we draw two lines that are perpendicular to the equator and therefore parallel there, and we extend the lines locally as straight as possible, these lines follow great circles that meet at the pole. Parallel lines do not remain parallel when they are extended, i.e., the sphere surface is not flat. This can also be seen if we again

© The Author(s), under exclusive license to Springer-Verlag GmbH,
DE, part of Springer Nature 2026

M. Ruhrländer, *Ascent to the Einstein Equations*,
https://doi.org/10.1007/978-3-662-72672-3_21

mark two points on the equator, which are a quarter of the equator length, or 90°, apart and then draw "straight" lines perpendicular to the North Pole, forming a triangle. This triangle has the sum of angles of $90° + 90° + 90° = 270°$, which also shows that the sphere surface cannot be flat. Another way to make the intrinsic curvature of the sphere surface visible is the so-called **parallel transport**. This method has the advantage that it can be easily transferred to higher-dimensional spaces.

21.1. Parallel Transport

In parallel transport, we take a vector and move it parallel around a closed curve. If the space is flat, the vector points in exactly the same direction after circling the curve as at the start.

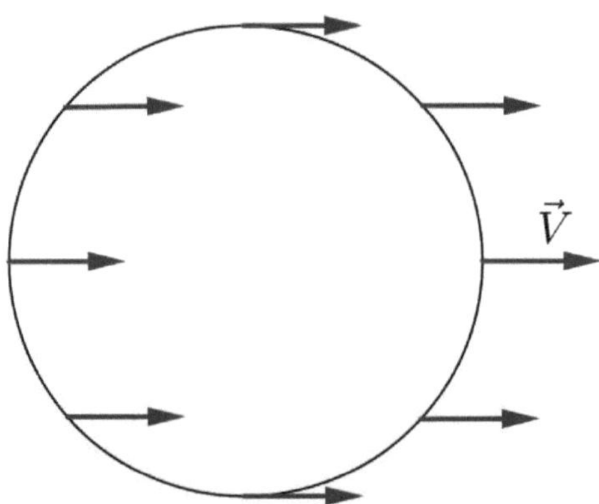

Figure 21.1.: Parallel transport in Euclidean plane

Fig. 21.1 demonstrates this with a circle around which the vector $\vec{V}$ is transported parallel. Here, parallel transport means that at each point along a circle, the vector is parallel to the one at the previous point. The situation is different on intrinsically curved surfaces. As an example, we again consider the sphere surface.

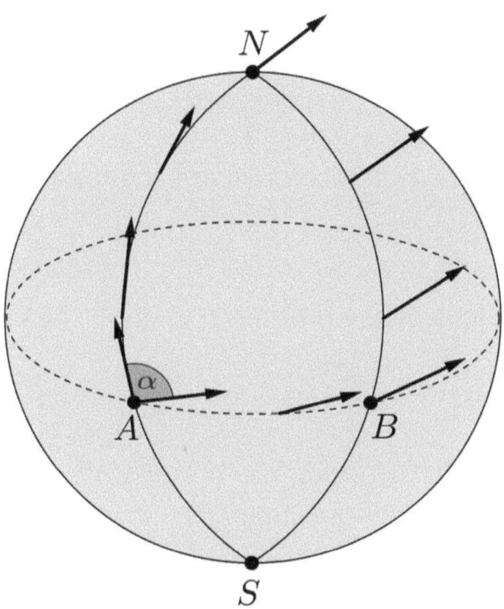

Figure 21.2.: Parallel transport on a spherical surface

A vector is transported from a point A over the North Pole N and the point B parallel back to A. On the first leg from A to the North Pole, the vector is shifted on the great circle so that it is always tangential to the great circle. Upon reaching the North Pole, the vector forms a certain angle with the great circle, which is formed by B and the North Pole. It is shifted from the North Pole to B in such a way that this angle always remains the same. Upon reaching B, the vector forms an angle with the great circle, which is formed by A and B (in Fig. 21.2 the equator), and it is pushed back to A while maintaining this angle. As can be seen from the graphic, the initial vector does not match the end vector at point A. Both vectors at point A form an angle that is different from zero, $\alpha \neq 0$. So there is a significant difference between flat space, where vectors remain parallel when parallel transported around a closed curve, and a curved space, where this is not the case. With the help of parallel transport, one can decide whether a section of space is curved, and one can also define a measure for the curvature, which we want to examine more closely in the following.

We first show that there is a connection between covariant derivation and parallel transport of a vector. We have defined the covariant derivative by a transformation of the ordinary derivative in the local inertial system to the Riemannian space, thus choosing a rather abstract definition. But there is also a more geometric approach to the covariant derivative, similar to the one we used to represent the ordinary derivative by limit values of difference quotients (see Sec. 2.1). If we wanted to define the derivative of a vector field $\vec{V} \to V^{\mu}(t, x, y, z)$ at a point $P \to (t, x, y, z)$ "classically", we would component-wise form the difference of two neighboring vectors

$$dV^{\mu} = V^{\mu}(t, x + dx, y, z) - V^{\mu}(t, x, y, z)$$

and then define the limit of the difference quotient as the derivative

$$\frac{\partial V^{\mu}}{\partial x} = \lim_{dx \to 0} \frac{dV^{\mu}}{dx}.$$

The problem now is that the difference between two vectors at different locations is *not* a vector. The components $V^{\mu}(t, x, y, z)$ transform with

$$V^{\mu'} = \left[\frac{\partial \mu'}{\partial \mu}\right]_{(t,x,y,z)} V^{\mu}$$

and the $V^{\mu}(t, x + dx, y, z)$ with

$$V^{\mu'} = \left[\frac{\partial \mu'}{\partial \mu}\right]_{(t,x+dx,y,z)} V^{\mu},$$

and in general, the transformation matrices from point (t, x, y, z) to point $(t, x + dx, y, z)$ are different, i.e., the difference between the two vectors has no well-defined transformation behavior, is thus not a vector, and thus the ordinary derivative is not a vector, which we have already shown by other means in Eq. (19.10). This fact, by the way, has nothing to do with the curvature of a space, as we have seen when using polar coordinates in the Euclidean plane. To construct a derivative that is a true vector, we must take the difference of two vectors *at the same* point. That means, we consider at the point

$$\{\alpha + d\alpha\} = (t + dt, x + dx, y + dy, z + dz)$$

two vectors, one of which is

$$\vec{V}\{\alpha + d\alpha\} = \vec{V}\{\alpha\} + d\vec{V}$$

and the other is a vector at the position $\{\alpha + d\alpha\}$, which is parallel to $\vec{V}$ at the position $\{\alpha\}$ and we denote it as $\vec{V} + \delta\vec{V}$. It is said, that the vector $\vec{V} + \delta\vec{V}$ is created by **parallel transport** or **parallel displacement** from $\{\alpha\}$ to $\{\alpha + d\alpha\}$.

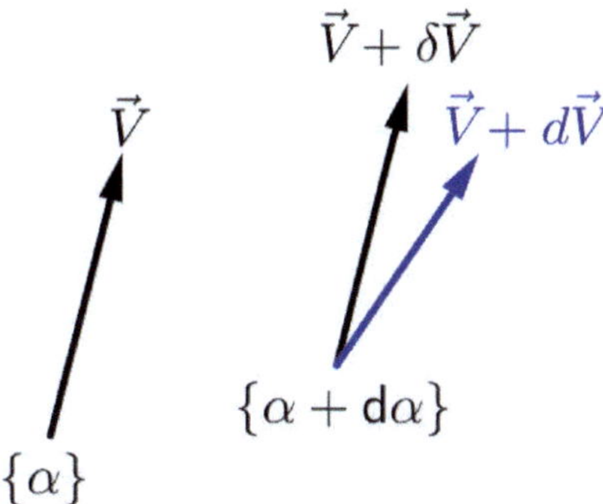

Figure 21.3.: Parallel transport of a vector

We thus obtain for the components of the two vectors

$$V^\mu\{\alpha\} + dV^\mu \;=\; V^\mu\{\alpha + d\alpha\}$$
$$V^\mu\{\alpha\} + \delta V^\mu \;=\; \text{vector parallel transported from } \{\alpha\} \text{ to } \{\alpha + d\alpha\},$$

see Fig. 21.3. The expression $\vec{V} + \delta\vec{V}$ is by definition a vector at the position $\{\alpha + d\alpha\}$, which in local Minkowski coordinates has the same components as the vector $\vec{V}$ at the position $\{\alpha\}$, i.e., in these coordinates $\delta V^\mu = 0$, which is not generally valid. Thus, $\delta\vec{V}$ is not a vector, but the sum $\vec{V} + \delta\vec{V}$ at the position $\{\alpha + d\alpha\}$ is. We thus obtain a very similar intermediate result as with the covariant derivative. There, the ordinary derivative and also the Christoffel symbols considered alone were not tensors, but their sum was.

We now form the difference of the two vectors, which we denote as $D\vec{V}$

$$D\vec{V} \to DV^\mu \;=\; V^\mu\{\alpha + d\alpha\} - (V^\mu\{\alpha\} + \delta V^\mu)$$
$$=\; V^\mu\{\alpha\} + dV^\mu - (V^\mu\{\alpha\} + \delta V^\mu) = dV^\mu - \delta V^\mu.$$

The question remains how to determine the quantities δV^μ. It is reasonable to assume that the δV^μ are proportional to both the vector components V^σ and the displacements $d\alpha$, i.e., there are numbers $\Gamma^\mu_{\sigma\tau}$, such that

$$\delta V^\mu = -\Gamma^\mu_{\sigma\tau} V^\sigma d\tau \tag{21.1}$$

holds, and of course these coefficients must be chosen so that $D\vec{V}$ becomes a vector. These properties are fulfilled by the Christoffel symbols and we obtain

$$DV^\mu = dV^\mu - \delta V^\mu = dV^\mu + \Gamma^\mu_{\sigma\tau} V^\sigma d\tau$$

and from this by division by $d\tau$

$$\frac{DV^\mu}{d\tau} = \frac{dV^\mu}{d\tau} + \Gamma^\mu_{\sigma\tau}\, V^\sigma = V^\mu_{,\tau} + \Gamma^\mu_{\sigma\tau}\, V^\sigma = V^\mu_{;\tau},$$

i.e., we end up again with the covariant derivative. One can write DV^μ as

$$DV^\mu = \frac{dV^\mu}{d\tau}\, d\tau + \Gamma^\mu_{\sigma\tau}\, V^\sigma d\tau = V^\mu_{;\tau}\, d\tau, \tag{21.2}$$

which is why DV^μ is also referred to as the **covariant differential**.

We want to extend the concept of covariant derivation to derivatives in the direction of a vector $\vec{U}$ and *define* the **covariant derivative of a vector field $\vec{V}$ in the direction of $\vec{U}$** by

$$\nabla_{\vec{U}}\,\vec{V} \to V^\mu_{;\nu}\, U^\nu = \left(V^\mu_{,\nu} + \Gamma^\mu_{\sigma\nu}\, V^\sigma\right) U^\nu.$$

Since the expression $\nabla_{\vec{U}}\,\vec{V}$ is derived from the contraction of two tensors, it is itself a tensor, specifically a vector. We use this derivative to define the **parallel transport along a curve**. Let $\{\mu\,(\lambda)\}$ be a curve with parameter λ and let

$$\vec{U} \to U^\mu = \frac{d\mu}{d\lambda}$$

be the not necessarily normalized tangent vector to the curve. If $\vec{V}$ is a vector field along the curve, i.e., $\vec{V}\,\{\mu\,(\lambda)\}$ is defined for each λ, then $\vec{V}$ is **parallel transported along $\{\mu\,(\lambda)\}$**, if

$$\nabla_{\vec{U}}\,\vec{V} \to V^\mu_{;\nu}\, U^\nu = \left(V^\mu_{,\nu} + \Gamma^\mu_{\sigma\nu}\, V^\sigma\right) U^\nu = 0 \tag{21.3}$$

holds. Or more intuitively: If the vectors $\vec{V}\,\{\alpha\}$ and $\vec{V}\,\{\alpha + d\alpha\}$ at infinitesimally close points on the curve are parallel and of the same length, then this is called parallel transport of $\vec{V}$ along the curve. Because, if we go to the local inertial system at a point P, we know that parallel vectors there have the same components. This means that the components of $\vec{V}$ are constant along the curve at P, which is nothing else than

$$\frac{\partial V^\mu}{d\lambda} = 0 \text{ at the point } P.$$

Now, with the chain rule,

$$\frac{\partial V^\mu}{d\lambda} = \frac{\partial V^\mu}{d\nu}\frac{\partial \nu}{d\lambda} = \frac{\partial V^\mu}{d\nu}\, U^\nu = V^\mu_{,\nu}\, U^\nu.$$

Since, according to Sec. 19.3, the Christoffel symbols are zero in local inertial systems, the ordinary and covariant derivatives coincide there,

$$V^\mu_{,\nu} = V^\mu_{;\nu} \text{ at the point } P,$$

and we get

$$\frac{\partial V^\mu}{d\lambda} = V^\mu_{,\nu}\, U^\nu = V^\mu_{;\nu}\, U^\nu = 0 \text{ at the point } P.$$

Now, the last equation is a tensor equation and thus holds not only in the local inertial system, but in any coordinate system.

Derivation of the Geodesic Equation using Parallel Transport

We want to use parallel transport to show a second way to derive the geodesic equation (20.2). The most important curves in flat space are straight lines. One of the Euclidean axioms states that two straight lines, which initially run parallel, remain parallel if they are extended. What does extension mean in this context? This does *not* mean that (only) the distance between the two lines remains unchanged, because the lines could curve equally upon extension. Rather, it means that the lines are extended in the direction they have run so far. Or to put it more precisely, that the tangent to the curves at one point is parallel to the tangent at a previous point. Indeed, a straight line in Euclidean space is the only curve that shifts its tangent vector parallel. In curved spaces, however, we can also define curves that are "as straight as possible", namely by requiring that the tangent vectors of these curves be transported parallel. Expressed in formulas, this means

$$\nabla_{\vec{U}}\, \vec{U} = 0,$$

or in component notation

$$U^\nu U^\mu_{;\nu} = U^\nu U^\mu_{,\nu} + \Gamma^\mu_{\sigma\nu}\, U^\sigma U^\nu = 0.$$

If we consider

$$U^\nu = \frac{d\nu}{d\lambda}$$

and with the chain rule

$$U^\nu U^\mu_{,\nu} = \frac{d\nu}{d\lambda}\frac{d}{d\nu}\left(\frac{d\mu}{d\lambda}\right) = \frac{d}{d\lambda}\left(\frac{d\mu}{d\lambda}\right) = \frac{d^2\mu}{d\lambda^2},$$

we obtain

$$\frac{d^2\mu}{d\lambda^2} + \Gamma^\mu_{\sigma\nu}\frac{d\sigma}{d\lambda}\frac{d\nu}{d\lambda} = 0,$$

which is again the geodesic equation (20.2). The geodesics are thus not only the curves of extreme proper time, but also those curves that run as straight as possible.

21.2. Riemannian Curvature Tensor

In the previous sections, we have already made plausible that the components of the metric $g_{\mu\nu}\,(x)$ express the curvature of a Riemannian space. Now we want to introduce a tensor that quantitatively describes the curvature. Since the curvature of space indicates the presence of a gravitational field, this curvature tensor plays a major role in Einstein's field equations. From the components of a metric, one cannot generally directly read whether these are only curvilinear coordinates in a flat space (like e.g. the polar coordinates in the Euclidean plane) or whether they "live" on a curved space (like e.g. the angular coordinates on a sphere surface). However, we will see that from the $g_{\mu\nu}\,(x)$ or the Christoffel symbols the curvature tensor can be calculated. And if this tensor is everywhere equal to zero, then the space is flat. If it is everywhere different from zero, then the space is curved.

The ordinary and the covariant derivatives have many comparable properties. But there is a fundamental difference, because the order of differentiation in second derivatives is not interchangeable in covariant derivatives, while it is in partial derivatives. If ϕ is a scalar field, then according to Schwarz's theorem (see Eq. 9.27 on page 202)

$$\frac{\partial}{\partial\mu}\left(\frac{\partial\phi}{\partial\nu}\right) = \frac{\partial}{\partial\nu}\left(\frac{\partial\phi}{\partial\mu}\right) \iff \phi_{,\nu,\mu} = \phi_{,\mu,\nu}.$$

First, we look at the covariant second derivative of the scalar field ϕ and note that for scalar fields $\phi_{;\mu} = \phi_{,\mu}$ holds, from which

$$\phi_{;\nu;\mu} = \phi_{,\nu;\mu} = \phi_{,\nu,\mu} - \Gamma^{\sigma}{}_{\nu\mu}\phi_{,\sigma}$$

and

$$\phi_{;\mu;\nu} = \phi_{,\mu;\nu} = \phi_{,\mu,\nu} - \Gamma^{\sigma}{}_{\mu\nu}\phi_{,\sigma}$$

follows. Since the Christoffel symbols are symmetric in the lower two indices, it also follows that

$$\phi_{;\nu;\mu} = \phi_{;\mu;\nu}.$$

For a vector $\vec{V} \to V^{\mu}$, however, this is generally not the case. We first calculate $V^{\mu}{}_{;\lambda;\nu}$ (see section 19.3) and note that $\nabla\vec{V} \to V^{\mu}{}_{;\lambda}$ is a tensor of rank $(1,1)$. It holds

$$
\begin{aligned}
V^{\mu}{}_{;\lambda;\nu} &= \left(V^{\mu}{}_{;\lambda}\right)_{,\nu} + \Gamma^{\mu}{}_{\rho\nu} V^{\rho}{}_{;\lambda} - \Gamma^{\tau}{}_{\lambda\nu} V^{\mu}{}_{;\tau} \\
&= \left(V^{\mu}{}_{,\lambda} + \Gamma^{\mu}{}_{\rho\lambda} V^{\rho}\right)_{,\nu} + \Gamma^{\mu}{}_{\rho\nu} V^{\rho}{}_{;\lambda} - \Gamma^{\tau}{}_{\lambda\nu} V^{\mu}{}_{;\tau}.
\end{aligned}
$$

We calculate the derivative of the bracket expression, taking into account the product rule

$$\left(V^{\mu}_{,\lambda} + \Gamma^{\mu}_{\rho\lambda} V^{\rho}\right)_{,\nu} = V^{\mu}_{,\lambda,\nu} + \Gamma^{\mu}_{\rho\lambda} V^{\rho}_{,\nu} + \Gamma^{\mu}_{\rho\lambda,\nu} V^{\rho}.$$

This results in

$$V^{\mu}_{;\lambda;\nu} = V^{\mu}_{,\lambda,\nu} + \Gamma^{\mu}_{\rho\lambda} V^{\rho}_{,\nu} + \Gamma^{\mu}_{\rho\lambda,\nu} V^{\rho} + \Gamma^{\mu}_{\rho\nu} V^{\rho}_{;\lambda} - \Gamma^{\tau}_{\lambda\nu} V^{\mu}_{;\tau}.$$

Now we replace $V^{\rho}_{;\lambda}$ with $V^{\rho}_{,\lambda} + \Gamma^{\rho}_{\sigma\lambda} V^{\sigma}$ and obtain

$$\begin{aligned}
V^{\mu}_{;\lambda;\nu} &= V^{\mu}_{,\lambda,\nu} + \Gamma^{\mu}_{\rho\lambda} V^{\rho}_{,\nu} + \Gamma^{\mu}_{\rho\lambda,\nu} V^{\rho} \\
&+ \Gamma^{\mu}_{\rho\nu}\left(V^{\rho}_{,\lambda} + \Gamma^{\rho}_{\sigma\lambda} V^{\sigma}\right) - \Gamma^{\tau}_{\lambda\nu}\left(V^{\mu}_{,\tau} + \Gamma^{\mu}_{\rho\tau} V^{\rho}\right) \\
&= V^{\mu}_{,\lambda,\nu} + \Gamma^{\mu}_{\rho\lambda} V^{\rho}_{,\nu} + \Gamma^{\mu}_{\rho\nu} V^{\rho}_{,\lambda} - \Gamma^{\tau}_{\lambda\nu} V^{\mu}_{,\tau} - \Gamma^{\tau}_{\lambda\nu} \Gamma^{\mu}_{\rho\tau} V^{\rho} \\
&+ \Gamma^{\mu}_{\rho\lambda,\nu} V^{\rho} + \Gamma^{\mu}_{\rho\nu} \Gamma^{\rho}_{\sigma\lambda} V^{\sigma}.
\end{aligned}$$

Now we swap the order of differentiation and obtain with $\lambda \leftrightarrow \nu$

$$\begin{aligned}
V^{\mu}_{;\nu;\lambda} &= V^{\mu}_{,\nu,\lambda} + \Gamma^{\mu}_{\rho\nu} V^{\rho}_{,\lambda} + \Gamma^{\mu}_{\rho\lambda} V^{\rho}_{,\nu} - \Gamma^{\tau}_{\nu\lambda} V^{\mu}_{,\tau} - \Gamma^{\tau}_{\nu\lambda} \Gamma^{\mu}_{\rho\tau} V^{\rho} \\
&+ \Gamma^{\mu}_{\rho\nu,\lambda} V^{\rho} + \Gamma^{\mu}_{\rho\lambda} \Gamma^{\rho}_{\sigma\nu} V^{\sigma}.
\end{aligned}$$

We form the difference of these two derivatives and, taking into account the symmetries in partial derivatives or Christoffel symbols, we find that the 1., (2. + 3.), 4. and 5. terms cancel out in the difference formation. If we also swap $\sigma \leftrightarrow \rho$ in the terms with the products of two Christoffel symbols, we get

$$\begin{aligned}
V^{\mu}_{;\lambda;\nu} - V^{\mu}_{;\nu;\lambda} &= \Gamma^{\mu}_{\rho\lambda,\nu} V^{\rho} + \Gamma^{\mu}_{\rho\nu} \Gamma^{\rho}_{\sigma\lambda} V^{\sigma} - \left(\Gamma^{\mu}_{\rho\nu,\lambda} V^{\rho} + \Gamma^{\mu}_{\rho\lambda} \Gamma^{\rho}_{\sigma\nu} V^{\sigma}\right) \\
&= V^{\rho}\left(\Gamma^{\mu}_{\rho\lambda,\nu} - \Gamma^{\mu}_{\rho\nu,\lambda} + \Gamma^{\mu}_{\sigma\nu} \Gamma^{\sigma}_{\rho\lambda} - \Gamma^{\mu}_{\sigma\lambda} \Gamma^{\sigma}_{\rho\nu}\right) \\
&= V^{\rho} R^{\mu}_{\rho\nu\lambda}
\end{aligned} \tag{21.4}$$

with

$$R^{\mu}_{\rho\nu\lambda} = \Gamma^{\mu}_{\rho\lambda,\nu} - \Gamma^{\mu}_{\rho\nu,\lambda} + \Gamma^{\mu}_{\sigma\nu} \Gamma^{\sigma}_{\rho\lambda} - \Gamma^{\mu}_{\sigma\lambda} \Gamma^{\sigma}_{\rho\nu}. \tag{21.5}$$

Since the components of a tensor are on the left side of Eq. (21.4), this must also be the case for the right side. Since $\vec{V}$ is a vector, the expression $R^{\mu}_{\rho\nu\lambda}$ also defines a tensor of rank $(1,3)$. This is called the **Riemann curvature tensor**. We can make an initial remark about the Riemann tensor.

If space is flat, a *global* transformation to Cartesian or Minkowski coordinates can be found for any coordinate system. However, this means that the Christoffel symbols are *everywhere* equal to zero and therefore also their derivative, i.e., in flat spaces in Cartesian or Minkowski coordinates,

$$R^{\mu}{}_{\rho\nu\lambda} = 0$$

and thus in all coordinate systems in flat space, since the $R^{\mu}{}_{\rho\nu\lambda}$ are the components of a tensor.

We want to derive the Riemann tensor again in a somewhat more illustrative way using parallel displacement. In curved spaces, parallel displacement results in different results when vectors are transported parallel from one point to another along different paths. The Riemann curvature tensor arises from the difference of covariant second derivatives, and we will see that these second derivatives represent the respective parallel displacements along the different paths. For simplicity, we consider a small surface element in a two-dimensional curved space with the coordinates x, y. In Fig. 21.4 on the next page, the infinitesimal surface element is bounded by the four coordinate lines

$$x = a, y = b, x = a + \delta a, y = b + \delta b.$$

We want to transport the vector V^{μ} at the point (a, b) once along the path $1 \to 2 \to 3$ from (a, b) via $(a + \delta a, b)$ to $(a + \delta a, b + \delta b)$ in parallel and secondly via the path $1 \to 2' \to 3'$ and calculate the difference

$$V^{\mu}(3) - V^{\mu}(3')$$

concretely, where we want to write for the positions for simplicity, e.g.

$$V^{\mu}(3) = V^{\mu}(a + \delta a, b + \delta b).$$

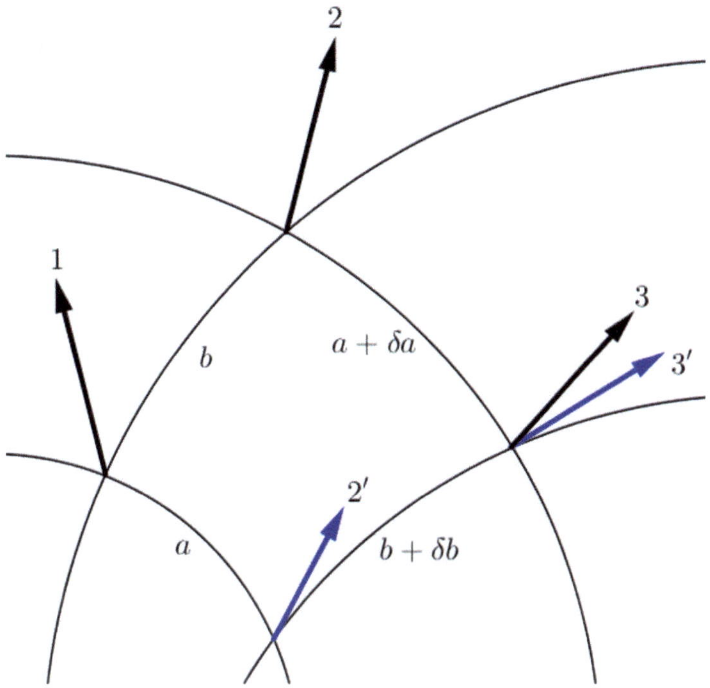

Figure 21.4.: Parallel shift of a vector along two paths

The curves along which the displacements occur are the coordinate lines, and we denote the individual path segments with $\vec{x}^{\,i\to j}(\lambda)$ i.e., it holds with suitable parametrizations

$$
\begin{aligned}
\vec{x}^{\,1\to 2}(\lambda) &\to (x(\lambda), b) \\
\vec{x}^{\,2\to 3}(\lambda) &\to (a+\delta a, y(\lambda)) \\
\vec{x}^{\,1\to 2'}(\lambda) &\to (a, y(\lambda)) \\
\vec{x}^{\,2'\to 3'}(\lambda) &\to (x(\lambda), b+\delta b),
\end{aligned}
$$

where the initial and final values of the parametrization λ_a and λ_e are chosen so that

$$
\begin{aligned}
x(\lambda_a) &= a, x(\lambda_e) = a+\delta a \\
y(\lambda_a) &= b, y(\lambda_e) = b+\delta b.
\end{aligned}
$$

For the tangential vectors

$$
\vec{U}^{\,i\to j} = \frac{d\vec{x}^{\,i\to j}(\lambda)}{d\lambda}
$$

it follows

$$
\begin{aligned}
\vec{U}^{\,1\to2} &\to \left(\frac{dx\,(\lambda)}{d\lambda},\frac{db}{d\lambda}\right)=\left(\frac{dx}{d\lambda},0\right)\\[2mm]
\vec{U}^{\,2\to3} &\to \left(\frac{d\,(a+\delta a)}{d\lambda},\frac{dy\,(\lambda)}{d\lambda}\right)=\left(0,\frac{dy}{d\lambda}\right)\\[2mm]
\vec{U}^{\,1\to2'} &\to \left(\frac{da}{d\lambda},\frac{dy\,(\lambda)}{d\lambda}\right)=\left(0,\frac{dy}{d\lambda}\right)\\[2mm]
\vec{U}^{\,2'\to3'} &\to \left(\frac{dx\,(\lambda)}{d\lambda},\frac{d\,(b+\delta b)}{d\lambda}\right)=\left(\frac{dx}{d\lambda},0\right).
\end{aligned}
$$

We use for the parallel shift the formula (21.3)

$$
\left(V^{\mu}_{\ ,\nu}+\Gamma^{\mu}_{\sigma\nu}V^{\sigma}\right)U^{\nu}=0
$$

with $\mu,\nu,\sigma=x,y$, which with the results for the components of the tangential vectors leads to the four equations

$$
\begin{aligned}
\left(V^{\mu}_{\ ,x}+\Gamma^{\mu}_{\sigma x}V^{\sigma}\right)\frac{dx}{d\lambda} &= 0\\[2mm]
\left(V^{\mu}_{\ ,y}+\Gamma^{\mu}_{\sigma y}V^{\sigma}\right)\frac{dy}{d\lambda} &= 0\\[2mm]
\left(V^{\mu}_{\ ,y}+\Gamma^{\mu}_{\sigma y}V^{\sigma}\right)\frac{dy}{d\lambda} &= 0\\[2mm]
\left(V^{\mu}_{\ ,x}+\Gamma^{\mu}_{\sigma x}V^{\sigma}\right)\frac{dx}{d\lambda} &= 0,
\end{aligned}
$$

which specify the two equations

$$
\begin{aligned}
V^{\mu}_{\ ,x} &= -\Gamma^{\mu}_{\sigma x}V^{\sigma}\\[2mm]
V^{\mu}_{\ ,y} &= -\Gamma^{\mu}_{\sigma y}V^{\sigma}.
\end{aligned}
$$

From this, we get the differences of the vectors with the differential

$$
\begin{aligned}
V^{\mu}\,(2)-V^{\mu}\,(1) &= V^{\mu}_{\ ,x}\,(1)\,\delta a = -\Gamma^{\mu}_{\sigma x}\,(1)\,V^{\sigma}\,(1)\,\delta a\\[2mm]
V^{\mu}\,(3)-V^{\mu}\,(2) &= V^{\mu}_{\ ,y}\,(2)\,\delta b = -\Gamma^{\mu}_{\sigma y}\,(2)\,V^{\sigma}\,(2)\,\delta b\\[2mm]
V^{\mu}\,(2')-V^{\mu}\,(1) &= V^{\mu}_{\ ,y}\,(1)\,\delta b = -\Gamma^{\mu}_{\sigma y}\,(1)\,V^{\sigma}\,(1)\,\delta b\\[2mm]
V^{\mu}\,(3')-V^{\mu}\,(2') &= V^{\mu}_{\ ,x}\,(2')\,\delta a = -\Gamma^{\mu}_{\sigma x}\,(2')\,V^{\sigma}\,(2')\,\delta a.
\end{aligned}
$$

We use the differential again to calculate the Christoffel symbols at the different locations. It applies

$$
\Gamma^{\mu}_{\sigma y}\,(2) = \Gamma^{\mu}_{\sigma y}\,(1)+\Gamma^{\mu}_{\sigma y,x}\,(1)\,\delta a
$$

$$\Gamma^{\mu}_{\sigma x}\left(2'\right) \;=\; \Gamma^{\mu}_{\sigma x}\left(1\right) + \Gamma^{\mu}_{\sigma x,y}\left(1\right)\delta b.$$

When this is inserted into the above equations, one obtains

$$
\begin{aligned}
V^{\mu}\left(2\right) - V^{\mu}\left(1\right) &= -\Gamma^{\mu}_{\sigma x}\left(1\right) V^{\sigma}\left(1\right)\delta a \\
V^{\mu}\left(3\right) - V^{\mu}\left(2\right) &= -\left[\Gamma^{\mu}_{\sigma y}\left(1\right) + \Gamma^{\mu}_{\sigma y,x}\left(1\right)\delta a\right] V^{\sigma}\left(2\right)\delta b \\
V^{\mu}\left(2'\right) - V^{\mu}\left(1\right) &= -\Gamma^{\mu}_{\sigma y}\left(1\right) V^{\sigma}\left(1\right)\delta b \\
V^{\mu}\left(3'\right) - V^{\mu}\left(2'\right) &= -\left[\Gamma^{\mu}_{\sigma x}\left(1\right) + \Gamma^{\mu}_{\sigma x,y}\left(1\right)\delta b\right] V^{\sigma}\left(2'\right)\delta a.
\end{aligned}
$$

We still set

$$
\begin{aligned}
V^{\sigma}\left(2\right) &= V^{\sigma}\left(1\right) - \Gamma^{\sigma}_{\tau x}\left(1\right) V^{\tau}\left(1\right)\delta a \\
V^{\sigma}\left(2'\right) &= V^{\sigma}\left(1\right) - \Gamma^{\sigma}_{\tau y}\left(1\right) V^{\tau}\left(1\right)\delta b
\end{aligned}
$$

and by inserting and simplifying and exchanging $\sigma \leftrightarrow \tau$ in the terms with the derivatives of the Christoffel symbols, we get

$$
\begin{aligned}
&V^{\mu}\left(3\right) - V^{\mu}\left(3'\right) \\
=\;& \left[V^{\mu}\left(3\right) - V^{\mu}\left(2\right)\right] + \left[V^{\mu}\left(2\right) - V^{\mu}\left(1\right)\right] \\
&- \left[V^{\mu}\left(2'\right) - V^{\mu}\left(1\right)\right] - \left[V^{\mu}\left(3'\right) - V^{\mu}\left(2'\right)\right] \\
=\;& \delta a\,\delta b\, V^{\tau}\left[\Gamma^{\mu}_{\tau x,y}\left(1\right) - \Gamma^{\mu}_{\tau y,x}\left(1\right) + \Gamma^{\mu}_{\sigma y}\left(1\right)\Gamma^{\sigma}_{\tau x}\left(1\right) - \Gamma^{\mu}_{\sigma x}\left(1\right)\Gamma^{\sigma}_{\tau y}\left(1\right)\right] \\
&- \delta a^{2}\,\delta b\, V^{\tau}\left[\Gamma^{\mu}_{\sigma y,x}\left(1\right)\Gamma^{\sigma}_{\tau x}\left(1\right)\right] - \delta a\,\delta b^{2}\, V^{\tau}\left[\Gamma^{\mu}_{\sigma x,y}\left(1\right)\Gamma^{\sigma}_{\tau y}\left(1\right)\right] \\
\approx\;& \delta a\,\delta b\, V^{\tau}\left[\Gamma^{\mu}_{\tau x,y}\left(1\right) - \Gamma^{\mu}_{\tau y,x}\left(1\right) + \Gamma^{\mu}_{\sigma y}\left(1\right)\Gamma^{\sigma}_{\tau x}\left(1\right) - \Gamma^{\mu}_{\sigma x}\left(1\right)\Gamma^{\sigma}_{\tau y}\left(1\right)\right],
\end{aligned}
$$

where in the last step we have neglected the (very small) third order terms $\delta a^{2}\,\delta b$ and $\delta a\,\delta b^{2}$. Overall, it is shown that with the above definition of the Riemann tensor (21.5)

$$dV^{\mu}\left(3\right) = V^{\mu}\left(3\right) - V^{\mu}\left(3'\right) = \delta a\,\delta b\, V^{\tau} R^{\mu}_{\ \tau yx}\left(1\right)$$

applies, i.e., one can measure the parallel displacement on different paths or on closed curves with the Riemann tensor, where the parallel displacement is also proportional to the area $\delta A = \delta a\,\delta b$ that the curve encloses.

The Riemannian tensor is composed of Christoffel symbols and their derivatives and is thus an intrinsic quantity. Since the Christoffel symbols themselves consist of the metric and its derivative, the Riemann tensor depends only on the metric and its first and second derivatives. Thus, the curvature of a space is determined by the metric. We already know that from Einstein's perspective, gravitation should be describable as a curvature of spacetime. Now we have derived how to calculate the curvature of a space, and have thus made a big step. However, there is still a lot to do to find out exactly how the curvature is incorporated in Einstein's theory.

21.3. Symmetries of the Riemannian Tensor

We want to derive some symmetries of the Riemannian tensor, which will lead us in total to the fact that out of the computational $4 \cdot 4 \cdot 4 \cdot 4 = 256$ components of the tensor only 20 are different. First, one can read directly from (21.5)

$$R^{\mu}{}_{\rho\nu\lambda} = \Gamma^{\mu}{}_{\rho\lambda,\nu} - \Gamma^{\mu}{}_{\rho\nu,\lambda} + \Gamma^{\mu}{}_{\sigma\nu}\,\Gamma^{\sigma}{}_{\rho\lambda} - \Gamma^{\mu}{}_{\sigma\lambda}\,\Gamma^{\sigma}{}_{\rho\nu},$$

that

$$R^{\mu}{}_{\rho\nu\lambda} = -R^{\mu}{}_{\rho\lambda\nu}$$

applies, the Riemann tensor is *antisymmetric* in the last two indices. This means that four of the 16 possible combinations of the last two indices yield zero:

$$R^{\mu}{}_{\rho00} = -R^{\mu}{}_{\rho00} \Rightarrow R^{\mu}{}_{\rho00} = 0$$

and similarly

$$R^{\mu}{}_{\rho11} = R^{\mu}{}_{\rho22} = R^{\mu}{}_{\rho33} = 0$$

Of the remaining twelve numbers, only six are independent, the others are obtained from these six by multiplication with -1.

Remark. The definition of the Riemannian tensor we have chosen is not unambiguous in the literature. Some books define the Riemann tensor with a minus sign compared to our definition. After the above shown symmetry in the last two indices, this means that the Riemann tensor in these books is defined by $R^{\mu}{}_{\rho\lambda\nu}$ instead of $R^{\mu}{}_{\rho\nu\lambda}$. This differing definition leads to the fact that in some tensor equations, like e.g. the Einstein equations, on the right side a minus sign appears. We follow in our definition the approach, which is chosen in most Anglo-Saxon books (e.g. Schutz [29], Ryder [26]). The Einstein equations still to be derived therefore come without the minus sign on the right side. For a more detailed presentation of the different conventions see [25]. $\square$

To uncover further symmetries of the Riemann tensor, we choose a local inertial system. We know from (19.18), that there the first derivatives of the metric and thus the Christoffel symbols are equal to zero, i.e., in the local inertial system

$$R^{\mu}{}_{\rho\nu\lambda} = \Gamma^{\mu}{}_{\rho\lambda,\nu} - \Gamma^{\mu}{}_{\rho\nu,\lambda}. \tag{21.6}$$

We use (19.16) to express the Christoffel symbols by the metric:

$$\Gamma^{\mu}{}_{\rho\lambda} = \frac{g^{\mu\sigma}}{2}\left(g_{\rho\sigma,\lambda} + g_{\lambda\sigma,\rho} - g_{\rho\lambda,\sigma}\right),$$

and derive using the product rule

$$
\begin{aligned}
\Gamma^{\mu}{}_{\rho\lambda,\nu} &= \frac{g^{\mu\sigma}{}_{,\nu}}{2}\left(g_{\rho\sigma,\lambda}+g_{\lambda\sigma,\rho}-g_{\rho\lambda,\sigma}\right)+\frac{g^{\mu\sigma}}{2}\left(g_{\rho\sigma,\lambda\nu}+g_{\lambda\sigma,\rho\nu}-g_{\rho\lambda,\sigma\nu}\right) \\
&= \frac{g^{\mu\sigma}}{2}\left(g_{\rho\sigma,\lambda\nu}+g_{\lambda\sigma,\rho\nu}-g_{\rho\lambda,\sigma\nu}\right),
\end{aligned}
$$

since in the local inertial system $g^{\mu\sigma}{}_{,\nu}=0$. Similarly follows

$$
\Gamma^{\mu}{}_{\rho\nu,\lambda}=\frac{g^{\mu\sigma}}{2}\left(g_{\rho\sigma,\nu\lambda}+g_{\nu\sigma,\rho\lambda}-g_{\rho\nu,\sigma\lambda}\right)
$$

and thus for the Riemann tensor in the local inertial system

$$
\begin{aligned}
R^{\mu}{}_{\rho\nu\lambda} &= \frac{g^{\mu\sigma}}{2}\left(g_{\rho\sigma,\lambda\nu}+g_{\lambda\sigma,\rho\nu}-g_{\rho\lambda,\sigma\nu}-\left(g_{\rho\sigma,\nu\lambda}+g_{\nu\sigma,\rho\lambda}-g_{\rho\nu,\sigma\lambda}\right)\right) \\
&= \frac{g^{\mu\sigma}}{2}\left(g_{\lambda\sigma,\rho\nu}-g_{\rho\lambda,\sigma\nu}-g_{\nu\sigma,\rho\lambda}+g_{\rho\nu,\sigma\lambda}\right), \tag{21.7}
\end{aligned}
$$

since according to Schwarz's theorem for the second derivatives

$$
g_{\rho\sigma,\lambda\nu}=g_{\rho\sigma,\nu\lambda}
$$

applies. We form (still in the local inertial system) from the Riemannian tensor and the metric the so-called **fully covariant Riemannian tensor**:

$$
R_{\mu\rho\nu\lambda}=g_{\mu\tau}\,R^{\tau}{}_{\rho\nu\lambda}
$$

and obtain

$$
\begin{aligned}
R_{\mu\rho\nu\lambda} &= g_{\mu\tau}\,R^{\tau}{}_{\rho\nu\lambda}=g_{\mu\tau}\,\frac{g^{\tau\sigma}}{2}\left(g_{\lambda\sigma,\rho\nu}-g_{\rho\lambda,\sigma\nu}-g_{\nu\sigma,\rho\lambda}+g_{\rho\nu,\sigma\lambda}\right) \\
&= \frac{1}{2}\left(g_{\lambda\mu,\rho\nu}-g_{\rho\lambda,\mu\nu}-g_{\nu\mu,\rho\lambda}+g_{\rho\nu,\mu\lambda}\right). \tag{21.8}
\end{aligned}
$$

From this representation we read off the above and further symmetries:

1.
$$
R_{\mu\rho\nu\lambda}=-R_{\mu\rho\lambda\nu}
$$

2.
$$
R_{\mu\rho\nu\lambda}=-R_{\rho\mu\nu\lambda}
$$

3.
$$
R_{\mu\rho\nu\lambda}=R_{\nu\lambda\mu\rho}
$$

4.
$$
R_{\mu\rho\nu\lambda}+R_{\mu\lambda\rho\nu}+R_{\mu\nu\lambda\rho}=0 \tag{21.9}
$$

The fourth equation is obtained by substituting (21.8) and simplifying. Since the four equations hold in the local inertial system and are tensor equations, they hold in all coordinate systems. $R_{\mu\rho\nu\lambda}$ is thus antisymmetric in the first and in the second pair of indices and symmetric in the exchange of the two pairs. The equations (1.) and (2.) thus state that there are six independent components, i.e., numbers, in the two pairs of indices. We can thus write

$$R_{\mu\rho\nu\lambda} = R_{AB},$$

where A denotes the first pair of indices and B the second pair of indices. R_{AB} is thus a (6×6)-matrix, which according to the third equation is symmetric, i.e.,

$$R_{AB} = R_{BA}.$$

A symmetric (6×6)-matrix has 21 independent components, since the elements below the main diagonal are created by reflection along the main diagonal. If we know the elements above and on the main diagonal, we know all $6 \times 6 = 36$ elements of the symmetric matrix. We calculate again: Six (on the main diagonal) plus five (on the diagonal above the main diagonal) plus four (on the one above that) plus three ditto plus two plus one, adds up to 21. The equation (4.) reduces this number by one condition, so that there are ultimately instead of computationally 256 only 20 independent components of the Riemann tensor in four dimensions.

We use Eq. (21.8) to show another property of the Riemann tensor. We calculate - still in the local inertial system - the expression

$$
\begin{aligned}
R_{\mu\rho\nu\lambda,\sigma} + R_{\mu\rho\sigma\nu,\lambda} + R_{\mu\rho\lambda\sigma,\nu} \;=\; & \frac{1}{2}\left(g_{\lambda\mu,\rho\nu\sigma} - g_{\rho\lambda,\mu\nu\sigma} - g_{\nu\mu,\rho\lambda\sigma} + g_{\rho\nu,\mu\lambda\sigma}\right) \\
+\; & \frac{1}{2}\left(g_{\nu\mu,\rho\sigma\lambda} - g_{\rho\nu,\mu\sigma\lambda} - g_{\sigma\mu,\rho\nu\lambda} + g_{\rho\sigma,\mu\nu\lambda}\right) \\
+\; & \frac{1}{2}\left(g_{\sigma\mu,\rho\lambda\nu} - g_{\rho\sigma,\mu\lambda\nu} - g_{\lambda\mu,\rho\sigma\nu} + g_{\rho\lambda,\mu\sigma\nu}\right).
\end{aligned}
$$

If we consider the symmetry of the metric tensor and the second partial derivatives, all terms on the right side cancel out. Since we are in the local inertial system and there the partial derivative coincides with the covariant one, we get

$$R_{\mu\rho\nu\lambda;\sigma} + R_{\mu\rho\sigma\nu;\lambda} + R_{\mu\rho\lambda\sigma;\nu} = 0, \tag{21.10}$$

and since this is a tensor equation, it holds in all coordinate systems. The equation is called the **second Bianchi identity**. The so-called **first Bianchi identity** is the equation (4.) from above, i.e.,

$$R_{\mu\rho\nu\lambda} + R_{\mu\lambda\rho\nu} + R_{\mu\nu\lambda\rho} = 0. \tag{21.11}$$

21.4. Ricci Tensor and Curvature Scalar

From the Riemann tensor two more tensors can be obtained by special contractions, which play a significant role in the theory of general relativity. First, the so-called **Ricci tensor** $R_{\mu\nu}$ is formed by

$$R_{\mu\nu} = R^{\rho}{}_{\mu\rho\nu} = g^{\rho\sigma} R_{\sigma\mu\rho\nu}. \tag{21.12}$$

The Ricci tensor is of rank $(0,2)$ and symmetric, since

$$R_{\nu\mu} = g^{\rho\sigma} R_{\sigma\nu\rho\mu} \underset{(3.)}{=} g^{\sigma\rho} R_{\rho\mu\sigma\nu} = R^{\sigma}{}_{\mu\sigma\nu} = R_{\mu\nu},$$

where we have used equation (3.) from (21.9) and the symmetry of the metric tensor. The components of the Ricci tensor thus form a symmetric (4×4)-matrix, so it has $4+3+2+1 = 10$ independent components. Other contractions of the Riemannian tensor make no sense, since

$$R^{\rho}{}_{\rho\mu\nu} = g^{\rho\sigma} R_{\sigma\rho\mu\nu}$$

all terms in the double sum cancel each other out due to

$$R_{\sigma\rho\mu\nu} = -R_{\rho\sigma\mu\nu}.$$

Because for all values that σ and ρ can take, the sum of the two terms yields

$$g^{\rho\sigma} R_{\sigma\rho\mu\nu} + g^{\sigma\rho} R_{\rho\sigma\mu\nu} = g^{\rho\sigma} \left(R_{\sigma\rho\mu\nu} + R_{\rho\sigma\mu\nu} \right) = g^{\rho\sigma} \left(R_{\sigma\rho\mu\nu} - R_{\sigma\rho\mu\nu} \right) = 0.$$

This is a general tensor law: if you contract a symmetric (here $g^{\rho\sigma}$) with an antisymmetric (here $R_{\sigma\rho\mu\nu}$ in the first two indices) tensor, the result is always zero. That means, we also get for the third possibility of contraction of the Riemann tensor

$$R^{\rho}{}_{\mu\nu\rho} = g^{\rho\sigma} R_{\mu\nu\sigma\rho} = 0.$$

The Ricci tensor can also be expressed by the Christoffel symbols, for this we use the formula (21.5)

$$R_{\mu\nu} = R^{\rho}{}_{\mu\rho\nu} = \Gamma^{\rho}_{\mu\nu,\rho} - \Gamma^{\rho}_{\mu\rho,\nu} + \Gamma^{\rho}_{\sigma\rho} \Gamma^{\sigma}_{\mu\nu} - \Gamma^{\rho}_{\sigma\nu} \Gamma^{\sigma}_{\mu\rho}. \tag{21.13}$$

The **curvature scalar** R is the contraction of the Ricci tensor with the metric

$$R = g^{\mu\nu} R_{\mu\nu}. \tag{21.14}$$

By construction, the curvature scalar is indeed a scalar and therefore takes the same value in every coordinate system, which the Riemannian and Ricci tensors do not do.

As examples, we calculate the curvatures of the two-dimensional plane with polar coordinates as well as the surface of the sphere.

Example 21.1.
First, we note that the indices in the two-dimensional plane can only take the values x and y. Due to the antisymmetry of the Riemannian tensor in the first two and last two indices the only non-zero components of the Riemannian tensor in two dimensions are those whose indices in the two index pairs are different, i.e., only four remain, namely $R_{xyxy}, R_{yxxy}, R_{xyyx}$ and R_{yxyx}. But with the symmetry properties of the Riemannian tensor we get:

$$\begin{aligned}
R_{xyxy} &= -R_{yxxy} \\
R_{xyxy} &= -R_{xyyx} \\
R_{xyxy} &= R_{yxyx},
\end{aligned}$$

so that the Riemannian tensor is determined by a single quantity, e.g. R_{xyxy}. We first consider polar coordinates and again set $\mu = r, \varphi$. The line element in polar coordinates was

$$ds^2 = dr^2 + r^2 d\varphi^2.$$

The Christoffel symbols are with Eq. (8.36):

$$\begin{aligned}
\Gamma^{\varphi}_{r\varphi} &= \Gamma^{\varphi}_{\varphi r} = \frac{1}{r} \\
\Gamma^{r}_{\varphi\varphi} &= -r,
\end{aligned}$$

all other Christoffel symbols are equal to zero. Now we calculate the Riemann tensor with Eq. (21.5)

$$\begin{aligned}
R^{r}{}_{\varphi r\varphi} &= \Gamma^{r}_{\varphi\varphi,r} - \Gamma^{r}_{\varphi r,\varphi} + \Gamma^{r}_{\sigma r}\,\Gamma^{\sigma}_{\varphi\varphi} - \Gamma^{r}_{\sigma\varphi}\,\Gamma^{\sigma}_{\varphi r} \\
&= \Gamma^{r}_{\varphi\varphi,r} - \Gamma^{r}_{\varphi r,\varphi} + \Gamma^{r}_{rr}\,\Gamma^{r}_{\varphi\varphi} + \Gamma^{r}_{\varphi r}\,\Gamma^{\varphi}_{\varphi\varphi} - \Gamma^{r}_{r\varphi}\,\Gamma^{r}_{\varphi r} - \Gamma^{r}_{\varphi\varphi}\,\Gamma^{\varphi}_{\varphi r} \\
&= \frac{\partial(-r)}{\partial r} - 0 + 0 + 0 - 0 - (-r)\cdot\frac{1}{r} = -1 + 1 = 0.
\end{aligned}$$

The Riemannian tensor is equal to zero and thus the space under consideration is flat, which we already know. The Ricci tensor and the curvature scalar are also equal to zero. $\square$

Example 21.2.
The surface of a sphere with radius a has according to Eq. (20.9) with $\mu = \vartheta, \varphi$ the line element

$$ds^2 = a^2\, d\vartheta^2 + a^2 \sin^2 \vartheta\, d\varphi^2,$$

the metric

$$(g_{\mu\nu}) = \begin{pmatrix} a^2 & 0 \\ 0 & a^2 \sin^2 \vartheta \end{pmatrix},$$

and the inverse metric

$$(g^{\mu\nu}) = \begin{pmatrix} \dfrac{1}{a^2} & 0 \\ 0 & \dfrac{1}{a^2 \sin^2 \vartheta} \end{pmatrix}$$

as well as according to Eq. (20.10) the Christoffel symbols

$$\Gamma^{\varphi}_{\vartheta\varphi} = \Gamma^{\varphi}_{\varphi\vartheta} = \cot \vartheta$$
$$\Gamma^{\vartheta}_{\varphi\varphi} = -\sin \vartheta \cos \vartheta,$$

all other Christoffel symbols vanish. Thus we obtain

$$\begin{aligned}
R^{\vartheta}_{\varphi\vartheta\varphi} &= \Gamma^{\vartheta}_{\varphi\varphi,\vartheta} - \Gamma^{\vartheta}_{\varphi\vartheta,\varphi} + \Gamma^{\vartheta}_{\sigma\vartheta}\,\Gamma^{\sigma}_{\varphi\varphi} - \Gamma^{\vartheta}_{\sigma\varphi}\,\Gamma^{\sigma}_{\varphi\vartheta} \\
&= \Gamma^{\vartheta}_{\varphi\varphi,\vartheta} - \Gamma^{\vartheta}_{\varphi\vartheta,\varphi} + \Gamma^{\vartheta}_{\vartheta\vartheta}\,\Gamma^{\vartheta}_{\varphi\varphi} + \Gamma^{\vartheta}_{\varphi\vartheta}\,\Gamma^{\varphi}_{\varphi\varphi} \\
&\quad - \Gamma^{\vartheta}_{\vartheta\varphi}\,\Gamma^{\vartheta}_{\varphi\vartheta} - \Gamma^{\vartheta}_{\varphi\varphi}\,\Gamma^{\varphi}_{\varphi\vartheta} \\
&= \frac{\partial(-\sin \vartheta \cos \vartheta)}{\partial\vartheta} - 0 + 0 + 0 - 0 - (-\sin \vartheta \cos \vartheta) \cdot \cot \vartheta \\
&= -\cos^2 \vartheta + \sin^2 \vartheta + \cos^2 \vartheta = \sin^2 \vartheta.
\end{aligned}$$

Therefore, the Riemannian tensor is not equal to zero, which means the space is curved. At the North and South poles, $\sin^2 \vartheta = 0$, but this is not an intrinsic property, but depends on how we have chosen the coordinate system. The surface of the sphere is homogeneous, and the North and South poles can be placed anywhere. The Ricci tensor is defined by

$$R_{\mu\nu} = g^{\rho\sigma} R_{\sigma\mu\rho\nu},$$

i.e., we first need to calculate the fully covariant Riemannian tensor

$$R_{\vartheta\varphi\vartheta\varphi} = g_{\vartheta\tau}\,R^{\tau}_{\varphi\vartheta\varphi} = g_{\vartheta\vartheta}\,R^{\vartheta}_{\varphi\vartheta\varphi} + g_{\vartheta\varphi}\,R^{\varphi}_{\varphi\vartheta\varphi} = g_{\vartheta\vartheta}\,R^{\vartheta}_{\varphi\vartheta\varphi} = a^2 \sin^2 \vartheta.$$

From this it follows

$$\begin{aligned}
R_{\vartheta\vartheta} &= g^{\rho\sigma} R_{\sigma\vartheta\rho\vartheta} = g^{\rho\vartheta} R_{\vartheta\vartheta\rho\vartheta} + g^{\rho\varphi} R_{\varphi\vartheta\rho\vartheta} \\
&= g^{\vartheta\vartheta} R_{\vartheta\vartheta\vartheta\vartheta} + g^{\varphi\vartheta} R_{\vartheta\vartheta\varphi\vartheta} + g^{\vartheta\varphi} R_{\varphi\vartheta\vartheta\vartheta} + g^{\varphi\varphi} R_{\varphi\vartheta\varphi\vartheta} \\
&= g^{\varphi\varphi} R_{\varphi\vartheta\varphi\vartheta} = \frac{1}{a^2 \sin^2 \vartheta} \left(a^2 \sin^2 \vartheta\right) = 1
\end{aligned}$$

and similarly

$$R_{\vartheta\varphi} = g^{\rho\sigma} R_{\sigma\vartheta\rho\varphi} = g^{\vartheta\vartheta} R_{\vartheta\vartheta\vartheta\varphi} + g^{\varphi\vartheta} R_{\vartheta\vartheta\varphi\varphi} + g^{\vartheta\varphi} R_{\varphi\vartheta\vartheta\varphi} + g^{\varphi\varphi} R_{\varphi\vartheta\varphi\varphi} = 0$$

$$
\begin{aligned}
&= R_{\varphi\vartheta} \\
R_{\varphi\varphi} &= g^{\rho\sigma} R_{\sigma\varphi\rho\varphi} = g^{\vartheta\vartheta} R_{\vartheta\varphi\vartheta\varphi} + g^{\varphi\vartheta} R_{\vartheta\varphi\varphi\varphi} + g^{\vartheta\varphi} R_{\varphi\varphi\vartheta\varphi} + g^{\varphi\varphi} R_{\varphi\varphi\varphi\varphi} \\
&= g^{\vartheta\vartheta} R_{\vartheta\varphi\vartheta\varphi} = \frac{1}{a^2} \left(a^2 \sin^2 \vartheta \right) = \sin^2 \vartheta.
\end{aligned}
$$

The curvature scalar is obtained with

$$
R = g^{\vartheta\vartheta} R_{\vartheta\vartheta} + g^{\varphi\varphi} R_{\varphi\varphi} = \frac{1}{a^2} + \frac{\sin^2 \vartheta}{a^2 \sin^2 \vartheta} = \frac{2}{a^2}.
$$

It is independent of the coordinate system and also independent of the position on the surface of the sphere. The surface of the sphere is equally curved everywhere, thus it is *homogeneously* curved. The curvature scalar is inversely proportional to the square of the radius a^2 of the sphere, which corresponds to our intuition. A spherical surface is more strongly curved the smaller the radius of the sphere is. $\square$

22. Riemannian Space and Einstein Equations

In the last chapters we have worked out, that there is no global inertial system in the presence of a non-homogeneous gravitational field and that spacetime is curved when gravity is effective. We have derived the quantities needed to describe the curvature of spacetime using mathematical arguments, and we want to focus more on the physical laws of spacetime in this chapter. We need to investigate how physical objects (such as particles or fluids) behave in curved spacetime, and find out how, conversely, the curvature of spacetime is generated and determined by these objects.

22.1. Principle of Covariance

The principle of covariance is a method for deriving physical laws with gravity from known laws of special relativity. We have already applied the procedure prescribed in this method several times in the last chapters: First, the physical phenomena are described in the local inertial system and then transformed into the actually used coordinate system by coordinate transformation. We also already know how to formulate equations so that they are form-invariant (i.e. *"covariant"*) in all coordinate systems, namely by using only tensors and covariant derivatives of these tensors in the equations. We can therefore define the principle of covariance for the equations valid in the gravitational field by the following two statements:

- The physical law must be in the form of a tensor equation, making it form-invariant / covariant in all coordinate systems.

- The law must be valid in the local inertial system. That is, if we replace the general metric $g_{\mu\nu}$ with the Minkowski metric $\eta_{\mu\nu}$, the corresponding law of special relativity must result.

Compared to the procedure in section 18.5, which was illustrated by Fig. 22.1 on the following page,

© The Author(s), under exclusive license to Springer-Verlag GmbH, DE, part of Springer Nature 2026
M. Ruhrländer, *Ascent to the Einstein Equations*,
https://doi.org/10.1007/978-3-662-72672-3_22

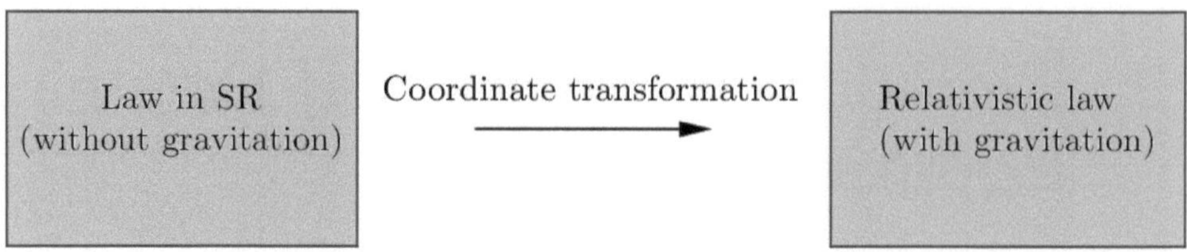

Figure 22.1.: Principle of Covariance

we can thus save ourselves the (often tedious) transformation of an SR equation by a general coordinate transformation by writing the SR law directly in covariant form. After that, it no longer changes its form by coordinate transformations, it is already the sought-after relativistic law with gravity.

As the first exemplary physical application of the principle of covariance, we consider the motion of a particle in the gravitational field, which we have already studied in Chap. 20. There we derived in Eq. (20.1) that in the local inertial system with the Minkowski coordinates $\hat{\alpha}$ and the four-velocity $u^{\hat{\alpha}}$ as well as the proper time t_E

$$\frac{du^{\hat{\alpha}}}{dt_E} = \frac{d^2\hat{\alpha}}{dt_E^2} = 0$$

applies. According to the principle of covariance, we are looking for a covariant equation that reduces to the above equation for $g_{\mu\nu} = \eta_{\mu\nu}$. If we define the general vector $\vec{U} \to U^\mu$ belonging to the four-velocity $u^{\hat{\alpha}}$ by

$$U^\mu = \frac{\partial\mu}{\partial\hat{\alpha}}\, u^{\hat{\alpha}} = \frac{\partial\mu}{\partial\hat{\alpha}}\frac{\partial\hat{\alpha}}{\partial t_E} = \frac{\partial\mu}{\partial t_E} = \frac{d\mu}{dt_E}, \tag{22.1}$$

where μ denotes any time-dependent coordinate system, we must covariantly derive $\vec{U}$ to set up the motion equation. The components of the covariant derivative are defined by

$$U^\mu_{\;;\nu} = \frac{\partial U^\mu}{\partial\nu} + \Gamma^\mu_{\sigma\nu} U^\sigma.$$

Contracting this with U^ν, we obtain the components of a (general) vector

$$U^\mu_{\;;\nu} U^\nu = U^\mu_{\;;\nu}\frac{d\nu}{dt_E} = \frac{\partial U^\mu}{\partial\nu}\frac{d\nu}{dt_E} + \Gamma^\mu_{\sigma\nu} U^\sigma \frac{d\nu}{dt_E} = \frac{\partial U^\mu}{dt_E} + \Gamma^\mu_{\sigma\nu}\frac{d\sigma}{dt_E}\frac{d\nu}{dt_E}.$$

We denote this vector with $D\vec{U}/dt_E$ and obtain for its components

$$\frac{DU^\mu}{dt_E} = \frac{\partial U^\mu}{dt_E} + \Gamma^\mu_{\sigma\nu}\frac{d\sigma}{dt_E}\frac{d\nu}{dt_E} = \frac{\partial^2\mu}{dt_E^2} + \Gamma^\mu_{\sigma\nu}\frac{d\sigma}{dt_E}\frac{d\nu}{dt_E},$$

so that as a covariant generalization of Eq. (20.1)

$$\frac{DU^\mu}{dt_E} = 0 \qquad (22.2)$$

is suggested. Of course, from this equation we again obtain the geodesic equation in the gravitational field

$$\frac{d^2\mu}{dt_E^2} = -\Gamma^\mu_{\sigma\nu}\frac{d\sigma}{dt_E}\frac{d\nu}{dt_E},$$

which we obtained in Chap. 20 by explicitly inserting a coordinate transformation. We can therefore state that the covariance principle often significantly simplifies the formulation of laws with gravity. And because the covariance principle is so important, we formulate it again in a slightly different way:

> Laws in the gravitational field $g_{\mu\nu}$ are covariant equations, which without a gravitational field, i.e., when $g_{\mu\nu} = \eta_{\mu\nu}$ applies, reduce to the laws of Special Relativity.

However, one must not forget that the covariance principle does not result from a strict mathematical derivation, but rather is based on physical experiences in experiments or astronomical observations. Let's take another vector $\vec{V}$, whose divergence in the co-moving inertial system is zero

$$V^\alpha_{,\alpha} = 0.$$

A *possible* generalization of this equation in curved spacetime could be the equation

$$V^\alpha_{;\alpha} = R,$$

where R denotes the curvature scalar. Because in the local inertial system the curvature scalar disappears and the covariant derivative is equal to the partial. The covariance principle, however, demands as the *"correct"* generalization the simplest case, and that is

$$V^\alpha_{;\alpha} = 0,$$

which was proven by physical experiments. We subsequently assume the validity of the covariance principle, especially since there are no indications, let alone physical experiments, that question the covariance principle.

Energy-Momentum Tensors in Riemannian Space

In Chap. 16 we examined energy-momentum tensors in Special Relativity and in particular proved the energy, momentum and particle number conservation laws by the divergence-free nature of the corresponding tensors for ideal fluids and dust. We now want to use the covariance principle to derive the corresponding generalizations. To do this, we reconsider Eq. (16.5) and (16.10), where we use small letters u^α for the components of the four-velocity in the co-moving inertial system:

$$N^\alpha_{\ ,\beta} = (nu^\alpha)_{,\alpha} = 0$$

and

$$T^{\alpha\beta}_{\ \ ,\beta} = \left[(\rho + p)u^\alpha u^\beta + p\eta^{\alpha\beta}\right]_{,\beta} = 0$$

The $T^{\alpha\beta}$ are the components of the energy-momentum tensor for fluids (and with $p = 0$ also for dust) and the N^α are the components of the particle flow vector. As in Eq. (22.1) we define the general vector of the four-velocity $\vec{U}$ in Riemannian space by

$$U^\mu = \frac{\partial\mu}{\partial\hat\alpha}u^{\hat\alpha} = \frac{\partial\mu}{\partial\hat\alpha}\frac{d\hat\alpha}{dt_E} = \frac{d\mu}{dt_E}$$

and obtain by applying the covariance principle as covariant equations

$$T^{\alpha\beta}_{\ \ ;\beta} = \left[(\rho + p)U^\alpha U^\beta + pg^{\alpha\beta}\right]_{;\beta} = 0$$

and

$$N^\alpha_{\ ;\beta} = (nU^\alpha)_{;\alpha} = 0.$$

Note that we have replaced the Minkowski metric $\eta^{\alpha\beta}$ with the general metric $g^{\alpha\beta}$. We thus also obtain the covariant energy-momentum tensor, whose components are given by

$$T^{\alpha\beta} = (\rho + p)U^\alpha U^\beta + pg^{\alpha\beta}. \tag{22.3}$$

22.2. Newtonian Limit Case

We want to show that the equation of motion (20.2) in the limit case of small velocities and a weak, static gravitational field reduces to the Newtonian equation of motion. This is given by formula (6.6):

$$m\ddot{\vec{r}} = -m\nabla\phi(\vec{r})$$

or in coordinate form after division by the mass m

$$\frac{d^2 i\,(t)}{dt^2} = -\frac{\partial\phi}{\partial i}, \tag{22.4}$$

where ϕ denotes the Newtonian gravitational potential and the index i can take the values x, y, z. Small velocity means that in natural units

$$|\vec{v}| \ll 1$$

and therefore the components of the four-velocity (see 14.8 on page 296):

$$U^\mu = \frac{\partial \mu}{\partial t_E} = \left(\frac{1}{\sqrt{1-v^2}}, \frac{v_x}{\sqrt{1-v^2}}, \frac{v_y}{\sqrt{1-v^2}}, \frac{v_z}{\sqrt{1-v^2}} \right)$$

satisfy the conditions

$$\left| \frac{di}{dt_E} \right| \ll \frac{dt}{dt_E}, \quad i = x, y, z$$

and

$$\frac{dt}{dt_E} = \frac{1}{\sqrt{1-v^2}} \approx 1.$$

From the last equation it also follows that

$$\frac{d\mu}{dt_E} = \frac{d\mu}{dt} \frac{dt}{dt_E} \approx \frac{d\mu}{dt}. \tag{22.5}$$

The assumption of weak fields means that the metric $g_{\mu\nu}$ does not deviate much from the Minkowski metric $\eta_{\mu\nu}$. We make the assumption

$$g_{\mu\nu} = \eta_{\mu\nu} + h_{\mu\nu}$$

with very small static $h_{\mu\nu}$, i.e.

$$|h_{\mu\nu}| \ll 1,$$

whose derivatives and thus also the Christoffel symbols may also be small. The coordinates α, γ, μ, ν are therefore, except for small deviations, the Minkowski coordinates and the geodesic equation reduces with (22.5) to

$$\frac{d^2\gamma}{dt_E^2} = -\Gamma^\gamma_{\mu\nu} \frac{d\mu}{dt_E} \frac{d\nu}{dt_E} \approx -\Gamma^\gamma_{tt} \frac{dt}{dt_E} \frac{dt}{dt_E} \approx -\Gamma^\gamma_{tt} \frac{dt}{dt} \frac{dt}{dt} = -\Gamma^\gamma_{tt}.$$

Now the components of $g_{\mu\nu}$ are temporally constant due to the assumed statics, so the temporal derivatives are zero, so with Eq. (19.16)

$$\Gamma^\gamma_{tt} = \frac{g^{\gamma\nu}}{2} \left(g_{t\nu,t} + g_{t\nu,t} - g_{tt,\nu} \right) \approx -\frac{\eta^{\gamma\nu}}{2} g_{tt,\nu} = -\frac{\eta^{\gamma\nu}}{2} h_{tt,\nu}$$

follows. In the last equation we have used

$$g_{tt,\nu} = (\eta_{tt} + h_{tt})_{,\nu} = h_{tt,\nu}.$$

The temporal derivative of h_{tt} is also zero and we obtain

$$
\begin{aligned}
\Gamma^x_{tt} &\approx -\frac{1}{2}\frac{\partial h_{tt}}{\partial x}\\
\Gamma^y_{tt} &\approx -\frac{1}{2}\frac{\partial h_{tt}}{\partial y}\\
\Gamma^z_{tt} &\approx -\frac{1}{2}\frac{\partial h_{tt}}{\partial z}\\
\Gamma^t_{tt} &= 0.
\end{aligned}
$$

In summary, for the spatial components of the particle path

$$
\frac{d^2\gamma}{dt^2} \approx \frac{d^2\gamma}{dt_E^2} \approx \frac{1}{2}\,h_{tt,\gamma},
$$

which can also be written as a three-dimensional vector equation in the approximation

$$
\frac{d^2\vec{r}}{dt^2} \to \frac{1}{2}\nabla h_{tt} = \frac{1}{2}\left(\frac{\partial h_{tt}}{\partial x},\frac{\partial h_{tt}}{\partial y},\frac{\partial h_{tt}}{\partial z}\right).
$$

This equation agrees for

$$
h_{tt} = -2\phi \tag{22.6}
$$

with the Newtonian equation of motion (22.4). If one also considers that

$$
g_{tt} = \eta_{tt} + h_{tt} = -(1+2\phi) \tag{22.7}
$$

applies, one obtains between the "time-time" component of the metric and the gravitational potential ϕ the same relationship as in Eq. (18.4) with the centrifugal potential. At a distance r from a body of mass M is

$$
\phi = -\frac{GM}{r},
$$

see formula 6.7 on page 89, i.e.

$$
g_{tt} = -\left(1 - \frac{2GM}{r}\right) \tag{22.8}
$$

and

$$
ds^2 = -\left(1 - \frac{2GM}{r}\right)dt^2 + \cdots .
$$

So we have found *one* component of the metric and that is all we can find when we compare the Newtonian theory with the Einsteinian, since the Newtonian gravitation is described by only *one* function, namely the gravitational potential ϕ.

22.3. Einstein's Field Equations

In this section, we establish the laws for the description of the gravitational field, i.e., we show how the sources of the gravitational field determine the metric. These field equations cannot be derived from Newtonian physics or from Special Relativity, for example, by the principle of covariance. Rather, they were found and postulated by Einstein after ten years of contemplation. The validity of General Relativity, i.e., the correctness of the field equations, must be confirmed by experiments and observations, like any other physical theory.

Even though the field equations came about by Einstein's creative insight, they can be understood and made plausible in retrospect. So let's imagine that we had the task of finding the general equations for the description of a gravitational field. Then it would not be unreasonable if we made three demands on these equations:

1. The equations should be covariant, i.e., they should be able to be set up as general tensor equations. This would give them the same form in any coordinate system.

2. In the Newtonian limit case (weak, static gravitational field, small velocities), the equations should reduce to Newton's law of gravitation, specifically to the Poisson equation (6.29)

$$\Delta \phi = 4\pi G \rho$$

   and thus contain the proven Newtonian theory of gravitation as a special case.

3. The equations should also be as simple as possible. This demand corresponds to a general scientific principle ("Ockham's law"), which states that the simplest theory is to be preferred over all others when there are several possible explanations for the same facts.

Let's first look at the right side of the Poisson equation. The source of the gravitational field in Newton's theory is the mass density ρ. In the sought-after relativistic theory, the sources must correspond to a relativistic approach, which is not given by mass alone. An obvious relativistic generalization is the total energy (see Eq. 12.18 on page 257), which includes the rest mass. In Sec. 16.2, we denoted the energy density of an ideal fluid in the current inertial system (also) with ρ, and it is obvious to use this as the source of the gravitational field. But this approach also fails, since ρ only represents the energy density for the co-moving observer, other observers measure the energy density as the component T^{tt} of the energy-momentum tensor in their own reference system (see Eq. 16.7 on page 338). So we discard the approach with the energy density ρ and consider T^{tt} as a generalization of Newtonian mass density. Now T^{tt} is only *one* component of the energy-momentum tensor, according to demand (1.) the equations should consist of (complete) tensors. Therefore, we choose for the right side of the sought equations the approach

$$k\,T^{\mu\nu},$$

with a constant k to be determined, where the $T^{\mu\nu}$ are the components of the energy-momentum tensor **T**.

When we now look at the left side of the Poisson equation, we find that there are second derivatives of the gravitational potential ϕ because of

$$\Delta\phi = \frac{\partial^2\phi}{\partial x^2} + \frac{\partial^2\phi}{\partial y^2} + \frac{\partial^2\phi}{\partial z^2}.$$

In the Newtonian limit case, according to Eq. (22.8)

$$g_{tt} = -(1 + 2\phi) \Rightarrow \phi = \frac{1}{2}\left(-1 - g_{tt}\right)$$

and thus

$$
\begin{aligned}
\Delta\phi &= -\frac{1}{2}\left(\frac{\partial^2(1+g_{tt})}{\partial x^2} + \frac{\partial^2(1+g_{tt})}{\partial y^2} + \frac{\partial^2(1+g_{tt})}{\partial z^2}\right) \\
&= -\frac{1}{2}\left(\frac{\partial^2 g_{tt}}{\partial x^2} + \frac{\partial^2 g_{tt}}{\partial y^2} + \frac{\partial^2 g_{tt}}{\partial z^2}\right) = -\frac{1}{2}\Delta g_{tt}
\end{aligned}
$$

holds, it follows from the Poisson equation

$$-\Delta g_{tt} = 8\pi G\rho.$$

From this we can derive two things. Firstly, for the constant k on the right side of our approach

$$k = 8\pi G,$$

and secondly, on the left side of the equation in the Newtonian limit, second derivatives of the metric must occur, specifically the tt-component of the sought $(2,0)$-tensor on the left side must be $-\Delta g_{tt}$. Now we know that the Ricci tensor (21.12) consists of second derivatives of the metric, so we now choose the simplest approach according to requirement (3.)

$$R^{\mu\nu} = 8\pi G\,T^{\mu\nu}.$$

To verify this approach, we remember that the (covariant) divergence of the energy-momentum tensor vanishes

$$T^{\mu\nu}_{;\nu} = 0,$$

i.e., if our approach is correct, the divergence of the Ricci tensor must also vanish. To calculate this, we start from the second Bianchi identity (21.10)

$$R_{\mu\rho\nu\lambda;\sigma} + R_{\mu\rho\sigma\nu;\lambda} + R_{\mu\rho\lambda\sigma;\nu} = 0$$

and multiply this equation by $g^{\mu\nu}$

$$g^{\mu\nu} R_{\mu\rho\nu\lambda;\sigma} + g^{\mu\nu} R_{\mu\rho\sigma\nu;\lambda} + g^{\mu\nu} R_{\mu\rho\lambda\sigma;\nu} = 0.$$

According to the definition of the Ricci tensor (21.12) the first term results in

$$g^{\mu\nu} R_{\mu\rho\nu\lambda;\sigma} = R_{\rho\lambda;\sigma},$$

since due to $g^{\mu\nu}_{;\sigma} = 0$ with the product rule

$$R_{\rho\lambda;\sigma} = (g^{\mu\nu} R_{\mu\rho\nu\lambda})_{;\sigma} = g^{\mu\nu}_{;\sigma} R_{\mu\rho\nu\lambda} + g^{\mu\nu} R_{\mu\rho\nu\lambda;\sigma} = g^{\mu\nu} R_{\mu\rho\nu\lambda;\sigma}$$

follows. Due to the antisymmetry of the Riemannian tensor in the last two indices, the second term also results in

$$g^{\mu\nu} R_{\mu\rho\sigma\nu;\lambda} = -g^{\mu\nu} R_{\mu\rho\nu\sigma;\lambda} = -R_{\rho\sigma;\lambda}.$$

The third term can be written as

$$g^{\mu\nu} R_{\mu\rho\lambda\sigma;\nu} = R^{\nu}_{\rho\lambda\sigma;\nu},$$

so that in total

$$R_{\rho\lambda;\sigma} - R_{\rho\sigma;\lambda} + R^{\nu}_{\rho\lambda\sigma;\nu} = 0$$

follows. We multiply this equation by $g^{\rho\lambda}$ and obtain with the definition of the curvature scalar (21.14):

$$R = g^{\rho\lambda} R_{\rho\lambda}$$

for the first term

$$g^{\rho\lambda} R_{\rho\lambda;\sigma} = R_{;\sigma},$$

for the second

$$-g^{\rho\lambda} R_{\rho\sigma;\lambda} = -R^{\lambda}_{\sigma;\lambda}$$

and again due to the antisymmetry of the Riemannian tensor in the last two indices for the third term

$$g^{\rho\lambda} R^{\nu}_{\rho\lambda\sigma;\nu} = -g^{\rho\lambda} R^{\nu}_{\rho\lambda\sigma;\nu} = -g^{\rho\lambda} g^{\nu\tau} R_{\tau\rho\sigma\lambda;\nu}.$$

We now swap both the first and the second pairs of indices of the Riemann tensor and obtain

$$
\begin{aligned}
-g^{\rho\lambda}g^{\nu\tau}R_{\tau\rho\sigma\lambda;\nu} &= -g^{\nu\tau}g^{\rho\lambda}R_{\rho\tau\lambda\sigma;\nu} = -g^{\nu\tau}R^{\lambda}{}_{\tau\lambda\sigma;\nu} \\
&= -g^{\nu\tau}R_{\tau\sigma;\nu} = -R^{\nu}{}_{\sigma;\nu} = -R^{\lambda}{}_{\sigma;\lambda},
\end{aligned}
$$

where in the last equation the summation index was swapped $\lambda \leftrightarrow \nu$. So we get in total

$$
R_{;\sigma} - R^{\lambda}{}_{\sigma;\lambda} - R^{\lambda}{}_{\sigma;\lambda} = R_{;\sigma} - 2R^{\lambda}{}_{\sigma;\lambda} = 0
$$

and because

$$
R_{;\sigma} = \delta^{\lambda}_{\sigma}\, R_{;\lambda}
$$

from this after multiplication with $g^{\tau\sigma}$

$$
R^{\tau\lambda}{}_{;\lambda} = g^{\tau\sigma}R^{\lambda}{}_{\sigma;\lambda} = \frac{1}{2}\,g^{\tau\sigma}\delta^{\lambda}_{\sigma}\,R_{;\lambda} = \frac{1}{2}\,g^{\tau\lambda}R_{;\lambda}.
$$

The covariant divergence of the Ricci tensor is therefore

$$
R^{\tau\lambda}{}_{;\lambda} = \frac{1}{2}\,g^{\tau\lambda}R_{;\lambda}
$$

and the right side is generally not zero. So we have to extend our approach. If one considers that the covariant derivative of the metric is zero, then for the derivative of the tensor $g^{\tau\lambda}R/2$

$$
\left(\frac{1}{2}\,g^{\tau\lambda}R\right)_{;\lambda} = \frac{1}{2}\,\underbrace{g^{\tau\lambda}{}_{;\lambda}}_{=0}\,R + \frac{1}{2}\,g^{\tau\lambda}R_{;\lambda} = \frac{1}{2}\,g^{\tau\lambda}R_{;\lambda}
$$

and thus with index change $\tau \to \mu, \lambda \to \nu$

$$
\left(R^{\mu\nu} - \frac{1}{2}\,g^{\mu\nu}R\right)_{;\nu} = 0. \tag{22.9}
$$

The (divergence-free) tensor

$$
G^{\mu\nu} = R^{\mu\nu} - \frac{1}{2}\,g^{\mu\nu}R
$$

is called **Einstein tensor**, this forms the left side of **Einstein's field equations**:

$$
\boxed{G^{\mu\nu} = R^{\mu\nu} - \frac{1}{2}\,g^{\mu\nu}R = 8\pi G\,T^{\mu\nu}} \tag{22.10}
$$

These equations were established by Einstein in 1915. They are, together with the equations of motion in the gravitational field (20.2)

$$\frac{d^2\gamma}{dt_E^2} = -\Gamma^\gamma{}_{\mu\nu}\frac{d\mu}{dt_E}\frac{d\nu}{dt_E}$$

the sought-after *fundamental equations of General Relativity*. Because

$$G_{\mu\nu} = g_{\mu\sigma}\,g_{\nu\tau}\,G^{\sigma\tau} = g_{\mu\sigma}\,g_{\nu\tau}\,8\pi G\,T^{\sigma\tau} = 8\pi G\,T_{\mu\nu}$$

one can also write Einstein's field equations in the form

$$G_{\mu\nu} = R_{\mu\nu} - \frac{1}{2}\,g_{\mu\nu}\,R = 8\pi G\,T_{\mu\nu}, \tag{22.11}$$

which is called the **covariant form**. Due to the symmetry of the Einstein and energy-momentum tensor and the metric initially gives ten instead of 16 independent Einstein's field equations. However, we must consider that the components of the metric depend on the chosen coordinate system. Since there are four coordinates, there are also four degrees of freedom for the ten $g_{\mu\nu}$, i.e., it is not possible to determine all ten components of the metric from ten given components of the energy-momentum tensor, since the coordinate system should be freely selectable. The problem is solved by the fact that due to the divergence-free nature of the Einstein tensor

$$G^{\mu\nu}{}_{;\nu} = 0$$

there are four (for each value of μ) conditions for the ten Einstein equations, so that ultimately six independent equations for the six components of the metric tensor, which describe the geometry independently of the coordinates, remain.

22.4. Interpretation of the Einstein Equations

We want to make it clear once again whether and how Einstein's field equations fulfill the three requirements set out at the beginning of the last section. First, we note that the Einstein equations are covariant, both the right and the left side of the field equations consist of tensors of the same rank.

From Einstein follows Newton

The second requirement was that the Poisson equation must result from the Einstein equations for the Newtonian limit case. To this end, we first multiply Eq. (22.11) by $g^{\mu\nu}$ and obtain for the left side

$$g^{\mu\nu}R_{\mu\nu} - \frac{1}{2}g^{\mu\nu}g_{\mu\nu}R = R - \frac{1}{2}\delta^\mu_\mu R = R - \frac{1}{2}4R = -R.$$

For the right side, we define T by

$$g^{\mu\nu}\, 8\pi G\, T_{\mu\nu} = 8\pi G\, T$$

and obtain

$$R = -8\pi G\, T.$$

With this, we can rewrite the field equations (22.11) to

$$R_{\mu\nu} = 8\pi G \left(T_{\mu\nu} - \frac{1}{2} g_{\mu\nu}\, T \right). \tag{22.12}$$

We assume as in Sect. 22.2 again that

$$g_{\mu\nu} = \eta_{\mu\nu} + h_{\mu\nu}$$

with very small static $h_{\mu\nu}$, i.e.

$$|h_{\mu\nu}| \ll 1$$

applies. Then follows

$$\left(\eta^{\mu\rho} - h^{\mu\rho}\right)\left(\eta_{\rho\nu} + h_{\rho\nu}\right) = \delta^{\mu}_{\nu} - h^{\mu}{}_{\nu} + h^{\mu}{}_{\nu} - h^{\mu\rho}h_{\rho\nu} \approx \delta^{\mu}_{\nu},$$

in linear approximation, i.e., when the quadratic term $h^{\mu\rho}h_{\rho\nu}$ is neglected. Because

$$\delta^{\mu}_{\nu} = g^{\mu\rho} g_{\rho\nu} = g^{\mu\rho}\left(\eta_{\rho\nu} + h_{\rho\nu}\right)$$

the last two equations in linear approximation result in

$$g^{\mu\nu} = \eta^{\mu\nu} - h^{\mu\nu}. \tag{22.13}$$

We now calculate the right side of (22.12) and remember that according to Eq. (16.11) for the energy-momentum tensor in the Newtonian limit case

$$(T^{\mu\nu}) = \begin{pmatrix} \rho & 0 & 0 & 0 \\ 0 & 0 & 0 & 0 \\ 0 & 0 & 0 & 0 \\ 0 & 0 & 0 & 0 \end{pmatrix} = (T_{\mu\nu})$$

applies. This results in

$$T = g^{\mu\nu}\, T_{\mu\nu} = \left(\eta^{\mu\nu} - h^{\mu\nu}\right) T_{\mu\nu} \approx \eta^{\mu\nu}\, T_{\mu\nu} = \eta^{tt}\, T_{tt} = -\rho$$

and from this

$$T_{\mu\nu} - \frac{1}{2} g_{\mu\nu}\, T \;=\; T_{\mu\nu} - \frac{1}{2}\left(\eta_{\mu\nu} + h_{\mu\nu}\right) T \approx T_{\mu\nu} - \frac{1}{2}\eta_{\mu\nu}\, T$$

$$= \begin{pmatrix} \rho & 0 & 0 & 0 \\ 0 & 0 & 0 & 0 \\ 0 & 0 & 0 & 0 \\ 0 & 0 & 0 & 0 \end{pmatrix} - \frac{-\rho}{2} \begin{pmatrix} -1 & 0 & 0 & 0 \\ 0 & 1 & 0 & 0 \\ 0 & 0 & 1 & 0 \\ 0 & 0 & 0 & 1 \end{pmatrix} = \frac{\rho}{2} \delta_{\mu\nu}.$$

Now we calculate R_{tt} and use the relation (21.13):

$$R_{\mu\nu} = R^{\rho}{}_{\mu\rho\nu} = \Gamma^{\rho}_{\mu\nu,\rho} - \Gamma^{\rho}_{\mu\rho,\nu} + \Gamma^{\rho}_{\sigma\rho}\Gamma^{\sigma}_{\mu\nu} - \Gamma^{\rho}_{\sigma\nu}\Gamma^{\sigma}_{\mu\rho}$$

The Christoffel symbols can be represented by the metric according to (19.16):

$$\Gamma^{\tau}_{\mu\lambda} = \frac{g^{\tau\nu}}{2}\left(g_{\mu\nu,\lambda} + g_{\lambda\nu,\mu} - g_{\mu\lambda,\nu}\right)$$

If you multiply two Christoffel symbols together, you get terms in h^2 and higher. In linear approximation, these fall away, and the Ricci tensor results in

$$R_{\mu\nu} = \Gamma^{\rho}_{\mu\nu,\rho} - \Gamma^{\rho}_{\mu\rho,\nu},$$

from which, just like in the derivation of the formula (21.8) and taking into account that the partial derivatives of $\eta_{\mu\nu}$ vanish,

$$R_{\mu\nu} = \frac{\eta^{\lambda\sigma}}{2}\left(h_{\nu\sigma,\mu\lambda} - h_{\mu\nu,\sigma\lambda} - h_{\lambda\sigma,\mu\nu} + h_{\mu\lambda,\sigma\nu}\right) \tag{22.14}$$

in linear approximation and thus

$$R_{tt} = \frac{\eta^{\lambda\sigma}}{2}\left(h_{t\sigma,t\lambda} - h_{tt,\sigma\lambda} - h_{\lambda\sigma,tt} + h_{t\lambda,\sigma t}\right)$$

follows. Since the $h_{\mu\nu}$ are assumed to be static, all derivatives with respect to time are zero, i.e., all terms that are derived at least once with respect to t vanish. So what remains is

$$\begin{aligned}
R_{tt} &= -\frac{\eta^{\lambda\sigma}}{2}h_{tt,\sigma\lambda} = -\frac{1}{2}\left(-\underbrace{h_{tt,tt}}_{=0} + h_{tt,xx} + h_{tt,yy} + h_{tt,zz}\right) \\
&= -\frac{1}{2}\left(\frac{\partial^2 h_{tt}}{\partial x^2} + \frac{\partial^2 h_{tt}}{\partial y^2} + \frac{\partial^2 h_{tt}}{\partial z^2}\right) = -\frac{1}{2}\Delta h_{tt}.
\end{aligned}$$

In Eq. (22.6) we derived that in the Newtonian limit

$$h_{tt} = -2\phi$$

applies. So we get

$$R_{tt} = -\frac{1}{2}\Delta h_{tt} = \Delta\phi$$

and thus overall for $\mu = \nu = t$

$$\Delta\phi = R_{tt} = 8\pi G \left(T_{tt} - \frac{1}{2} g_{tt} T \right) = 8\pi G \frac{\rho}{2} \delta_{tt} = 4\pi G\rho.$$

In the Newtonian approximation, the Einstein equations therefore yield the Poisson equation and thus the Newtonian theory of gravitation.

Cosmological Constant

We consider the third requirement we have made of the field equations of General Relativity: They should be as simple as possible. The above derivation has shown that the Einstein equations

$$R^{\mu\nu} - \frac{1}{2} g^{\mu\nu} R = 8\pi G\, T^{\mu\nu}$$

represent almost the minimum that can be achieved based on the other requirements, especially the divergence-free condition. However, the structure of the equations allows for an additional term to be added that depends linearly on the metric. The equations then take the form

$$G^{\mu\nu} + \Lambda g^{\mu\nu} = R^{\mu\nu} - \frac{1}{2} g^{\mu\nu} R + \Lambda g^{\mu\nu} = 8\pi G\, T^{\mu\nu} \qquad (22.15)$$

where Λ is a Riemann scalar. The divergence-free condition of the left-hand side is guaranteed, as the covariant derivative of the metric equals zero.

It can be shown that $G^{\mu\nu} + \Lambda g^{\mu\nu}$ is the only $(2, 0)$-tensor that consists of the first and second derivatives of the metric and is simultaneously divergence-free. This means that the extension of the Einstein equations by the term $\Lambda g^{\mu\nu}$ is also the only physically meaningful one.

Einstein himself added this term to his equations in 1917 to ensure that his equations guarantee that the universe remains in a static state. For this reason, the constant Λ is also called the **cosmological constant**. When it became clear in the late twenties of the last century, particularly by observations by the astronomer *Hubble*, that the universe is expanding, Einstein retracted the extension of his field equations and spoke of the "biggest blunder" he had ever made. However, today it is believed that the cosmological constant could be a way to account for the so-called **vacuum energy** in GR. In other words: Even if the cosmological constant represents a wrong extension of his equations from Einstein's point of view, it seems to be of essential importance from today's perspective.

The Einstein equations including the cosmological constants no longer meet the requirement that they can be reduced to the Poisson equation in the Newtonian limit. Since the Ricci tensor $R^{\mu\nu}$ contains second derivatives with respect to the coordinates, the dimension of the cosmological constant (just like that of the curvature scalar) is $1/\text{length}^2$. The Newtonian theory of gravitation is perfectly suitable for describing almost all phenomena in our solar system, i.e., the additional term with the cosmological constant must therefore be very small. Specifically, the length $\Lambda^{-1/2}$ must be much larger than the diameter of our solar system

$$\Lambda^{-1/2} \gg 10^5 \text{ lightyears.}$$

Because of this restriction, the cosmological constant only plays a role in the large-scale consideration of the universe, the current models favor a cosmological constant of the order $\Lambda^{-1/2} \sim 10^{10} \text{ Ly}$. Since we do not want to deal with cosmological world models in this book, but want to investigate phenomena of our solar system or individual objects in the universe, we will work with the Einstein equations without the additional term, i.e., with (22.10):

$$R^{\mu\nu} - \frac{1}{2} g^{\mu\nu} R = 8\pi G \, T^{\mu\nu}$$

Energy Conservation as a Consequence of Spacetime Geometry

We have derived the special form of the Einstein tensor $G^{\mu\nu}$ from the requirement that the law of energy conservation, i.e. the divergence-free nature of the energy-momentum tensor

$$T^{\mu\nu}_{\;;\nu} = 0,$$

should apply. If we consider the Einstein equations

$$G^{\mu\nu} = 8\pi G \, T^{\mu\nu}$$

and only assume that the Einstein tensor (like the energy-momentum tensor) is symmetric, without presupposing energy conservation, then the Einstein equations consist of ten equations, from which the ten independent components of the metric tensor could be determined. As we have already explained above, the components of the metric can only be determined from six of the ten equations, as we need four equations for the arbitrary choice of a coordinate system. We therefore need four more (internal) equations, which according to the assumption made, cannot be derived from the divergence-free nature of the energy-momentum tensor in order to determine the ten components of the metric. In deriving the Einstein equations, we derived the divergence-free nature of the Einstein tensor

$$G^{\mu\nu}_{\;;\nu} = 0$$

solely from the second Bianchi identity (21.10), i.e. from purely geometric facts. This means, if we assume the Einstein equations to be valid, then it follows conversely, that due to the divergence-free nature of the Einstein tensor, the energy-momentum tensor must also be divergence-free.

Geodesics as a Consequence of the Einstein Equations

As we have seen, the conservation of energy

$$T^{\mu\nu}_{\ ;\nu} = 0$$

follows from the divergence-free nature of the Einstein tensor. We now want to show that this also leads to the geodesic equation, where for simplicity we assume that the object under investigation is a dust particle. The energy-momentum tensor is then due to $p = 0$ according to Eq. (22.3):

$$T^{\mu\nu} = (\rho + p)\, U^\mu U^\nu + pg^{\mu\nu} = \rho U^\mu U^\nu$$

With the product rule it follows that

$$0 = T^{\mu\nu}_{\ ;\nu} = (\rho U^\mu U^\nu)_{;\nu} = U^\mu \left(\rho U^\nu\right)_{;\nu} + \rho U^\nu \left(U^\mu_{\ ;\nu}\right) \quad (*).$$

Now the relationship

$$-1 = g_{\mu\alpha}\, U^\mu U^\alpha,$$

applies to the four-velocity $\mathbf{U} \to U^\mu$, from which with the product rule due to $g_{\mu\alpha;\nu} = 0$

$$0 = (g_{\mu\alpha}\, U^\mu U^\alpha)_{;\nu} = g_{\mu\alpha}\, U^\mu U^\alpha_{\ ;\nu} + g_{\mu\alpha}\, U^\alpha U^\mu_{\ ;\nu}$$

follows. Due to the symmetry of the metric tensor and after index swapping $\mu \leftrightarrow \alpha$ this results in

$$0 = (g_{\mu\alpha} U^\mu U^\alpha)_{;\nu} = 2g_{\mu\alpha}\, U^\alpha U^\mu_{\ ;\nu}$$

and thus

$$0 = g_{\mu\alpha} U^\alpha U^\mu_{\ ;\nu}.$$

Now we multiply equation $(*)$ with $g_{\mu\alpha}\, U^\alpha$ and get

$$0 = g_{\mu\alpha}\, U^\alpha \left(\rho U^\mu U^\nu\right)_{;\nu} = \underbrace{g_{\mu\alpha}\, U^\alpha U^\mu}_{=-1} \left(\rho U^\nu\right)_{;\nu} + \rho U^\nu \underbrace{g_{\mu\alpha}\, U^\alpha U^\mu_{\ ;\nu}}_{=0},$$

thus

$$0 = (\rho U^\nu)_{;\nu}.$$

We insert this into equation $(*)$, and with the definition of the covariant derivative we get

$$0 = U^\nu \left(U^\mu_{\ ;\nu}\right) = U^\nu \left(\frac{\partial U^\mu}{\partial \nu} + \Gamma^\mu_{\beta\nu}\, U^\beta\right).$$

Now it holds

$$U^\nu = \frac{d\nu}{dt_E}$$

and thus, following the chain rule, we get

$$0 = \frac{\partial U^\mu}{\partial \nu}\frac{d\nu}{dt_E} + \Gamma^\mu_{\beta\nu}\frac{d\beta}{dt_E}\frac{d\nu}{dt_E} = \frac{dU^\mu}{dt_E} + \Gamma^\mu_{\beta\nu}\frac{d\beta}{dt_E}\frac{d\nu}{dt_E} = \frac{d^2\mu}{dt_E^2} + \Gamma^\mu_{\beta\nu}\frac{d\beta}{dt_E}\frac{d\nu}{dt_E},$$

which is the geodesic equation, see formula (20.2).

The next task will be to find the ten functions $g_{\mu\nu}$ for a given energy-momentum tensor, i.e., for given sources. Now, the Ricci tensor and also the curvature scalar R consist of a *nonlinear* combination of the $g_{\mu\nu}$ and their derivatives. By nonlinear, we mean that in the definition of the tensors on the left side of the Einstein equations, for example, products of the $g_{\mu\nu}$ and their derivatives can occur. For example, the Ricci tensor

$$R_{\mu\nu} = \frac{g^{\lambda\sigma}}{2}\left(g_{\nu\sigma,\mu\lambda} - g_{\mu\nu,\sigma\lambda} - g_{\lambda\sigma,\mu\nu} + g_{\mu\lambda,\sigma\nu}\right)$$

contains products of $g^{\mu\nu}$ and second derivatives of the metric. For such nonlinear (differential) equations, there is no closed solution theory and no standard procedure with which one could derive (special) solutions of the Einstein equations. To still arrive at solutions, one often proceeds in such a situation by trying to find solutions for certain special cases (as we have already done above for the Newtonian limit case) and to gain knowledge from these special solutions for more general ones.

23. Static Spherical Gravitational Fields

In this chapter, we want to assume that the gravitational fields we are investigating are time-independent (static) and spherically symmetric (spherical), which is approximately true for the gravitational fields of the Sun and Earth.

Many applications of General Relativity refer to physical phenomena in our solar system. For the investigation of these effects, we want to disregard the slow ($v \ll c$) rotation and the flattening at the north and south poles of the Sun (and the planets), so we assume that the Sun is a spherical and static, spatially limited mass distribution M. In this chapter, we want to investigate physical phenomena outside this mass distribution, the gravitational effects inside the star are not considered here. Due to the spatially limited mass distribution, we expect that with increasing distance r from this central body, we will encounter an ever-weakening gravitational field, so that asymptotically, i.e., for $r \to \infty$, Minkowski spacetime can be assumed.

23.1. Coordinate Systems for Static Spherical Spacetimes

We start with the selection of a suitable coordinate system that reflects the assumed temporal and spatial symmetries. To do this, we first generalize the spherical coordinates introduced in Remark 19.2 in three-dimensional Euclidean space to the Minkowski spacetime. Specifically, we define the coordinate transformation to the *spatial* spherical coordinates again by

$$
\begin{aligned}
x &= r \sin \vartheta \cos \varphi \\
y &= r \sin \vartheta \sin \varphi \\
z &= r \cos \vartheta,
\end{aligned}
$$

the temporal coordinate in spherical coordinates should remain unchanged

$$
t = t.
$$

We obtain for the differentials

$$
dt = \frac{\partial t}{\partial t} dt + \frac{\partial t}{\partial r} dr + \frac{\partial t}{\partial \vartheta} d\vartheta + \frac{\partial t}{\partial \varphi} d\varphi = dt
$$

© The Author(s), under exclusive license to Springer-Verlag GmbH, DE, part of Springer Nature 2026
M. Ruhrländer, *Ascent to the Einstein Equations*,
https://doi.org/10.1007/978-3-662-72672-3_23

$$
\begin{aligned}
dx &= \frac{\partial x}{\partial t}\,dt + \frac{\partial x}{\partial r}\,dr + \frac{\partial x}{\partial \vartheta}\,d\vartheta + \frac{\partial x}{\partial \varphi}\,d\varphi \\
&= \sin\vartheta\cos\varphi\,dr + r\cos\vartheta\cos\varphi\,d\vartheta - r\sin\vartheta\sin\varphi\,d\varphi \\
dy &= \frac{\partial y}{\partial t}\,dt + \frac{\partial y}{\partial r}\,dr + \frac{\partial y}{\partial \vartheta}\,d\vartheta + \frac{\partial y}{\partial \varphi}\,d\varphi \\
&= \sin\vartheta\sin\varphi\,dr + r\cos\vartheta\sin\varphi\,d\vartheta + r\sin\vartheta\cos\varphi\,d\varphi \\
dz &= \frac{\partial z}{\partial t}\,dt + \frac{\partial z}{\partial r}\,dr + \frac{\partial z}{\partial \vartheta}\,d\vartheta + \frac{\partial z}{\partial \varphi}\,d\varphi = \cos\vartheta\,dr - r\sin\vartheta\,d\vartheta.
\end{aligned}
$$

And thus for the line element in flat space

$$
\begin{aligned}
ds^2 &= -dt^2 + dx^2 + dy^2 + dz^2 \\
&= -dt^2 + \sin^2\vartheta\cos^2\varphi\,dr^2 + r^2\cos^2\vartheta\cos^2\varphi\,d\vartheta^2 + r^2\sin^2\vartheta\sin^2\varphi\,d\varphi^2 \\
&+ \ \sin^2\vartheta\sin^2\varphi\,dr^2 + r^2\cos^2\vartheta\sin^2\varphi\,d\vartheta^2 + r^2\sin^2\vartheta\cos^2\varphi\,d\varphi^2 \\
&+ \ \cos^2\vartheta\,dr^2 + r^2\sin^2\vartheta\,d\vartheta^2 + \mathit{mixed\ terms} \\
&= -dt^2 + dr^2 + r^2 d\vartheta^2 + r^2\sin^2\vartheta\,d\varphi^2,
\end{aligned}
$$

where we have not written down the *mixed terms* for clarity, which all cancel each other out. The line element in Minkowski space can thus be written as

$$
ds^2 = -dt^2 + dr^2 + r^2 d\vartheta^2 + r^2\sin^2\vartheta\,d\varphi^2. \tag{23.1}
$$

One can see that the surface with constant r and t (i.e. $dt = dr = 0$) corresponds as expected to the surface of a three-dimensional sphere with radius r. If only the time coordinate t is kept constant and the distance r is varied, the Euclidean line element in three-dimensional space results:

$$
dl^2 = dr^2 + r^2\left(d\vartheta^2 + \sin^2\vartheta\,d\varphi^2\right) = dr^2 + r^2 d\Omega^2,
$$

whereby the **surface element** $d\Omega$ is defined. We now want to work out how the assumed time and space symmetries affect the general line element

$$
ds^2 = g_{\mu\nu}\,d\mu\,d\nu, \tag{23.2}
$$

where the components of the metric (initially) can be arbitrary functions dependent on the coordinates:

$$
g_{\mu\nu} = g_{\mu\nu}\left(t, r, \vartheta, \varphi\right)
$$

The assumed time-independence means on the one hand that the $g_{\mu\nu}$ do not depend on the time coordinate t, and on the other hand, that ds^2 must be invariant under time reversal ($t \to -t$). If we rewrite Eq. (23.2), it follows with the symmetry of the metric tensor ($g_{\mu\nu} = g_{\nu\mu}$) and $i, j = r, \vartheta, \varphi$

$$
ds^2 = g_{tt}\,dt^2 + 2g_{ti}\,dt\,di + g_{ij}\,di\,dj, \tag{23.3}
$$

which due to the invariance of time reversal must be equal to

$$ds^2 = g_{tt}\, dt^2 - 2g_{ti}\, dt\, di + g_{ij}\, di\, dj. \tag{23.4}$$

Subtraction of the two equations leads to

$$4g_{ti}\, dt\, di = 0,$$

i.e.

$$g_{ti}\, di = 0,$$

from which

$$g_{ti} = g_{it} = 0$$

follows, since the di can be chosen arbitrarily.

The assumed spherical symmetry states on the one hand that the components of the metric only depend on the radial component r and not on a specific direction, i.e. $g_{tt} = g_{tt}(r)$ and $g_{ij} = g_{ij}(r)$. On the other hand, no spatial direction should be distinguished, i.e., we require that the line element does not change when we replace $d\vartheta$ with $-d\vartheta$ or $d\varphi$ with $-d\varphi$. Just like above in (23.3) and (23.4) for the invariant time reversal, this implies that in the line element all mixed terms containing $d\vartheta$ or $d\varphi$ (i.e. $dr\, d\vartheta, dr\, d\varphi, d\vartheta\, d\varphi$), disappear, so

$$g_{r\vartheta} = g_{r\varphi} = g_{\vartheta\varphi} = 0.$$

The line element is thus reduced by the assumed symmetries to

$$ds^2 = g_{tt}(r)\, dt^2 + g_{rr}(r)\, dr^2 + g_{\vartheta\vartheta}(r)\, r^2 d\vartheta^2 + g_{\varphi\varphi}(r)\, r^2 \sin^2\vartheta\, d\varphi^2.$$

We further assume that the "deformation" of the two-dimensional surface element $d\Omega$ by the functions $g_{\vartheta\vartheta}$ and $g_{\varphi\varphi}$ occurs equally in the ϑ-direction as well as in the φ-direction, i.e., we assume that $g_{\vartheta\vartheta} = g_{\varphi\varphi}$, and we get

$$ds^2 = g_{tt}(r)\, dt^2 + g_{rr}(r)\, dr^2 + g_{\vartheta\vartheta}(r)\, r^2 d\Omega^2.$$

So, by the assumed symmetries, we have arrived at the point where we only need to determine three of the ten metric components. In fact, by redefining the radial parameter r, the number can be reduced to two. For this, we set

$$\bar{r}^2 = g_{\vartheta\vartheta}(r)\, r^2 \Rightarrow \bar{r} = \sqrt{g_{\vartheta\vartheta}}\, r \tag{23.5}$$

and obtain with the product rule

$$\frac{d\bar{r}}{dr} = \sqrt{g_{\vartheta\vartheta}} + r\, \frac{1}{2\sqrt{g_{\vartheta\vartheta}}}\, \frac{dg_{\vartheta\vartheta}}{dr} = \sqrt{g_{\vartheta\vartheta}}\left(1 + \frac{r}{2\, g_{\vartheta\vartheta}}\, \frac{dg_{\vartheta\vartheta}}{dr}\right),$$

from which

$$dr^2 = \frac{1}{g_{\vartheta\vartheta}}\left(1 + r\,\frac{1}{2\,g_{\vartheta\vartheta}}\frac{dg_{\vartheta\vartheta}}{dr}\right)^{-2} d\bar{r}^2$$

follows. With this, we define the quantities $g_{\bar{r}\bar{r}}$ and $g_{\bar{t}\bar{t}}$ by

$$g_{rr}\,dr^2 = \frac{g_{rr}}{g_{\vartheta\vartheta}}\left(1 + r\,\frac{1}{2\,g_{\vartheta\vartheta}}\frac{dg_{\vartheta\vartheta}}{dr}\right)^{-2} d\bar{r}^2 = g_{\bar{r}\bar{r}}\,d\bar{r}^2$$

and

$$g_{tt}\left(r\right) = g_{\bar{t}\bar{t}}\left(\bar{r}\right).$$

The effect of this redefinition is that the function $g_{\bar{\vartheta}\bar{\vartheta}}$ can be replaced by 1, because then with (23.5)

$$g_{\bar{\vartheta}\bar{\vartheta}}\,\bar{r}^2 = \bar{r}^2 = g_{\vartheta\vartheta}\,r^2$$

and thus

$$\begin{aligned}
ds^2 &= g_{tt}\,dt^2 + g_{rr}\,dr^2 + g_{\vartheta\vartheta}\,r^2 d\Omega^2 \\
&= g_{\bar{t}\bar{t}}\,d\bar{t}^2 + g_{\bar{r}\bar{r}}\,d\bar{r}^2 + \bar{r}^2 d\Omega^2.
\end{aligned}$$

We write (for simplicity) again r instead of $\bar{r}$ and t instead of $\bar{t}$ and finally obtain for the line element

$$ds^2 = g_{tt}\left(r\right)dt^2 + g_{rr}\left(r\right)dr^2 + r^2 d\Omega^2. \tag{23.6}$$

This is a general approach for a spherical and static metric, which is called the **standard form**. When we compare the standard form with the Minkowski metric in spherical coordinates (23.1) and assume that at a great distance from the central body the space becomes flat, then for the two components of the metric it must hold:

$$g_{tt}\left(r\right) \to -1 \ (r \to \infty)$$

and

$$g_{rr}\left(r\right) \to 1 \ (r \to \infty)$$

To ensure this and to simplify the following calculations, we set

$$g_{tt}\left(r\right) = -e^{2\nu(r)}, \ g_{rr}\left(r\right) = e^{2\lambda(r)} \tag{23.7}$$

with two unknown functions $\nu\left(r\right)$ and $\lambda\left(r\right)$, which

$$\nu\left(r\right), \lambda\left(r\right) \to 0 \ (r \to \infty)$$

fulfill. The metric tensor now results in

$$g_{\mu\nu} = \begin{pmatrix} -e^{2\nu} & 0 & 0 & 0 \\ 0 & e^{2\lambda} & 0 & 0 \\ 0 & 0 & r^2 & 0 \\ 0 & 0 & 0 & r^2 \sin^2 \vartheta \end{pmatrix} \tag{23.8}$$

and the inverse metric to

$$g^{\mu\nu} = \begin{pmatrix} -e^{-2\nu} & 0 & 0 & 0 \\ 0 & e^{-2\lambda} & 0 & 0 \\ 0 & 0 & r^{-2} & 0 \\ 0 & 0 & 0 & r^{-2} \sin^{-2} \vartheta \end{pmatrix}, \tag{23.9}$$

which can be easily verified by matrix multiplication $g_{\mu\rho}\, g^{\rho\nu} = \delta_\mu^\nu$.

23.2. (Outer) Schwarzschild Metric

We want to concretely calculate the functions $\nu\,(r)$ and $\lambda\,(r)$ introduced in the last section and use Einstein's field equations for this. Since we limit ourselves to finding solutions that are valid outside the limited area of mass distribution, we set the energy-momentum tensor $T_{\mu\nu} = 0$. The Einstein equations are thus reduced to

$$R_{\mu\nu} - \frac{1}{2}\, g_{\mu\nu}\, R = 0. \tag{23.10}$$

If we multiply this equation by $g^{\mu\nu}$, we get

$$\begin{aligned} 0 = g^{\mu\nu} \cdot 0 \;&=\; g^{\mu\nu}\left(R_{\mu\nu} - \frac{1}{2} g_{\mu\nu}\, R \right), \\ &=\; \underbrace{g^{\mu\nu} R_{\mu\nu}}_{=R} - \frac{1}{2} \underbrace{g^{\mu\nu} g_{\mu\nu}}_{=\delta_\mu^\mu}\, R \\ &=\; R - \frac{1}{2} \underbrace{\delta_\mu^\mu}_{=4}\, R \\ &=\; R - 2R = -R \end{aligned}$$

i.e., the curvature scalar R is zero and Eq. (23.10) is reduced to

$$R_{\mu\nu} = 0. \tag{23.11}$$

These equations are called **vacuum field equations**. If we are in a flat Minkowski space, i.e. $g_{\mu\nu} = \eta_{\mu\nu}$, then the left side of the Einstein equations equals zero ($R_{\mu\nu} = 0$, since the Riemannian tensor is zero everywhere), i.e., the Minkowski metric is a solution to the vacuum field equations. We will subsequently see that there are other metrics that also satisfy the vacuum equations.

To determine the functions $\nu\,(r)$ and $\lambda\,(r)$ we calculate (in a very long calculation) first the Ricci tensor using the Christoffel symbols, for which we use the formula (19.16)

$$\Gamma^{\tau}_{\mu\lambda} = \frac{g^{\tau\nu}}{2}\left(g_{\mu\nu,\lambda} + g_{\lambda\nu,\mu} - g_{\mu\lambda,\nu}\right).$$

On the right side is a sum (over ν), which but due to the diagonal shape of $g^{\tau\nu}$ only for $\nu = \tau$ has a non-zero summand.

$$\Gamma^{\tau}_{\mu\lambda} = \frac{g^{\tau\tau}}{2}\left(g_{\mu\tau,\lambda} + g_{\lambda\tau,\mu} - g_{\mu\lambda,\tau}\right).$$

It follows for $\tau = t$ with $i, j = r, \vartheta, \varphi$

$$
\begin{aligned}
\Gamma^{t}_{tt} &= \frac{g^{tt}}{2}\left(g_{tt,t} + g_{tt,t} - g_{tt,t}\right) = \frac{-e^{-2\nu}}{2}\frac{d}{dt}\left(-e^{2\nu}\right) = 0 \\
\Gamma^{t}_{rt} &= \Gamma^{t}_{tr} = \frac{g^{tt}}{2}\left(g_{rt,t} + g_{tt,r} - g_{rt,t}\right) = \frac{-e^{-2\nu}}{2}\frac{d}{dr}\left(-e^{2\nu}\right) \\
&= \frac{-e^{-2\nu}}{2}\left(-2e^{2\nu}\right)\frac{d\nu}{dr} = \nu' \\
\Gamma^{t}_{\vartheta t} &= \Gamma^{t}_{t\vartheta} = \frac{g^{tt}}{2}\left(g_{\vartheta t,t} + g_{tt,\vartheta} - g_{\vartheta t,t}\right) = \frac{-e^{-2\nu}}{2}\frac{d}{d\vartheta}\left(-e^{2\nu}\right) = 0 \\
\Gamma^{t}_{\varphi t} &= \Gamma^{t}_{t\varphi} = \frac{g^{tt}}{2}\left(g_{\varphi t,t} + g_{tt,\varphi} - g_{\varphi t,t}\right) = \frac{-e^{-2\nu}}{2}\frac{d}{d\varphi}\left(-e^{2\nu}\right) = 0 \\
\Gamma^{t}_{ij} &= \Gamma^{t}_{ji} = \frac{g^{tt}}{2}\left(g_{it,j} + g_{jt,i} - g_{ij,t}\right) = 0,
\end{aligned}
$$

where the prime $'$ is intended to represent the derivative with respect to the variable r:

$$\nu' = \frac{d\nu}{dr}$$

Furthermore, for $\tau = r$

$$
\begin{aligned}
\Gamma^{r}_{tt} &= \frac{g^{rr}}{2}\left(g_{tr,t} + g_{tr,t} - g_{tt,r}\right) = \frac{e^{-2\lambda}}{2}\frac{d}{dr}\left(e^{2\nu}\right) = \frac{e^{-2\lambda}}{2}\left(2e^{2\nu}\right)\frac{d\nu}{dr} \\
&= \nu'\,e^{2\nu-2\lambda} \\
\Gamma^{r}_{rr} &= \frac{g^{rr}}{2}\left(g_{rr,r} + g_{rr,r} - g_{rr,r}\right) = \frac{e^{-2\lambda}}{2}\frac{d}{dr}\left(e^{2\lambda}\right) = \frac{e^{-2\lambda}}{2}\left(2e^{2\lambda}\right)\frac{d\lambda}{dr} = \lambda'
\end{aligned}
$$

$$\Gamma^{r}_{\vartheta\vartheta} = \frac{g^{rr}}{2}\left(g_{\vartheta r,\vartheta} + g_{\vartheta r,\vartheta} - g_{\vartheta\vartheta,r}\right) = \frac{e^{-2\lambda}}{2}\frac{d}{dr}\left(-r^{2}\right) = -re^{-2\lambda}$$

$$\Gamma^{r}_{\varphi\varphi} = \frac{g^{rr}}{2}\left(g_{\varphi r,\varphi} + g_{\varphi r,\varphi} - g_{\varphi\varphi,r}\right) = \frac{e^{-2\lambda}}{2}\frac{d}{dr}\left(-r^{2}\sin^{2}\vartheta\right) = -r\sin^{2}\vartheta\, e^{-2\lambda}$$

$$\Gamma^{r}_{\mu\lambda} = \frac{g^{rr}}{2}\left(g_{\mu r,\lambda} + g_{\lambda r,\mu} - g_{\mu\lambda,r}\right) = 0 \; f\ddot{u}r\, \mu \neq \lambda.$$

For $\tau = \vartheta$ we get

$$\Gamma^{\vartheta}_{t\mu} = \Gamma^{\vartheta}_{\mu t} = \frac{g^{\vartheta\vartheta}}{2}\left(g_{t\vartheta,\mu} + g_{\mu\vartheta,t} - g_{t\mu,\vartheta}\right) = 0$$

$$\Gamma^{\vartheta}_{rr} = \frac{g^{\vartheta\vartheta}}{2}\left(g_{r\vartheta,r} + g_{r\vartheta,r} - g_{rr,\vartheta}\right) = 0$$

$$\Gamma^{\vartheta}_{\vartheta r} = \Gamma^{\vartheta}_{r\vartheta} = \frac{g^{\vartheta\vartheta}}{2}\left(g_{\vartheta\vartheta,r} + g_{r\vartheta,\vartheta} - g_{\vartheta r,\vartheta}\right) = \frac{1}{2r^{2}}\frac{d}{dr}\left(r^{2}\right) = \frac{1}{r}$$

$$\Gamma^{\vartheta}_{\varphi r} = \Gamma^{\vartheta}_{r\varphi} = \frac{g^{\vartheta\vartheta}}{2}\left(g_{\varphi\vartheta,r} + g_{r\vartheta,\varphi} - g_{\varphi r,\vartheta}\right) = 0$$

$$\Gamma^{\vartheta}_{\vartheta\vartheta} = \frac{g^{\vartheta\vartheta}}{2}\left(g_{\vartheta\vartheta,\vartheta} + g_{\vartheta\vartheta,\vartheta} - g_{\vartheta\vartheta,\vartheta}\right) = 0$$

$$\Gamma^{\vartheta}_{\varphi\vartheta} = \Gamma^{\vartheta}_{\vartheta\varphi} = \frac{g^{\vartheta\vartheta}}{2}\left(g_{\varphi\vartheta,\vartheta} + g_{\vartheta\vartheta,\varphi} - g_{\varphi\vartheta,\vartheta}\right) = 0$$

$$\Gamma^{\vartheta}_{\varphi\varphi} = \frac{g^{\vartheta\vartheta}}{2}\left(g_{\varphi\vartheta,\varphi} + g_{\varphi\vartheta,\varphi} - g_{\varphi\varphi,\vartheta}\right) = -\frac{1}{2r^{2}}\frac{d}{d\vartheta}\left(r^{2}\sin^{2}\vartheta\right) = -\sin\vartheta\cos\vartheta.$$

And finally for $\tau = \varphi$

$$\Gamma^{\varphi}_{t\mu} = \Gamma^{\varphi}_{\mu t} = \frac{g^{\varphi\varphi}}{2}\left(g_{t\varphi,\mu} + g_{\mu\varphi,t} - g_{t\mu,\varphi}\right) = 0$$

$$\Gamma^{\varphi}_{rr} = \frac{g^{\varphi\varphi}}{2}\left(g_{r\varphi,r} + g_{r\varphi,r} - g_{rr,\varphi}\right) = 0$$

$$\Gamma^{\varphi}_{\vartheta r} = \Gamma^{\varphi}_{r\vartheta} = \frac{g^{\varphi\varphi}}{2}\left(g_{\vartheta\varphi,r} + g_{r\varphi,\vartheta} - g_{\vartheta r,\varphi}\right) = 0$$

$$\Gamma^{\varphi}_{\varphi r} = \Gamma^{\varphi}_{r\varphi} = \frac{g^{\varphi\varphi}}{2}\left(g_{\varphi\varphi,r} + g_{r\varphi,\varphi} - g_{\varphi r,\varphi}\right) = \frac{1}{2r^{2}\sin^{2}\vartheta}\frac{d}{dr}\left(r^{2}\sin^{2}\vartheta\right) = \frac{1}{r}$$

$$\Gamma^{\varphi}_{\vartheta\vartheta} = \frac{g^{\varphi\varphi}}{2}\left(g_{\vartheta\varphi,\vartheta} + g_{\vartheta\varphi,\vartheta} - g_{\vartheta\vartheta,\varphi}\right) = 0$$

$$\Gamma^{\varphi}_{\varphi\vartheta} = \Gamma^{\varphi}_{\vartheta\varphi} = \frac{g^{\varphi\varphi}}{2}\left(g_{\varphi\varphi,\vartheta} + g_{\vartheta\varphi,\varphi} - g_{\varphi\vartheta,\varphi}\right) = \frac{1}{2r^{2}\sin^{2}\vartheta}\frac{d}{d\vartheta}\left(r^{2}\sin^{2}\vartheta\right)$$

$$= \frac{\sin\vartheta\cos\vartheta}{\sin^{2}\vartheta} = \cot\vartheta$$

$$\Gamma^{\varphi}_{\varphi\varphi} = \frac{g^{\varphi\varphi}}{2}\left(g_{\varphi\varphi,\varphi} + g_{\varphi\varphi,\varphi} - g_{\varphi\varphi,\varphi}\right) = 0.$$

We summarize again which Christoffel symbols are different from zero:

$$
\begin{aligned}
\Gamma^{t}_{rt} &= \Gamma^{t}_{tr} = \nu' \\
\Gamma^{r}_{tt} &= \nu'\, e^{2\nu - 2\lambda} \\
\Gamma^{r}_{rr} &= \lambda' \\
\Gamma^{r}_{\vartheta\vartheta} &= -r e^{-2\lambda} \\
\Gamma^{r}_{\varphi\varphi} &= -r \sin^2\vartheta\, e^{-2\lambda} \\
\Gamma^{\vartheta}_{\vartheta r} &= \Gamma^{\vartheta}_{r\vartheta} = \frac{1}{r} \\
\Gamma^{\vartheta}_{\varphi\varphi} &= -\sin\vartheta\cos\vartheta \\
\Gamma^{\varphi}_{\varphi r} &= \Gamma^{\varphi}_{r\varphi} = \frac{1}{r} \\
\Gamma^{\varphi}_{\varphi\vartheta} &= \Gamma^{\varphi}_{\vartheta\varphi} = \cot\vartheta
\end{aligned}
\tag{23.12}
$$

To calculate the components of the Ricci tensor, we use the formula (21.13):

$$
R_{\mu\nu} = \Gamma^{\rho}_{\mu\nu,\rho} - \Gamma^{\rho}_{\mu\rho,\nu} + \Gamma^{\rho}_{\sigma\rho}\,\Gamma^{\sigma}_{\mu\nu} - \Gamma^{\rho}_{\sigma\nu}\,\Gamma^{\sigma}_{\mu\rho}
$$

We treat the four terms on the right side separately and define

$$
R^{(1)}_{\mu\nu} = \Gamma^{\rho}_{\mu\nu,\rho} = \Gamma^{t}_{\mu\nu,t} + \Gamma^{r}_{\mu\nu,r} + \Gamma^{\vartheta}_{\mu\nu,\vartheta} + \Gamma^{\varphi}_{\mu\nu,\varphi}
$$

and

$$
R^{(2)}_{\mu\nu} = \Gamma^{\rho}_{\mu\rho,\nu} = \Gamma^{t}_{\mu t,\nu} + \Gamma^{r}_{\mu r,\nu} + \Gamma^{\vartheta}_{\mu\vartheta,\nu} + \Gamma^{\varphi}_{\mu\varphi,\nu}
$$

and

$$
\begin{aligned}
R^{(3)}_{\mu\nu} = \Gamma^{\rho}_{\sigma\rho}\,\Gamma^{\sigma}_{\mu\nu} ={}& \Gamma^{t}_{tt}\,\Gamma^{t}_{\mu\nu} + \Gamma^{t}_{rt}\,\Gamma^{r}_{\mu\nu} + \Gamma^{t}_{\vartheta t}\,\Gamma^{\vartheta}_{\mu\nu} + \Gamma^{t}_{\varphi t}\,\Gamma^{\varphi}_{\mu\nu} \\
+{}& \Gamma^{r}_{tr}\,\Gamma^{t}_{\mu\nu} + \Gamma^{r}_{rr}\,\Gamma^{r}_{\mu\nu} + \Gamma^{r}_{\vartheta r}\,\Gamma^{\vartheta}_{\mu\nu} + \Gamma^{r}_{\varphi r}\,\Gamma^{\varphi}_{\mu\nu} \\
+{}& \Gamma^{\vartheta}_{t\vartheta}\,\Gamma^{t}_{\mu\nu} + \Gamma^{\vartheta}_{r\vartheta}\,\Gamma^{r}_{\mu\nu} + \Gamma^{\vartheta}_{\vartheta\vartheta}\,\Gamma^{\vartheta}_{\mu\nu} + \Gamma^{\vartheta}_{\varphi\vartheta}\,\Gamma^{\varphi}_{\mu\nu} \\
+{}& \Gamma^{\varphi}_{t\varphi}\,\Gamma^{t}_{\mu\nu} + \Gamma^{\varphi}_{r\varphi}\,\Gamma^{r}_{\mu\nu} + \Gamma^{\varphi}_{\vartheta\varphi}\,\Gamma^{\vartheta}_{\mu\nu} + \Gamma^{\varphi}_{\varphi\varphi}\,\Gamma^{\varphi}_{\mu\nu}
\end{aligned}
$$

as well as

$$
\begin{aligned}
R^{(4)}_{\mu\nu} = \Gamma^{\rho}_{\sigma\nu}\,\Gamma^{\sigma}_{\mu\rho} ={}& \Gamma^{t}_{t\nu}\,\Gamma^{t}_{\mu t} + \Gamma^{t}_{r\nu}\,\Gamma^{r}_{\mu t} + \Gamma^{t}_{\vartheta\nu}\,\Gamma^{\vartheta}_{\mu t} + \Gamma^{t}_{\varphi\nu}\,\Gamma^{\varphi}_{\mu t} \\
+{}& \Gamma^{r}_{t\nu}\,\Gamma^{t}_{\mu r} + \Gamma^{r}_{r\nu}\,\Gamma^{r}_{\mu r} + \Gamma^{r}_{\vartheta\nu}\,\Gamma^{\vartheta}_{\mu r} + \Gamma^{r}_{\varphi\nu}\,\Gamma^{\varphi}_{\mu r} \\
+{}& \Gamma^{\vartheta}_{t\nu}\,\Gamma^{t}_{\mu\vartheta} + \Gamma^{\vartheta}_{r\nu}\,\Gamma^{r}_{\mu\vartheta} + \Gamma^{\vartheta}_{\vartheta\nu}\,\Gamma^{\vartheta}_{\mu\vartheta} + \Gamma^{\vartheta}_{\varphi\nu}\,\Gamma^{\varphi}_{\mu\vartheta} \\
+{}& \Gamma^{\varphi}_{t\nu}\,\Gamma^{t}_{\mu\varphi} + \Gamma^{\varphi}_{r\nu}\,\Gamma^{r}_{\mu\varphi} + \Gamma^{\varphi}_{\vartheta\nu}\,\Gamma^{\vartheta}_{\mu\varphi} + \Gamma^{\varphi}_{\varphi\nu}\,\Gamma^{\varphi}_{\mu\varphi}.
\end{aligned}
$$

We now calculate

$$R_{\mu\nu} = R^{(1)}_{\mu\nu} - R^{(2)}_{\mu\nu} + R^{(3)}_{\mu\nu} - R^{(4)}_{\mu\nu}$$

for all ten independent combinations of μ, ν. It holds

$$
\begin{aligned}
R^{(1)}_{tt} &= \Gamma^t_{tt,t} + \Gamma^r_{tt,r} + \Gamma^\vartheta_{tt,\vartheta} + \Gamma^\varphi_{tt,\varphi} = \Gamma^r_{tt,r} \\
&= \frac{d}{dr}\left(\nu'\, e^{2\nu-2\lambda}\right) = e^{2\nu-2\lambda}\left(\nu'' + 2\left(\nu'\right)^2 - 2\nu'\lambda'\right)
\end{aligned}
$$

and

$$
R^{(2)}_{tt} = \Gamma^t_{tt,t} + \Gamma^r_{tr,t} + \Gamma^\vartheta_{t\vartheta,t} + \Gamma^\varphi_{t\varphi,t} = 0.
$$

Similarly

$$
\begin{aligned}
R^{(3)}_{tt} &= \Gamma^t_{tt}\,\Gamma^t_{tt} + \Gamma^t_{rt}\,\Gamma^r_{tt} + \Gamma^t_{\vartheta t}\,\Gamma^\vartheta_{tt} + \Gamma^t_{\varphi t}\,\Gamma^\varphi_{tt} \\
&+ \Gamma^r_{tr}\,\Gamma^t_{tt} + \Gamma^r_{rr}\,\Gamma^r_{tt} + \Gamma^r_{\vartheta r}\,\Gamma^\vartheta_{tt} + \Gamma^r_{\varphi r}\,\Gamma^\varphi_{tt} \\
&+ \Gamma^\vartheta_{t\vartheta}\,\Gamma^t_{tt} + \Gamma^\vartheta_{r\vartheta}\,\Gamma^r_{tt} + \Gamma^\vartheta_{\vartheta\vartheta}\,\Gamma^\vartheta_{tt} + \Gamma^\vartheta_{\varphi\vartheta}\,\Gamma^\varphi_{tt} \\
&+ \Gamma^\varphi_{t\varphi}\,\Gamma^t_{tt} + \Gamma^\varphi_{r\varphi}\,\Gamma^r_{tt} + \Gamma^\varphi_{\vartheta\varphi}\,\Gamma^\vartheta_{tt} + \Gamma^\varphi_{\varphi\varphi}\,\Gamma^\varphi_{tt} \\
&= \Gamma^t_{rt}\,\Gamma^r_{tt} + \Gamma^r_{rr}\,\Gamma^r_{tt} + \Gamma^\vartheta_{r\vartheta}\,\Gamma^r_{tt} + \Gamma^\varphi_{r\varphi}\,\Gamma^r_{tt} \\
&= \left(\nu'\right)^2 e^{2\nu-2\lambda} + \lambda'\,\nu'\, e^{2\nu-2\lambda} + \frac{\nu'}{r}\, e^{2\nu-2\lambda} + \frac{\nu'}{r}\, e^{2\nu-2\lambda}
\end{aligned}
$$

as well as

$$
\begin{aligned}
R^{(4)}_{tt} &= \Gamma^t_{tt}\,\Gamma^t_{tt} + \Gamma^t_{rt}\,\Gamma^r_{tt} + \Gamma^t_{\vartheta t}\,\Gamma^\vartheta_{tt} + \Gamma^t_{\varphi t}\,\Gamma^\varphi_{tt} \\
&+ \Gamma^r_{tt}\,\Gamma^t_{tr} + \Gamma^r_{rt}\,\Gamma^r_{tr} + \Gamma^r_{\vartheta t}\,\Gamma^\vartheta_{tr} + \Gamma^r_{\varphi t}\,\Gamma^\varphi_{tr} \\
&+ \Gamma^\vartheta_{tt}\,\Gamma^t_{t\vartheta} + \Gamma^\vartheta_{rt}\,\Gamma^r_{t\vartheta} + \Gamma^\vartheta_{\vartheta t}\,\Gamma^\vartheta_{t\vartheta} + \Gamma^\vartheta_{\varphi t}\,\Gamma^\varphi_{t\vartheta} \\
&+ \Gamma^\varphi_{tt}\,\Gamma^t_{t\varphi} + \Gamma^\varphi_{rt}\,\Gamma^r_{t\varphi} + \Gamma^\varphi_{\vartheta t}\,\Gamma^\vartheta_{t\varphi} + \Gamma^\varphi_{\varphi t}\,\Gamma^\varphi_{t\varphi} \\
&= \Gamma^t_{rt}\,\Gamma^r_{tt} + \Gamma^r_{tt}\,\Gamma^t_{tr} \\
&= \left(\nu'\right)^2 e^{2\nu-2\lambda} + \left(\nu'\right)^2 e^{2\nu-2\lambda}.
\end{aligned}
$$

Summarized, we get

$$
\begin{aligned}
R_{tt} &= R^{(1)}_{tt} - R^{(2)}_{tt} + R^{(3)}_{tt} - R^{(4)}_{tt} \\
&= e^{2\nu-2\lambda}\left(\nu'' + 2\left(\nu'\right)^2 - 2\nu'\lambda'\right) + \left(\nu'\right)^2 e^{2\nu-2\lambda} + \lambda'\,\nu'\, e^{2\nu-2\lambda} \\
&+ \frac{2\nu'}{r}\, e^{2\nu-2\lambda} - 2\left(\nu'\right)^2 e^{2\nu-2\lambda} \\
&= e^{2\nu-2\lambda}\left(\nu'' + \left(\nu'\right)^2 - \nu'\lambda' + \frac{2\nu'}{r}\right).
\end{aligned}
$$

Analogously, R_{rr} is calculated, we only write down the intermediate results:

$$R_{rr}^{(1)} = \frac{d}{dr}\left(\lambda'\right) = \lambda''$$

and

$$R_{rr}^{(2)} = \nu'' + \lambda'' - \frac{1}{r^2} - \frac{1}{r^2}$$

and

$$R_{rr}^{(3)} = \nu'\lambda' + (\lambda')^2 + \frac{\lambda'}{r} + \frac{\lambda'}{r}$$

as well as

$$R_{rr}^{(4)} = (\nu')^2 + (\lambda')^2 + \frac{1}{r^2} + \frac{1}{r^2}$$

Summarized, this results in

$$\begin{aligned}
R_{rr} &= R_{rr}^{(1)} - R_{rr}^{(2)} + R_{rr}^{(3)} - R_{rr}^{(4)} \\
&= \lambda'' - \left(\nu'' + \lambda'' - \frac{2}{r^2}\right) + \nu'\lambda' + (\lambda')^2 + \frac{2\lambda'}{r} - \left((\nu')^2 + (\lambda')^2 + \frac{2}{r^2}\right) \\
&= -\nu'' + \nu'\lambda' + \frac{2\lambda'}{r} - (\nu')^2.
\end{aligned}$$

For $R_{\vartheta\vartheta}$, it follows accordingly

$$R_{\vartheta\vartheta}^{(1)} = \frac{d}{dr}\left(-re^{-2\lambda}\right) = e^{-2\lambda}\left(-1 + 2r\lambda'\right)$$

and

$$R_{\vartheta\vartheta}^{(2)} = \frac{d}{d\vartheta}\left(\cot\vartheta\right) = \frac{d}{d\vartheta}\left(\frac{\cos\vartheta}{\sin\vartheta}\right) = -\frac{1}{\sin^2\vartheta}$$

and

$$R_{\vartheta\vartheta}^{(3)} = \nu'\left(-re^{-2\lambda}\right) + \lambda'\left(-re^{-2\lambda}\right) + \frac{1}{r}\left(-re^{-2\lambda}\right) + \frac{1}{r}\left(-re^{-2\lambda}\right)$$

as well as

$$R_{\vartheta\vartheta}^{(4)} = \frac{1}{r}\left(-re^{-2\lambda}\right) + \frac{1}{r}\left(-re^{-2\lambda}\right) + \cot^2\vartheta.$$

Summarized, this results in

$$R_{\vartheta\vartheta} = R_{\vartheta\vartheta}^{(1)} - R_{\vartheta\vartheta}^{(2)} + R_{\vartheta\vartheta}^{(3)} - R_{\vartheta\vartheta}^{(4)}$$

$$
\begin{aligned}
&= e^{-2\lambda}\left(-1+2r\lambda'\right)-\left(-\frac{1}{\sin^2\vartheta}\right)-e^{-2\lambda}\left(r\nu'+r\lambda'+2\right) \\
&\quad - \left(\frac{2}{r}\left(-re^{-2\lambda}\right)+\cot^2\vartheta\right) \\
&= e^{-2\lambda}\left(-1+r\lambda'-r\nu'\right)+\frac{1}{\sin^2\vartheta}-\frac{\cos^2\vartheta}{\sin^2\vartheta} \\
&= e^{-2\lambda}\left(-1+r\lambda'-r\nu'\right)+1.
\end{aligned}
$$

For $R_{\varphi\varphi}$ it holds

$$
\begin{aligned}
R^{(1)}_{\varphi\varphi} &= \frac{d}{dr}\left(-r\sin^2\vartheta\,e^{-2\lambda}\right)+\frac{d}{d\vartheta}\left(-\sin\vartheta\cos\vartheta\right) \\
&= \sin^2\vartheta\,e^{-2\lambda}\left(-1+2r\lambda'\right)-\cos^2\vartheta+\sin^2\vartheta
\end{aligned}
$$

and

$$
R^{(2)}_{\varphi\varphi} = 0
$$

and

$$
\begin{aligned}
R^{(3)}_{\varphi\varphi} &= \nu'\left(-r\sin^2\vartheta\,e^{-2\lambda}\right)+\lambda'\left(-r\sin^2\vartheta\,e^{-2\lambda}\right) \\
&\quad + \frac{2}{r}\left(-r\sin^2\vartheta\,e^{-2\lambda}\right)+\cot\vartheta\left(-\sin\vartheta\cos\vartheta\right)
\end{aligned}
$$

as well as

$$
\begin{aligned}
R^{(4)}_{\varphi\varphi} &= \frac{1}{r}\left(-r\sin^2\vartheta\,e^{-2\lambda}\right)+\cot\vartheta\left(-\sin\vartheta\cos\vartheta\right) \\
&\quad + \frac{1}{r}\left(-r\sin^2\vartheta\,e^{-2\lambda}\right)+\cot\vartheta\left(-\sin\vartheta\cos\vartheta\right).
\end{aligned}
$$

Summarizing, we get

$$
\begin{aligned}
R_{\varphi\varphi} &= R^{(1)}_{\varphi\varphi}-R^{(2)}_{\varphi\varphi}+R^{(3)}_{\varphi\varphi}-R^{(4)}_{\varphi\varphi} \\
&= \sin^2\vartheta\,e^{-2\lambda}\left(-1+2r\lambda'\right)-\cos^2\vartheta+\sin^2\vartheta \\
&\quad + \sin^2\vartheta\,e^{-2\lambda}\left(-\nu'r-\lambda'r-2\right) \\
&\quad + \cot\vartheta\left(-\sin\vartheta\cos\vartheta\right)-\left(\left(-2\sin^2\vartheta\,e^{-2\lambda}\right)+2\cot\vartheta\left(-\sin\vartheta\cos\vartheta\right)\right) \\
&= \sin^2\vartheta\,e^{-2\lambda}(-1+r\lambda'-r\nu')-\cos^2\vartheta+\sin^2\vartheta+\cot\vartheta\left(\sin\vartheta\cos\vartheta\right) \\
&= \sin^2\vartheta\,e^{-2\lambda}(-1+r\lambda'-r\nu')-\cos^2\vartheta+\sin^2\vartheta+\cos^2\vartheta \\
&= \sin^2\vartheta\left[e^{-2\lambda}(-1+r\lambda'-r\nu')+1\right]=\sin^2\vartheta\,R_{\vartheta\vartheta}.
\end{aligned}
$$

Similarly, it is shown that

$$R_{\mu\nu} = 0 \; f\ddot{u}r \; \mu \neq \nu.$$

The vacuum field equations thus provide four equations, three of which are independent, to determine the unknown functions $\nu(r)$ and $\lambda(r)$:

$$
\begin{aligned}
R_{tt} &= e^{2\nu - 2\lambda}\left(\nu'' + (\nu')^2 - \nu'\lambda' + \frac{2\nu'}{r}\right) = 0 \\
R_{rr} &= -\nu'' + \nu'\lambda' + \frac{2\lambda'}{r} - (\nu')^2 = 0 \\
R_{\vartheta\vartheta} &= e^{-2\lambda}\left(-1 + r\lambda' - r\nu'\right) + 1 = 0 \\
R_{\varphi\varphi} &= \sin^2\vartheta \, R_{\vartheta\vartheta}
\end{aligned}
\tag{23.13}
$$

The factor $e^{2\nu - 2\lambda}$ in the first equation is non-zero, so we get

$$\nu'' + (\nu')^2 - \nu'\lambda' + \frac{2\nu'}{r} = 0.$$

Adding this equation to the above second one, we get

$$\frac{2\nu'}{r} + \frac{2\lambda'}{r} = 0 \Leftrightarrow \nu' + \lambda' = 0.$$

If we integrate this equation, we get

$$\nu(r) + \lambda(r) = const.$$

We have already shown above that the two functions

$$\nu(r), \lambda(r) \to 0 \; (r \to \infty)$$

must satisfy, i.e., the constant is zero, from which

$$\lambda(r) = -\nu(r) \tag{23.14}$$

follows. If we insert this into the equation for $R_{\vartheta\vartheta}$, we get

$$e^{2\nu}\left(-1 - r\nu' - r\nu'\right) + 1 = 0$$

so

$$e^{2\nu}\left(1 + 2r\nu'\right) = 1.$$

Now the left side is the derivative of the function $re^{2\nu}$ with respect to r, so we get

$$e^{2\nu}\left(1 + 2r\nu'\right) = \left(re^{2\nu}\right)' = 1,$$

and by integration we get

$$re^{2\nu} = r - 2m \Rightarrow e^{2\nu} = 1 - \frac{2m}{r},$$

where $-2m$ is a suitable integration constant, whose physical meaning becomes immediately clear. So, with our approach (23.7) and with (23.14)

$$g_{tt} = -\left(1 - \frac{2m}{r}\right)$$

$$g_{rr} = \left(1 - \frac{2m}{r}\right)^{-1}.$$

Now we know that according to Eq. (22.8) in the Newtonian limit

$$g_{tt} = -\left(1 - \frac{2GM}{r}\right)$$

applies, i.e., we can determine the integration constant as $m = GM$, where M denotes the total mass of the central body under consideration, and obtain for the sought line element

$$ds^2 = -\left(1 - \frac{2GM}{r}\right) dt^2 + \frac{1}{\left(1 - \frac{2GM}{r}\right)} dr^2 + r^2 \left(d\vartheta^2 + \sin^2 \vartheta \, d\varphi^2\right). \quad (23.15)$$

This is the famous **(outer) Schwarzschild solution**, named after the German astronomer Karl Schwarzschild, who found this metric as early as 1916. It is *exact* under the assumed symmetries and not an approximation. Since it covers the Newtonian limit, the approximation solution developed earlier for this limit is also exact. As required, the Schwarzschild solution approaches the Minkowski metric with $r \to \infty$ and it applies outside a body with mass M. It does not matter *how* the mass is distributed in the body. A very similar result we have derived in Sect. 6.3 for the Newtonian gravitational field. There, the gravitational field outside a body with arbitrary mass distribution is the same, as if the mass is concentrated point-like at the center of gravity.

23.3. Physical Interpretation of the Schwarzschild Solution

The Schwarzschild metric characterizes events in spacetime by the coordinates t, r, ϑ, φ, which look somewhat like the spherical coordinates for flat spacetime, see (23.1). To find out what physical meaning these coordinates have, we must examine the metric, for only it provides information about how distances or time intervals can be physically measured.

The Radial Coordinate

We consider a circle with constant r, i.e. $dr = 0$ in the equatorial plane, i.e.

$$\vartheta = \frac{\pi}{2} \Rightarrow \sin\vartheta = 1, d\vartheta = 0$$

at a certain fixed time t, i.e. $dt = 0$, and want to calculate the circumference U of this circle. To do this, we integrate the infinitesimal spacetime interval ds along the circumference curve. It applies under the made assumptions

$$ds^2 = 0 + 0 + 0 + r^2 d\varphi^2$$

and thus

$$ds = r d\varphi.$$

Integration yields

$$U = \int_U ds = \int_0^{2\pi} r\, d\varphi = 2\pi r \implies r = \frac{U}{2\pi}.$$

This means that the Schwarzschild metric defines the radial coordinate r of such a circle as the circumference divided by 2π, and since the orientation of the equatorial plane can be chosen arbitrarily, this applies to every circle around the considered object. The Schwarzschild solution thus yields the same result as the Euclidean geometry of flat space. But the r coordinate is - unlike for a sphere in three dimensions - *not equal* to the radial distance from the center to the circle line. To see this, we integrate ds along a radial line, i.e., we assume

$$dt = d\vartheta = d\varphi = 0$$

and obtain

$$ds^2 = \frac{1}{\left(1 - \dfrac{2GM}{r}\right)} dr^2 \implies ds = \frac{1}{\sqrt{1 - \dfrac{2GM}{r}}} dr.$$

The distance between two points with coordinates r_A and r_E results from this as

$$\Delta s = \int ds = \int_{r_A}^{r_E} \frac{1}{\sqrt{1 - \dfrac{2GM}{r}}} dr.$$

To approximate the integral, we estimate the integrand with the differential:

$$\Delta f = f(x_0 + dx) - f(x_0) \approx f'(x_0)\, dx$$

(see formula 4.7 on page 58). We assume that the coordinate r is very large compared to $2GM$ (it will be explained below that this can be assumed for most objects in the universe outside their mass distribution), then the term $2GM/r$ is very small and we can set $x_0 = 0$ in the formula for the differential. We define the function f by

$$f(x) = \frac{1}{\sqrt{1-x}} = (1-x)^{-1/2}\,,$$

so with

$$f'(x) = \frac{1}{2}\,(1-x)^{-3/2}$$

the approximation results

$$f\left(\frac{2GM}{r}\right) - f(0) \approx f'(0)\,\frac{2GM}{r}$$

and from this

$$\frac{1}{\sqrt{1 - \dfrac{2GM}{r}}} \approx 1 + \frac{1}{2}\,\frac{2GM}{r} = 1 + \frac{GM}{r}.$$

Now we approximate the above integral:

$$
\begin{aligned}
\Delta s &= \int_{r_A}^{r_E} \frac{1}{\sqrt{1 - \dfrac{2GM}{r}}}\, dr \\[2mm]
&\approx \int_{r_A}^{r_E} \left(1 + \frac{GM}{r}\right) dr \\[2mm]
&= \left[r + GM \ln r\right]_{r_A}^{r_E} \\[2mm]
&= r_E - r_A + GM\left(\ln\left(r_E\right) - \ln\left(r_A\right)\right)
\end{aligned}
$$

We obtain a larger expression than $r_E - r_A$, which would be the corresponding distance in flat space. The coordinate r therefore does *not* provide the radial distance of a point to the coordinate origin, and this in turn is an indication that the Schwarzschild metric describes a curved spacetime.

The Time Coordinate

We consider a clock at rest at a fixed position r, i.e.

$$dr = d\vartheta = d\varphi = 0.$$

In Eq. (20.3) we derived the proper time between two events A and B and obtained the expression

$$\Delta t_E = t_E(B) - t_E(A) = \int \sqrt{-ds^2} = \int_{\lambda_A}^{\lambda_B} \sqrt{-g_{\mu\nu}\, d\mu\, d\nu}.$$

If we substitute the Schwarzschild metric for $g_{\mu\nu}$, we obtain with the above assumptions

$$\begin{aligned}
\Delta t_E &= \int_{t_A}^{t_B} \sqrt{\left(1 - \frac{2GM}{r}\right) dt^2} = \sqrt{1 - \frac{2GM}{r}} \int_{t_A}^{t_B} dt \\
&= \sqrt{1 - \frac{2GM}{r}}\, (t_B - t_A) = \sqrt{1 - \frac{2GM}{r}}\, \Delta t.
\end{aligned}$$

The proper time t_E of the clock only coincides with the coordinate time t when the root is equal to 1, i.e., when $r = \infty$. The time difference between two events measured with the coordinate time therefore only corresponds to the proper time when the measuring clock is located far away "at infinity". But how can the coordinate time actually be measured at the locations of the events if the clock is located very far away? Well, one can imagine that clocks are installed at each point in space, which receive signals from the distant clock, which these that sends out every second. The local clock advances by one second whenever it receives a signal from the distant clock.

The last equation also states that a clock at rest measures less time than the coordinate time and that this difference becomes greater the smaller r becomes. This in turn is also an indication that the spacetime described by the Schwarzschild metric is curved.

The Schwarzschild Radius

Looking more closely at the Schwarzschild metric, one sees that the term

$$\frac{1}{\left(1 - \dfrac{2GM}{r}\right)}$$

tends towards infinity for $r \to 2GM$. The quantity

$$r_S = 2GM \tag{23.16}$$

is called **Schwarzschild radius**. The question of what happens as one approaches the Schwarzschild radius will be explored in more detail in the following chapters and will lead us to black holes. In this chapter, we want to exclusively consider the physical consequences of the Schwarzschild solution in the solar system and test the validity of the General Theory of Relativity there. The **singularity** (the divergence) of the Schwarzschild metric at $2GM$ is irrelevant in the solar system, as the Schwarzschild radius of the Sun is much smaller than the radius of the Sun and we only want to conduct investigations in the outer field of the Sun. To do this, we calculate the Schwarzschild radius r_S for the Sun in km and must first convert from natural units $(c = 1)$ back to SI units $(c = 3 \cdot 10^8 \, \text{m})$. To decide at which point in the Schwarzschild radius the speed of light c occurs, we make a dimensional consideration. The quantity $2GM/r$ has with the solar mass $M = 1.99 \cdot 10^{30} \, \text{kg}$ and the Newtonian gravitational constant

$$G = 6.67 \cdot 10^{-11} \, \frac{\text{m}^3}{\text{kg}\,\text{s}^2}$$

the unit

$$\left[\frac{2GM}{r} \right] = \frac{\text{m}^3 \, \text{kg}}{\text{kg}\,\text{s}^2\,\text{m}} = \frac{\text{m}^2}{\text{s}^2},$$

but should be dimensionless in SI units. Therefore, the Schwarzschild radius in SI units becomes

$$r_S = \frac{2GM}{c^2}. \tag{23.17}$$

We insert the numbers and get

$$r_S = \frac{2GM}{c^2} = \frac{2 \cdot 6.67 \cdot 10^{-11} \cdot 1.99 \cdot 10^{30}}{9 \cdot 10^{16}} \approx 2.95 \, \text{km}.$$

Since the solar radius is $R = 696,000 \, \text{km}$, the Schwarzschild radius lies within the Sun. So we do not approach the Schwarzschild radius of the Sun, as we only investigate phenomena in the outer field of the Sun. There $r > R$ and thus

$$\frac{2GM}{rc^2} = \frac{r_S}{r} < 4.2 \cdot 10^{-6},$$

i.e., the Schwarzschild metric deviates only slightly from the Minkowski metric, and the assumption made earlier, that in most calculations $r \gg r_S$ can be assumed, is thus explained.

The coefficients of the Schwarzschild metric become singular at $r = r_S$, but this does not necessarily mean that spacetime also becomes singular there. For example, if one looks at the spherical coordinates in Euclidean space, it follows

$$g^{\varphi\varphi} = \frac{1}{\sin^2 \vartheta} \to \infty \, (\vartheta \to 0, \pi),$$

i.e., the metric is singular at the poles, although nothing unusual happens at the poles. The surface of the sphere is homogeneous and no point differs from another. Such singularities are called **coordinate singularities**. In contrast to "real" singularities, we will see in the next chapters that the singularity of the Schwarzschild metric at $r = r_S$ is also only a coordinate singularity, which can be resolved by suitable coordinate transformations.

Nevertheless, the Schwarzschild radius has a special physical significance. In Sect. 6.1 we calculated how large the escape velocity must be, so that a particle can overcome the gravity of a spherical body with radius R and mass M, namely (see 6.8 on page 92):

$$v_F > \sqrt{\frac{2GM}{R}}$$

Assuming that light consists of particles (photons) that move at the speed of light c, it follows that light can escape from a body when

$$c^2 > \frac{2GM}{R}$$

or conversely, that light cannot escape when

$$R < \frac{2GM}{c^2} = r_S.$$

Therefore, stars whose radius is smaller than the Schwarzschild radius ($R < r_S$), are called **black holes**. From the surface of a black hole, no photon can escape (this only applies if quantum effects like the so-called **Hawking radiation** are ignored). More details can be found in [25]. If ρ is the density of a spherical homogeneous star, it follows

$$M = \frac{4}{3} \pi R^3 \rho$$

and thus for a star that emits no light

$$R < \frac{2GM}{c^2} = \frac{8\pi G R^3 \rho}{3c^2},$$

so

$$R^2 > \frac{3c^2}{8\pi G \rho},$$

i.e., if the star is large enough given the density ρ, it will also not shine.

23.4. Gravitational Redshift in Schwarzschild Spacetime

We continue the considerations from Sect. 18.2 and now assume that we are in Schwarzschild spacetime. The metric tensor

$$
g_{\mu\nu} = \begin{pmatrix} -\left(1 - \dfrac{2GM}{r}\right) & 0 & 0 & 0 \\ 0 & \left(1 - \dfrac{2GM}{r}\right)^{-1} & 0 & 0 \\ 0 & 0 & r^2 & 0 \\ 0 & 0 & 0 & r^2 \sin^2 \vartheta \end{pmatrix}
$$

is time-constant, i.e. independent of the coordinate time t. This again refers to the time displayed on the clock of the observer assumed to be far away. The proper time t_E of a particle in the gravitational field is the time that the particle itself measures on its own clock, and is calculated according to the formula (19.8) by

$$
-dt_E^2 = ds^2 = g_{\mu\nu}\, d\mu\, d\nu,
$$

from which

$$
dt_E = \sqrt{-g_{\mu\nu}\, d\mu\, d\nu} \tag{23.18}
$$

follows. If one wants to measure the proper time between two events, each taking place at a fixed point in space, then

$$
dx = dy = dz = 0
$$

and one obtains

$$
dt_E = \sqrt{-g_{tt}\, dt^2} = \sqrt{1 - \frac{2GM}{r}}\, dt.
$$

Since the root is less than 1, it follows that $dt_E < dt$, which, with the same argument as in Sect. 18.2, means that clocks and thus time in a Schwarzschild field run slower. We derived the same result in Sect. 22.2 for a weak, static gravitational field. There it was

$$
g_{tt} = -(1 + 2\phi),
$$

which with

$$
\phi = -\frac{GM}{r}
$$

in the Newtonian limit also leads to

$$dt_E = \sqrt{1 - \frac{2GM}{r}}\, dt.$$

To calculate the redshift of light in Schwarzschild spacetime, we need to compare the time differences at two different points in the gravitational field. To do this, we imagine that we emit light of a certain frequency ν_1 at a point r_1, which is received at a second point r_2 with a frequency ν_2 measured there. The wavelength of the light corresponds to a certain proper length ds and this in turn to the proper time $dt_E \sim ds$. In Fig. 23.1, the emissions of two successive light waves are shown. The dashed lines can be interpreted as wave crests of the two light waves. We define the coordinate time interval Δt_1 as the time span between the emission of two successive wave crests. Due to the time independence of the Schwarzschild metric, it follows that both wave crests take the same time to get from r_1 to r_2.

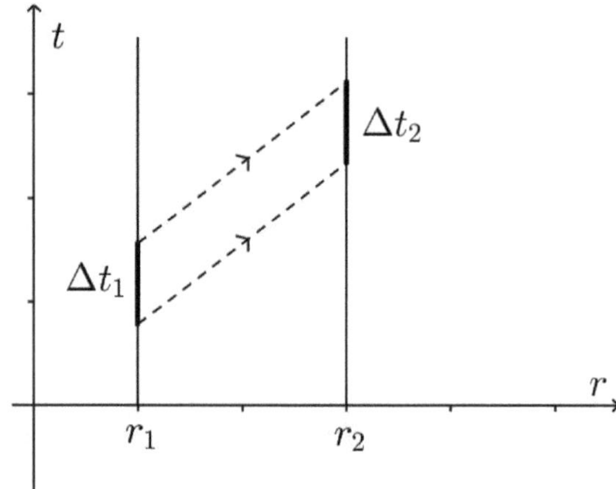

Figure 23.1.: Light waves in Schwarzschild spacetime

This means that the time interval for reception at r_2 is also Δt_1, so

$$\Delta t_2 = \Delta t_1.$$

The proper times $(dt_E)_1$ and $(dt_E)_2$ measure the time between the two successive wave crests and thus the period T of the light waves. Since the period is inversely proportional to the frequency $(T = 1/f)$, it follows

$$(dt_E)_1 = \frac{1}{f_1} = \Delta t_1 \sqrt{-g_{tt}(r_1)}$$

and

$$(dt_E)_2 = \frac{1}{f_2} = \Delta t_2 \sqrt{-g_{tt}(r_2)} = \Delta t_1 \sqrt{-g_{tt}(r_2)}.$$

Division yields

$$\frac{f_2}{f_1} = \sqrt{\frac{-g_{tt}(r_1)}{-g_{tt}(r_2)}} = \sqrt{\frac{1 - 2GM/r_1}{1 - 2GM/r_2}}. \tag{23.19}$$

To make the last term a bit clearer, we estimate the numerator and denominator each with the differential. To do this, we first consider the function $f(x) = \sqrt{1-x}$ and get for small x

$$f(x) \approx f(0) + x \cdot f'(0) = 1 + x \left[-\frac{1}{2\sqrt{1-x}} \right]_{x=0} = 1 - \frac{x}{2}.$$

So for the numerator

$$\sqrt{1 - \frac{2GM}{r_1}} \approx 1 - \frac{GM}{r_1}.$$

Similarly, as already shown in section 23.3 with $f(x) = 1/\sqrt{1-x}$:

$$f(x) \approx f(0) + x \cdot f'(0) = 1 + x \left[-\frac{-1}{2(1-x)^{3/2}} \right]_{x=0} = 1 + \frac{x}{2}$$

for the denominator

$$\frac{1}{\sqrt{1 - \dfrac{2GM}{r_2}}} \approx 1 + \frac{GM}{r_2}$$

With this estimate, we get

$$\frac{f_2}{f_1} \approx \left(1 - \frac{GM}{r_1} \right) \left(1 + \frac{GM}{r_2} \right)$$

and in linear approximation, i.e. without the quadratic term $\dfrac{G^2 M^2}{r_1 r_2}$,

$$\frac{f_2}{f_1} \approx 1 - GM \left(\frac{1}{r_1} - \frac{1}{r_2} \right). \tag{23.20}$$

Since $r_1 < r_2$, it follows that $1/r_1 > 1/r_2$ and the bracket on the right side is positive, i.e. $f_2 < f_1$. The light is redshifted. We want to calculate the redshift in the Sun-Earth system specifically and consider a photon that is emitted from the surface of the Sun $r_1 = 6.96 \cdot 10^8$ m with a frequency f_1. Since the distance Sun-Earth is $r_2 = 1.5 \cdot 10^{11}$ m and thus $r_2 \gg r_1$, it follows with (23.20)

$$\frac{f_2}{f_1} \approx 1 - GM \left(\frac{1}{r_1} - \frac{1}{r_2} \right) \approx 1 - GM \frac{1}{r_1}.$$

Now the Schwarzschild radius of the Sun is $2GM = 2950$ m, from which

$$\frac{f_2}{f_1} \approx 1 - \frac{GM}{r_1} = 1 - \frac{1475}{6.96 \cdot 10^8} = 1 - 2.12 \cdot 10^{-6}$$

follows. If $\Delta f = f_2 - f_1$ denotes the difference between the two frequencies, then the relative frequency change is

$$\frac{\Delta f}{f_1} = -2.12 \cdot 10^{-6}.$$

This is the prediction for the redshift of sunlight resulting from the Schwarzschild metric. The effect is quite small and its determination is complicated by the relative speed between Earth and Sun, by the thermal motion of the atoms and the convection currents in the solar atmosphere. Nevertheless a good agreement of the measured values with the theoretical predictions in the order of magnitude

$$\Delta f_{exp} = \Delta f_{theor} \cdot (1.01 \pm 0.06)$$

could be demonstrated, see [39].

Moving Clocks and the Global Positioning System (GPS)

So far, we have considered the time differences between stationary clocks in a gravitational field. For moving clocks, the relation between the proper and coordinate time also results from (23.18):

$$dt_E = \sqrt{-g_{\mu\nu} \, d\mu \, d\nu},$$

where the relative speeds of the clocks must be taken into account. As an example, we take a clock that rests on the surface of the Earth, and another that orbits the Earth in a satellite. If you compare both clocks, you can see two opposing effects. On the one hand, the clock in the satellite runs faster because the gravitational field of the Earth is weaker there. On the other hand, it runs slower because the satellite is moving relative to the Earth. Modern satellite systems like the Global Positioning System have to take both effects into account. We want (albeit under significantly simplified assumptions, see the article "The GPS" by Grafarend and Schwarze in the journal Physics Journal

1) to calculate the total time differences for a fictional satellite approximately. For this, we first assume that the Earth is an instantaneous inertial system, i.e., the orbital speed of the Earth around the Sun and also the rotation of the Earth is neglected. The satellite including the clock may be at a height of approx. $20,000\,\mathrm{km}$ and with a speed $v = 3.87\,\mathrm{km/s}$ move relative to the clock on the Earth. To calculate the time differences due to the relative movement of the clocks, we use Minkowski coordinates, i.e. $g_{\mu\nu} = \eta_{\mu\nu}$, then it follows with $\dfrac{di}{dt} = v^i$

$$
\begin{aligned}
dt_E &= \sqrt{-g_{\mu\nu}\,d\mu\,d\nu} = \sqrt{-\eta_{\mu\nu}\left(\frac{d\mu}{dt}\frac{d\nu}{dt}\right)dt^2} = \sqrt{1 - \delta_{ij}\left(v^i v^j\right)}\,dt \\
&= \sqrt{1 - v^2}\,dt,
\end{aligned}
$$

so the well-known formula for time dilation in Special Relativity. We insert the numbers and obtain in *SI units* for the relative time difference per second

$$
\frac{dt_E - dt}{dt} = \frac{\sqrt{1 - \dfrac{v^2}{c^2}}\,dt - dt}{dt} = \sqrt{1 - \frac{v^2}{c^2}} - 1 = -8.3 \cdot 10^{-11}.
$$

This negative time difference accumulates in one day to approximately seven microseconds $\left(= -7 \cdot 10^{-6}\,\mathrm{s}\right)$. For the calculation of the gravitational effect, we use the formula

$$
dt_E = \sqrt{1 - \frac{2GM}{r}}\,dt
$$

and again calculate in SI units the proper time for the clock at the Earth's surface with the Earth's radius $r_E = 6.378 \cdot 10^6\,\mathrm{m}$, the Earth's mass $M_E = 5.97 \cdot 10^{24}\,\mathrm{kg}$ and the gravitational constant $G = 6.67 \cdot 10^{-11}\,\mathrm{m^3\,kg^{-1}\,s^{-2}}$

$$
dt_{Earth} = \sqrt{1 - \frac{2GM_E}{c^2 r_E}}\,dt = \sqrt{1 - \frac{2 \cdot 6.67 \cdot 10^{-11} \cdot 5.97 \cdot 10^{24}}{(3 \cdot 10^8)^2\, 6.378 \cdot 10^6}}\,dt.
$$

For the satellite, the following applies accordingly

$$
\begin{aligned}
dt_{Satellite} &= \sqrt{1 - \frac{2GM_E}{c^2(r_E + 20,000,000)}}\,dt \\
&= \sqrt{1 - \frac{2 \cdot 6.67 \cdot 10^{-11} \cdot 5.97 \cdot 10^{24}}{(3 \cdot 10^8)^2\,(6.378 + 20)\,10^6}}\,dt.
\end{aligned}
$$

For the relative difference of the two clocks per second, this results in

$$\frac{dt_{Satellite} - dt_{Earth}}{dt_{Earth}} \approx 52.7 \cdot 10^{-11},$$

and per day there is a positive deviation of about 45 microseconds. If both effects are taken into account, the clocks in the satellites run about 38 microseconds faster per day. Since the satellites are equipped with precise atomic clocks, these small time differences can be measured and corrected. This correction is necessary, as otherwise the accuracy of the position indication ($10\,$m) can no longer be guaranteed.

23.5. Movements in Schwarzschild Spacetime

In this section, we work again with natural units, unless otherwise stated. We derive the equations of motion of planets and light in the gravitational field of the Sun, i.e., we consider the planets as test bodies and neglect their own gravitational field. According to Eq. (20.2) this movement is given by the geodesic equation:

$$\frac{d^2\mu}{d\lambda^2} + \Gamma^\mu_{\nu\rho}\frac{d\nu}{d\lambda}\frac{d\rho}{d\lambda} = 0,$$

where λ is an initially arbitrary path parameter $\mu = \mu(\lambda)$. The non-zero Christoffel symbols are obtained according to Eq. (23.12) and with $\lambda = -\nu$ and

$$e^{2\nu} = 1 - \frac{2GM}{r} \Rightarrow \nu = \frac{1}{2}\ln\left(1 - \frac{2GM}{r}\right)$$

$$\Rightarrow \nu' = \frac{1}{2}\frac{1}{1 - \dfrac{2GM}{r}}\frac{2GM}{r^2} = \frac{GM}{r\,(r - 2GM)}$$

to

$$\Gamma^t_{rt} = \Gamma^t_{tr} = \nu' = \frac{GM}{r\,(r - 2GM)}$$

$$\Gamma^r_{tt} = \nu'\,e^{2\nu - 2\lambda} = \frac{GM}{r\,(r - 2GM)}\left(1 - \frac{2GM}{r}\right)^2 = \frac{GM\,(r - 2GM)}{r^3}$$

$$\Gamma^r_{rr} = \lambda' = -\frac{GM}{r\,(r - 2GM)}$$

$$\Gamma^r_{\vartheta\vartheta} = -re^{-2\lambda} = -(r - 2GM)$$

$$\Gamma^r_{\varphi\varphi} = -r\sin^2\vartheta\,e^{-2\lambda} = -(r - 2GM)\sin^2\vartheta$$

$$
\begin{aligned}
\Gamma^{\vartheta}_{\vartheta r} &= \Gamma^{\vartheta}_{r\vartheta} = \frac{1}{r} \\
\Gamma^{\vartheta}_{\varphi\varphi} &= -\sin\vartheta\cos\vartheta \\
\Gamma^{\varphi}_{\varphi r} &= \Gamma^{\varphi}_{r\varphi} = \frac{1}{r} \\
\Gamma^{\varphi}_{\varphi\vartheta} &= \Gamma^{\varphi}_{\vartheta\varphi} = \cot\vartheta.
\end{aligned}
$$

For $\mu = t$, the geodesic equation yields

$$
\frac{d^2 t}{d\lambda^2} + \Gamma^{t}_{rt}\frac{dr}{d\lambda}\frac{dt}{d\lambda} + \Gamma^{t}_{tr}\frac{dt}{d\lambda}\frac{dr}{d\lambda} = \ddot{t} + \frac{2GM}{r\,(r-2GM)}\dot{r}\,\dot{t} = 0,
$$

where we have chosen the "dot notation" for the derivative with respect to λ, e.g. $\left(dt/d\lambda = \dot{t}\right)$. Now, with the product and chain rule

$$
\frac{d}{d\lambda} = \frac{d}{dr}\frac{dr}{d\lambda}
$$

and thus

$$
\begin{aligned}
\frac{d}{d\lambda}\left(\left(1 - \frac{2GM}{r}\right)\frac{dt}{d\lambda}\right) &= \frac{2GM}{r^2}\frac{dr}{d\lambda}\frac{dt}{d\lambda} + \frac{d^2 t}{d\lambda^2}\left(1 - \frac{2GM}{r}\right) \\
&= \left(1 - \frac{2GM}{r}\right)\Big[\underbrace{\ddot{t} + \frac{2GM}{r\,(r-2GM)}\dot{r}\,\dot{t}}_{=0}\Big] = 0,
\end{aligned}
$$

applies, i.e., it also holds that

$$
\frac{d}{d\lambda}\left(\left(1 - \frac{2GM}{r}\right)\dot{t}\right) = 0
$$

or after integration,

$$
\left(1 - \frac{2GM}{r}\right)\dot{t} = b = const. \tag{23.21}
$$

with an integration constant b. For $\mu = \vartheta$, the geodesic equation yields

$$
\frac{d^2\vartheta}{d\lambda^2} + \Gamma^{\vartheta}_{\vartheta r}\frac{d\vartheta}{d\lambda}\frac{dr}{d\lambda} + \Gamma^{\vartheta}_{r\vartheta}\frac{dr}{d\lambda}\frac{d\vartheta}{d\lambda} + \Gamma^{\vartheta}_{\varphi\varphi}\frac{d\varphi}{d\lambda}\frac{d\varphi}{d\lambda} = \ddot{\vartheta} + \frac{2}{r}\dot{\vartheta}\,\dot{r} - \sin\vartheta\cos\vartheta\,\dot{\varphi}^2 = 0. \tag{23.22}
$$

And for $\mu = \varphi$, similarly, we get

$$
\ddot{\varphi} + \frac{2}{r}\dot{\varphi}\,\dot{r} + 2\cot\vartheta\,\dot{\vartheta}\,\dot{\varphi} = 0. \tag{23.23}
$$

Instead of $\mu = r$, we consider the line element

$$ds^2 = -\left(1 - \frac{2GM}{r}\right) dt^2 + \frac{1}{\left(1 - \frac{2GM}{r}\right)} dr^2 + r^2 \left(d\vartheta^2 + \sin^2\vartheta \, d\varphi^2\right)$$

and divide it by $ds^2 = -dt_E^2$:

$$1 = \left(1 - \frac{2GM}{r}\right)\left(\frac{dt}{dt_E}\right)^2 - \frac{1}{\left(1 - \frac{2GM}{r}\right)}\left(\frac{dr}{dt_E}\right)^2$$
$$- r^2\left(\left(\frac{d\vartheta}{dt_E}\right)^2 + \sin^2\vartheta\left(\frac{d\varphi}{dt_E}\right)^2\right)$$

The advantage of this equation is that for light $ds^2 = 0$ applies, so that the left side of the last equation is 0 instead of 1. If m is the mass of the test body, one can therefore write with $\lambda = t_E$ for a planet or $\lambda \neq t_E$ for light summarily

$$\begin{cases} \left(1 - \dfrac{2GM}{r}\right)\dot{t}^2 - \dfrac{1}{\left(1 - \dfrac{2GM}{r}\right)}\dot{r}^2 - r^2\left(\dot{\vartheta}^2 + \sin^2\vartheta\,\dot{\varphi}^2\right) = 1 \quad m \neq 0 \\[4ex] \left(1 - \dfrac{2GM}{r}\right)\dot{t}^2 - \dfrac{1}{\left(1 - \dfrac{2GM}{r}\right)}\dot{r}^2 - r^2\left(\dot{\vartheta}^2 + \sin^2\vartheta\,\dot{\varphi}^2\right) = 0 \quad m = 0 \end{cases}$$

$$(23.24)$$

(23.22) can apparently be solved by

$$\vartheta = \frac{\pi}{2} = const. \tag{23.25}$$

This is not a restriction of the variety of solutions, because if one sets the coordinate system as the initial condition of the planet's motion so that $\vartheta(0) = \pi/2$ and $\dot{\vartheta}(0) = 0$ applies (i.e., the initial location and velocity vector lie in the equatorial plane), then it follows from Eq. (23.22) that $\ddot{\vartheta}(0) = 0$ is. If you differentiate this equation once more, $\dddot{\vartheta}(0) = 0$ follows, and if you continue this process, all higher derivatives of the function ϑ at the point 0 are also zero. But if such a situation exists, then a theorem of mathematics ("**Taylor's theorem**"), which we do not want to prove here, says that $\vartheta(\lambda) = \pi/2$ holds for all λ, i.e., ϑ is a constant function, so $\dot{\vartheta}(\lambda) = 0$ follows also for all λ. This means that the entire trajectory lies in the equatorial plane. We get the same result as in Newtonian mechanics, there too the orbits of the planets around the Sun lie in a plane.

But if $\dot{\vartheta}(\lambda) = 0$, then from Eq. (23.23) we get

$$\ddot{\varphi} + \frac{2}{r}\,\dot{\varphi}\,\dot{r} = 0.$$

Now is

$$\frac{d}{d\lambda}\left(r^2\dot{\varphi}\right) = 2r\dot{r}\dot{\varphi} + r^2\ddot{\varphi} = r^2\left(\ddot{\varphi} + \frac{2}{r}\,\dot{\varphi}\,\dot{r}\right),$$

i.e., it also follows that

$$\frac{d}{d\lambda}\left(r^2\dot{\varphi}\right) = 0.$$

By integration, we get

$$r^2\dot{\varphi} = l = const. \tag{23.26}$$

We can justify Eq. (23.25) and (23.26) directly with the isotropy, as is common in the non-relativistic Kepler problem (see formula 6.14 on page 97 and the explanations there). Because of the isotropy, the angular momentum $\vec{l}$ is preserved, so the direction of $\vec{l}$ is constant, i.e., the coordinate system can be chosen so that $\vec{l}$ is parallel to the z-axis, from which directly follows Eq. (23.25). Since the magnitude of $\vec{l}$ is also constant, Eq. (23.26) applies. The integration constant l there corresponds to the angular momentum from (6.14) divided by the mass.

Taking into account that from (23.26) and with the chain rule

$$\dot{r} = \frac{dr}{d\lambda} = \frac{dr}{d\varphi}\frac{d\varphi}{d\lambda} = \frac{dr}{d\varphi}\dot{\varphi} = \frac{dr}{d\varphi}\frac{l}{r^2} \tag{23.27}$$

follows, we obtain by substituting (23.21) and (23.27) into (23.24)

$$
\begin{aligned}
1 &= \left(1 - \frac{2GM}{r}\right)\dot{t}^2 - \frac{1}{\left(1 - \dfrac{2GM}{r}\right)}\dot{r}^2 - r^2\left(\dot{\vartheta}^2 + \sin^2\vartheta\,\dot{\varphi}^2\right) \\
&= \left(1 - \frac{2GM}{r}\right)^{-1}b^2 - \left(1 - \frac{2GM}{r}\right)^{-1}\left(\frac{dr}{d\varphi}\frac{l}{r^2}\right)^2 - r^2\left(\frac{l}{r^2}\right)^2 \\
&= \left(1 - \frac{2GM}{r}\right)^{-1}b^2 - \left(1 - \frac{2GM}{r}\right)^{-1}\frac{l^2}{r^4}\left(\frac{dr}{d\varphi}\right)^2 - \frac{l^2}{r^2},
\end{aligned}
$$

where in the second equation we have exploited $\vartheta = \pi/2, \dot{\vartheta} = 0$ and Eq. (23.26). Now it holds

$$\left[\frac{d}{d\varphi}\left(\frac{1}{r}\right)\right]^2 = \left[-\frac{1}{r^2}\frac{dr}{d\varphi}\right]^2 = \frac{1}{r^4}\left(\frac{dr}{d\varphi}\right)^2. \tag{23.28}$$

Substituting this into the last equation yields

$$1 = \left(1 - \frac{2GM}{r}\right)^{-1} b^2 - \left(1 - \frac{2GM}{r}\right)^{-1} l^2 \left[\frac{d}{d\varphi}\left(\frac{1}{r}\right)\right]^2 - \frac{l^2}{r^2}.$$

Multiplication with $\left(1 - \dfrac{2GM}{r}\right)/l^2$ results in

$$\left(1 - \frac{2GM}{r}\right)/l^2 = b^2/l^2 - \left[\frac{d}{d\varphi}\left(\frac{1}{r}\right)\right]^2 - \frac{\left(1 - \dfrac{2GM}{r}\right)}{r^2}.$$

Rearranging and simplifying lead to

$$\left[\frac{d}{d\varphi}\left(\frac{1}{r}\right)\right]^2 + \frac{1}{r^2} = \frac{b^2 - 1}{l^2} + \frac{2GM}{rl^2} + \frac{2GM}{r^3}. \tag{23.29}$$

We differentiate this equation with respect to φ and obtain with

$$\frac{d}{d\varphi}\left[\frac{d}{d\varphi}\left(\frac{1}{r}\right)\right]^2 = 2\frac{d}{d\varphi}\left(\frac{1}{r}\right)\frac{d^2}{d\varphi^2}\left(\frac{1}{r}\right)$$

$$\frac{d}{d\varphi}\left(\frac{1}{r}\right)^2 = \frac{2}{r}\frac{d}{d\varphi}\left(\frac{1}{r}\right)$$

$$\frac{d}{d\varphi}\left(\frac{b^2 - 1}{l^2}\right) = 0$$

$$\frac{d}{d\varphi}\left(\frac{2GM}{rl^2}\right) = \frac{2GM}{l^2}\frac{d}{d\varphi}\left(\frac{1}{r}\right)$$

$$\frac{d}{d\varphi}\left(\frac{2GM}{r^3}\right) = -\frac{6GM}{r^4}\frac{dr}{d\varphi} = \frac{6GM}{r^2}\left(\frac{-1}{r^2}\frac{dr}{d\varphi}\right) = \frac{6GM}{r^2}\frac{d}{d\varphi}\left(\frac{1}{r}\right)$$

and with $\sigma = 1/r$ the equation

$$2\frac{d\sigma}{d\varphi}\frac{d^2\sigma}{d\varphi^2} + 2\sigma\frac{d\sigma}{d\varphi} = \frac{2GM}{l^2}\frac{d\sigma}{d\varphi} + 6GM\sigma^2\frac{d\sigma}{d\varphi}.$$

First, we notice that this equation is solved by

$$\frac{d\sigma}{d\varphi} = 0,$$

which by integration leads to

$$\sigma(\varphi) = const. \Rightarrow r(\varphi) = const.$$

So, this solution is a circle. Planetary orbits in Schwarzschild spacetime can therefore be circular orbits around the Sun, just like in Newtonian gravitational theory. If we exclude the circular orbits, we can divide the above equation by $d\sigma/d\varphi$ and obtain

$$\frac{d^2\sigma}{d\varphi^2} + \sigma = \frac{GM}{l^2} + 3GM\sigma^2 \tag{23.30}$$

as the determining equation for the planetary orbits. In the case of light, we obtain with exactly the same derivation instead of (23.30) the equation of motion

$$\frac{d^2\sigma}{d\varphi^2} + \sigma = 3GM\sigma^2. \tag{23.31}$$

We want to compare Eq. (23.30) with the equation of motion of a planet in Newtonian gravitational theory and start from Eq. 6.15 on page 99 from the section on Kepler orbits. There it was:

$$-\frac{d\sigma}{d\varphi} = \frac{1}{r^2}\frac{dr}{d\varphi} = \sqrt{\frac{2m\,(E + A\sigma)}{L^2} - \sigma^2}$$

with

$$
\begin{aligned}
m &= \textit{mass of the planet,}\\
E &= \textit{total energy} = \textit{const.}\\
A &= GMm\\
L &= \textit{angular momentum} = l \cdot m
\end{aligned}
$$

If you square the last equation and insert the values, you get

$$\frac{1}{r^4}\left(\frac{dr}{d\varphi}\right)^2 + \frac{1}{r^2} = \frac{2m\left(E + \dfrac{GMm}{r}\right)}{m^2 l^2} = \frac{2E/m}{l^2} + \frac{2GM}{l^2 r}.$$

If you consider that as above in Eq. (23.28)

$$\frac{1}{r^4}\left(\frac{dr}{d\varphi}\right)^2 = \left[\frac{d}{d\varphi}\left(\frac{1}{r}\right)\right]^2$$

holds, it follows

$$\left[\frac{d}{d\varphi}\left(\frac{1}{r}\right)\right]^2 + \frac{1}{r^2} = \frac{2E/m}{l^2} + \frac{2GM}{l^2 r}.$$

So you get an equation that is very similar to (23.29) and is treated in the same way. Differentiation with respect to φ gives with $\sigma = 1/r$ as above

$$\frac{d\sigma}{d\varphi}\frac{d^2\sigma}{d\varphi^2} + \sigma\frac{d\sigma}{d\varphi} = \frac{GM}{l^2}\frac{d\sigma}{d\varphi}.$$

Again, it should be noted that the equation is solved by $d\sigma/d\varphi = 0$, i.e., by a circular orbit. The other solutions are obtained after division by $d\sigma/d\varphi$:

$$\frac{d^2\sigma}{d\varphi^2} + \sigma = \frac{GM}{l^2} \tag{23.32}$$

If you compare this Newtonian orbit equation with the one from Schwarzschild spacetime (23.30), the equations differ only by the term $3GM\sigma^2$, which additionally appears on the right side in the Schwarzschild solution.

23.6. Perihelion Precession of Mercury

In Sect. 6.2 on planetary orbits in the Newtonian gravitational field, it was pointed out that there is an anomaly in the orbit of Mercury known since the 19th century, which could not be explained by Newton's law of gravitation. In this section, we want to show that the Schwarzschild solution is suitable for explaining this small deviation (43 arcseconds per century) of Mercury's orbit. Since we want to work with SI units again, we set the Schwarzschild radius to

$$r_S = \frac{2GM}{c^2} = 2m,$$

which defines the quantity m. In Sect. 6.2 we have found the solution of (23.32) (see Eq. 6.17 on page 101). It was

$$\sigma\left(\varphi\right) = \frac{1}{r\left(\varphi\right)} = \frac{1 + \varepsilon\cos\varphi}{p} \tag{23.33}$$

with $p = l^2/GM = l^2/mc^2$. The polar coordinates are chosen so that the point closest to the Sun (the **perihelion**) is at $\varphi = 0$. The quantity ε is the eccentricity of the elliptical orbit and is related to p via the relationship

$$p = a\left(1 - \varepsilon^2\right),$$

where a denotes the semi-major axis of the elliptical orbit. The additional term $3m\sigma^2$ in Eq. (23.30) is a small perturbation term in our planetary system. For the planet Mercury, this is estimated by inserting the unperturbed solution (23.33) into the additional term. It then results with

$$r_S = 2m = 2.95\,\text{km}, a = 5.8 \cdot 10^7\,\text{km}, \varepsilon = 0.21 \tag{23.34}$$

(see Sect. 1.1) for the relative size of the two terms on the right side of Eq. (23.30)

$$\frac{2.\,Term}{1.\,Term} = \frac{3m\sigma^2}{mc^2/l^2} = \frac{3m\left(1 + \varepsilon\cos\varphi\right)^2}{\left(mc^2/l^2\right)p^2} \tag{23.35}$$

$$= \frac{3m\left(1 + \varepsilon \cos\varphi\right)^2}{p} \approx \frac{3m}{p} = \frac{3m}{a\left(1 - \varepsilon^2\right)}$$

$$= \frac{3 \cdot 2.95/2}{5.8 \cdot 10^7 \left(1 - 0.0441\right)} = 8 \cdot 10^{-8}.$$

In the second line, we have approximated the term $3m\left(1 + \varepsilon \cos\varphi\right)^2/p$ by $3m/p$. If you insert the values from (23.34) into the first term, you get

$$\frac{3m\left(1 + \varepsilon \cos\varphi\right)^2}{p} \leq \frac{4.4 \cdot \left(1 + 0.21\right)^2}{5.8 \cdot 10^7 \left(1 - 0.0441\right)} \approx \frac{6.4}{5.5 \cdot 10^7} \approx 1.4 \cdot 10^{-7}$$

i.e., the difference of the two terms is at most $6 \cdot 10^{-8}$.

Since the ratio in Eq. (23.35) is small, one can determine the changes of the Kepler orbits by **perturbation calculation**. In the perturbation calculation, one makes the following approach. One writes the sought solution $\sigma(\varphi)$ as the sum of the Kepler solution (23.33) and a **perturbation function** $\delta(\varphi)$:

$$\sigma(\varphi) = \sigma^0\left(\varphi\right) + \delta\left(\varphi\right)$$

with

$$\sigma^0 = \frac{1 + \varepsilon \cos\varphi}{p}$$

If we insert this approach into Eq. (23.30), we get

$$\left(\sigma^0 + \delta\right)'' + \left(\sigma^0 + \delta\right) = \frac{m}{l^2} + 3m\left(\sigma^0 + \delta\right)^2,$$

where for simplicity we have again marked the derivative with respect to φ with a prime $'$. Now we use the relationship (23.32):

$$\frac{d^2\sigma^0}{d\varphi^2} + \sigma^0 = \frac{m}{l^2}$$

and obtain a determining equation for $\delta(\varphi)$

$$\delta'' + \delta = 3m(\sigma^0 + \delta)^2.$$

If we neglect the terms $6m\sigma^0\delta$ and $3m\delta^2$, which result from expanding the right side, we get for δ the approximate differential equation

$$\delta'' + \delta = 3m(\sigma^0)^2 = 3m\left(\frac{1 + \varepsilon \cos\varphi}{p}\right)^2.$$

Since ε^2/p^2 is a small quantity, we neglect the quadratic term on the right side and obtain

$$\delta'' + \delta = \frac{3m}{p^2}\left(1 + 2\varepsilon\cos\varphi\right) \approx \frac{6m\varepsilon}{p^2}\cos\varphi, \qquad (23.36)$$

where we have also neglected the term $3m/p^2$, which is small according to (23.35). This is a second order linear differential equation, which can be solved using special mathematical methods. However, since we do not want to deal with these methods here, we check whether the function

$$\delta(\varphi) = \frac{3m\varepsilon}{p^2}\,\varphi\cdot\sin\varphi$$

is a solution of the last equation. We calculate the first two derivatives using the product rule

$$\delta' \;=\; \frac{3m\varepsilon}{p^2}\left(\sin\varphi + \varphi\cos\varphi\right)$$

$$\delta'' \;=\; \frac{3m\varepsilon}{p^2}\left(\cos\varphi + \cos\varphi - \varphi\sin\varphi\right)$$

and add δ'' and δ:

$$\frac{3m\varepsilon}{p^2}\left(\cos\varphi + \cos\varphi - \varphi\sin\varphi\right) + \frac{3m\varepsilon}{p^2}\,\varphi\cdot\sin\varphi = \frac{6m\varepsilon}{p^2}\cos\varphi$$

This means that the function

$$\delta(\varphi) = \frac{3m\varepsilon}{p^2}\,\varphi\cdot\sin\varphi$$

solves the differential equation (23.36). In total, it follows that

$$\sigma(\varphi) \;=\; \sigma^0(\varphi) + \delta(\varphi) = \frac{1 + \varepsilon\cos\varphi}{p} + \frac{3m\varepsilon}{p^2}\,\varphi\cdot\sin\varphi$$

$$\;=\; \frac{1}{p}\left(1 + \varepsilon\left(\cos\varphi + \frac{3m}{p}\,\varphi\cdot\sin\varphi\right)\right)$$

is an approximate solution for the planetary motion of Mercury. Note that the last term on the right side ensures that the function is not periodic, and therefore not an ellipse. Since the term $3m/p$ is a small quantity according to (23.35), we can estimate

$$\left(\cos\varphi + \frac{3m}{p}\,\varphi\cdot\sin\varphi\right)$$

in linear approximation by the differential. It holds with

$$x = \varphi - \frac{3m}{p}\,\varphi,\; x_0 = \varphi$$

according to formula 4.8 on page 59

$$\cos x \approx \cos\left(x_0\right) - \sin\left(x_0\right)\left(x - x_0\right),$$

if x_0 is close to x. If you substitute the values for x and x_0, you get

$$\cos\left(\varphi - \frac{3m}{p}\,\varphi\right) \approx \cos\varphi - \sin\varphi\left(-\frac{3m}{p}\,\varphi\right) = \cos\varphi + \frac{3m}{p}\,\varphi \cdot \sin\varphi.$$

So we can approximate the above equation linearly and get

$$\sigma(\varphi) = \frac{1}{p}\left(1 + \varepsilon\cos\left[\varphi\left(1 - \frac{3m}{p}\right)\right]\right).$$

This function becomes maximal when

$$\cos\left[\varphi\left(1 - \frac{3m}{p}\right)\right] = 1$$

is, then $r = 1/\sigma$ is minimal and the planet is in perihelion. When is the cosine equal to 1? Well, when φ takes the values

$$\varphi \;=\; 0, \frac{2\pi}{1 - \dfrac{3m}{p}}, \frac{4\pi}{1 - \dfrac{3m}{p}}, \cdots$$

The term $1/\left(1 - 3m/p\right)$ can again be estimated with the differential:

$$\frac{1}{1 - \dfrac{3m}{p}} \approx 1 + \frac{3m}{p},$$

and you get for the relevant angles

$$\varphi \;=\; 0, 2\pi + 2\pi\frac{3m}{p}, \cdots$$

i.e., the deviation $\Delta\varphi$ compared to the period 2π is per revolution

$$\Delta\varphi = 2\pi\frac{3m}{p} = 2\pi \cdot 8 \cdot 10^{-8} = 5.0265 \cdot 10^{-7}\,\text{rad} = 0.10368''.$$

Mercury has an orbital period of about 88 Earth days. Over the course of a century (on Earth), its perihelion therefore moves by the amount

$$\Delta\varphi \cdot \frac{100 \cdot 365}{88} \;=\; 43''.$$

The actually observed perihelion rotation of Mercury is considerably larger, it is about 574 arcseconds per century, see Fig. 23.2. Astronomical calculations, which take into account the influence of the other planets and other causes on the original Kepler ellipse of Mercury, however, only yield a partial amount of about $531''$. The observed difference of $\sim 43''$ is thus conclusively explained by the General Theory of Relativity.

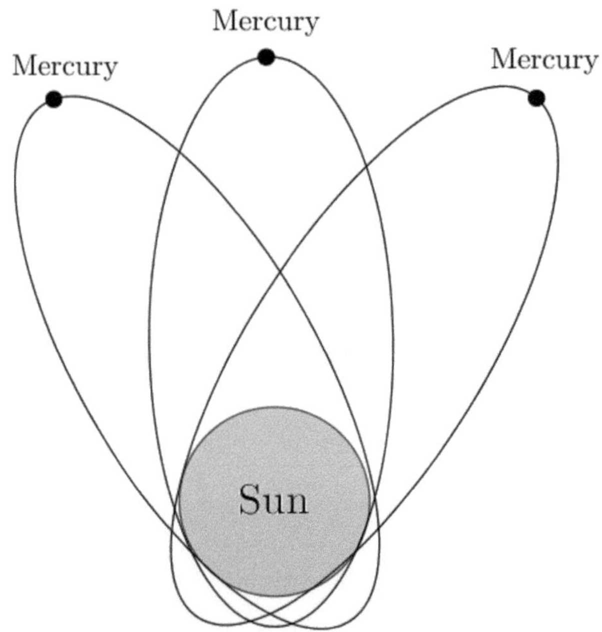

Figure 23.2.: Perihelion Precession of Mercury

23.7. Light Deflection in Schwarzschild Spacetime

In addition to the perihelion rotation, the deflection of a light beam by the Sun is the second "classic" test of the General Theory of Relativity. We start from the equation of motion for light (23.31) in Schwarzschild spacetime:

$$\frac{d^2\sigma}{d\varphi^2} + \sigma = 3m\sigma^2$$

In this equation, the term on the right side is very small in the case of scattering of light at the Sun. Because if you compare $3m\sigma^2$ with σ for a light beam that grazes the Sun at its edge, then with the solar radius $R = 696,000\,\text{km}$ and with $\sigma = 1/r$ it holds

$$\frac{3m\sigma^2}{\sigma} = \frac{3m}{r} \leq \frac{3r_S}{2R} = \frac{3 \cdot 2.95}{2 \cdot 696,000} \approx 6.4 \cdot 10^{-6}.$$

Without the term on the right side of the equation of motion for light, we get

$$\frac{d^2\sigma}{d\varphi^2} + \sigma = 0,$$

the general solution for this (homogeneous) equation is

$$\sigma(\varphi) = c_1 \sin\varphi + c_2 \cos\varphi.$$

Because if we insert $\sigma(\varphi)$ into the equation, we get

$$\frac{d^2\sigma}{d\varphi^2} + \sigma = -c_1 \sin\varphi - c_2 \cos\varphi + c_1 \sin\varphi + c_2 \cos\varphi = 0.$$

If we set the coordinate system so that $\sigma(\pi/2) = 1/R$, then it follows

$$1/R = \sigma(\pi/2) = c_1 \sin(\pi/2) + c_2 \cos(\pi/2) = c_1,$$

so $c_1 = 1/R$, i.e., the solution for the homogeneous equation is

$$\sigma(\varphi) = \frac{1}{r(\varphi)} = \frac{1}{R} \sin\varphi.$$

Rearranging gives

$$R = r \sin\varphi = y,$$

i.e., the homogeneous solution is the straight line $y = R$.

Fig. 23.3 illustrates our task. The Sun is at the origin of the coordinates. The homogeneous solution is the horizontal dashed line that passes through the point $(0, R)$. We are looking for the formula for the dotted line, i.e., the solution of Eq. (23.31).

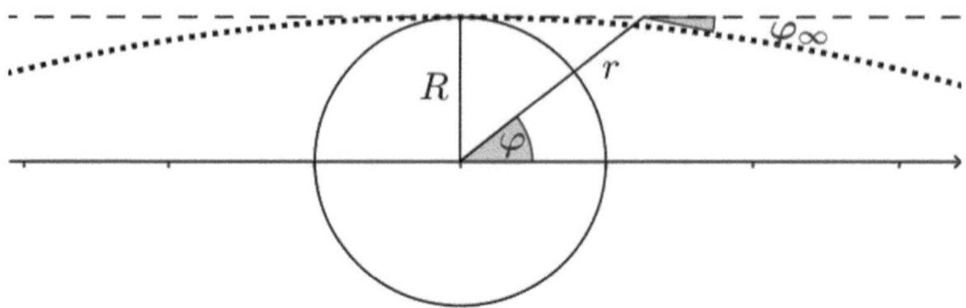

Figure 23.3.: Light deflecion at the Sun in Schwarzschild spacetime

Once we have found this, we can also determine the angle φ_∞, which is formed by the dotted line (at infinity) and the straight line $y = R$. The light beam is deflected to the lower right at the edge of the Sun by this angle.

To obtain the solution of Eq. (23.31), we again apply the method of perturbation calculation. So we make the following assumption again:

$$\sigma(\varphi) = \sigma^0(\varphi) + \delta(\varphi),$$

with $\sigma^0 = (\sin\varphi)/R$. If we insert this assumption into (23.31), we get

$$\left(\sigma^0 + \delta\right)'' + \left(\sigma^0 + \delta\right) = 3m\left(\sigma^0 + \delta\right)^2.$$

If we again neglect the terms $6m\sigma^0\delta$ and $3m\delta^2$, which result from multiplying out the right side, and note that because

$$\left(\sigma^0\right)'' = -\frac{\sin\varphi}{R}$$

the σ^0 terms on the left side cancel out, we get the approximate differential equation for δ:

$$\delta'' + \delta = \frac{3m}{R^2}\sin^2\varphi = \frac{3m}{R^2}\left(1 - \cos^2\varphi\right) = \frac{3m}{2R^2}\left(1 - \cos\left(2\varphi\right)\right), \qquad (23.37)$$

where we have used the addition theorem (see Chap. 29):

$$\cos 2\varphi = 2\cos^2\varphi - 1 \Rightarrow \cos^2\varphi = \frac{\cos\left(2\varphi\right) + 1}{2}.$$

The function

$$\delta(\varphi) = \frac{3m}{2R^2}\left(1 + \frac{1}{3}\cos\left(2\varphi\right)\right)$$

solves Eq. (23.37), as can be seen as follows. It holds

$$\delta'(\varphi) = \frac{3m}{2R^2} \left(-\frac{2}{3} \sin(2\varphi) \right)$$

and

$$\delta''(\varphi) = \frac{3m}{2R^2} \left(-\frac{4}{3} \cos(2\varphi) \right),$$

thus

$$\begin{aligned}
\delta''(\varphi) + \delta(\varphi) &= \frac{3m}{2R^2} \left(-\frac{4}{3} \cos(2\varphi) \right) + \frac{3m}{2R^2} \left(1 + \frac{1}{3} \cos(2\varphi) \right) \\
&= \frac{3m}{2R^2} \left(1 - \cos(2\varphi) \right).
\end{aligned}$$

The entire solution is thus approximately equal to

$$\sigma(\varphi) = \sigma^0(\varphi) + \delta(\varphi) = \frac{\sin\varphi}{R} + \frac{3m}{2R^2} \left(1 + \frac{1}{3} \cos(2\varphi) \right).$$

With this solution, one can calculate the deflection of a light beam coming from infinity at the Sun, which would run straight without disturbance. With $r \to \infty$, i.e. $\varphi \to 0$, $\sin\varphi \approx \varphi$ and $\cos(2\varphi) \approx 1$ as well as $\sigma \approx 0$ and thus

$$\frac{\varphi}{R} + \frac{3m}{2R^2} \frac{4}{3} \approx 0,$$

from which

$$\varphi_\infty = -\frac{2m}{R}$$

follows. The total light deflection between $-\infty$ and ∞ is

$$\Delta = |2\varphi_\infty| = \frac{4m}{R} = \frac{4GM}{R} = \frac{2 \cdot 2,95}{696.000} \text{ rad} = 1,75'',$$

which is twice as large as the one predicted by Newton's theory of gravitation in section 18.3 and thus eliminates the discrepancy between the measured values and the Newtonian prediction.

If one wants to measure the bending of light on Earth, one needs light sources that are close to the edge of the Sun from the perspective of the Earth. However, stars that fulfill this are not visible during the day due to the brightness of the Sun. This is only possible during a solar eclipse. If one measures their position relative to the background of other stars during a solar eclipse, it appears shifted compared to the one measured at night. The stars appear further away from the Sun and thus further apart from each other during a solar eclipse than on a usual night. This is further illustrated in Fig. 23.4.

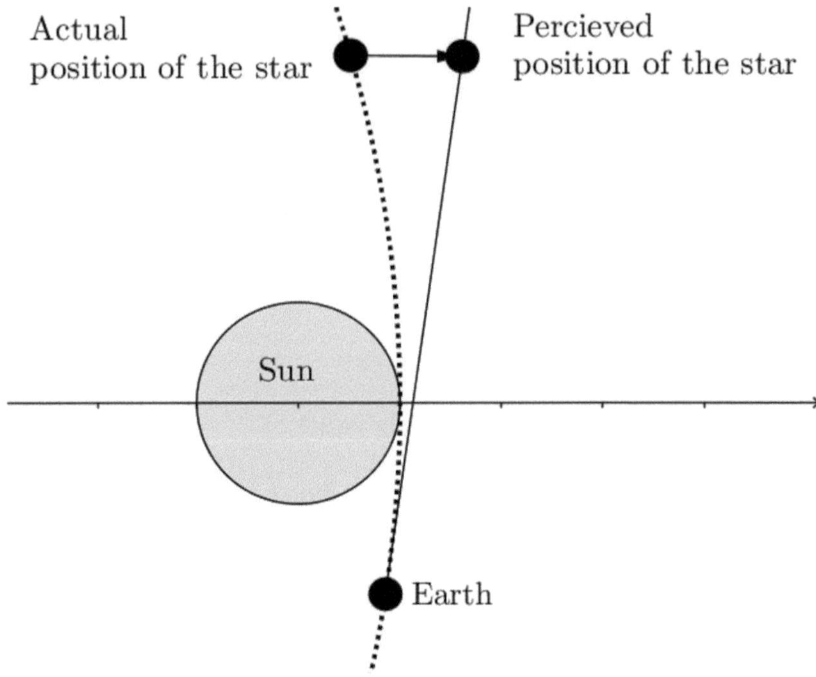

Figure 23.4.: Light deflection during Sun eclipse

This prediction of the General Theory of Relativity was first verified in 1919, when two different expeditions during a solar eclipse detected light deflections of about $1.98''$ and $1.61''$ respectively. This factual agreement with the predictions of the general theory of relativity made Einstein famous overnight. Measurements in subsequent years have also confirmed the theoretical value of light deflection, see e.g. [39].

24. Literature References and Further Reading on Part IV

The first popular science book recommended here on the general theory of relativity is a "classic", probably the most widely read popular science book on relativity theory (and quantum physics) and its significance for the development of our universe. It is Stephen Hawking's A Brief History of Time [16]. Although somewhat dated, the first approximately 100 pages provide a good impression of what Einstein's General Theory of Relativity can and cannot achieve.

The book by Schutz, Gravity from the Ground Up [28], which was already mentioned in the literature references to part I, is also one of the predominantly everyday language presentations of the General Theory of Relativity and is - as already mentioned - very worth reading because it explains topics such as the development of cosmological models in a generally understandable way, beyond the questions dealt with here.

Also worth reading is the popular science book by Will, Was Einstein right? [39], in which the status of the verification of the General Theory of Relativity by experiments up to around 1990 is well presented for laypeople. In particular, the sometimes big difficulties of the experimenters are described, who often have to clearly attribute the very small effects of measurements in the solar system (e.g., the perihelion rotation of Mercury or the light deflection at the Sun) to GR.

Since the first *German* edition of this book was published at the end of 2014, to my knowledge two books have been published that follow roughly the same approach as in this text. The first book by P. Collier (2017) [6] describes Einstein's theories of relativity, including a chapter on cosmology, on 274 pages and therefore does not reach the level of detail of the present text. The author focuses on the presentation of the mathematical foundations of the theories of relativity, which are introduced very thoroughly and carefully. However, the physics is somewhat neglected; many important results, such as the introduction of the Einstein equations, the consequences of the Schwarzschild solution in the solar system, black holes, or gravitational waves, are only superficially treated or presented without detailed introductions.

The second *German* book by T. Blankenheim (2018) [2], which comprises

M. Ruhrländer, *Ascent to the Einstein Equations*,
https://doi.org/10.1007/978-3-662-72672-3_24

approximately 630 pages, covers (except for the topic of gravitational waves) the most important phenomena of Special and General Relativity, including charged and rotating black holes and two chapters on the relativistic foundations of cosmology. The author pursues a slightly different approach than this book, as he places great emphasis on detailed descriptions of the physical relationships and therefore sometimes omits deriving results from what has been learned so far and simply writes down the results. The mathematics necessary for a deeper understanding of the General Theory of Relativity (including tensor calculus and differential geometry) is briefly presented. For the study of this second book, mathematical and physical knowledge about the Maxwell equations of electrodynamics is helpful.

All other "simple" books on the General Theory of Relativity that I am aware of, which simply put "get by with fewer formulas", do not meet the requirement of this book, namely that the theory is fully and comprehensibly presented. In many such treatises, "the physical laws (e.g., the Einstein equations or the Schwarzschild metric) simply fall from the sky", without the reader having the chance to understand *why* these laws are as they are.

In drafting part IV, I had to resort to "proper" specialist literature and try to simplify the procedures used there, which all presuppose mathematical knowledge beyond what is conveyed in this book, in such a way that - as I hope - something comprehensible has come out. After reading this book, however, one should be able to work through the books mentioned below (with some effort) and thus deepen one's knowledge of the General Theory of Relativity. The following textbooks, which all contributed to making the part what it is, are ordered by degree of difficulty; in my view the simpler ones are listed first.

- The book by Schutz [29], which has been quoted several times, is the source I relied on the most during the creation of this entire book. The derivation of tensor calculus in Riemannian space and the introduction to the Einstein equations are particularly well done in this book.

- The textbook by Thomas A. Moore, A General Relativity Workbook [21], contains many calculated passages in the application of the Einstein equations (e.g., in the determination of the geodesic equation or also on the topic of gravitational waves) and is particularly suitable for self-study to delve deeper into the subject.

- The treatise by Lewis Ryder, Introduction to General Relativity [26], is "student-friendly" (as stated in the introduction), as the calculations are presented and explained in detail.

Part V.

Application of the General Theory of Relativity to Selected Cosmological Phenomena

In this part we go beyond our solar system and look at some phenomena observed in the universe. We first show that Einstein's theory of gravity predicts the existence of **gravitational waves**. Gravitational waves are distortions of spacetime generated by rapidly changing relativistic masses. The Einstein equations for weak non-stationary gravitational fields can be reduced to wave equations, the solutions of which are gravitational waves. We investigate the propagation of gravitational waves that propagate freely through empty space and show what effect gravitational waves have on test particles. The quadrupole formula, already found by Einstein, represents the total energy radiated per unit of time by a mass source. For the detection of gravitational waves, we describe two different methods. The first is to measure the consequences of the energy loss of a system due to gravitational wave radiation, e.g., by recording changes in the rotation speed of a binary system over a long period of time. The second method involves detecting gravitational waves with measuring instruments on Earth. The main difficulty in measuring gravitational waves on Earth is that the amplitudes of the incoming waves are on the order of only about 10^{-21} meters. Since the 1980s, the necessary measuring instruments have been continuously improved, until in September 2015 gravitational waves were detected on Earth for the first time. For this success, three of the main scientists involved (Kip Thorne, Rainer Weiss, Barry Barish) were awarded the Nobel Prize in Physics in 2017.

In the following chapter, we describe the basic mechanisms that can lead to a gravitational collapse of a star. We derive the **interior Schwarzschild solution** for spherically symmetric stationary stars and calculate the pressure and density inside the star. The resulting Tolman-Oppenheimer-Volkoff equation describes the relativistic hydrostatic equilibrium inside a star.

The last chapter is dedicated to the most mysterious objects in the universe, namely black holes. If the radius of a collapsing star falls below the Schwarzschild radius, it is referred to as a black hole. Using the outer Schwarzschild metric, we derive some properties of non-rotating black holes and introduce Eddington-Finkelstein and Kruskal coordinates to also make physical statements about the area within the Schwarzschild radius.

This part of the book is the most mathematically challenging. At the one or at other points, we will use solutions that cannot be (simply) derived with the prerequisites we have assumed. For each of these cases, there are references to literature where the reader should be able to follow the derivations omitted here with some effort if they want to delve deeper into mathematics.

25. Gravitational Waves

In this chapter, we demonstrate that rapidly changing relativistic masses can be the sources of distortions in spacetime. These distortions are called **gravitational waves** and propagate similarly to electromagnetic waves at the speed of light. There are many important sources of gravitational waves in the universe, e.g., binary star systems or black holes, supernova explosions, and the Big Bang itself. We assume that an observer who wants to measure gravitational waves is far away from their source and the gravitational field around him is weak, but not static.

In this chapter, for the sake of clarity, we denote the partial derivative of a physical quantity A^μ with respect to the coordinate ν by $\partial_\nu A^\mu$ instead of $A^\mu_{,\nu}$ as before. The presentations in this chapter are based on my *German* book [25].

25.1. Einstein Equations for Weak Gravitational Fields

In sections 22.2 and 22.4 we have dealt with the Newtonian limit case of the Einstein equations and assumed there that the gravitational field should not only be weak but also static, and that the particles move slowly. Here, the gravitational field should also be weak, but the gravitational field may change over time, and there are no restrictions on the movement of the particles. In a weak gravitational field, spacetime is "almost flat", i.e., there exist coordinates such that the metric components have the form

$$g_{\mu\nu} = \eta_{\mu\nu} + h_{\mu\nu}, \quad |h_{\mu\nu}| \ll 1, \tag{25.1}$$

where $h_{\mu\nu}$ denotes a **perturbation function**, whose derivatives may also be small, i.e.

$$|\partial_\rho h_{\mu\nu}| \ll 1,$$

see section 22.2.

© The Author(s), under exclusive license to Springer-Verlag GmbH, DE, part of Springer Nature 2026
M. Ruhrländer, *Ascent to the Einstein Equations*,
https://doi.org/10.1007/978-3-662-72672-3_25

Background Lorentz Transformation

A **background Lorentz transformation** for weak gravitational fields is defined by

$$\mu' = L^{\mu'}_{\ \mu}\, \mu,$$

where $(L^{\mu}_{\ \nu})$ represents a (constant) Lorentz matrix (see formula 14.4 on page 292) as in the Special Theory of Relativity. Of course, we are not in SR, i.e., the background Lorentz transformations only form *one* class of all possible transformations. When we transform the metric, we get

$$g_{\mu'\nu'} = L^{\mu}_{\ \mu'}\, L^{\nu}_{\ \nu'}\, g_{\mu\nu} = L^{\mu}_{\ \mu'}\, L^{\nu}_{\ \nu'}\, \eta_{\mu\nu} + L^{\mu}_{\ \mu'}\, L^{\nu}_{\ \nu'}\, h_{\mu\nu}.$$

Now $L^{\mu}_{\ \nu}$ is a Lorentz matrix, i.e., it holds with (15.14)

$$\eta_{\mu'\nu'} = L^{\mu}_{\ \mu'}\, L^{\nu}_{\ \nu'}\, \eta_{\mu\nu}$$

and it results in

$$g_{\mu'\nu'} = \eta_{\mu'\nu'} + h_{\mu'\nu'}$$

with

$$h_{\mu'\nu'} = L^{\mu}_{\ \mu'}\, L^{\nu}_{\ \nu'}\, h_{\mu\nu}.$$

Under a background Lorentz transformation $h_{\mu\nu}$ transforms itself like a Lorentz tensor. Even though the general tensor property of $h_{\mu\nu}$ (i.e., invariance under *arbitrary* coordinate transformations) cannot be guaranteed, we interpret a weak gravitational field as a *flat spacetime* with a Lorentz-invariant symmetric "tensor" $h_{\mu\nu}$. Then all physical fields such as the Riemann tensor are defined by terms of $h_{\mu\nu}$ and look like fields in a flat spacetime. Such a perspective simplifies many of the following calculations, but should not lead us to forget that real spacetime is curved.

Gauge Transformation

We want to show that there are degrees of freedom in the choice of perturbation functions. For this, we consider the coordinate transformation

$$\mu \to \tilde{\mu} := \mu + \xi^{\mu} \tag{25.2}$$

with a function ξ^{μ}, which may depend on the coordinates α and like $|h_{\mu\nu}|$ fulfills the condition

$$|\partial_{\nu}\, \xi^{\mu}| \ll 1.$$

The relationship (25.2) is called a **gauge transformation**. For the *transformation matrix* we then get

$$\frac{\partial \tilde{\mu}}{\partial \nu} = \frac{\partial\,(\mu + \xi^{\mu})}{\partial \nu} = \delta^{\mu}_{\nu} + \partial_{\nu}\, \xi^{\mu}.$$

For the calculation of the inverse, we use a similar approach as in (22.13) and obtain in linear order in $\partial_\nu \xi^\mu$:

$$\left(\delta^\mu_\rho - \partial_\rho \xi^\mu\right)\left(\delta^\rho_\nu + \partial_\nu \xi^\rho\right) = \delta^\mu_\rho \delta^\rho_\nu + \delta^\mu_\rho \partial_\nu \xi^\rho - \delta^\rho_\nu \partial_\rho \xi^\mu - \underbrace{\partial_\rho \xi^\mu \partial_\nu \xi^\rho}_{\approx 0}$$

$$= \delta^\mu_\nu + \partial_\nu \xi^\mu - \partial_\nu \xi^\mu = \delta^\mu_\nu,$$

from which for the inverse matrix in linear approximation

$$\frac{\partial \mu}{\partial \tilde\nu} = \delta^\mu_\nu - \partial_\nu \xi^\mu$$

follows. For the transformed metric, we also get in linear approximation

$$\tilde g_{\mu\nu} = \frac{\partial \rho}{\partial \tilde\mu}\frac{\partial \sigma}{\partial \tilde\nu}\, g_{\rho\sigma} = \left(\delta^\rho_\mu - \partial_\mu \xi^\rho\right)\left(\delta^\sigma_\nu - \partial_\nu \xi^\sigma\right) g_{\rho\sigma}$$

$$= \left(\delta^\rho_\mu \delta^\sigma_\nu - \delta^\rho_\mu \partial_\nu \xi^\sigma - \delta^\sigma_\nu \partial_\mu \xi^\rho + \underbrace{\partial_\mu \xi^\rho \partial_\nu \xi^\sigma}_{\approx 0}\right) g_{\rho\sigma}$$

$$= g_{\mu\nu} - \partial_\nu \xi^\sigma g_{\mu\sigma} - \partial_\mu \xi^\rho g_{\rho\nu}.$$

We now insert (25.1) into the last equation and obtain with $|h_{\mu\nu}| \ll 1$ and $|\partial_\nu \xi^\mu| \ll 1$:

$$\tilde g_{\mu\nu} = \eta_{\mu\nu} + h_{\mu\nu} - \left(\partial_\nu \xi^\sigma \eta_{\mu\sigma} + \underbrace{\partial_\nu \xi^\sigma h_{\mu\sigma}}_{\approx 0}\right) - \left(\partial_\mu \xi^\rho \eta_{\rho\nu} + \underbrace{\partial_\mu \xi^\rho h_{\rho\nu}}_{\approx 0}\right)$$

$$= \eta_{\mu\nu} + h_{\mu\nu} - \partial_\nu \xi_\mu - \partial_\mu \xi_\nu$$

with $\xi_\mu = \eta_{\mu\sigma} \xi^\sigma$. If we redefine $h_{\mu\nu}$ by

$$h_{\mu\nu} \to \tilde h_{\mu\nu} = h_{\mu\nu} - \partial_\nu \xi_\mu - \partial_\mu \xi_\nu, \tag{25.3}$$

then $\left|\tilde h_{\mu\nu}\right| \ll 1$ still applies, i.e., after the transformation we are again in a weak gravitational field.

Einstein Equations

We first want to determine the left side of the Einstein equations and for this purpose first calculate the Christoffel symbols. These take according to (19.16)

$$\Gamma^\tau_{\mu\lambda} = \frac{g^{\tau\nu}}{2}\left(\partial_\lambda g_{\mu\nu} + \partial_\mu g_{\lambda\nu} - \partial_\nu g_{\mu\lambda}\right)$$

in a weak gravitational field in linear approximation the simple form

$$\Gamma^\tau_{\mu\lambda} = \frac{1}{2}\,\eta^{\tau\nu}\left(\partial_\lambda\,h_{\mu\nu,\lambda} + \partial_\mu\,h_{\lambda\nu} - \partial_\nu\,h_{\mu\lambda}\right),\qquad(25.4)$$

since the partial derivatives of $\eta_{\mu\nu}$ vanish and the products of the disturbance functions $h_{\mu\nu}$ are omitted. We want to calculate the Riemann tensor in a weak gravitational field and can use the formula (21.6) since the products of the Christoffel symbols in Eq. (21.5)

$$R^\mu_{\ \rho\nu\lambda} = \partial_\nu\,\Gamma^\mu_{\rho\lambda} - \partial_\lambda\,\Gamma^\mu_{\rho\nu} + \Gamma^\mu_{\sigma\nu}\,\Gamma^\sigma_{\rho\lambda} - \Gamma^\mu_{\sigma\lambda}\,\Gamma^\sigma_{\rho\nu}$$

vanish in linear approximation. With formula (21.8)

$$R_{\mu\rho\nu\lambda} = \frac{1}{2}\left(\partial_\nu\,\partial_\rho\,g_{\lambda\mu} - \partial_\nu\,\partial_\mu\,g_{\rho\lambda} - \partial_\lambda\,\partial_\rho\,g_{\nu\mu} + \partial_\lambda\,\partial_\mu\,g_{\rho\nu}\right)$$

it follows that

$$R_{\mu\rho\nu\lambda} = \frac{1}{2}\left(\partial_\nu\,\partial_\rho\,h_{\lambda\mu} - \partial_\nu\,\partial_\mu\,h_{\rho\lambda} - \partial_\lambda\,\partial_\rho\,h_{\nu\mu} + \partial_\lambda\,\partial_\mu\,h_{\rho\nu}\right).$$

The same result is obtained if we replace $h_{\mu\nu}$ with $h_{\mu\nu} - \partial_\nu\,\xi_\mu - \partial_\mu\,\xi_\nu$ because

$$\begin{aligned}
\partial_\nu\,\partial_\rho\,h_{\lambda\mu} - \partial_\nu\,\partial_\rho\,\partial_\mu\,\xi_\lambda - \partial_\nu\,\partial_\rho\,\partial_\lambda\,\xi_\mu - \partial_\nu\,\partial_\mu\,h_{\rho\lambda} + \partial_\nu\,\partial_\mu\,\partial_\lambda\,\xi_\rho + \partial_\nu\,\partial_\mu\,\partial_\rho\,\xi_\lambda\quad &- \\
\partial_\lambda\,\partial_\rho\,h_{\nu\mu} + \partial_\lambda\,\partial_\rho\,\partial_\mu\,\xi_\nu + \partial_\lambda\,\partial_\rho\,\partial_\nu\,\xi_\mu + \partial_\lambda\,\partial_\mu\,h_{\rho\nu} - \partial_\lambda\,\partial_\mu\,\partial_\nu\,\xi_\rho - \partial_\lambda\,\partial_\mu\,\partial_\rho\,\xi_\nu\quad &= \\
\partial_\nu\,\partial_\rho\,h_{\lambda\mu} - \partial_\nu\,\partial_\mu\,h_{\rho\lambda} - \partial_\lambda\,\partial_\rho\,h_{\nu\mu} + \partial_\lambda\,\partial_\mu\,h_{\rho\nu},&
\end{aligned}$$

since according to the Schwarz theorem (9.27) all mixed derivatives of the ξ_μ cancel each other out. We define some quantities by

$$h^\mu_{\ \nu} := \eta^{\mu\rho}h_{\rho\nu},\quad h^{\mu\nu} := \eta^{\nu\rho}h^\mu_{\ \rho},\quad h := h^\mu_{\ \mu},$$

where we have raised the indices with the inverse Minkowski metric. This is allowed in linear approximation, since with (22.13)

$$g^{\mu\nu} = \eta^{\mu\nu} - h^{\mu\nu}$$

the relationship

$$g^{\mu\rho}h_{\rho\nu} = \left(\eta^{\mu\rho} - h^{\mu\rho}\right)h_{\rho\nu} = \eta^{\mu\rho}h_{\rho\nu} - \underbrace{h^{\mu\rho}h_{\rho\nu}}_{\approx 0} = \eta^{\mu\rho}h_{\rho\nu}$$

follows. With these definitions, we can represent the also gauge-invariant Ricci tensor, which we have already derived in (22.14), as follows:

$$R_{\mu\nu} \;=\; \frac{\eta^{\lambda\sigma}}{2}\left(\partial_\lambda\,\partial_\mu\,h_{\nu\sigma} - \partial_\lambda\,\partial_\sigma\,h_{\mu\nu} - \partial_\nu\,\partial_\mu\,h_{\lambda\sigma} + \partial_\nu\,\partial_\sigma\,h_{\mu\lambda}\right)$$

$$
\begin{aligned}
&= \frac{1}{2}\left(\partial^{\sigma}\,\partial_{\mu}\,h_{\nu\sigma} + \partial_{\nu}\,\partial^{\lambda}\,h_{\mu\lambda} - \partial_{\lambda}\,\partial^{\lambda}\,h_{\mu\nu} - \partial_{\nu}\,\partial_{\mu}\,h^{\sigma}{}_{\sigma}\right)\\
&\underset{\sigma\leftrightarrow\lambda}{=} \frac{1}{2}\left(\partial_{\mu}\,\partial^{\lambda}\,h_{\nu\lambda} + \partial_{\nu}\,\partial^{\lambda}\,h_{\mu\lambda} - \Box h_{\mu\nu} - \partial_{\nu}\,\partial_{\mu}\,h\right),
\end{aligned}
\qquad (25.5)
$$

where the **D'Alembert operator** $\Box$ is the four-dimensional generalization of the *Laplace operator* ∇^2 (see 6.28 on page 124) and in Cartesian coordinates is defined by

$$
\Box := \nabla^2 - \frac{\partial^2}{\partial t^2} = -\frac{\partial^2}{\partial t^2} + \frac{\partial^2}{\partial x^2} + \frac{\partial^2}{\partial y^2} + \frac{\partial^2}{\partial z^2} = \eta_{\mu\nu}\,\frac{\partial}{\partial\mu}\,\frac{\partial}{\partial\nu} =: \partial^{\mu}\partial_{\mu}. \qquad (25.6)
$$

The curvature scalar in linear order results from (25.5) to

$$
\begin{aligned}
R &= \frac{1}{2}\,\eta^{\mu\nu}\left(\partial_{\mu}\,\partial^{\lambda}\,h_{\nu\lambda} + \partial_{\nu}\,\partial^{\lambda}\,h_{\mu\lambda} - \Box h_{\mu\nu} - \partial_{\nu}\,\partial_{\mu}\,h\right)\\
&= \frac{1}{2}\left(\partial^{\nu}\,\partial^{\lambda}\,h_{\nu\lambda} + \partial^{\mu}\,\partial^{\lambda}\,h_{\mu\lambda} - \Box h^{\mu}{}_{\mu} - \partial_{\nu}\,\partial^{\nu}\,h\right)\\
&= \partial^{\mu}\,\partial^{\lambda}\,h_{\mu\lambda} - \Box h,
\end{aligned}
\qquad (25.7)
$$

and we obtain for the Einstein tensor

$$
\begin{aligned}
G_{\mu\nu} &= R_{\mu\nu} - \frac{1}{2}\,\eta_{\mu\nu}\,R\\
&= \frac{1}{2}\left(\partial_{\mu}\,\partial^{\lambda}\,h_{\nu\lambda} + \partial_{\nu}\,\partial^{\lambda}\,h_{\mu\lambda} - \Box h_{\mu\nu}\right)\\
&\quad - \frac{1}{2}\left(\partial_{\nu}\,\partial_{\mu}\,h + \eta_{\mu\nu}\,\partial^{\rho}\,\partial^{\lambda}\,h_{\rho\lambda} - \eta_{\mu\nu}\,\Box h\right),
\end{aligned}
\qquad (25.8)
$$

thus a somewhat confusing expression. We want to simplify this by defining new variables by

$$
H_{\mu\nu} := h_{\mu\nu} - \frac{1}{2}\,\eta_{\mu\nu}\,h. \qquad (25.9)
$$

These fulfill

$$
H := \eta^{\mu\nu}\,H_{\mu\nu} = \eta^{\mu\nu}\,h_{\mu\nu} - \frac{1}{2}\,\eta^{\mu\nu}\,\eta_{\mu\nu}\,h = h - \frac{4}{2}\,h = -h
$$

as well as

$$
h_{\mu\nu} = H_{\mu\nu} - \frac{1}{2}\,\eta_{\mu\nu}\,H \qquad (25.10)
$$

and are therefore also called **trace reverse** of $h_{\mu\nu}$. We calculate the Ricci and Einstein tensor again, using the variables $H_{\mu\nu}$. We insert (25.10) into (25.5) and obtain for the Ricci tensor

$$
R_{\mu\nu} = \frac{1}{2}\left(\partial_{\mu}\,\partial^{\lambda}\,h_{\nu\lambda} + \partial_{\nu}\,\partial^{\lambda}\,h_{\mu\lambda} - \Box h_{\mu\nu} - \partial_{\nu}\,\partial_{\mu}\,h\right)
$$

$$
\begin{aligned}
&= \frac{1}{2}\left(\partial_\mu \partial^\lambda \left(H_{\nu\lambda} - \frac{1}{2}\eta_{\nu\lambda} H \right) + \partial_\nu \partial^\lambda \left(H_{\mu\lambda} - \frac{1}{2}\eta_{\mu\lambda} H \right) \right) \\
&\quad - \frac{1}{2}\left(\Box\left(H_{\mu\nu} - \frac{1}{2}\eta_{\mu\nu} H \right) + \partial_\nu \partial_\mu H \right) \\
&= \frac{1}{2}\left(\partial_\mu \partial^\lambda H_{\nu\lambda} + \partial_\nu \partial^\lambda H_{\mu\lambda} - \Box H_{\mu\nu} + \frac{1}{2}\eta_{\mu\nu}\Box H \right),
\end{aligned}
$$

since in the penultimate line the second, fourth and last term cancel out. For the curvature scalar, this results in

$$
\begin{aligned}
\mathcal{R} &= \frac{1}{2}\eta^{\mu\nu}\left(\partial_\mu \partial^\lambda H_{\nu\lambda} + \partial_\nu \partial^\lambda H_{\mu\lambda} - \Box H_{\mu\nu} + \frac{1}{2}\eta_{\mu\nu}\Box H \right) \\
&= \frac{1}{2}\left(\partial^\nu \partial^\lambda H_{\nu\lambda} + \partial^\mu \partial^\lambda H_{\mu\lambda} - \Box H + \frac{4}{2}\Box H \right) \\
&= \partial^\mu \partial^\lambda H_{\mu\lambda} + \frac{1}{2}\Box H,
\end{aligned}
$$

and the Einstein tensor is then

$$
G_{\mu\nu} = \frac{1}{2}\left(\partial_\mu \partial^\lambda H_{\nu\lambda} + \partial_\nu \partial^\lambda H_{\mu\lambda} - \Box H_{\mu\nu} - \eta_{\mu\nu}\partial^\rho \partial^\sigma H_{\rho\sigma} \right). \tag{25.11}
$$

The Einstein tensor has simplified somewhat, but is still complicated. The Einstein equations corresponding to (25.11) are

$$
\Box H_{\mu\nu} - \partial_\mu \partial^\lambda H_{\nu\lambda} - \partial_\nu \partial^\lambda H_{\mu\lambda} + \eta_{\mu\nu}\partial^\rho \partial^\sigma H_{\rho\sigma} = -16\pi G T_{\mu\nu}. \tag{25.12}
$$

The left side of (25.12) contains three terms, each containing a term of the form $\partial^\lambda H_{\mu\lambda}$, and simplifies considerably if we require that

$$
\partial^\lambda H_{\mu\lambda} = 0 \tag{25.13}
$$

holds. These are four equations, and we will see that we can choose the four free gauge functions ξ^μ such that (25.13) is fulfilled. The gauge (25.13) is called **Lorenz gauge** after the physicist L.V. Lorenz (1829 - 1891), not to be confused with the Dutchman H.A. Lorentz!

We show with (25.9) and the gauge transformation (25.3), that $H_{\mu\nu}$ transforms according to

$$
H_{\mu\nu} \to \tilde{H}_{\mu\nu} = H_{\mu\nu} - \partial_\nu \xi_\mu - \partial_\mu \xi_\nu + \eta_{\mu\nu}\partial^\rho \xi_\rho. \tag{25.14}
$$

It holds:

$$
H_{\mu\nu} \to \tilde{H}_{\mu\nu} = \tilde{h}_{\mu\nu} - \frac{1}{2}\eta_{\mu\nu}\tilde{h} = h_{\mu\nu} - \partial_\nu \xi_\mu - \partial_\mu \xi_\nu - \frac{1}{2}\eta_{\mu\nu}\eta^{\rho\sigma}\tilde{h}_{\rho\sigma}
$$

$$\begin{aligned} &= h_{\mu\nu} - \partial_\nu \xi_\mu - \partial_\mu \xi_\nu - \frac{1}{2} \eta_{\mu\nu} \eta^{\rho\sigma} \left(h_{\rho\sigma} - \partial_\sigma \xi_\rho - \partial_\rho \xi_\sigma \right) \\ &= h_{\mu\nu} - \frac{1}{2} \eta_{\mu\nu} h - \partial_\nu \xi_\mu - \partial_\mu \xi_\nu + \frac{1}{2} \eta_{\mu\nu} \left(\partial^\rho \xi_\rho + \partial^\sigma \xi_\sigma \right) \\ &= H_{\mu\nu} - \partial_\nu \xi_\mu - \partial_\mu \xi_\nu + \eta_{\mu\nu} \partial^\rho \xi_\rho \end{aligned}$$

We calculate the *four-divergence* (see Eq. 16.1 on page 330) and obtain

$$\begin{aligned} \partial^\nu \tilde{H}_{\mu\nu} &= \partial^\nu H_{\mu\nu} - \partial^\nu \partial_\nu \xi_\mu - \partial_\mu \partial^\nu \xi_\nu + \eta_{\mu\nu} \partial^\nu \partial^\rho \xi_\rho \\ &= \partial^\nu H_{\mu\nu} - \partial^\nu \partial_\nu \xi_\mu - \partial_\mu \partial^\nu \xi_\nu + \partial_\mu \partial^\rho \xi_\rho \\ &= \partial^\nu H_{\mu\nu} - \partial^\nu \partial_\nu \xi_\mu . \end{aligned}$$

So if we want $\partial^\nu \tilde{H}_{\mu\nu} = 0$ to hold, the gauge functions ξ_μ must be determined by the equation

$$\Box \xi_\mu = \partial^\nu \partial_\nu \xi_\mu = \partial^\nu H_{\mu\nu} . \tag{25.15}$$

Eq. (25.15) represents a system of four (for each value of μ) inhomogeneous differential equations, the solutions of which exist even under weak assumptions about the *disturbance function* on the right-hand side, which we cannot elaborate further here due to the necessary mathematics. For further details, see e.g. [25]. The gauge functions ξ_μ are not unique, we can add arbitrary functions χ_μ, which satisfy

$$\Box \chi_\mu = 0, \tag{25.16}$$

and then it holds

$$\Box \left(\xi_\mu + \chi_\mu \right) = \Box \xi_\mu + \Box \chi_\mu = \partial^\nu H_{\mu\nu},$$

the Lorenz gauge is thus preserved.

In summary, we can state that the Einstein equations in the Lorenz gauge $\partial^\nu H_{\mu\nu} = 0$ simplify to

$$\Box H_{\mu\nu} = -16\pi G T_{\mu\nu} . \tag{25.17}$$

They are then called **linear** (due to the linearity in $H_{\mu\nu}$) or **weak** Einstein equations.

25.2. Propagation of Gravitational Waves

In this section, we consider gravitational waves that propagate freely through empty space, but we do not investigate whether or how the gravitational waves are generated by changes in an object in a gravitational field. Our approach assumes that we have a weak gravitational field and are far away from sources where the energy-momentum tensor vanishes, i.e., we solve the linear Einstein equations (in the Lorenz gauge) in vacuum:

$$\Box H_{\mu\nu} = \left(-\frac{\partial}{\partial t^2} + \nabla^2\right) H_{\mu\nu} = 0 \tag{25.18}$$

An equation of the form (25.18) is also called a **homogeneous wave equation**.

Approach with a Plane Wave

We want to show that the linear Einstein equations in vacuum (25.18) are solved by plane waves and generalize the wave equation represented in Eq. 12.10 on page 240 to the case of an arbitrary propagation direction.

Remark 25.1. **MT: Plane wave with arbitrary propagation direction** The propagation direction is determined by the (constant) **wave vector** $\vec{k} \to (k^x, k^y, k^z)$. Then the equation of a plane wave (12.10) generalizes to

$$A\left(\vec{x}, t\right) = A_0 \sin\left(\vec{k} \cdot \vec{x} - \omega t\right) = A_0 \sin\left(k^x x + k^y y + k^z z - \omega t\right). \tag{25.19}$$

We extend the wave vector to the **wave four-vector**

$$k \to k^\rho = (\omega, k^x, k^y, k^z)$$

and obtain with Eq. 15.16 on page 318 for the one-form corresponding to k

$$\tilde{k} \to k_\rho = (-\omega, k_x, k_y, k_z).$$

This allows us to briefly write Eq. (25.19) as

$$A\left(t, x, y, z\right) = A_0 \sin\left(-\omega t + k^x x + k^y y + k^z z\right) = A_0 \sin\left(k_\rho \rho\right). \Box$$

For the homogeneous wave equation (25.18), we choose plane waves as the *solution approach*:

$$H_{\mu\nu}\left(t, x, y, z\right) = A_{\mu\nu} \sin\left(k_\rho \rho\right) = A_{\mu\nu} \sin\left(\vec{k} \cdot \vec{x} - \omega t\right), \tag{25.20}$$

where $A_{\mu\nu}$ are constant tensor components, and we determine the necessary properties of the constants $A_{\mu\nu}$ and k. Such a plane wave moves with phase velocity $v = \omega/\left|\vec{k}\right|$ in the direction of $\vec{k}$.

Substituting the approach (25.20) into the linear Einstein equation (25.18) yields

$$
\begin{aligned}
0 &= \Box H_{\mu\nu} = \partial^\alpha \partial_\alpha \, A_{\mu\nu} \, \sin\left(k_\rho \, \rho\right) = \eta^{\alpha\beta} \partial_\beta \, \partial_\alpha \, A_{\mu\nu} \, \sin\left(k_\rho \, \rho\right) \\
&= \eta^{\alpha\beta} \partial_\beta \left(k_\alpha \, A_{\mu\nu} \, \cos\left(k_\rho \, \rho\right)\right) = -A_{\mu\nu} \, \eta^{\alpha\beta} k_\alpha \, k_\beta \, \sin\left(k_\rho \, \rho\right),
\end{aligned}
$$

from which

$$
\eta^{\alpha\beta} k_\alpha \, k_\beta = k^\alpha k_\alpha = 0 \tag{25.21}
$$

follows. From the Lorenz gauge (25.13) we obtain

$$
\begin{aligned}
0 &= \partial^\nu \left(A_{\mu\nu} \, \sin\left(k_\rho \, \rho\right)\right) = \eta^{\nu\sigma} \partial_\sigma \left(A_{\mu\nu} \, \sin\left(k_\rho \, \rho\right)\right) \\
&= \eta^{\nu\sigma} k_\sigma \, A_{\mu\nu} \, \cos\left(k_\rho \, \rho\right) = k^\nu \, A_{\mu\nu} \, \cos\left(k_\rho \, \rho\right) \\
&\Rightarrow k^\nu A_{\mu\nu} = 0 \tag{25.22}
\end{aligned}
$$

as well as with the symmetry of $H_{\mu\nu}$

$$
A_{\mu\nu} = A_{\nu\mu}. \tag{25.23}
$$

From Eq. (25.21) it follows

$$
\begin{aligned}
0 &= k^\alpha k_\alpha = \eta^{\alpha\beta} k_\alpha \, k_\beta \\
&= (-\omega)^2 \, \eta^{tt} + k_x^2 \, \eta^{xx} + k_y^2 \, \eta^{yy} + k_z^2 \, \eta^{zz} \\
&= -\omega^2 + k_x^2 + k_y^2 + k_z^2 = -\omega^2 + \left|\vec{k}\right| \\
&\Rightarrow \omega^2 = \left|\vec{k}\right|^2 \Rightarrow v = \omega/\left|\vec{k}\right| = 1,
\end{aligned}
$$

i.e., a plane gravitational wave moves like an electromagnetic wave at the speed of light ($v = c = 1$).

Transverse-Traceless Gauge

We now use the freedom that we can additively extend the Lorenz gauge by another gauge function

$$
\chi_\mu := B_\mu \, \cos\left(k_\rho \, \rho\right),
$$

where the k_ρ are chosen as in (25.20) and the B_μ are constant. The prerequisite (25.16) is satisfied due to (25.21) because

$$
\partial^\alpha \partial_\alpha \, \chi_\mu = -k^\alpha k_\alpha \, B_\mu \, \cos\left(k_\rho \, \rho\right) = 0.
$$

We define

$$
\tilde{H}_{\mu\nu} = \tilde{A}_{\mu\nu} \, \sin\left(k_\rho \, \rho\right),
$$

then it follows with (25.14):

$$
\begin{aligned}
\tilde{H}_{\mu\nu} &= H_{\mu\nu} - \partial_\nu \chi_\mu - \partial_\mu \chi_\nu + \eta_{\mu\nu}\, \partial^\rho \chi_\rho \\
&= A_{\mu\nu}\, \sin\left(k_\rho\, \rho\right) - \partial_\nu B_\mu\, \cos\left(k_\rho\, \rho\right) - \partial_\mu B_\nu\, \cos\left(k_\rho\, \rho\right) \\
&\quad + \eta_{\mu\nu}\, \partial^\alpha B_\alpha\, \cos\left(k_\rho\, \rho\right) \\
&= A_{\mu\nu}\, \sin\left(k_\rho\, \rho\right) + k_\nu B_\mu\, \sin\left(k_\rho\, \rho\right) + k_\mu B_\nu\, \sin\left(k_\rho\, \rho\right) \\
&\quad - \eta_{\mu\nu}\, k^\alpha B_\alpha\, \sin\left(k_\rho\, \rho\right)
\end{aligned}
$$

When we factor out the terms with the sine function, we get

$$
\tilde{A}_{\mu\nu} = A_{\mu\nu} + k_\nu B_\mu + k_\mu B_\nu - \eta_{\mu\nu}\, k^\alpha B_\alpha. \tag{25.24}
$$

We show that $\tilde{A}_{\mu\nu}$ also satisfies the orthogonality relation (25.22)

$$
k^\nu \tilde{A}_{\mu\nu} = 0. \tag{25.25}
$$

It holds

$$
\begin{aligned}
k^\nu \tilde{A}_{\mu\nu} &= \underbrace{k^\nu A_{\mu\nu}}_{=0} + \underbrace{k^\nu k_\nu}_{=0} B_\mu + k^\nu k_\mu B_\nu - \eta_{\mu\nu}\, k^\nu k^\alpha B_\alpha \\
&= k_\mu k^\nu B_\nu - k_\mu k^\alpha B_\alpha = 0.
\end{aligned}
$$

The components B_μ are arbitrarily selectable, and we want to specify them to get further restrictions for $\tilde{A}_{\mu\nu}$ in addition to (25.25). It follows from (25.24):

$$
\begin{aligned}
\tilde{A}^\mu_\mu &= \eta^{\mu\nu} \tilde{A}_{\mu\nu} = A^\mu_\mu + \eta^{\mu\nu} k_\nu B_\mu + \eta^{\mu\nu} k_\mu B_\nu - \eta^{\mu\nu} \eta_{\mu\nu}\, k^\alpha B_\alpha \\
&= A^\mu_\mu + k^\mu B_\mu + k^\nu B_\nu - 4 k^\alpha B_\alpha \\
&= A^\mu_\mu - 2 k^\mu B_\mu
\end{aligned}
$$

We choose

$$
k^\mu B_\mu = \frac{1}{2}\, A^\mu_\mu
$$

and obtain the restriction

$$
\tilde{A}^\mu_\mu = 0. \tag{25.26}
$$

Let $\vec{U} \to U^\nu$ be a constant four-velocity, i.e., a constant timelike unit vector, then we define as further conditions

$$
\tilde{A}_{\mu\nu}\, U^\nu = 0. \tag{25.27}
$$

At first glance, it seems as if this specification requires four degrees of freedom (for each value of μ), but it holds quite analogously to (25.25), that for *every* choice of B_μ the relationship

$$
k^\mu \tilde{A}_{\mu\nu}\, U^\nu = 0
$$

holds, i.e., only *three* degrees of freedom are needed. From now on, we write the transformed quantities again without the tilde ˜. The two conditions (25.26) and (25.27) are called **transverse-traceless (TT) gauge**, and according to (25.10)

$$h_{\mu\nu}^{TT} = H_{\mu\nu}^{TT} - \frac{1}{2}\eta_{\mu\nu}\,H^{TT} = H_{\mu\nu}^{TT} \tag{25.28}$$

because

$$H^{TT} = \left(H^{TT}\right)_{\mu}^{\mu} = \underbrace{A_{\mu}^{\mu}}_{=0}\cos\left(k_{\rho}\,\rho\right) = 0,$$

i.e., in the TT gauge there is no difference between the disturbance functions $h_{\mu\nu}$ and the inverse trace functions $H_{\mu\nu}$.

We now want to show that in the TT gauge for the components $A_{\mu\nu}$ of the original ten (due to symmetry) only two degrees of freedom remain, and for the sake of clarity we write again

$$H_{\mu\nu} = H_{\mu\nu}^{TT} = h_{\mu\nu}^{TT} = h_{\mu\nu}.$$

In Eq. (25.27) we choose the vector $\vec{U} = \vec{e}_t \rightarrow (1,0,0,0)$ and obtain

$$A_{\mu\nu}\,U^{\nu} = A_{\mu t} = A_{t\mu} = 0. \tag{25.29}$$

In this coordinate system, the axes are oriented such that the wave propagates in the z-direction, i.e.

$$k^{\nu} = (\omega, 0, 0, \omega), \tag{25.30}$$

which guarantees $k_{\nu}\,k^{\nu} = 0$. From (25.30) it follows in the Lorenz gauge according to (25.22)

$$0 = k^{\nu}A_{\mu\nu} = \omega\,\underbrace{A_{\mu t}}_{=0} + \omega A_{\mu z} \Rightarrow A_{\mu z} = A_{z\mu} = 0.$$

So in total only

$$A_{xx}, A_{yy}, \quad und\ \ A_{\times} := A_{xy} = A_{yx}$$

are different from zero. From the tracelessness of $A_{\mu\nu}$ (25.26) it follows that

$$A_{xx} = -A_{yy} =: A_{+},$$

so that $A_{\mu\nu}$ can be represented in the following matrix form:

$$A_{\mu\nu} = \begin{pmatrix} 0 & 0 & 0 & 0 \\ 0 & A_{+} & A_{\times} & 0 \\ 0 & A_{\times} & -A_{+} & 0 \\ 0 & 0 & 0 & 0 \end{pmatrix} = A_{+}\begin{pmatrix} 0 & 0 & 0 & 0 \\ 0 & 1 & 0 & 0 \\ 0 & 0 & -1 & 0 \\ 0 & 0 & 0 & 0 \end{pmatrix} + A_{\times}\begin{pmatrix} 0 & 0 & 0 & 0 \\ 0 & 0 & 1 & 0 \\ 0 & 1 & 0 & 0 \\ 0 & 0 & 0 & 0 \end{pmatrix} \tag{25.31}$$

with the two independent constants A_+ and $A_\times$. A plane gravitational wave in the TT-gauge thus consists of a linear combination of two different solutions, which are called the **polarizations** of the wave. The part that is proportional to A_+ is called + polarization, the second $\times$ polarization.

25.3. Observation of Gravitational Waves

When a wave packet propagates in the z-direction, we can align all plane waves so that they attain the form (25.31), i.e., each wave in the TT-gauge has only the two independent components $H_{xx} = h_{xx}$ and $H_{xy} = h_{xy}$. We consider a particle that is hit by a gravitational wave in an otherwise wave-free region, and choose a coordinate system in which the particle is at rest, and the corresponding TT-gauge. The vector U from (25.27) denotes the four-velocity in the rest system of the particle

$$U = \left(\sqrt{-g_{xx}}, 0, 0, 0 \right) \approx (1, 0, 0, 0)$$

in linear approximation. For a freely falling particle, the geodesic equation (20.2) applies:

$$\frac{d^2\mu}{dt_E^2} + \Gamma^\mu_{\nu\rho} \frac{d\nu}{dt_E} \frac{d\rho}{dt_E} = \frac{dU^\mu}{dt_E} + \Gamma^\mu_{\nu\rho} U^\nu U^\rho = 0,$$

and it follows with (25.4):

$$0 = \frac{dU^\mu}{dt_E} + \Gamma^\mu_{\nu\rho} U^\nu U^\rho = \frac{dU^\mu}{dt_E} + \Gamma^\mu_{tt} = \frac{dU^\mu}{dt_E} + \frac{1}{2} \eta^{\mu\nu} \underbrace{\left(\partial_t h_{t\nu} + \partial_t h_{t\nu} - \partial_\nu h_{tt} \right)}_{=0},$$

since all $h_{t\mu}$ vanish. During the passage of the wave, the four-acceleration of the particle is zero, i.e., the particle remains at rest as if the gravitational wave had no influence on the particle. However, we must be careful with such an interpretation, as we have chosen a co-moving coordinate system that leaves the coordinates of the particle unchanged. The result therefore has *no geometrically invariant* meaning.

To get a better estimate of the effects of a gravitational wave, we consider at $t = 0$ two spatially infinitesimally separated particles, one at the origin and the other at

$$x = L^*, y = z = 0.$$

Then for the distance $L(t)$ between these particles we get

$$L(t) = \int_0^{L^*} dx \sqrt{g_{xx}} = \int_0^{L^*} dx \sqrt{\eta_{xx} + h_{xx}}$$

$$\approx \; L^* \sqrt{1 + h_{xx}\left(t, x = 0\right)} \approx L^*\left(1 + \frac{1}{2}\, h_{xx}\left(t, x = 0\right)\right). \quad (25.32)$$

We define the **length change** δL by

$$\delta L := L\left(t\right) - L^*$$

and from (25.32) we get for the **relative length change**

$$\frac{\delta L}{L^*} = \frac{1}{2}\, h_{xx}\left(t, x = 0\right). \quad (25.33)$$

The disturbance function h_{xx} is time-dependent, i.e., the invariant proper distance of the particles changes with time. To investigate this more closely, we first look at how a $+$ polarized gravitational wave affects a set of particles arranged in a ring in the (x, y) plane. We consider with (25.30)

$$h_{\mu\nu} = \begin{pmatrix} 0 & 0 & 0 & 0 \\ 0 & A_+ & 0 & 0 \\ 0 & 0 & -A_+ & 0 \\ 0 & 0 & 0 & 0 \end{pmatrix} \sin\left(\omega\left(z - t\right)\right).$$

Since we have anchored the particle ring in the (x, y)-plane, we set $z = 0$, and the disturbance function simplifies to

$$h_{\mu\nu} = \begin{pmatrix} 0 & 0 & 0 & 0 \\ 0 & A_+ & 0 & 0 \\ 0 & 0 & -A_+ & 0 \\ 0 & 0 & 0 & 0 \end{pmatrix} \sin\left(-\omega t\right).$$

The spatial line element of the weak gravitational field results from this with (25.1) and because $z = 0$ to:

$$ds^2 = \left(1 + A_+ \sin\left(-\omega t\right)\right) dx^2 + \left(1 - A_+ \sin\left(-\omega t\right)\right) dy^2 \quad (25.34)$$

The distances between a particle in the center of the ring and the particles on the ring change over time according to the metric (25.34). To calculate these, we introduce a new coordinate system:

$$X: \;=\; \left(1 - \frac{A_+}{2}\sin\left(-\omega t\right)\right) x = \left(1 + \frac{A_+}{2}\sin \omega t\right) x$$

$$Y: \;=\; \left(1 - \frac{A_+}{2}\sin \omega t\right) y$$

Then in first order in A_+

$$dX^2 = \left(1 + A_+ \sin\omega t + \left(\underbrace{\frac{A_+^2}{4} \sin^2\omega t}_{\approx 0}\right)\right) dx^2 = \left(1 + A_+ \sin\omega t\right) dx^2$$

$$dY^2 = \left(1 - A_+ \sin\omega t + \left(\underbrace{\frac{A_+^2}{4} \sin^2\omega t}_{\approx 0}\right)\right) dy^2 = \left(1 - A_+ \sin\omega t\right) dy^2,$$

so the spatial line element (25.34) in first order is that of the Euclidean (X, Y) plane:

$$ds^2 = dX^2 + dY^2$$

Fig. 25.1 illustrates the temporal behavior of the positions of the eight particles on the ring, which is perpendicular to the propagation direction of the gravitational wave. The general pattern is an ellipse, whose axes oscillate periodically in time. In the first quarter of a period, the ring is continuously squeezed in the Y direction and stretched in the X direction. In the second quarter, the Y direction expands and the X direction contracts until the ring is formed again. A quarter further, it goes the other way round: The Y direction is stretched, the X direction squeezed. And finally, in the last quarter, the ring shape is formed again, and everything starts from the beginning.

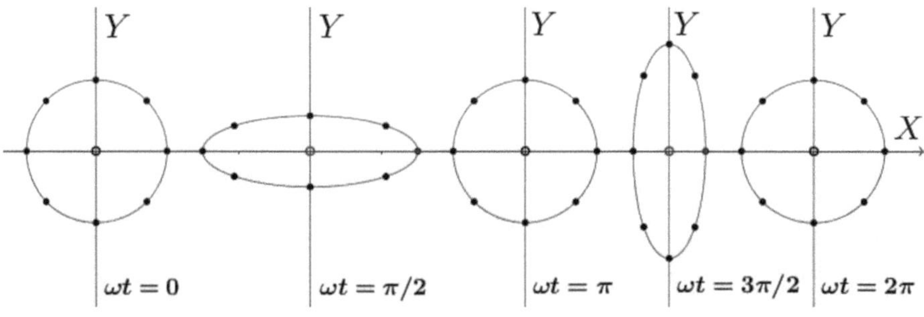

Figure 25.1.: Deformation during the passage of a $+$ gravitational wave

Such a pattern is characteristic of all gravitational waves and can be used as a recognition feature of such a wave.

We now deal with the case of a $\times$ polarized wave, i.e., we set

$$h_{\mu\nu} = \begin{pmatrix} 0 & 0 & 0 & 0 \\ 0 & 0 & A_\times & 0 \\ 0 & A_\times & 0 & 0 \\ 0 & 0 & 0 & 0 \end{pmatrix} \sin\left(-\omega t\right)$$

and obtain for the spatial line element

$$ds^2 = dx^2 + 2A_\times \sin\left(-\omega t\right) dx\, dy + dy^2. \tag{25.35}$$

To see what movements of the particles on the ring are triggered by such a gravitational wave, it is easiest to rotate the axes by 45° in the plane. The new coordinates are then with the rotation matrix 8.13:

$$\tilde{x} = x \cos 45° + y \sin 45° = \frac{1}{\sqrt{2}}\left(x + y\right)$$

$$\tilde{y} = -x \sin 45° + y \cos 45° = \frac{1}{\sqrt{2}}\left(-x + y\right),$$

and it follows

$$d\tilde{x}^2 + d\tilde{y}^2 = dx^2 + dy^2, \quad d\tilde{x}^2 - d\tilde{y}^2 = 2dx\, dy.$$

This makes the line element (25.35) to

$$\begin{aligned} ds^2 &= d\tilde{x}^2 + d\tilde{y}^2 + A_\times \sin\left(-\omega t\right)\left(d\tilde{x}^2 - d\tilde{y}^2\right) \\ &= \left(1 + A_+ \sin\left(-\omega t\right)\right) d\tilde{x}^2 + \left(1 - A_+ \sin\left(-\omega t\right)\right) d\tilde{y}^2, \end{aligned}$$

and we again obtain the metric (25.34). The corresponding Fig. 25.2 shows a similar temporal behavior of the eight particles on the ring, but the ellipses are rotated by 45°.

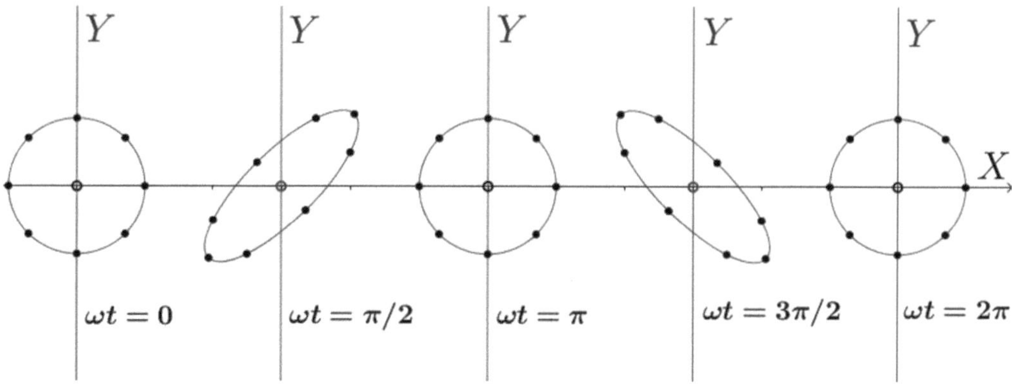

Figure 25.2.: Deformation during the passage of a × gravitational wave

Typical Amplitudes of Astrophysical Sources

A large number of astrophysical sources produce gravitational waves with an amplitude of A_+ or $A_\times$ in an order of magnitude between 10^{-18} and 10^{-22}

meters (as measured on Earth). This means that free particles, which are at a distance of $R = 10^6$ m $= 1000$ km from each other, oscillate back and forth with amplitudes of the order of 10^{-14} m (which is approximately the size of ten atomic nuclei) when they are hit by a gravitational wave with amplitude $A_{+/\times} \approx 10^{-20}$. This is the main reason why the detection of gravitational waves on Earth is so difficult. The deflections in Figures 25.1 and 25.2 are therefore greatly exaggerated. We will deal with the detection of gravitational waves in more detail in Sect. 25.5.

25.4. Generation of Gravitational Waves

We now assume the presence of sources (e.g., stars rotating around each other) that are characterized by the energy-momentum tensor $T_{\mu\nu} \neq 0$. The sources may be located in a limited spatial area, so that the energy-momentum tensor vanishes outside.

Our task is to examine the ten independent inhomogeneous linear Einstein equations (25.17)

$$\Box H_{\mu\nu} = -16\pi G T_{\mu\nu}$$

more closely and find solutions. For this, we use the general solution formula

$$A_\mu (t, \vec{x}) = \frac{1}{4\pi} \int d^3x' \, \frac{J_\mu (t_r, \vec{x}\,')}{|\vec{x} - \vec{x}\,'|} \tag{25.36}$$

for inhomogeneous wave equations of the form

$$\Box A_\mu = -J_\mu,$$

whose derivation exceeds the mathematical prerequisites for this book (a good and not too abstract introduction to the solution theory of differential equations can be found in the book by Borthwick [4]). In the solution formula (25.36), $|\vec{x} - \vec{x}\,'|$ represents the distance between the source point $\vec{x}\,'$ and the field point $\vec{x}$ (e.g., a location on Earth where the gravitational wave is measured). Since gravitational effects propagate at the speed of light, it does not depend on the state of the source at the current time t, but on the state the source had at the earlier time

$$t_r = t - \left|\vec{x} - \vec{x}\,'\right|.$$

Since we have set $c = 1$, $|\vec{x} - \vec{x}\,'|$ is the time it takes for light to travel the distance $|\vec{x} - \vec{x}\,'|$. The quantity t_r is called **retarded time**. We transfer the solution formula (25.36) to the inhomogeneous linear Einstein equations and obtain

$$H_{\mu\nu}(t,\vec{x}) = 16\pi G \frac{1}{4\pi} \int d^3x' \frac{T_{\mu\nu}(t_r,\vec{x}')}{|\vec{x}-\vec{x}'|} = 4G \int d^3x' \frac{T_{\mu\nu}(t_r,\vec{x}')}{|\vec{x}-\vec{x}'|}. \qquad (25.37)$$

For further treatment, we will consider some special cases in the following.

Long-Distance Approximation

If we are far away from the source area and assume that the extension of the source is small, then

$$r := |\vec{x}| \gg |\vec{x}'| =: r',$$

and we can approximate $|\vec{x}-\vec{x}'|$ by r and thus (25.37) by

$$H_{\mu\nu}(t,\vec{x}) \xrightarrow[r\to\infty]{} 4G \frac{1}{r} \int d^3x' \, T_{\mu\nu}(t-r,\vec{x}') \qquad (25.38)$$

see 25.3. The symbol $\xrightarrow[r\to\infty]{}$ represents in this section the statement that the following term is to be calculated at a large distance r from the source; the arrow thus symbolizes *no* mathematical limit.

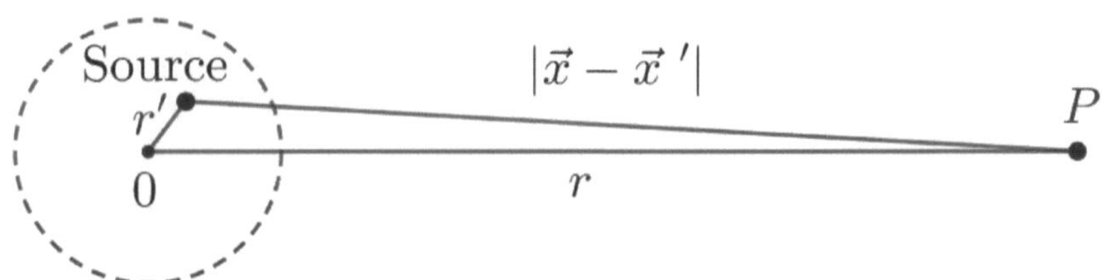

Figure 25.3.: Long-Distance Approximation

We further assume that a weak gravitational field is present and that the sources move slowly (compared to the speed of light). The energy-momentum tensor is then dominated by the (rest) mass density ρ:

$$T^{\mu\nu} = \rho U^\mu U^\nu,$$

i.e., the pressure p is very small compared to the mass density ρ and is neglected, see (16.6). According to (16.10), the energy-momentum tensor satisfies the conservation law of flat space:

$$\partial_\nu T^{\mu\nu} = 0 \qquad (25.39)$$

For $\mu = t$ and $i = x, y, z$ it follows from this and with the Schwarz theorem (9.27)

$$\partial_t\, \partial_t\, T^{tt} = -\partial_t\, \partial_i\, T^{ti} = -\partial_i\, \partial_t\, T^{ti}. \tag{25.40}$$

If we set $\mu = k$ in (25.39), we get

$$\partial_t\, T^{kt} + \partial_i\, T^{ki} = 0 \Rightarrow \partial_i\, T^{ki} = -\partial_t\, T^{kt}$$

and after differentiating again with respect to ∂_k, we get

$$\partial_i\, T^{ki} = -\partial_t\, T^{kt} \Rightarrow \partial_k\, \partial_i\, T^{ki} = -\partial_k\, \partial_t\, T^{kt}$$

and from this, with (25.40), the relationship

$$\partial_t\, \partial_t\, T^{tt} = \partial_k\, \partial_i\, T^{ki}. \tag{25.41}$$

We multiply Eq. (25.41) with $x^m x^n$ and integrate the right side partially over the whole space

$$
\begin{aligned}
\int d^3x\, \partial_k\, \partial_i\, T^{ki} x^m x^n &= \left[\partial_i\, T^{ki} x^m x^n\right]_\Sigma - \int d^3x \left(\partial_i\, T^{ki}\right)\left(\partial_k\left(x^m x^n\right)\right) \\
&= -\int d^3x \left(\partial_i\, T^{ki}\right)\left((\partial_k\, x^m)\, x^n + x^m\,(\partial_k\, x^n)\right) \\
&= -\int d^3x\, \partial_i\, T^{ki}\left(\delta^m_k\, x^n + \delta^n_k\, x^m\right),
\end{aligned}
$$

since the surface term vanishes when Σ is larger than the source region. Another partial integration leads to

$$
\begin{aligned}
\int d^3x\, \partial_k\, \partial_i\, T^{ki} x^m x^n &= \int d^3x\, T^{ki}\left(\delta^m_k\, \partial_i\, x^n + \delta^n_k\, \partial_i\, x^m\right) \\
&= \int d^3x\, T^{ki}\left(\delta^m_k\, \delta^n_i + \delta^n_k\, \delta^m_i\right) \\
&= \int d^3x\,\left(T^{mn} + T^{nm}\right) = 2\int d^3x\, T^{mn}.
\end{aligned}
$$

Together with (25.41), we obtain

$$\int d^3x\, T^{mn} = \frac{1}{2}\frac{\partial^2}{\partial t^2}\int d^3x\, T^{tt} x^m x^n \tag{25.42}$$

In our non-relativistic limit case T^{tt} is determined by the mass density ρ (see Eq. 16.11 on page 341), and the integral on the right side of (25.42) defines the (symmetric) **quadrupole moment tensor** $I^{mn}(t)$ of the energy density ρ by

$$I^{mn}(t) := \int d^3x\, \rho(t, \vec{x})\, x^m x^n. \tag{25.43}$$

From (25.43) and (25.38) we obtain for the *spatial* components of $H^{\mu\nu}$

$$H^{mn}(t, \vec{x}) \xrightarrow[r \to \infty]{} \frac{2G}{r}\, \ddot{I}^{mn}(t - r), \tag{25.44}$$

where a dot represents a derivative with respect to time t.

Gravitational Radiation of a Binary Star System

We want to calculate the quadrupole moment for the case of a binary star system and consider two (neutron) stars with approximately the same mass M, which move on a circular orbit with radius R around their center of mass, see Fig. 25.4.

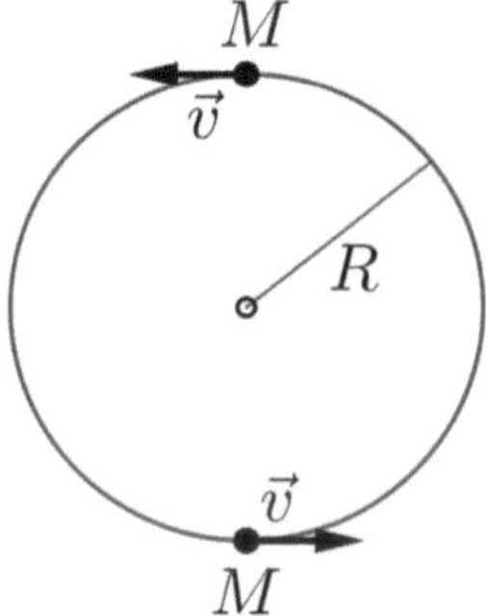

Figure 25.4.: Binary star system

We assume that the stars move slowly enough that their (equal) velocity v is non-relativistic and that they are far enough apart so that Newton's theory of gravitation is suitable to describe their movements. From Newton's law of gravitation (6.2) the magnitude of the gravitational acceleration is given by

$$a = \frac{GM}{(2R)^2},$$

and according to the second law of Newton, the magnitude of the *centripetal acceleration* (2.11) of a star is

$$a = \frac{GM}{(2R)^2} = \frac{v^2}{R} = R\left(\frac{v}{R}\right)^2 = R\omega^2 \Rightarrow \omega^2 = \frac{GM}{4R^3} \tag{25.45}$$

with the *angular frequency* (Eq. 1.5 on page 11) $\omega = v/R$. The last equation reminds us of the *third law of Kepler*, see Eq. 1.5 on page 11. At time t the two stars may be at the positions

$$
\begin{aligned}
(x(t), y(t), z(t)) &= (R\cos\omega t, R\sin\omega t, 0) \quad \text{und} \\
(x(t), y(t), z(t)) &= (-R\cos\omega t, -R\sin\omega t, 0)
\end{aligned}
$$

respectively. For the quadrupole moments, due to

$$
\int d^3x\,\rho(t, \vec{x}) = 2M
$$

we then have

$$
I^{mn} = 2M x^m(t)\, x^n(t).
$$

With the addition theorems of the angle functions in section 29.1 we get

$$
\begin{aligned}
I^{xx} &= 2MR^2\cos^2\omega t = MR^2(1 + \cos 2\omega t) \\
I^{xy} &= I^{yx} = 2MR^2\sin\omega t\cos\omega t = MR^2(\sin 2\omega t) \\
I^{yy} &= 2MR^2\sin^2\omega t = MR^2(1 - \cos 2\omega t) \\
I^{xz} &= I^{yz} = I^{zz} = 0.
\end{aligned}
$$

We obtain for the second time derivatives of the quadrupole moments:

$$
\ddot{I}^{mn}(t) = -4\omega^2 MR^2
\begin{pmatrix}
\cos(2\omega t) & \sin(2\omega t) & 0 \\
\sin(2\omega t) & -\cos(2\omega t) & 0 \\
0 & 0 & 0
\end{pmatrix}
\tag{25.46}
$$

and from this with (25.44)

$$
H^{mn}(t, \vec{x}) \xrightarrow[r\to\infty]{} -\frac{8GMR^2\omega^2}{r}
\begin{pmatrix}
\cos(2\omega(t - r)) & \sin(2\omega(t - r)) & 0 \\
\sin(2\omega(t - r)) & -\cos(2\omega(t - r)) & 0 \\
0 & 0 & 0
\end{pmatrix}.
\tag{25.47}
$$

The frequency of the emitted radiation is twice as large as the orbit frequency, which was also to be expected, since after half a period the two (equal) masses are quasi exchanged and thus the initial situation is recreated. For the polarization amplitudes, we get from (25.47) with (25.45):

$$
A_+ = A_\times = \frac{8GM\omega^2}{r}R^2 = \frac{2G^2M^2}{rR} = \frac{8GM\omega^2}{r}\frac{(GM)^{2/3}}{4^{2/3}\omega^{4/3}} = \frac{8GM}{r}\left(\frac{GM\omega}{4}\right)^{2/3}
\tag{25.48}
$$

The faster the stars rotate around each other, the larger the amplitudes become, the further one is from the source, the smaller the amplitudes become. We see that the disturbance functions

$$H_{ST}^{\mu\nu} = H^{\mu\nu} \xrightarrow[r\to\infty]{} -\frac{8GMR^2\omega^2}{r} \begin{pmatrix} 0 & 0 & 0 & 0 \\ 0 & \cos\left(2\omega\left(t-r\right)\right) & \sin\left(2\omega\left(t-r\right)\right) & 0 \\ 0 & \sin\left(2\omega\left(t-r\right)\right) & -\cos\left(2\omega\left(t-r\right)\right) & 0 \\ 0 & 0 & 0 & 0 \end{pmatrix}$$

$$(25.49)$$

are already in TT gauge (25.31), i.e., the $H^{\mu\nu}$ in (25.49) describe a wave propagation in z-direction perpendicular to the orbital plane of the two stars. The wave consists of an equally weighted linear combination of the two linear polarizations in (25.31).

Quadrupole Formula

For the total radiated energy per unit of time L_{GW}, also known as **(gravitational wave) luminosity**, we get for our binary star system

$$L_{GW} = -\frac{dE}{dt} = \frac{G}{5}\left\langle \dddot{I}_{ij}\, \dddot{I}^{ij} \right\rangle, \qquad (25.50)$$

where the angle brackets denote the averaging over all frequencies ω. This is Einstein's famous **quadrupole formula** for the radiated energy of a gravitational wave, which he derived from his theory of gravitation as early as 1916 and improved in 1918. The derivation of this formula exceeds the mathematical and physical prerequisites of this book, so we only quote it here. If you want to delve deeper, you can find an exhaustive derivation with all detailed calculations in my book [25].

The radiated power represents a loss for the considered masses, hence the negative sign before the term dE/dt. From the quadrupole formula, we can also see that the third time derivative of the quadrupole moment must be different from zero in order for a gravitational wave to be generated at all. This is also the reason why static spherically symmetric and uniformly rotating spherically symmetric objects *do not* emit gravitational waves.

For the binary star system considered above, we get with (25.46)

$$\dddot{I}_{ij} = \frac{d}{dt}\left(-4\omega^2 MR^2 \begin{pmatrix} \cos\left(2\omega t\right) & \sin\left(2\omega t\right) & 0 \\ \sin\left(2\omega t\right) & -\cos\left(2\omega t\right) & 0 \\ 0 & 0 & 0 \end{pmatrix}\right)$$

$$= 8\omega^3 MR^2 \begin{pmatrix} \sin\left(2\omega t\right) & -\cos\left(2\omega t\right) & 0 \\ -\cos\left(2\omega t\right) & -\sin\left(2\omega t\right) & 0 \\ 0 & 0 & 0 \end{pmatrix} \qquad (25.51)$$

and from this

$$
\left\langle \dddot{I}_{ij}\, \dddot{I}^{ij} \right\rangle = \left\langle 64\omega^6 M^2 R^4 \left(2\sin^2(2\omega t) + 2\cos^2(2\omega t)\right) \right\rangle
$$
$$
= 128\omega^6 M^2 R^4 \left\langle \sin^2(2\omega t) + \cos^2(2\omega t) \right\rangle.
$$

The functions $\sin^2(2\omega t)$ and $\cos^2(2\omega t)$ take all values between 0 and 1, so we get for the average

$$
\left\langle \sin^2(2\omega t) + \cos^2(2\omega t) \right\rangle = \frac{1}{2} + \frac{1}{2} = 1
$$

and thus

$$
\left\langle \dddot{I}_{ij}\, \dddot{I}^{ij} \right\rangle = 128\omega^6 M^2 R^4.
$$

We use Kepler's law (25.45):

$$
\omega^2 = \frac{GM}{4R^3} \Rightarrow R = \frac{(GM)^{1/3}}{4^{1/3}\,\omega^{2/3}}
$$

and with (25.50) we get for the luminosity of the binary star system

$$
L_{GW} = -\frac{dE}{dt} = \frac{128G}{5}\omega^6 M^2 R^4 = \frac{2}{5G}\left(\frac{GM}{R}\right)^5 = \frac{2}{5G}(2GM\omega)^{10/3}. \quad (25.52)
$$

This is stronger the smaller the distance between the two stars or the larger the orbital frequency ω is. We want to write the formula (25.52) in SI units and use the Schwarzschild radius proportional to the mass M (23.17):

$$
r_S = \frac{2GM}{c^2}
$$

as well as the distance $D = 2R$ between the two stars. Then follows

$$
L_{GW} = -\frac{dE}{dt} = \frac{2}{5Gc^5}\left(\frac{GM}{R}\right)^5 = \frac{2}{5}\frac{c^5}{G}\left(\frac{r_S}{D}\right)^5 \approx \left(\frac{r_S}{D}\right)^5 \cdot 10^{52}\,\text{Watt}. \quad (25.53)
$$

For the radiation power, it depends on the ratio of the Schwarzschild radius to the distance of the stars. We want to illustrate the possible orders of magnitude of the radiation with three examples.

Example 25.1.

1. As a first example, we calculate the gravitational wave luminosity of the Earth orbiting the Sun. For this, we need to extend the formula (25.52)

to the case of different masses M_1 and M_2. In [21] one can find a detailed derivation, which is very similar to the above and results in

$$L_{GW} = \frac{32}{5} \frac{G}{c^5} \omega^6 \frac{(M_1 M_2)^2}{(M_1 + M_2)^2} D^4. \tag{25.54}$$

If the two masses are equal $M_1 = M_2 = M$, then from (25.54) it follows

$$L_{GW} = \frac{32 G}{5 c^5} \omega^6 \frac{M^4}{(2M)^2} (2R)^4 = \frac{128 G}{5} \omega^6 M^2 R^4,$$

so the special case (25.52). The Kepler's law is then

$$D^3 = \frac{G(M_1 + M_2)}{\omega^2}, \tag{25.55}$$

and we obtain from (25.54) with (25.55)

$$L_{GW} = \frac{32}{5} \frac{G^4}{c^5} \frac{(M_1 + M_2)(M_1 M_2)^2}{D^5} = \frac{32}{5} \frac{G^{7/3}}{c^5} \frac{(M_1 M_2)^2}{(M_1 + M_2)^{2/3}} \omega^{10/3}. \tag{25.56}$$

We use the following numerical values (in SI units) for the physical quantities, also compare Tab. 30.3.

Physical quantity	Symbol	Numerical value
Mass of Sun	$M_S, M_\odot$	$1.99 \cdot 10^{30}\,\mathrm{kg}$
Mass of Earth	M_E	$5.974 \cdot 10^{24}\,\mathrm{kg}$
Newton's gravitational const.	G	$6.67 \cdot 10^{-11}\,\mathrm{m^3\,kg^{-1}\,s^{-2}}$
Speed of light	c	$3 \cdot 10^8\,\mathrm{m\,s^{-1}}$
Astronomical Unit	AU	$1.5 \cdot 10^{11}\,\mathrm{m}$

Table 25.1.: Physical quantities

This results in

$$\begin{aligned} L_{GW} &= \frac{32}{5} \frac{G^4}{c^5} \frac{(M_S + M_E)(M_S M_E)^2}{D^5} \frac{32}{5} \\ &\approx \frac{2}{10^{41}} \frac{1}{2.4 \cdot 10^{42}} \frac{\left(1.99 \cdot 10^{30}\right)\left(1.41 \cdot 10^{110}\right)}{7.6 \cdot 10^{55}} \\ &\approx 200\,\mathrm{watt}, \end{aligned}$$

which is a very modest number. Therefore, it would theoretically take a very long time (much longer than the universe has existed) for the Earth to crash into the Sun due to the loss of energy by gravitational wave radiation. However, this is not much consolation, as the Sun will have burned out billions of years before this event.

2. In this example, we want to estimate the luminosity for two neutron stars, each with 1.4 solar masses. The Schwarzschild radius of such a star, according to Tab. 25.1, is

$$r_S = \frac{2GM}{c^2} = \frac{2 \cdot 6.67 \cdot 1.4 \cdot 1.99 \cdot 10^{30}}{9 \cdot 10^{16} \cdot 10^{11}} \approx 4\,\mathrm{km},$$

which is about a third of its radius. We initially assume that the neutron stars orbit at a distance of one astronomical unit (= average distance Sun - Earth, see Tab. 25.1), and obtain with (25.53)

$$L_{GW} \approx \left(\frac{4 \cdot 10^3}{1.5 \cdot 10^{11}}\right)^5 \cdot 10^{52} \approx 1.35 \cdot 10^{14}\,\mathrm{watt}.$$

This is already a considerable number, about 6000 times the power of one of the largest power plant in the world (Three Gorges Dam in China). However, considering that the electromagnetic radiation power of the Sun is about $4 \cdot 10^{26}$ watt, the gravitational radiation of neutron stars on an astronomical scale is rather low. However, the radiation power of the binary system increases considerably when the distance between the two decreases. Thus, we obtain a radiation power of

$$L_{GW} \approx \left(\frac{4 \cdot 10^3}{5 \cdot 10^8}\right)^5 \cdot 10^{52} \approx 3.37 \cdot 10^{26}\,\mathrm{watt},$$

i.e., on the order of magnitude of the Sun, when the neutron stars orbit each other at a distance of $500,000\,\mathrm{km}$. If we further assume that the distance of the binary system to Earth is (only) about 40 light years (like, for example, the binary system i Boo), then the size of the gravitational wave amplitude with

$$\begin{aligned}
M^2 &\approx\ 8 \cdot 10^{60},\, G^2 \approx 45/10^{22},\, r = 40\,\mathrm{Lj} = 3.8 \cdot 10^{17}, \\
R &=\ 2.5 \cdot 10^8,\, c^4 = 81 \cdot 10^{32}
\end{aligned}$$

according to formula (25.48) in SI units

$$A_{+/\times} = \frac{2G^2 M^2}{c^4 r R} \approx 9.4 \cdot 10^{-20},$$

is of a similar order of magnitude as shown at the end of section 25.3.

3. The greatest radiation effect occurs when the two neutron stars are just before their merger, i.e., when their distance just equals their diameter:

$$L_{GW} \approx \left(\frac{4 \cdot 10^3}{24 \cdot 10^3} \right)^5 \cdot 10^{52} \approx 1.29 \cdot 10^{48} \, \text{watt}$$

This is an unimaginably large number, equivalent to the luminosity of more than 10^{21} suns. In fact, an even greater luminosity, which lasted only fractions of a second, was observed in September 2015 during the merger of two black holes, which we will describe in more detail in section 25.5. $\square$

25.5. Detection of Gravitational Waves

There are essentially two, albeit fundamentally different, methodological approaches to the detection of gravitational waves. One involves measuring the consequences of the energy loss of a system due to gravitational wave radiation, without detecting the radiated gravitational waves themselves (on Earth). This method is called **indirect detection** of gravitational waves. The detection of gravitational waves with measuring devices on Earth (and possibly in space in the future), which was first achieved in September 2015, is called **direct detection** of gravitational waves. We first discuss some results obtained by the indirect method.

Indirect Detection of Gravitational Waves

One consequence of energy loss due to gravitational waves is the temporal change in the orbital period of a binary star system. If it is determined by concrete measurements that this is just as large as the theoretical prediction of energy loss due to gravitational waves, and if it can be ruled out that there are other influencing effects, then one can speak of an indirect detection of gravitational waves.

As in section 25.4, we consider two (neutron) stars with approximately the same mass M, which move on a circular orbit with radius R around their center of mass, see Fig. 25.4. These assumptions are not very realistic, as the masses are usually different and the orbital motions take place on elliptical orbits, but they do give a good idea of the essential concepts.

In our Newtonian approximation, the total energy E of the binary star system is

$$E = E_{kin} + E_{pot} = \frac{1}{2} \left(2M \right) v^2 - \frac{GM^2}{2R} = M\omega^2 R^2 - \frac{GM^2}{2R}.$$

With the "Kepler's law" (25.45)

$$\omega^2 = \frac{GM}{4R^3} \Rightarrow R = \frac{(GM)^{1/3}}{4^{1/3}\,\omega^{2/3}}$$

(25.57)

we get from this

$$E = MR^2 \frac{GM}{4R^3} - \frac{GM^2}{2R} = -\frac{GM^2}{4R} = -\frac{GM^2 4^{1/3}\,\omega^{2/3}}{4\,(GM)^{1/3}} = -\frac{M\,(GM\omega)^{2/3}}{4^{2/3}}.$$

(25.58)

The two stars in the binary system neither have a constant distance nor a constant circular frequency. Due to the radiation of gravitational wave energy, the distance decreases over time, and the orbital frequency increases. To explain this in more detail, we write the time derivatives again with a dot and get with (25.58)

$$\frac{\dot{E}}{E} = \frac{\dfrac{GM^2}{4R^2}\dot{R}}{-\dfrac{GM^2}{4R}} = -\frac{\dot{R}}{R}$$

(25.59)

as well as with (25.45) and the orbital period $T = 2\pi/\omega$:

$$-\frac{\dot{R}}{R} = -\frac{2}{3}\frac{\dfrac{(GM)^{1/3}}{4^{1/3}\,\omega^{5/3}}\dot{\omega}}{\dfrac{(GM)^{1/3}}{4^{1/3}\,\omega^{2/3}}} = -\frac{2}{3}\frac{\dot{\omega}}{\omega} = -\frac{2}{3}\frac{\dfrac{2\pi}{T^2}}{\dfrac{2\pi}{T}}\dot{T} = -\frac{2}{3}\frac{\dot{T}}{T}$$

(25.60)

Since we assume that the energy loss is exclusively a result of the radiation of gravitational waves, we can use the luminosity L_{GW} of the binary star system in SI units (25.53)

$$\dot{E} = \frac{dE}{dt} = -\frac{2}{5Gc^5}\left(\frac{GM}{R}\right)^5$$

and get

$$\frac{\dot{E}}{E} = \frac{-\dfrac{2}{5Gc^5}\left(\dfrac{GM}{R}\right)^5}{-\dfrac{GM^2}{4R}} = \frac{8}{5c^5}\frac{G^3 M^3}{R^4} = -\frac{\dot{R}}{R} \Rightarrow \dot{R} = -\frac{8}{5c^5}\left(\frac{GM}{R}\right)^3.$$

(25.61)

We combine the terms with R on the left side and get

$$\dot{R}R^3 = -\frac{8}{5c^5}(GM)^3.$$

This differential equation is solved with the initial condition $R(0) = R_0$ by integrating both sides:

$$\int dt\, \dot{R}R^3 = \frac{1}{4}R^4 = -\int dt\, \frac{8}{5c^5}(GM)^3 \Rightarrow R(t) = \left(const. - \frac{32}{5c^5}(GM)^3 t\right)^{1/4}$$

with

$$R(0) = const.^{1/4} \Rightarrow const. = R_0^4.$$

We define

$$t_e := \frac{5c^5 R_0^4}{32\,(GM)^3} \tag{25.62}$$

and get as a solution of (25.61)

$$R(t) = R_0 \left(1 - \frac{t}{t_e}\right)^{1/4}, \tag{25.63}$$

i.e., at $t = t_e$ is $R(t) = 0$, t_e is therefore the time until the collision of the two stars. Analogously, using Kepler's law (25.45) for the period T:

$$R = \frac{(GM)^{1/3}}{4^{1/3}\,\omega^{2/3}} = \frac{(GM)^{1/3}\,T^{2/3}}{4^{1/3}\,(2\pi)^{2/3}}$$

and because

$$\frac{\dot{T}}{T} = -\frac{3}{2}\frac{\dot{E}}{E} = -\frac{12}{5c^5}\frac{G^3 M^3}{R^4} = -\frac{12}{5c^5}\frac{(GM)^3\,4^{4/3}\,(2\pi)^{8/3}}{(GM)^{4/3}\,T^{8/3}}$$

$$= -\frac{24}{5c^5}(2\pi)^{8/3}\frac{(2GM)^{5/3}}{T^{8/3}},$$

we get the following differential equation for T:

$$\dot{T} = -\frac{24}{5c^5}(2\pi)^{8/3}(2GM)^{5/3}\frac{1}{T^{5/3}} \tag{25.64}$$

The solution is analogous to above

$$T(t) = \left(const. - \frac{64}{5c^5}(2\pi)^{8/3}(2GM)^{5/3} t\right)^{3/8}$$

and can be written with the initial condition $T(0) = T_0$ and with

$$t_e = \frac{5c^5\,T_0^{8/3}}{64\,(2\pi)^{8/3}(2GM)^{5/3}} \tag{25.65}$$

as

$$T(t) = T_0 \left(1 - \frac{t}{t_e}\right)^{3/8}, \tag{25.66}$$

where t_e is again the decay time of the system, this time expressed by T instead of R.

The Hulse-Taylor Pulsar

The prime example for an indirect detection of gravitational waves is provided by the double star system **PSR B1913+16**, which consists of two orbiting neutron stars, one of which is a **pulsar** (i.e., a star that emits very regular recurring electromagnetic signals) and the other is invisible to us. Such a double star system (also called **binary pulsar**) was first discovered in 1974 by the astronomers Joseph Taylor and Russell Hulse, and instead of astronomically correct PSR B1913+16 it is also simply called **Hulse-Taylor Pulsar**. This binary pulsar emits gravitational waves of such strength that its orbital parameters have noticeably changed over a period of now more than 50 years. The exact astronomical data of the Hulse-Taylor Pulsar (as of: 2016) are according to [19]:

- Distance to Earth: $\approx 22,500\,\text{Lj}$.

- Pulse period, i.e. average rotation time of the pulsar: $59\,\text{milliseconds}$.

- Mass of the pulsar: 1.44 solar masses, $m_p = 1.44\,M_\odot$.

- Mass of the invisible companion: $m_c = 1.39\,M_\odot$.

- Orbit time of the two stars: $T = 7\,\text{h}\,45\,\text{min}$.

- Since the orbit deviates strongly from a circular orbit (the eccentricity of the elliptical orbit (see formula 1.1 on page 8) is $\varepsilon = 0.62$), the distance of the neutron stars in the **periastron** (closest point to the star) is 1.1 solar radii $\left(R_\odot = 696,342\,\text{km}\right)$ and in the **apastron** (furthest point from the star) 4.8 solar radii.

We calculate with (25.64) the period change for the case $M = 1.44\,M_\odot$ and $T = 7.75\,\text{h}$ and obtain with

$$
\begin{aligned}
c^5 &= 2.4 \cdot 10^{42}, (2\pi)^{8/3} = 134.4, M^{5/3} = 5.8 \cdot 10^{50}, \\
G^{5/3} &= \frac{23.8}{10^{55/3}}, T^{5/3} = 2.6 \cdot 10^7
\end{aligned}
$$

for the change of T per second:

$$
\dot{T} = -\frac{24}{5c^5}\,(2\pi)^{8/3}\,(2GM)^{5/3}\,\frac{1}{T^{5/3}} \approx -2.3 \cdot 10^{-13}
$$

This value is only a rough estimate, as we have assumed equal masses and a circle as the orbit in the derivation of (25.64). The equality of masses is almost fulfilled, but the circular orbit hypothesis is not. The value calculated with the "correct" formula

$$\dot{T} = -\frac{192\pi G^{5/3}}{5c^5} \left(\frac{T}{2\pi}\right)^{-5/3} \frac{1 + \frac{73}{24}\varepsilon^2 + \frac{37}{96}\varepsilon^4}{(1 - \varepsilon^2)^{7/2}} \frac{m_p m_c}{(m_p + m_c)^{1/3}} \tag{25.67}$$

(see [19]) is

$$\dot{T} = -2.4025 \cdot 10^{-12} \tag{25.68}$$

and agrees to 0.2 % with the measured values. If the orbit is a circle, i.e., the eccentricity is $\varepsilon = 0$, and the masses are equal, then the differential equation (25.64) immediately results from (25.67)).

The decrease in orbital times in (25.68) amounts to 75 millionths of a second per year. In Fig. 25.5 the cumulative reduction of the orbital period from 1975 to 2013 for the Hulse-Taylor pulsar is shown, the difference in these 38 years is more than a minute.

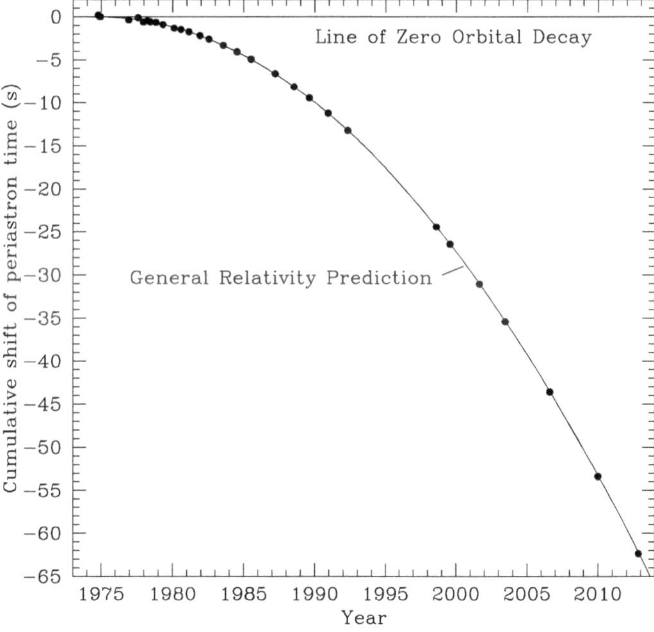

Figure 25.5.: Decrease of the orbital period in the Hulse-Taylor pulsar system. (Source: [38])

The solid line, which represents the prediction of the general theory of relativity, is not a straight line, but falls off more and more steeply, which again documents the amplification effect of the radiation power of gravitational waves with decreasing orbital times.

For this first indirect evidence of gravitational waves and the associated confirmation of the general theory of relativity, the discoverers of the binary pulsar PRS B1913+16, Hulse and Taylor, received the Nobel Prize in Physics in 1993.

Direct Detection of Gravitational Waves

In Sect. 25.3 we saw that physical distances in the plane perpendicular to the direction of propagation of gravitational waves change periodically. The *relative length change* is according to (25.33) and (25.48)

$$\frac{\delta L}{L} \approx \frac{1}{2}\, h = \frac{G^2 M^2}{c^4 r R} = \frac{r_S^2}{4 r R}, \tag{25.69}$$

where h denotes the amplitude of the gravitational wave and r_S the Schwarzschild radius. In (25.49) it was shown that the frequency f_{GW} of a gravitational wave is twice the orbital frequency of a binary star system, i.e., with (25.45) it follows

$$f_{GW} = \frac{2\omega}{2\pi} = \frac{(GM)^{1/2}}{2\pi R^{3/2}} = \frac{c\,(r_S)^{1/2}}{\pi\,(2R)^{3/2}}.$$

Resonance Detectors

The search for gravitational waves has a long history, which began in the 1960s with the experiments of **Joseph Weber** on a cylindrical **resonance detector** (also called **Weber cylinder**) developed by him. If a gravitational wave hits such a cylinder, it is set into oscillations. These can be converted into electrical impulses with piezoelectric measuring devices and detected. The measurement accuracy of modern resonance detectors is about a relative length change of about 10^{-18} to 10^{-19}.

Example 25.2.
To get a feel for such an order of magnitude, we calculate the relative length change for the Hulse-Taylor pulsar, assuming that

- the masses of the two stars are equal,

$$M = m_p = m_c = 1.4\, M_\odot,$$

- they orbit each other on a circular path with radius $R = 10^9\,\text{m}$,

- they are at a distance of $r = 25,000\,\text{Lj} = 2.1 \cdot 10^{20}\,\text{m}$ from Earth

- and on Earth the free test particles $L = 10^3$ m are placed apart from each other.

Then it follows

$$\frac{\delta L}{L} \approx \frac{1}{2} h \approx \frac{G^2 M^2}{c^4 r R} \approx 2.1 \cdot 10^{-23}. \ \square$$

Since the length of resonance detectors is $1-2$ m, a detector for the detection of such a gravitational wave requires a sensitivity of 10^{-23} m, i.e., one that must be several orders of magnitude better than the above-mentioned measurement accuracy. For the detection of gravitational waves with a Weber cylinder, extraordinarily strong gravitational sources are required.

Example 25.3.
We consider the merger of two (fictional) black holes each with ten solar masses in our Milky Way. We assume that the two black holes are at a distance that is ten times their Schwarzschild radius, and that they are $r = 25,000$ Ly away from Earth. Then with (25.69)

$$\frac{\delta L}{L} \approx \frac{R_s^2}{4rR} = \frac{\left(1.5 \cdot 10^4\right)^2}{4 \cdot 2.1 \cdot 10^{20} \cdot 1.5 \cdot 10^5} = 1.8 \cdot 10^{-18}.$$

Thus, the merger of two not too distant black holes would in principle be measurable with a Weber cylinder. $\square$

Interferometers

Since such events are extremely rare and there are also other difficulties (such as the suppression of acoustic and thermal noise) with the resonance detector, people began to build so-called **laser interferometers** as gravitational wave detectors in the 1970s.

Figure 25.6 shows a (greatly simplified) interferometer, as used by the **LIGO** (Laser Interferometer Gravitational-Wave Observatory) project for the detection of gravitational waves. The basic experimental setup of interferometers has been known since the 19th century (think of the famous **Michelson-Morley experiment** for the detection of the ether).

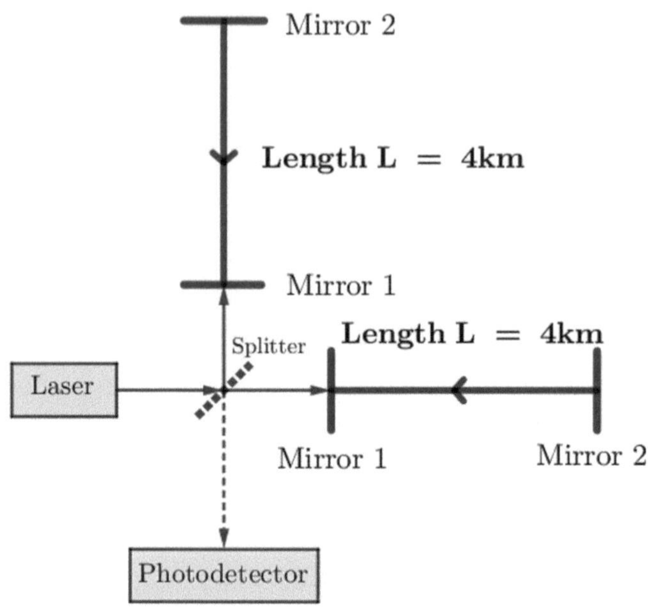

Figure 25.6.: Laser interferometer

The LIGO interferometer is based on the principle that laser light is split into two partial beams at a beam splitter, which move perpendicularly to each other in vacuum tubes of length L to two freely movable "mirrors 2", where they are reflected and hit also freely movable semi-transparent "mirrors 1", where they are reflected and directed back to mirror 2. This process step is repeated approximately 280 times, which is represented in Fig. 25.6 by the bold "light rays" between the two mirrors 1 and 2. This significant extension compared to a normal Michelson interferometer increases the arm length L to 1120 km. In the next step, the partial beams are returned to the beam splitter, superimposed there, and jointly directed to the photodetector for the measurement of their interference. When a gravitational wave hits the interferometer arms, the distances between mirrors 1 and 2 are lengthened in one direction and shortened in the other, and then vice versa, as we have described in Sect. 25.3. This changes the interference pattern (the two partial beams are no longer "in phase") in a typical way, and the gravitational wave can be identified with the photodetector.

In addition to extending the arms, refining the optical devices, and increasing the laser power, the LIGO project also prioritized damping the noise present on

Earth. To detect gravitational waves, other possible sources of vibration must be excluded when measuring the deflections of the mirrors. There are many sources of noise that have different effects at different frequencies. Below 10 Hertz, seismic vibrations dominate. Gravitational waves with such frequencies cannot be detected by LIGO. At the other end in the kilohertz range, the so-called **quantum noise** is the limiting factor. The LIGO interferometer has achieved a sensitivity of about 10^{-19} m over the more than 25 years of its existence by constant improvements and is thus able to detect gravitational waves with amplitudes up to 10^{-23} m and frequencies in the range $50-1.500\,\mathrm{Hz}$.

Example 25.4.
We calculate the respective frequencies of the emitted gravitational waves for examples 25.2 and 25.3.

1. For the Hulse-Taylor pulsar, the current value is

$$f_{GW} = \frac{c\,(r_S)^{1/2}}{\pi\,(2R)^{3/2}} = \frac{3\cdot 10^8 \cdot \left(4\cdot 10^3\right)^{1/2}}{\pi\,(2\cdot 10^9)^{3/2}} \approx \frac{1.9\cdot 10^{10}}{8.9\cdot 3.1\cdot 10^{13}} \approx 7\cdot 10^{-5}\,\mathrm{s}^{-1},$$

thus, the frequencies are outside the LIGO band. Only when the two neutron stars are about to collide and their distance is approximately their Schwarzschild radius (e.g. $R = 3\cdot 10^4\,\mathrm{m}$), the frequency of the emitted gravitational waves is

$$f_{GW} = \frac{c\,(r_S)^{1/2}}{\pi\,(2R)^{3/2}} = \frac{3\cdot 10^8 \cdot \left(4\cdot 10^3\right)^{1/2}}{\pi\,(2\cdot 3\cdot 10^4)^{3/2}} \approx \frac{1,9\cdot 10^{10}}{8.9\cdot 5.2\cdot 10^6} \approx 4\cdot 10^2\,\mathrm{s}^{-1}.$$

The merging of the two neutron stars can therefore (theoretically) be measured by the LIGO interferometer, but the researchers have to wait for this event for about 300 million years.

2. In example 25.3 about the merging of two (fictional) black holes, the frequency of the emitted gravitational waves with the above example data is:

$$f_{GW} = \frac{c\,(r_S)^{1/2}}{\pi\,(2R)^{3/2}} = \frac{3\cdot 10^8 \cdot \left(1.5\cdot 10^4\right)^{1/2}}{\pi\,(3\cdot 10^5)^{3/2}} \approx \frac{3.7\cdot 10^{10}}{16.3\cdot 3.1\cdot 10^7} = 72\,\mathrm{s}^{-1},$$

i.e., the gravitational waves emitted during the merger could be measured with the LIGO interferometer. $\square$

GW150914

On September 14, 2015, a gravitational wave was measured on Earth for the
first time.

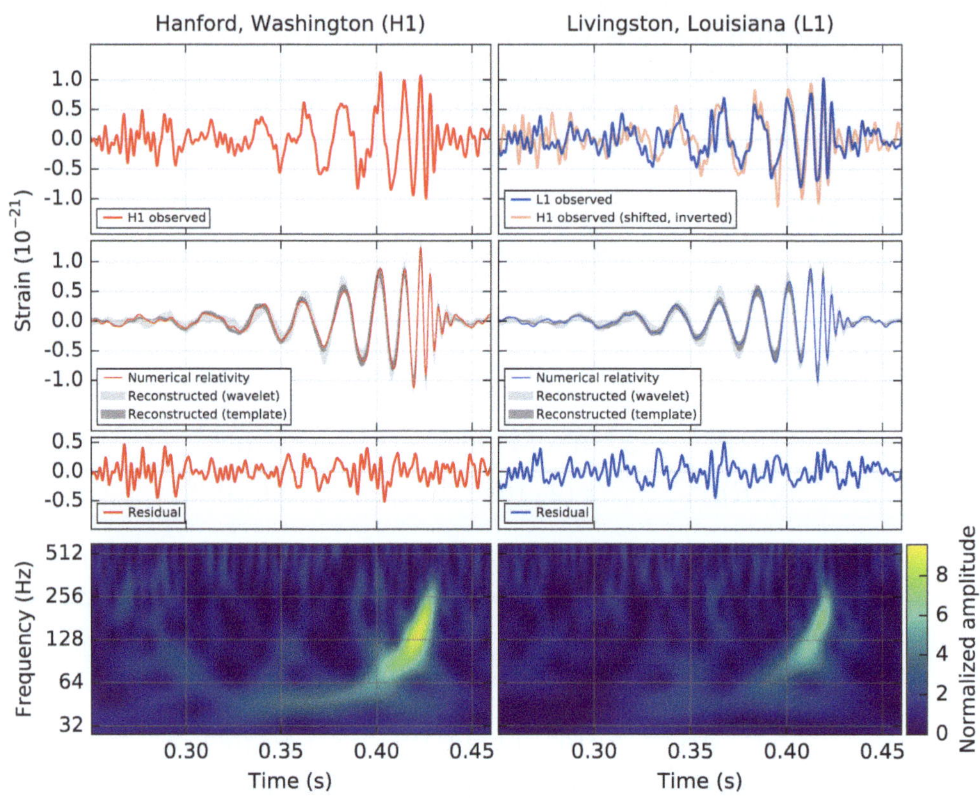

Figure 25.7.: GW150914. (Source [1])

The 4 km long arms of the two LIGO detectors, which are identical in construction and located in Hanford and Livingston in the USA, were hit by a
gravitational wave and were only deflected by a thousandth part of a proton
radius $\left(\approx 10^{-18}\,\mathrm{m}\right)$. This event is named **GW150914** and was announced on
February 11, 2016, after months of careful checks by many of the around a
thousand scientists involved in the LIGO project. The signal was registered
by the two LIGO detectors with a time difference of 6.9 ms, which is exactly
the time it takes for light to travel the distance between the two laboratories.
Thus, it was also measured for the first time that gravitational waves propagate
- as predicted by theory - at the speed of light. Fig. 25.7, which comes from
the initial publication of the LIGO project [1], illustrates various properties of

the gravitational wave signal measured with the detectors.

In the upper third of the figure, we see the signal measured in Hanford on the left. In the upper right box, the recording from Livingston is plotted in *dark*; the Hanford signal, shifted by 6.9 ms, is also shown in *light* underneath, so that one can clearly see that the course of both signals roughly matches. Local terrestrial triggers can thus be almost completely ruled out. The time in seconds is plotted horizontally. We see that the whole event only lasted 0.2 seconds. On the vertical axis in the upper area is the relative change in length $\delta L/L$ (called **strain**) in units of 10^{-21}. If we substitute the arm length $L = 4 \cdot 10^3$ m, then it follows

$$\frac{\delta L}{L} \sim \frac{10^{-18}}{10^3} = 10^{-21},$$

so that this vertical scaling makes sense. In the two boxes below, the signals are entered, which come from a theoretical numerical calculation, which due to the strength of the gravitational effects is based on the *full* (and *not* on the *linear*) Einstein equations. Below that, you can see the difference between the measured and the theoretical numerical signals. At the very bottom, you can (somewhat blurred) read off the frequency development on the bright stripes, the frequencies increase from 35 Hz to 150 Hz and drop rapidly at about 0.42 s. This suggests a system of two objects that orbit each other more and more closely and thus faster and then merge.

If we look at the time course in a little more detail, we can distinguish three phases, see also [1]:

1. **Inspiral**: Frequencies and amplitudes increase, i.e., the two objects are approaching each other.

2. **Merging**: The amplitude becomes maximum and chaotic, i.e., the two objects collide and merge.

3. **Ringdown**: Amplitudes and frequencies drop rapidly and then disappear completely. The newly formed object emits only low and finally no gravitational waves anymore.

Such a signal is also called **chirp**, which in technology means an increasing frequency. Since in the inspiral phase of a chirp signal the (still small) frequencies increase, the masses M_1 and M_2 in the binary system can be calculated from this frequency behavior by expressing the so-called **chirp mass**

$$M_{chirp} = \frac{(M_1 M_2)^{3/5}}{(M_1 + M_2)^{1/5}}$$

as a function of the frequency and its time derivative:

$$M_{chirp} = \frac{c^3}{G} \left(\frac{5 \dot{f}_{GW}}{96 \pi^{8/3} f_{GW}^{11/3}} \right)^{3/5} , \tag{25.70}$$

see [1]. We do not prove this here, but note the similarity of (25.70) with
the differential equation for the period T (25.64) in the special case of equal
masses. For the event GW150914, the chirp mass is about 30 solar masses.
With the following consideration, we can estimate the total mass $M_1 + M_2$ of
the two objects. It applies

$$0 \; \leq \; (M_1 - M_2)^2 \Leftrightarrow 4M_1 M_2 \leq (M_1 + M_2)^2 \Leftrightarrow \frac{4M_1 M_2}{(M_1 + M_2)^2} \leq 1$$

$$\Rightarrow \; \frac{(4M_1 M_2)^{3/5}}{(M_1 + M_2)^{6/5}} \leq 1 \Leftrightarrow \frac{(M_1 M_2)^{3/5}}{(M_1 + M_2)^{1/5}} \leq \frac{M_1 + M_2}{2^{6/5}} ,$$

from which

$$M_1 + M_2 \geq 2^{6/5} M_{Chirp} \approx 70 \, M_\odot \tag{25.71}$$

follows. If we assume equal masses $M_1 = M_2 = 35 \, M_\odot$, then the Schwarzschild
radius of an object is at

$$r_S = \frac{2GM_1}{c^2} \approx 105 \, \text{km}.$$

If two objects orbit each other with the orbital frequency of $\omega = 2\pi \cdot 75 \, \text{Hz}$ (half
the maximum of the gravitational wave frequency), the distance D between
the two objects on a (circular) orbit applies to the Kepler law (25.57)

$$D = 2R = 2 \frac{(GM)^{1/3}}{4^{1/3} \, \omega^{2/3}} \approx 2 \left(\frac{6.7 \cdot 35 \cdot 2 \cdot 10^{30}}{4 \cdot 39.5 \cdot 5625 \cdot 10^{11}} \right)^{1/3} \approx 350 \, \text{km} \cdot$$

The two objects must therefore be very compact, i.e., have a radius on the
order of their Schwarzschild radius. Therefore, only neutron stars or black
holes are possible.

- Neutron stars have a mass upper limit of about three solar masses, so a
  binary system with neutron stars is ruled out.

- A binary star system consisting of a neutron star and a black hole can
  also be ruled out, as due to the large mass of the black hole the orbital
  frequency would have been lower (siehe [1]).

- So, on 14.09.2015, the merger of two black holes was observed and from the measured data the masses $M_1 = 29\,M_\odot$ and $M_2 = 36\,M_\odot$ were determined by numerical calculation of the full Einstein equations. Their sum is 65 solar masses, i.e., we have overestimated the total mass calculated in the linear approximation in (25.71).

The newly formed object is a rotating black hole and has a mass of 62 solar masses, i.e., in only 0.2 seconds the energy of three solar masses was radiated in the form of gravitational waves! The luminosity achieved according to (25.53) is an unimaginable

$$L_{GW} \approx \left(\frac{R_s}{D}\right)^5 \cdot 10^{52}\,\text{Watt} \approx 2.4 \cdot 10^{49}\,\text{Watt}.$$

It is also astonishing that the merger of the two black holes occurred 1.3 billion light years away from us, i.e., the gravitational wave took 1.3 billion years to reach us.

Recent Discoveries and Outlook

After the event GW150914 nearly 100 other mergers of binary systems were recorded until 2024. Among these events, most were mergers of black holes, some of black holes and neutron stars and of binary neutron stars. In the event **GW170817** the union of two neutron stars was observed *for the first time* in 2017.

In addition to the further expansion of the existing detectors, more will be added in the future. We only list a few examples here, a detailed description of the current plans can be found in [22].

- The **LIGO-Virgo-KERGA Collaboration** consists of the two observatories in the USA, the interferometer **Virgo** in Italy and the Japanese gravitational wave detector **KAGRA**. The construction of a forth detector **LIGO-India** in India is planned (scheduled start 2030). With this worldwide distribution the measurements become more reliable, and the place of origin of gravitational waves in the universe can be determined much more precisely. Fig. 25.8 gives an overview of the already detected mergers sorted by solar masses. Each merger is illustrated by two lower dots and an arrow to a third dot, which is the final object after the merger. Blue dots represent black holes, yellow dots neutron stars.

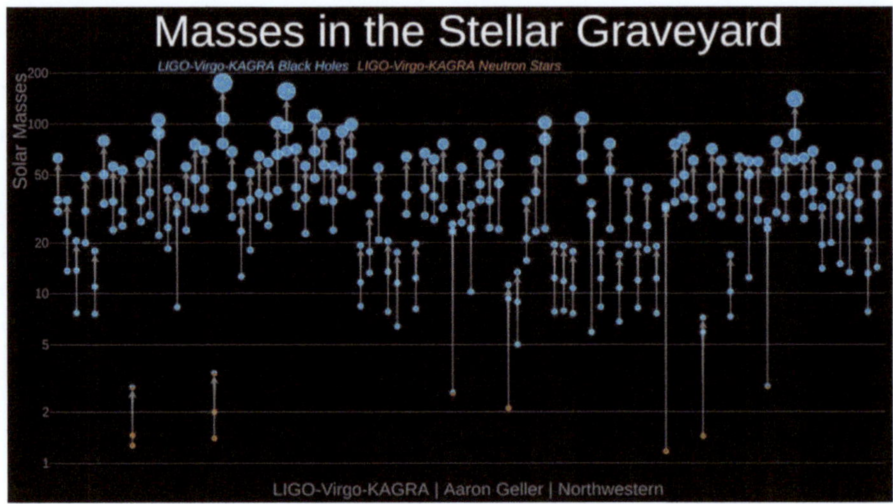

Figure 25.8.: Mergers of LIGO-Virgo-Kagra Collaboration (Source [18])

- The **Laser Interferometer Space Antenna** (LISA) is a gravitational wave detector planned by the ESA in space. Three satellites are supposed to form a triangle with 2.5 million kilometers of side length each. Due to the absence of seismic disturbances, LISA will be able to measure gravitational waves with frequencies between 0.001 and 1 Hz. This can, for example, detect two orbiting supermassive black holes (planned launch 2037).

26. Gravitational Collapse and the Interior Schwarzschild Metric

In this chapter, we want to set up the equations for calculating the pressure and density of matter inside a star. How stars can theoretically form at all and what stages of development they go through, we cannot describe in detail in this book, as this requires in-depth knowledge from, among other things, the physical fields of thermodynamics and quantum mechanics. We therefore limit ourselves to a general description of specific development models of stars.

26.1. Star Collapse

We want to understand a star as a collection of matter that is held together by its own gravity. This gravitational force tends to compress the star further and further. The star can only prevent the collapse as long as its matter can resist the gravitational pressure. If we also consider planets as stars, the Earth can do this because the atoms inside can build up enough counterpressure. For stars that have about the mass of our Sun, the atomic structure is no longer sufficient to resist gravity, here the required matter pressure is produced by nuclear fusions of various kinds. Once the fusion material is exhausted, the star cools down by radiation and the atoms created by the fusions, such as helium or iron, cannot withstand the gravitational pressure. They are virtually "crushed" and for stars of the size of our Sun, an electron gas is created, whose so-called **Fermi pressure** counteracts gravity. Stars that have reached a state of equilibrium in this way are called **white dwarfs**. So white dwarfs have roughly the mass of our Sun $M \approx M_S$ and a radius of about 6000 km, i.e., about one hundredth of the current solar radius.

If the mass of the star is greater than that of our Sun (the theoretical upper limit is (only!) about $1.4\,M_S$ and is called the **Chandrasekhar limit mass**), the Fermi pressure cannot withstand gravity. Through further contraction, electrons and protons can be converted into neutrons and again the Fermi pressure of the neutrons can come into equilibrium with the gravitational pressure. Stars with this property are called **neutron stars**. They can also have smaller masses in the final state, e.g., if a **supernova** has previously been created by the collapse of the Fermi pressure. A (fictitious) neutron star

M. Ruhrländer, *Ascent to the Einstein Equations*, https://doi.org/10.1007/978-3-662-72672-3_26

with solar mass has only a radius r of about $10\,\text{km}$, i.e., the radius is not too far from the Schwarzschild radius $r \approx 3\,r_S$). But the conversion into neutron stars is not the only possibility for even larger masses of stars (here too there is an estimate of the limit mass, namely the range from 1.3 to 3.2 solar masses, the so-called **Tolman-Oppenheimer-Volkoff limit**). If the mass is greater than the limit mass, neutron stars also contract further and eventually become black holes.

26.2. Calculation of the Interior Schwarzschild Solution

We assume in our considerations that the star to be examined is spherically symmetric and static. In Chap. 23 we derived the standard form of the metric (see 23.8 on page 451) for such a case:

$$
g_{\mu\nu} = \begin{pmatrix}
-e^{2\nu(r)} & 0 & 0 & 0 \\
0 & e^{2\lambda(r)} & 0 & 0 \\
0 & 0 & r^2 & 0 \\
0 & 0 & 0 & r^2 \sin^2 \vartheta
\end{pmatrix}
$$

The inverse metric to this is according to Eq. (23.9):

$$
g^{\mu\nu} = \begin{pmatrix}
-e^{-2\nu(r)} & 0 & 0 & 0 \\
0 & e^{-2\lambda(r)} & 0 & 0 \\
0 & 0 & r^{-2} & 0 \\
0 & 0 & 0 & r^{-2} \sin^{-2} \vartheta
\end{pmatrix}
$$

The two unknown functions $\nu(r)$ and $\lambda(r)$ cannot be calculated in the same way as in the above-mentioned chapter, as we now want to venture into the interior of a star. To set up the field equations for this, we need the energy-momentum tensor for the star matter, which we want to assume exists as an ideal fluid. According to Eq. 22.3 on page 432, the energy-momentum tensor for such a fluid is

$$
T^{\alpha\beta} = (\rho + p)\, U^{\alpha} U^{\beta} + p g^{\alpha\beta}
$$

or, after multiplication with $g_{\mu\alpha}\, g_{\nu\beta}$ on both sides of the equation,

$$
T_{\mu\nu} = (\rho + p)\, U_{\mu} U_{\nu} + p g_{\mu\nu}.
$$

Since we are studying static stars, the fluid is time-constant, so it has no velocity, i.e., the spatial coordinates of the proper velocity $\vec{U}$ are zero in the current inertial system

$$
\vec{U} \to \left(U^t, 0, 0, 0\right).
$$

Now the identity applies

$$g_{\alpha\beta}\, U^\alpha U^\beta = U^\alpha U_\alpha = g^{\alpha\beta} U_\alpha\, U_\beta,$$

and from the normalization condition

$$-1 = g_{\alpha\beta}\, U^\alpha U^\beta = g_{tt}\, U^t U^t = -e^{2\nu(r)} U^t U^t = U^t U_t$$

it follows that

$$U^t = e^{-\nu(r)} \ \text{and from that} \ \ U_t = -e^{\nu(r)}.$$

With this, we can calculate the components of the energy-momentum tensor specifically

$$
\begin{aligned}
T_{tt} &= (\rho + p)\, U_t\, U_t + pg_{tt} = (\rho + p)\, e^{2\nu(r)} + p\left(-e^{2\nu(r)}\right) = \rho e^{2\nu(r)} \\
T_{rr} &= pg_{rr} = pe^{2\lambda(r)} \\
T_{\vartheta\vartheta} &= pr^2 \\
T_{\varphi\varphi} &= pr^2 \sin^2 \vartheta = T_{\vartheta\vartheta} \sin^2 \vartheta.
\end{aligned}
$$

And $T_{\mu\nu} = 0$ for $\mu \neq \nu$. In the further course, we want to use the Einstein's field equations in the form (22.12)

$$R_{\mu\nu} = 8\pi G \left(T_{\mu\nu} - \frac{1}{2} g_{\mu\nu} T \right)$$

and calculate the scalar

$$
\begin{aligned}
T &= T^\mu_\mu = g^{\mu\nu} T_{\mu\nu} = g^{tt} T_{tt} + g^{rr} T_{rr} + g^{\vartheta\vartheta} T_{\vartheta\vartheta} + g^{\varphi\varphi} T_{\varphi\varphi} \\
&= -e^{-2\nu(r)} \rho e^{2\nu(r)} + e^{-2\lambda(r)} pe^{2\lambda(r)} + r^{-2} pr^2 + r^{-2} \sin^{-2} \vartheta \cdot pr^2 \sin^2 \vartheta \\
&= -\rho + 3p.
\end{aligned}
$$

This results in the right side of the Einstein equations to

$$
\begin{aligned}
8\pi G \left(T_{tt} - \frac{1}{2} g_{tt} (-\rho + 3p) \right) &= 8\pi G \left(\rho e^{2\nu} - \frac{1}{2} e^{2\nu} \rho + \frac{3}{2} pe^{2\nu} \right) \\
&= 4\pi G e^{2\nu} (\rho + 3p) \\
8\pi G \left(T_{rr} - \frac{1}{2} g_{rr} (-\rho + 3p) \right) &= 8\pi G \left(pe^{2\lambda} + \frac{1}{2} e^{2\lambda} \rho - \frac{3}{2} pe^{2\lambda} \right) \\
&= 4\pi G e^{2\lambda} (\rho - p) \\
8\pi G \left(T_{\vartheta\vartheta} - \frac{1}{2} g_{\vartheta\vartheta} (-\rho + 3p) \right) &= 8\pi G \left(pr^2 + \frac{1}{2} r^2 \rho - \frac{3}{2} pr^2 \right)
\end{aligned}
$$

$$8\pi G \left(T_{\varphi\varphi} - \frac{1}{2}g_{\varphi\varphi}\left(-\rho+3p\right)\right) \begin{aligned} &= 4\pi Gr^2\left(\rho-p\right) \\ &= 8\pi G\sin^2\vartheta\left(pr^2+\frac{1}{2}r^2\rho-\frac{3}{2}pr^2\right) \\ &= 4\pi Gr^2\sin^2\vartheta\left(\rho-p\right), \end{aligned}$$

where we have omitted the variable r in the functions $\lambda\left(r\right)$ and $\nu\left(r\right)$ for simplicity. We have already calculated the left side of the Einstein equations in Eq. (23.13):

$$\begin{aligned} R_{tt} &= e^{2\nu-2\lambda}\left(\nu''+\left(\nu'\right)^2-\nu'\lambda'+\frac{2\nu'}{r}\right) \\ R_{rr} &= -\nu''+\nu'\lambda'+\frac{2\lambda'}{r}-\left(\nu'\right)^2 \\ R_{\vartheta\vartheta} &= e^{-2\lambda}\left(-1+r\lambda'-r\nu'\right)+1 \\ R_{\varphi\varphi} &= \sin^2\vartheta\,R_{\vartheta\vartheta}, \end{aligned}$$

where the prime $'$ denotes the derivative with respect to the variable r. We thus obtain three relevant determining equations, as the fourth for $T_{\varphi\varphi}, R_{\varphi\varphi}$ follows directly from the third. After dividing the first equation by $e^{2\nu}$ and the second by $e^{2\lambda}$, the following three equations result:

$$\begin{aligned} e^{-2\lambda}\left(\nu''+\left(\nu'\right)^2-\nu'\lambda'+\frac{2\nu'}{r}\right) &= 4\pi G\left(\rho+3p\right) \\ e^{-2\lambda}\left(-\nu''+\nu'\lambda'+\frac{2\lambda'}{r}-\left(\nu'\right)^2\right) &= 4\pi G\left(\rho-p\right) \\ e^{-2\lambda}\left(-1+r\lambda'-r\nu'\right)+1 &= 4\pi Gr^2\left(\rho-p\right) \end{aligned}$$

If you divide the first two equations by 2 and the third by r^2, you get

$$\begin{aligned} \frac{e^{-2\lambda}}{2}\left(\nu''+\left(\nu'\right)^2-\nu'\lambda'+\frac{2\nu'}{r}\right) &= 2\pi G\left(\rho+3p\right) \\ \frac{e^{-2\lambda}}{2}\left(-\nu''+\nu'\lambda'+\frac{2\lambda'}{r}-\left(\nu'\right)^2\right) &= 2\pi G\left(\rho-p\right) \\ \frac{e^{-2\lambda}}{r^2}\left(-1+r\lambda'-r\nu'\right)+\frac{1}{r^2} &= 2\pi G\left(2\rho-2p\right). \end{aligned}$$

Subtracting the second equation from the first results in

$$e^{-2\lambda}\left(\nu''+\left(\nu'\right)^2-\nu'\lambda'+\frac{\nu'}{r}-\frac{\lambda'}{r}\right) = 8\pi Gp. \tag{26.1}$$

Now we add the three equations and get

$$\frac{e^{-2\lambda}}{r^2}\left(2r\lambda' - 1\right) + \frac{1}{r^2} = 8\pi G\rho. \tag{26.2}$$

If we add the first two equations, we get

$$e^{-2\lambda}\left(\frac{2\nu'}{r} + \frac{2\lambda'}{r}\right) = 8\pi G\left(\rho + p\right), \tag{26.3}$$

and if we subtract the third from this, we finally get

$$\frac{e^{-2\lambda}}{r^2}\left(2r\nu' + 1\right) - \frac{1}{r^2} = 8\pi Gp. \tag{26.4}$$

The last equations can now be used to find the unknown functions $\nu\left(r\right)$ and $\lambda\left(r\right)$. However, additional information about the pressure p and the mass density ρ is required. In general, one assumes an **equation of state** of the form

$$\frac{p}{\rho} = f(\rho, T)$$

for the star matter, where the function f can depend on the mass density ρ and the temperature T. However, we want to take a simpler approach here and assume as a special case that the matter is **incompressible**, i.e., we assume that the density ρ is constant. This is an acceptable assumption for lighter stars like the Sun and simplifies the following (not entirely simple) derivation of the metric considerably.

We write Eq. (26.2) in the form

$$e^{-2\lambda}\left(1 - 2r\lambda'\right) = 1 - 8\pi G\rho r^2$$

and use the product rule

$$\frac{d}{dr}\left(re^{-2\lambda}\right) = e^{-2\lambda} - 2\lambda' re^{-2\lambda}$$

to get

$$\frac{d}{dr}\left(re^{-2\lambda}\right) = 1 - 8\pi G\rho r^2.$$

This equation is integrated ($\rho = const.!$)

$$re^{-2\lambda} = \int \frac{d}{dr}\left(re^{-2\lambda}\right) dr = \int \left(1 - 8\pi G\rho r^2\right) dr = r - \frac{8\pi G\rho r^3}{3} + C$$

with an integration constant C. If we set $r = 0$, it follows that $C = 0$. So after division by r we get

$$e^{-2\lambda} = 1 - \frac{8\pi G\rho r^2}{3} = 1 - Ar^2, \tag{26.5}$$

where the constant A is defined, i.e., we have found a solution for

$$g_{rr} = e^{2\lambda} = \left(1 - Ar^2\right)^{-1}.$$

Now we look at Eq. (26.4) and write it in the form

$$e^{-2\lambda}\left(2r\nu' + 1\right) = 1 + 8\pi Gpr^2.$$

We differentiate both sides with respect to r, noting that the pressure $p = p(r)$ depends on r, and again using the product rule we get

$$-2\lambda'e^{-2\lambda}\left(2r\nu' + 1\right) + e^{-2\lambda}\left(2\nu' + 2r\nu''\right) = 8\pi Gp'r^2 + 16\pi Gpr.$$

From Eq. (26.1) we get by rearrangement

$$e^{-2\lambda}r\nu'' = 8\pi Gpr - e^{-2\lambda}\left(r\left(\nu'\right)^2 - r\nu'\lambda' + \nu' - \lambda'\right)$$

and insert this expression into the penultimate equation, which we also divide by 2:

$$-\lambda'e^{-2\lambda}\left(2r\nu' + 1\right) + e^{-2\lambda}\nu' + 8\pi Gpr - e^{-2\lambda}\left(r\left(\nu'\right)^2 - r\nu'\lambda' + \nu' - \lambda'\right) =$$
$$4\pi Gp'r^2 + 8\pi Gpr$$

Some terms cancel out and we obtain

$$-\frac{2e^{-2\lambda}}{r}\left(\nu'\left(\nu' + \lambda'\right)\right) = 8\pi Gp'.$$

Now we use Eq. (26.3), to replace the term

$$e^{-2\lambda}\left(\frac{2\nu'}{r} + \frac{2\lambda'}{r}\right),$$

and we get

$$p' = -\nu'(\rho + p) \Leftrightarrow \frac{p'}{(\rho + p)} = -\nu'. \tag{26.6}$$

We integrate the last equation, considering the substitution rule

$$p'\,dr = \frac{dp}{dr}\,dr = dp$$

and for the left side we get

$$\int \frac{p'}{(\rho + p)}\, dr = \int \frac{1}{(\rho + p)}\, dp = \ln\left(\rho + p(r)\right) + C_1$$

and for the right side

$$-\int \nu'\, dr = -\nu(r) + C_2.$$

So after combining the two integration constants C_1, C_2 and by exponentiating we get

$$\ln\left(\rho + p(r)\right) = -\nu(r) + C \implies \rho + p(r) = e^{-\nu(r)+C} = Be^{-\nu(r)} \tag{26.7}$$

with the integration constant $B = e^C$. To eliminate the pressure $p(r)$ from the last equation, we multiply it by $8\pi G$ and use Eq. (26.4) again

$$8\pi G\rho + 8\pi Gp = 8\pi G\rho + \frac{e^{-2\lambda}}{r^2}\left(2r\nu' + 1\right) - \frac{1}{r^2} = 8\pi GBe^{-\nu}.$$

Now we replace the term $e^{-2\lambda}$ according to Eq. (26.5) and we get

$$8\pi G\rho + \frac{1 - Ar^2}{r^2}\, 2r\nu' + \frac{1 - Ar^2}{r^2} - \frac{1}{r^2} = 8\pi GBe^{-\nu}.$$

Since $A = 8\pi G\rho/3$, we get

$$2A + \frac{1 - Ar^2}{r}\, 2\nu' = 8\pi GBe^{-\nu}$$

or after rearranging

$$Are^{\nu} + \left(1 - Ar^2\right)\nu' e^{\nu} = 4\pi GBr. \tag{26.8}$$

To solve this (differential) equation, we set

$$u(r) = e^{\nu(r)},$$

then we get

$$u'(r) = \nu'(r)\, e^{\nu(r)}$$

and we obtain the differential equation

$$Aru(r) + \left(1 - Ar^2\right)u'(r) = 4\pi GBr.$$

A special solution is obtained if we set $u(r)$ as a constant function, then $u'(r) = 0$ and we get

$$u(r) = 4\pi G\, \frac{B}{A}.$$

For the general solution, we still need the solution of the homogeneous differential equation, i.e. of

$$Aru\left(r\right) + \left(1 - Ar^2\right) u'\left(r\right) = 0.$$

We divide this equation by $u\left(r\right)$ and $\left(1 - Ar^2\right)$ and we get

$$\frac{u'\left(r\right)}{u\left(r\right)} = -\frac{Ar}{1 - Ar^2}.$$

Integration on both sides again with

$$u'\,dr = \frac{du}{dr}\,dr = du$$

for the left side gives

$$\int \frac{u'\left(r\right)}{u\left(r\right)}\,dr = \int \frac{1}{u}\,du = \ln u + C_1$$

and for the right side

$$\int \left(-\frac{Ar}{1 - Ar^2}\right) dr = \frac{1}{2}\ln\left(1 - Ar^2\right) + C_2 = \ln\left(1 - Ar^2\right)^{\frac{1}{2}} + C_2,$$

which can be verified by differentiation. We again combine the integration constants, exponentiate both sides and obtain

$$u\left(r\right) = D\left(1 - Ar^2\right)^{\frac{1}{2}}$$

with an integration constant D. The general solution of (26.8) is now the sum of the special solution and the solution of the homogeneous differential equation, so

$$e^{\nu(r)} = u\left(r\right) = 4\pi G\frac{B}{A} + D\left(1 - Ar^2\right)^{\frac{1}{2}}, \qquad (26.9)$$

and thus we have found the solution for g_{tt}:

$$g_{tt} = -e^{2\nu(r)} = -u\left(r\right)^2 = -\left(4\pi G\frac{B}{A} + D\left(1 - Ar^2\right)^{\frac{1}{2}}\right)^2 \qquad (26.10)$$

The line element of the metric results with (26.5) to

$$ds^2 = -\left(4\pi G\frac{B}{A} + D\left(1 - Ar^2\right)^{\frac{1}{2}}\right)^2 dt^2 + \left(1 - Ar^2\right)^{-1} dr^2 + r^2 d\Omega^2.$$

This metric is called the **interior Schwarzschild metric**, it was also found in 1916 by the German astronomer Karl Schwarzschild. The metric still contains the two constants B and D (A is determined by the density ρ), which we want to determine by the following requirements. First, the matter pressure at the surface of the star should be zero. If we denote the radius of the star with R, then $p(R) = 0$ should apply. And secondly, the interior Schwarzschild metric at the surface of the star should match the outer Schwarzschild metric.

If we set $p(R) = 0$ in Eq. (26.7), we get

$$\rho + p(R) = \rho = Be^{-\nu(R)} \quad \Longrightarrow \quad B = \rho e^{\nu(R)}.$$

Now we replace the term $e^{\nu(R)}$ according to (26.9) and get with $A = 8\pi G\rho/3$:

$$
\begin{aligned}
B &= \rho\left(4\pi G\frac{B}{A} + D\left(1 - AR^2\right)^{\frac{1}{2}}\right) = 4\pi G\rho\frac{B}{\frac{8\pi G\rho}{3}} + \rho D\left(1 - AR^2\right)^{\frac{1}{2}} \\
&= \frac{3}{2}B + \rho D\left(1 - AR^2\right)^{\frac{1}{2}},
\end{aligned}
$$

so

$$B = -2\rho D\left(1 - AR^2\right)^{\frac{1}{2}}.$$

This inserted into (26.9) yields again with $A = 8\pi G\rho/3$

$$
\begin{aligned}
e^{\nu(r)} &= 4\pi G\rho\frac{-2D\left(1 - AR^2\right)^{\frac{1}{2}}}{A} + D\left(1 - Ar^2\right)^{\frac{1}{2}} \\
&= -3D\left(1 - AR^2\right)^{\frac{1}{2}} + D\left(1 - Ar^2\right)^{\frac{1}{2}} \\
&= 8\pi G\rho\frac{-D\left(1 - AR^2\right)^{\frac{1}{2}}}{\frac{8\pi G\rho}{3}} + D\left(1 - Ar^2\right)^{\frac{1}{2}} \\
&= -3D\left(1 - AR^2\right)^{\frac{1}{2}} + D\left(1 - Ar^2\right)^{\frac{1}{2}},
\end{aligned}
$$

and the line element of the interior Schwarzschild metric is thus

$$ds^2 = -D^2\left(-3\left(1 - AR^2\right)^{\frac{1}{2}} + \left(1 - Ar^2\right)^{\frac{1}{2}}\right)^2 dt^2 + \left(1 - Ar^2\right)^{-1} dr^2 + r^2 d\Omega^2.$$

If we compare this line element with that of the outer Schwarzschild metric

$$ds^2 = -\left(1 - \frac{2MG}{r}\right) dt^2 + \left(1 - \frac{2MG}{r}\right)^{-1} dr^2 + r^2 d\Omega^2$$

and demand that both should coincide at $r = R$, we first find that the total mass M of the (spherical) star due to the constant density ρ is equal to

$$M = V \cdot \rho = \frac{4}{3}\pi R^3 \rho$$

and thus

$$\frac{2MG}{R} = \frac{\frac{8}{3}\pi R^3 \rho G}{R} = AR^2$$

provides, so that at $r = R$ the terms before dr^2 coincide. To determine the constant D, we set $r = R$ and obtain

$$D^2\left(-3\left(1-AR^2\right)^{\frac{1}{2}} + \left(1-AR^2\right)^{\frac{1}{2}}\right)^2 \overset{!}{=} 1 - \frac{2MG}{R} = 1 - AR^2.$$

The left side can be further calculated

$$D^2\left(-3\left(1-AR^2\right)^{\frac{1}{2}} + \left(1-AR^2\right)^{\frac{1}{2}}\right)^2 = D^2 4\left(1-AR^2\right),$$

so it follows

$$D = \pm\frac{1}{2},$$

and the line element of the interior Schwarzschild metric results in

$$ds^2 = -\left(-3D\left(1-AR^2\right)^{\frac{1}{2}} + D\left(1-Ar^2\right)^{\frac{1}{2}}\right)^2 dt^2 + \left(1-Ar^2\right)^{-1} dr^2 + r^2 d\Omega^2.$$

Since at $r = R$ the outer Schwarzschild metric should come out, $D = -1/2$ must be chosen and we obtain the final expression

$$ds^2 = -\left(\frac{3}{2}\left(1-AR^2\right)^{\frac{1}{2}} - \frac{1}{2}\left(1-Ar^2\right)^{\frac{1}{2}}\right)^2 dt^2 + \left(1-Ar^2\right)^{-1} dr^2 + r^2 d\Omega^2$$

$$(26.11)$$

for the line element of the interior Schwarzschild metric with

$$A = \frac{8\pi G\rho}{3}.$$

We calculate the value AR^2 for the Sun and use the equation from above

$$\frac{2MG}{R} = AR^2.$$

The solar radius is $R_s = 6.69 \cdot 10^8$ m, and the Schwarzschild radius of the Sun is $r_S = 2M_S G/c^2 = 2.96 \cdot 10^3$ m, so it follows

$$AR_S^2 = \frac{2.96 \cdot 10^3}{6.69 \cdot 10^8} = 4,42 \cdot 10^{-6},$$

i.e. $AR_S^2 \ll 1$, and also the interior Schwarzschild metric (just like the outer one) does not deviate very much from the Minkowski metric near the edge of the Sun.

26.3. The Tolman-Oppenheimer-Volkoff Equation

We derive another important relationship between the increase in pressure and density, assuming again that the density ρ is constant and therefore we can use the results of the last sections. We start from Eq. (26.6) as follows:

$$\frac{p'}{(\rho + p)} = -\nu'$$

and want to further calculate the right side using Eq. (26.4):

$$\frac{e^{-2\lambda}}{r^2}\left(2r\nu' + 1\right) - \frac{1}{r^2} = 8\pi G p$$

First, we replace the term $e^{-2\lambda}$ according to Eq. (26.5) by

$$e^{-2\lambda} = 1 - \frac{8\pi G \rho r^2}{3} = 1 - Ar^2$$

and get

$$\frac{1 - Ar^2}{r^2}\left(2r\nu' + 1\right) - \frac{1}{r^2} = 8\pi G p.$$

Multiplying out and summarizing the left side leads to

$$-A + \frac{2\nu'}{r} - 2Ar\nu' = 8\pi G p.$$

Multiplication by r, division by 2, substitution of A and rearrangement results in

$$\nu'\left(1 - Ar^2\right) = 4\pi G p r + \frac{4\pi G \rho r}{3} = 4\pi G r\left(p + \frac{\rho}{3}\right).$$

So

$$\nu' = \frac{4\pi G r}{1 - Ar^2}\left(p + \frac{\rho}{3}\right).$$

This inserted into Eq. (26.6)

$$\frac{p'}{(\rho + p)} = -\nu'$$

leads to

$$p' = \frac{dp}{dr} = -\frac{4\pi G r}{1 - Ar^2}\left(p + \frac{\rho}{3}\right)(\rho + p). \tag{26.12}$$

If one replaces

$$1 - Ar^2 = \frac{r - Ar^3}{r} = \frac{r - 2\left(\frac{4}{3}\pi\rho r^3\right)G}{r} = \frac{r - 2M\left(r\right)G}{r},$$

one obtains the usual form of the **Tolman-Oppenheimer-Volkoff equation (TOV equation)**

$$\frac{dp}{dr} = -\frac{4\pi G r^2 \left(p + \frac{\rho}{3}\right)(\rho + p)}{r - 2M(r)G} \tag{26.13}$$

or the special case $\rho = const$. The equation shows that a higher density ρ leads to a stronger decrease in pressure, which favors the gravitational collapse of the star. This is also intuitively expected, but what is surprising is the fact that an increase in pressure p also promotes collapse. This is again a major difference between general relativity, where pressure increases gravity, while in Newtonian theory, pressure not only has no enhancing effect on gravity, but often prevents a collapse, which we now want to make plausible with a simple example.

Non-relativistic Limit Case, Emden's Equation

For this, we consider a spherical star with density $\rho(r)$ and a shell of thickness dr that may be located at the length r from the center.

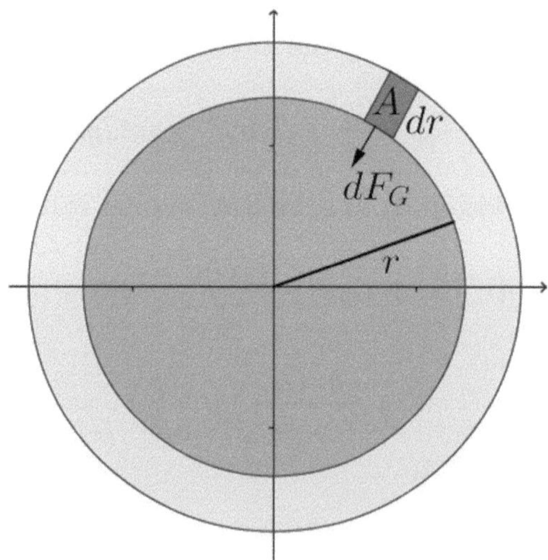

Figure 26.1.: Equilibrium between pressure and gravity

A (small) section of the spherical shell in Fig. 26.1 with area A and thickness dr has the mass $\rho A\, dr$ and experiences a (Newtonian) gravitational force dF_G of the size

$$dF_G = -\frac{GM(r)}{r^2}\,\rho(r)\, A\, dr,$$

where $M(r)$ is the mass of the star from the center to the spherical shell. In order for the spherical star not to collapse, this force directed towards the center must be counteracted by the pressure difference $dp = p(r + dr) - p(r)$, i.e.

$$dF_G = A\,dp.$$

So the equilibrium condition results in

$$\frac{dp}{dr} = -\frac{GM(r)}{r^2}\,\rho(r). \tag{26.14}$$

This equation is called **Emden's equation** and reduces in the case of constant density due to

$$M(r) = \frac{4}{3}\pi\rho r^3$$

to

$$\frac{dp}{dr} = -\frac{4\pi G}{3}\rho^2 r. \tag{26.15}$$

If we look again at the TOV equation (26.13) and consider the non-relativistic case, i.e. $r \gg 2MG$ and $\rho \gg p$, the equation reduces to

$$\frac{dp}{dr} = -\frac{4\pi G r^2 \left(p + \frac{\rho}{3}\right)(\rho + p)}{r - 2M(r)G} = -\frac{4\pi G r^2 \left(\frac{\rho}{3}\right)\rho}{r} = -\frac{4\pi G}{3}\rho^2 r,$$

i.e., the Tolman-Oppenheimer-Volkoff equation in the non-relativistic limit becomes the Emden's equation. If we integrate the Emden's equation (26.15), we get

$$\int_0^r dp = p(r) - p(0) = \int_0^r -\frac{4\pi G}{3}\rho^2 r\,dr = -\frac{2\pi G}{3}\rho^2 r^2,$$

so

$$p(r) = p(0) - \frac{2\pi G}{3}\rho^2 r^2,$$

where the integration constant $p(0) = p_0$ is the pressure at the center of the star. The pressure is therefore highest at the center of the star and decreases with increasing r. Since for $r = R$ (star radius) the pressure $p(R)$ must be zero, we get

$$p_0 = \frac{2\pi G}{3}\rho^2 R^2$$

and thus

$$p(r) = \frac{2\pi G}{3}\rho^2 \left(R^2 - r^2\right), \tag{26.16}$$

or

$$\frac{p_0}{\rho} = \frac{2\pi G}{3}\rho R^2 = \frac{2\left(\frac{4\pi\rho}{3}R^3\right)G}{4R} = \frac{2MG}{4R} = \frac{r_S}{4R},$$

where r_S denotes the Schwarzschild radius of the star. Thus, the equilibrium of the star is determined in the non-relativistic case and at constant density ρ by a certain ratio of Schwarzschild radius to star radius. In "normal" stars like our Sun, the kinetic pressure of the temperature movement of the gas molecules balances gravity. We consider this gas as ideal, so the pressure can be determined from the **ideal gas law** (see Chap. 29):

$$p = \frac{Nk_B T}{V}.$$

where N denotes the number of particles in the volume V, T the temperature and k_B the so-called **Boltzmann constant** (see Chap. 30). If one considers that the mass M can be calculated as the sum of the masses m of all individual particles

$$M = Nm$$

and that for the density

$$\rho = \frac{M}{V} = \frac{Nm}{V}$$

applies, one obtains

$$p = \frac{k_B \rho T}{m}.$$

The quantity $k_B T/m$ is very small for our Sun

$$\frac{p}{\rho} = \frac{k_B T}{M} \approx 10^{-6},$$

so that

$$\frac{r_S}{4R} \approx \frac{p}{\rho} \approx \frac{k_B T}{m} \approx 10^{-6}$$

follows. If we consider that the Schwarzschild radius of the Sun is approximately $r_S = 2.96 \cdot 10^3$ m, we can estimate the radius R of the Sun by the above equilibrium condition

$$R \approx \frac{2.96}{4} 10^9 \, \text{m} = 740{,}000 \, \text{km},$$

which is close to the actual radius of the Sun of about $700{,}000 \, \text{km}$.

Consequences for Relativistic Stars

We now leave the non-relativistic limit case and derive some results for relativistic stars from the TOV equation (26.12), i.e. for stars with a strong gravitational field. The TOV equation can be rewritten as

$$\frac{p'}{\left(p + \frac{\rho}{3}\right)(\rho + p)} = -\frac{4\pi G r}{1 - Ar^2}.$$

We integrate both sides and obtain for the right side

$$\int \frac{p'}{\left(p + \frac{\rho}{3}\right)(\rho + p)}\, dr = -\frac{3}{2\rho} \ln \left(\frac{p + \rho}{p + \frac{\rho}{3}}\right) + const.$$

To see this, we differentiate the right side with the chain and quotient rule and obtain because $\rho = const.$

$$\frac{d}{dr}\left[-\frac{3}{2\rho} \ln \left(\frac{p + \rho}{p + \frac{\rho}{3}}\right)\right] = \left(-\frac{3}{2\rho}\right)\left(\frac{1}{\frac{p + \rho}{p + \frac{\rho}{3}}}\right)\left(\frac{p'\left(p + \frac{\rho}{3}\right) - (p + \rho)\, p'}{\left(p + \frac{\rho}{3}\right)^2}\right)$$

$$= \left(-\frac{3}{2\rho}\right)\left(\frac{p + \frac{\rho}{3}}{p + \rho}\right)\left(\frac{-\frac{2\rho}{3}\, p'}{\left(p + \frac{\rho}{3}\right)^2}\right)$$

$$= \frac{p'}{\left(p + \frac{\rho}{3}\right)(\rho + p)}.$$

For the left side we get

$$\int -\frac{4\pi Gr}{1 - Ar^2}\, dr = \frac{4\pi G}{2A} \ln \left(1 - Ar^2\right) + const.$$

Here too, we differentiate the right side to check

$$\frac{d}{dr}\left[\frac{4\pi G}{2A} \ln \left(1 - Ar^2\right)\right] = \left(\frac{4\pi G}{2A}\right)\left(\frac{1}{1 - Ar^2}\right)(-2Ar) = -\frac{4\pi Gr}{1 - Ar^2}.$$

We combine the two integration constants and finally obtain from Eq. (26.12) as a result

$$\ln \left(\frac{p + \rho}{p + \frac{\rho}{3}}\right) = -\frac{4\pi G\rho}{3A} \ln \left(1 - Ar^2\right) + const.$$

Now $A = 8\pi G\rho/3$, so it follows

$$\ln \left(\frac{p + \rho}{p + \frac{\rho}{3}}\right) = -\frac{1}{2} \ln \left(1 - Ar^2\right) + const.$$

We exponentiate and obtain

$$\frac{p + \rho}{p + \frac{\rho}{3}} = C \left(1 - Ar^2\right)^{-\frac{1}{2}}. \tag{26.17}$$

Since the pressure at the edge of the star is zero, i.e. $p(R) = 0$, the constant C is determined by

$$\frac{p(R) + \rho}{p(R) + \frac{\rho}{3}} = C\left(1 - AR^2\right)^{-\frac{1}{2}} \Longleftrightarrow 3 = C\left(1 - AR^2\right)^{-\frac{1}{2}} \Longleftrightarrow C = 3\sqrt{1 - AR^2}.$$

We solve Eq. (26.17) for p :

$$
\begin{aligned}
\frac{p + \rho}{p + \frac{\rho}{3}} &= C\left(1 - Ar^2\right)^{-\frac{1}{2}} \Longleftrightarrow \\
p &= \left(p + \frac{\rho}{3}\right)\left(C\left(1 - Ar^2\right)^{-\frac{1}{2}}\right) - \rho \Longleftrightarrow \\
p\left(1 - C\left(1 - Ar^2\right)^{-\frac{1}{2}}\right) &= -\rho\left(1 - \frac{C}{3}\left(1 - Ar^2\right)^{-\frac{1}{2}}\right) \Longleftrightarrow \\
p &= \frac{-\rho\left(1 - \frac{C}{3}\left(1 - Ar^2\right)^{-\frac{1}{2}}\right)}{1 - C\left(1 - Ar^2\right)^{-\frac{1}{2}}}.
\end{aligned}
$$

We bring numerator and denominator to the common denominator, insert the constant C and get

$$p = \rho \,\frac{\sqrt{1 - Ar^2} - \sqrt{1 - AR^2}}{3\sqrt{1 - AR^2} - \sqrt{1 - Ar^2}}.$$

With

$$Ar^2 = \frac{8\pi G\rho r^2}{3} = 2\,\frac{4\pi G\rho R^3}{3}\,\frac{Gr^2}{R^3} = 2MG\,\frac{r^2}{R^3} = \frac{r_S r^2}{R^3}$$

we introduce the Schwarzschild radius and finally obtain the final equation for the pressure in the relativistic case

$$p = \rho \,\frac{\sqrt{1 - \dfrac{r_S\, r^2}{R^3}} - \sqrt{1 - \dfrac{r_S}{R}}}{3\sqrt{1 - \dfrac{r_S}{R}} - \sqrt{1 - \dfrac{r_S\, r^2}{R^3}}}. \tag{26.18}$$

For $r \geq R$, $p(r) = 0$.

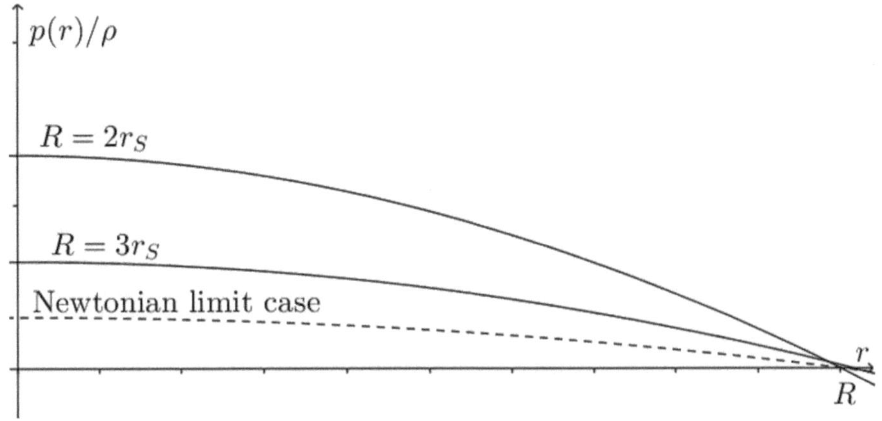

Figure 26.2.: Relative pressure curve p/ρ

Fig. 26.2 shows some pressure curves and also includes a comparison with the non-relativistic limit case (26.16), which can be written with the Schwarzschild radius r_S as

$$p\left(r\right) = \frac{2\pi G}{3}\, \rho^2 \left(R^2 - r^2\right) = \rho\, \frac{r_S}{4R^3} \left(R^2 - r^2\right).$$

As the radius of the star R approaches the Schwarzschild radius, the relativistic effects increase. For the pressure p_0 in the center of the star, it follows from Eq. (26.18)

$$p_0 = p\left(0\right) = \rho\, \frac{1 - \sqrt{1 - \dfrac{r_S}{R}}}{3\sqrt{1 - \dfrac{r_S}{R}} - 1},$$

i.e., the gravitational pressure is maximal in the center and it diverges for $R \to 9/8\, r_S$, since for the denominator

$$3\sqrt{1 - \frac{r_S}{\dfrac{9}{8}\, r_S}} - 1 = 3\sqrt{1 - \frac{8}{9}} - 1 = 0$$

holds, so

$$p_0 \to \infty \quad \text{für} \quad R \to \frac{9}{8}\, r_S.$$

If the gravitational pressure in the center diverges, then no equilibrium with the material pressure is possible anymore, i.e., the star collapses. A stellar equilibrium therefore requires

$$R > \frac{9}{8}\, r_S.$$

Since the Schwarzschild radius $r_S = 2MG$, this stability condition means that the mass M must not exceed an upper limit for a given stellar radius, or if the mass is given, the radius must not be too small. In the non-relativistic case and for incompressible matter ($\rho = const.$), as shown above, the maximum pressure is

$$p_0 = \frac{2\pi G}{3}\,\rho^2 R^2,$$

so always finite. However, if we abandon the assumption of incompressibility, an instability can also arise here, which, for example, leads to a mass upper limit for white dwarfs.

The instability for $R \to 9/8\,r_S$, on the other hand, is of a fundamental nature, as it is caused by the relativistic effects in the TOV equation. We have already explained above that high pressures, regardless of the type of matter, lead to a self-amplifying increase in pressure towards the center. The Einstein's field equations have thus led to the result that the central gravitational pressure of a star (with constant mass density) diverges as the radius of the star approaches the Schwarzschild radius. The consequence of this is that the star undergoes a gravitational collapse, which in the extreme case can lead to a black hole, which we want to deal with in the following final chapter 27 of this book.

27. Black Holes

No object in our universe stimulates the imagination more than a black hole. Many science fiction authors have written about it and attributed the most fantastic properties to it, such as enabling time travel, wormholes, superluminal speed, and much more. We treat black holes here like all other gravitational phenomena described so far, namely soberly, and ask ourselves what can be quantitatively stated or specifically measured about black holes.

27.1. Mass Density of Black Holes

We have defined black holes in the previous chapters as accumulations of matter of mass M whose expansion radius r is smaller than the Schwarzschild radius ($r \leq r_S = 2GM$). And we have also already proven that no light can escape from such objects, hence the name "black holes". If one assumes that a spherical evenly distributed accumulation of matter M is about to form a black hole, one can say something about its mass density ρ. The volume of the accumulation of matter is then

$$V = \frac{4\pi r_S^3}{3} = \frac{4\pi \left(2GM\right)^3}{3} = \frac{32\pi G^3 M^3}{3},$$

and for the density it follows

$$\rho = \frac{M}{V} = \frac{3}{32\pi G^3 M^2},$$

i.e., the matter density needed to form a black hole is inversely proportional to the square of the mass of the accumulation of matter. Objects that only have a small mass must therefore have a high mass density to form a black hole. Conversely, a low mass density is sufficient if the mass of the accumulation of matter is large enough. We want to take a closer look at the last equation and do some calculations with it. First, we represent the density ρ in SI units

$$\rho = \frac{M}{V} = \frac{3M}{32\pi \left(\dfrac{GM}{c^2}\right)^3} = \frac{3c^6}{32\pi G^3 M^2} = \frac{7.3 \cdot 10^{79}}{M^2}\,\mathrm{kg\,m^{-3}}.$$

M. Ruhrländer, *Ascent to the Einstein Equations*, https://doi.org/10.1007/978-3-662-72672-3_27

This quantity alone is difficult to interpret, so we relate it to the mass of the Sun $M_S = 1.99 \cdot 10^{30}\,\text{kg}$ and get

$$\rho = \frac{3c^6}{32\pi G^3 M^2} = 1.84 \cdot 10^{19} \left(\frac{M_S}{M}\right)^2 \text{kg}\,\text{m}^{-3}.$$

Now we know from atomic physics that the typical mass density in a nucleus is about $2 \cdot 10^{17}\,\text{kg}\,\text{m}^{-3}$ and this density also prevails in a neutron star. The density calculated above for a black hole *with solar mass* is therefore about 100 times higher than the nuclear density, and that is the reason why we cannot expect black holes with solar mass in the universe.

Black holes with a mass ten times that of the Sun have about nuclear density, so it is no surprise that the black holes observed in binary star systems typically have this size. The huge black holes in galaxy centers can arise with very low mass densities. If one assumes such black holes to have a mass a billion times greater than that of our Sun, such a black hole is created with a mass density of about $40\,\text{kg}\,\text{m}^{-3}$, which is significantly lower than the density of water $\left(1000\,\text{kg}\,\text{m}^{-3}\right)$.

27.2. Redshift at the Schwarzschild Radius

In the following sections, we assume the existence of a black hole and use the external Schwarzschild solution

$$ds^2 = -\left(1 - \frac{2m}{r}\right) dt^2 + \frac{1}{\left(1 - \dfrac{2m}{r}\right)} dr^2 + r^2 \left(d\vartheta^2 + \sin^2\vartheta\, d\varphi^2\right)$$

to examine the area outside the black hole. Since the radius of the black hole R is smaller than the Schwarzschild radius, this means that we must also move into the region $R < r \leq 2GM = 2m$. We have already shown that

$$g_{rr} = \frac{1}{\left(1 - \dfrac{2m}{r}\right)}$$

becomes singular at $r = 2m$. If one looks at g_{tt} at $r = 2m$, then

$$g_{tt} = 0.$$

This value is mathematically ok, but it leads to the redshift becoming infinite on the surface $r = 2m$. To this end, we look again at the equation of the redshift in the external Schwarzschild field (23.19):

$$\frac{f_2}{f_1} = \sqrt{\frac{g_{00}(r_1)}{g_{00}(r_2)}} = \sqrt{\frac{1 - 2GM/r_1}{1 - 2GM/r_2}},$$

where r_1 is a place "at infinity", i.e., far away from the Schwarzschild radius ($r_1 \to \infty$) and r_2 may approach the Schwarzschild radius. Then for the numerator

$$\sqrt{1 - \frac{2GM}{r_1}} \approx 1$$

and for the denominator

$$\sqrt{1 - \frac{2GM}{r_2}} \to 0 \ \text{ für } \ r_2 \to 2GM,$$

so overall

$$\frac{f_2}{f_1} = \sqrt{\frac{1 - 2GM/r_1}{1 - 2GM/r_2}} \to \infty \ \text{ für } \ r_2 \to 2GM.$$

As an object approaches the Schwarzschild radius from the outside, signals sent by the object are increasingly redshifted "at infinity", until finally no light signal can be observed by a distant observer. We want to investigate this more closely and first consider objects that fall radially into a black hole in Schwarzschild spacetime.

27.3. Radial Fall in Schwarzschild Spacetime

We consider a particle that at time $t = 0$ from a distance $r = R > 2m$ with initial velocity

$$\left(\frac{dr}{dt}\right)_{r=R} = 0$$

falls radially onto a black hole, as shown in Fig. 27.1. By radial, we mean that the angles ϑ and φ do not change, i.e.,

$$d\vartheta = d\varphi = 0$$

applies. The time t is read on a clock "at infinity", so it is the coordinate time of a distant observer.

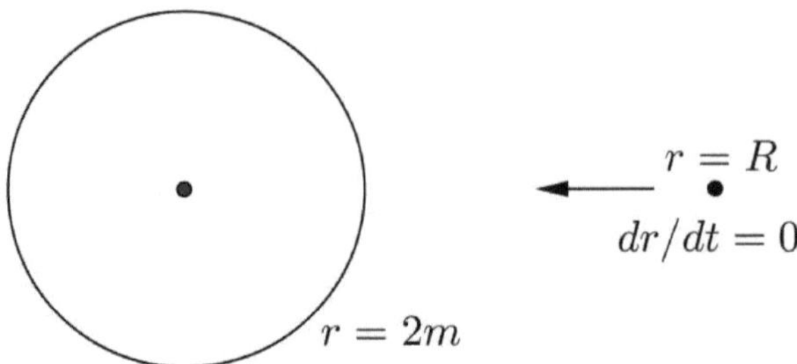

Figure 27.1.: Radial fall of a particle into a Black Hole

This results in the line element of the Schwarzschild metric as

$$ds^2 = -\left(1 - \frac{2m}{r}\right) dt^2 + \frac{1}{\left(1 - \frac{2m}{r}\right)} dr^2.$$

With

$$ds^2 = -dt_E^2$$

and the definitions

$$\dot{t} = \frac{dt}{dt_E}, \dot{r} = \frac{dr}{dt_E}$$

it follows after division by ds^2

$$\left(1 - \frac{2m}{r}\right) \dot{t}^2 - \frac{1}{\left(1 - \frac{2m}{r}\right)} \dot{r}^2 = 1.$$

Now with the chain rule

$$\dot{r} = \frac{dr}{dt_E} = \frac{dr}{dt}\frac{dt}{dt_E} = \frac{dr}{dt}\dot{t},$$

so it follows

$$\left[\left(1 - \frac{2m}{r}\right) - \frac{1}{\left(1 - \frac{2m}{r}\right)}\left(\frac{dr}{dt}\right)^2\right] \dot{t}^2 = 1. \tag{27.1}$$

With the initial condition $(dr/dt)_{r=R} = 0$ it follows

$$\left[\left(1 - \frac{2m}{r}\right)_{r=R} - \frac{1}{\left(1 - \frac{2m}{r}\right)}\left(\frac{dr}{dt}\right)^2_{r=R}\right]\left(\frac{dt}{dt_E}\right)^2_{r=R} = $$

$$\left[\left(1-\frac{2m}{R}\right)\right]\left(\frac{dt}{dt_E}\right)^2_{r=R} = 1,$$

from which by rearranging and taking the square root

$$\left(\frac{dt}{dt_E}\right)_{r=R} = \left(\frac{R}{R-2m}\right)^{1/2}$$

follows. According to Eq. (23.21) for all r:

$$\left(1-\frac{2m}{r}\right)\frac{dt}{dt_E} = b = const.$$

so in particular also for $r=R$, from which

$$b = \left(1-\frac{2m}{R}\right)\left(\frac{dt}{dt_E}\right)_{r=R} = \left(1-\frac{2m}{R}\right)\left(\frac{R}{R-2m}\right)^{1/2} = \left(\frac{R-2m}{R}\right)^{1/2}$$

follows. So for all r

$$\frac{dt}{dt_E} = \left(1-\frac{2m}{r}\right)^{-1} b = \left(\frac{r}{r-2m}\right)\left(\frac{R-2m}{R}\right)^{1/2}. \tag{27.2}$$

We insert this into Eq. (27.1) and get

$$1 = \left[\left(1-\frac{2m}{r}\right) - \frac{1}{\left(1-\frac{2m}{r}\right)}\left(\frac{dr}{dt}\right)^2\right]\dot{t}^2$$

$$= \left[\left(\frac{r-2m}{r}\right) - \frac{1}{\left(\frac{r-2m}{r}\right)}\left(\frac{dr}{dt}\right)^2\right]\left(\frac{r}{r-2m}\right)^2\left(\frac{R-2m}{R}\right).$$

We rearrange for dr/dt and first get

$$\left[\left(\frac{r-2m}{r}\right) - \frac{1}{\left(\frac{r-2m}{r}\right)}\left(\frac{dr}{dt}\right)^2\right] = \left(\frac{r-2m}{r}\right)^2\left(\frac{R}{R-2m}\right),$$

from which

$$\left(\frac{dr}{dt}\right)^2 = -\left(\frac{r-2m}{r}\right)\left(\frac{r-2m}{r}\right)^2\left(\frac{R}{R-2m}\right) + \left(\frac{r-2m}{r}\right)\left(\frac{r-2m}{r}\right)$$

$$= -\left(\frac{r-2m}{r}\right)^2\left(\left(\frac{r-2m}{r}\right)\left(\frac{R}{R-2m}\right)-1\right)$$

$$= -\left(\frac{r-2m}{r}\right)^2\left(\left(\frac{r-2m}{r}\right)\left(\frac{R}{R-2m}\right)-\frac{r\,(R-2m)}{r\,(R-2m)}\right)$$

$$= -\left(\frac{r-2m}{r}\right)^2\left(\frac{Rr-2Rm-Rr+2rm}{r(R-2m)}\right)$$

$$= \left(\frac{r-2m}{r}\right)^2\frac{2m\,(R-r)}{r(R-2m)}$$

follows. It further results in

$$\frac{dr}{dt} = \pm\left(\frac{r-2m}{r}\right)\sqrt{\frac{2m\,(R-r)}{r(R-2m)}}.$$

We must choose the negative sign in front of the term on the right side, since we are considering a fall into the black hole, i.e., with increasing t, r becomes smaller and smaller. Thus we obtain

$$\frac{dr}{dt} = -\sqrt{\frac{2m}{R-2m}}\left(\frac{(r-2m)\,(R-r)^{1/2}}{r^{3/2}}\right) \tag{27.3}$$

and from this

$$dt = -\sqrt{\frac{R-2m}{2m}}\left(\frac{r^{3/2}}{(r-2m)\,(R-r)^{1/2}}\right)dr.$$

Both sides are integrated, where we use t' and r' as integration variables

$$\int_0^t dt' = -\sqrt{\frac{R-2m}{2m}}\int_R^r \frac{(r')^{3/2}}{(r'-2m)\,(R-r')^{1/2}}\,dr'.$$

The left side gives t, so it follows

$$t = -\sqrt{\frac{R-2m}{2m}}\int_R^r \frac{(r')^{3/2}}{(r'-2m)\,(R-r')^{1/2}}\,dr'.$$

This is the time it takes for the particle to travel from the starting point $r = R$ to any point $r > 2m$. The integrand tends towards infinity when $r' \to 2m$. To investigate this more closely, we set

$$r' = 2m + \varepsilon$$

with a small $\varepsilon > 0$. Then it follows with this substitution, that $d\varepsilon = dr'$ and the integration limits change to $R \to R - 2m$ and $r \to r - 2m$. We obtain

$$\begin{aligned}
t &= -\sqrt{\frac{R-2m}{2m}} \int_{R-2m}^{r-2m} \frac{(2m+\varepsilon)^{3/2}}{\varepsilon\,(R-2m-\varepsilon)^{1/2}} \, d\varepsilon \\
&\approx -\sqrt{\frac{R-2m}{2m}} \int_{R-2m}^{r-2m} \frac{(2m)^{3/2}}{\varepsilon\,(R-2m)^{1/2}} \, d\varepsilon \\
&= -2m \int_{R-2m}^{r-2m} \frac{1}{\varepsilon} \, d\varepsilon = -2m \ln\left(\frac{r-2m}{R-2m}\right).
\end{aligned}$$

We exponentiate this equation and it follows

$$r - 2m = (R - 2m)\, e^{-t/2m}.$$

The term on the right side is always positive because $R > 2m$ and converges to zero when $t \to \infty$. Therefore, the term on the left side is greater than zero for every time point t. In other words: From the perspective of the distant observer, the particle falls towards the black hole and approaches the Schwarzschild radius more and more over time, but never reaches it, let alone: crosses it.

What does an observer falling with the particle see on his own clock? To calculate this, we calculate the particle's proper time it takes to reach the Schwarzschild radius. We start from dr/dt_E. It applies with Eq. (27.2) and (27.3)

$$\begin{aligned}
\frac{dr}{dt_E} &= \frac{dr}{dt}\frac{dt}{dt_E} \\
&= -\sqrt{\frac{2m}{R-2m}} \left(\frac{(r-2m)\,(R-r)^{1/2}}{r^{3/2}}\right) \left(\frac{r}{r-2m}\right) \left(\frac{R-2m}{R}\right)^{1/2} \\
&= -\left(\frac{2m\,(R-r)}{Rr}\right)^{1/2} = -\sqrt{\frac{2m}{R}}\left(\frac{R-r}{r}\right)^{1/2},
\end{aligned}$$

from which

$$dt_E = -\sqrt{\frac{R}{2m}} \left(\frac{R}{r}-1\right)^{-1/2} dr$$

follows. Integration on both sides yields

$$\int_0^{t_E} dt'_E = -\sqrt{\frac{R}{2m}} \int_R^r \frac{dr'}{\sqrt{\dfrac{R}{r'}-1}}.$$

The integral on the left side yields t_E, the one on the right side we transform by using the substitution

$$\rho = \frac{r'}{R} \;\Rightarrow\; \frac{d\rho}{dr'} = \frac{1}{R} \;\Rightarrow\; \frac{dr'}{d\rho} = R.$$

Under consideration of the change of the integration limits (cf. Remark 6.4) we find

$$-\sqrt{\frac{R}{2m}} \int_R^r \frac{dr'}{\sqrt{\dfrac{R}{r'} - 1}} = -\sqrt{\frac{R}{2m}} \int_1^{r/R} \frac{R\,d\rho}{\sqrt{\dfrac{1}{\rho} - 1}} = -\sqrt{\frac{R^3}{2m}} \int_1^{r/R} \frac{\sqrt{\rho}\,d\rho}{\sqrt{1-\rho}}.$$

We now substitute

$$\rho = u^2 \;\Rightarrow\; \frac{d\rho}{du} = 2u$$

and thus obtain for the integral on the right side

$$-\sqrt{\frac{R^3}{2m}} \int_1^{r/R} \frac{\sqrt{\rho}\,d\rho}{\sqrt{1-\rho}} \;=\; -\sqrt{\frac{R^3}{2m}} \int_1^{\sqrt{r/R}} \frac{u\,2u\,du}{\sqrt{1-u^2}}$$

$$= \; -2\sqrt{\frac{R^3}{2m}} \int_1^{\sqrt{r/R}} \frac{u^2\,du}{\sqrt{1-u^2}}.$$

Finally, we substitute

$$u = \cos v \;\Rightarrow\; \frac{du}{dv} = -\sin v$$

and obtain with

$$\sqrt{1-u^2} = \sqrt{1-\cos^2 v} = \sin v$$

for the right integral

$$-2\sqrt{\frac{R^3}{2m}} \int_1^{\sqrt{r/R}} \frac{u^2\,du}{\sqrt{1-u^2}} \;=\; -2\sqrt{\frac{R^3}{2m}} \int_{\arccos(1)}^{\arccos\left(\sqrt{r/R}\right)} \frac{\cos^2 v\,(-\sin v)\,dv}{\sin v}$$

$$= \; 2\sqrt{\frac{R^3}{2m}} \int_0^{\arccos\left(\sqrt{r/R}\right)} \cos^2 v\,dv.$$

Since

$$\left(\frac{\cos v \sin v + v}{2}\right)' \;=\; \frac{1}{2}(-\sin^2 v + \cos^2 v + 1) = \frac{1}{2}(1 - \sin^2 v + \cos^2 v)$$

$$= \; \frac{1}{2}(2\cos^2 v) = \cos^2 v,$$

is

$$\frac{\cos v \sin v + v}{2}$$

the sought primitive function for the last integral. We obtain

$$2\sqrt{\frac{R^3}{2m}} \int_0^{\arccos\left(\sqrt{r/R}\right)} \cos^2 v\, dv \;=\;$$

$$\sqrt{\frac{R^3}{2m}} \left[\cos v \sin v + v\right]_0^{\arccos\left(\sqrt{r/R}\right)} \;=\;$$

$$\sqrt{\frac{R^3}{2m}} \left[\cos v \sqrt{1 - \cos^2 v} + v\right]_0^{\arccos\left(\sqrt{r/R}\right)} \;=\;$$

$$\sqrt{\frac{R^3}{2m}} \left[\left(\sqrt{r/R}\right)\sqrt{1 - r/R} + \arccos\left(\sqrt{r/R}\right)\right].$$

So, the total proper time sought is

$$t_E = \sqrt{\frac{R^3}{2m}} \left[\left(\sqrt{r/R}\right)\sqrt{1 - r/R} + \arccos\left(\sqrt{r/R}\right)\right],$$

which is a well-defined expression for all $r \leq R$. The point $r = 2m$ can therefore be reached by the particle in finite time, even more: If you set $r = 0$, the proper time until the "impact" is

$$t_E = \sqrt{\frac{R^3}{2m}} \arccos(0) = \frac{\pi}{2}\sqrt{\frac{R^3}{2m}},$$

also a finite proper time span. In summary, a distant observer sees a particle falling towards the Schwarzschild radius, but he never sees the particle reach it. For this reason, the location (the area) $r = 2m$ is also called the **event horizon**. However, if the particle measures its own time, it reaches the horizon in finite time, crosses it, and falls into the black hole in finite time as well. The horizon has no physical significance for the falling particle. This is an important hint that the singularity of the Schwarzschild metric at $r = 2m$ is "only" a coordinate singularity, which can be circumvented by cleverly choosing other coordinate systems. We will show this in the following sections.

27.4. Eddington-Finkelstein Coordinates

We recall that the General Theory of Relativity allows free choice of the coordinate system. The Einstein equations are formulated covariantly, i.e., they are valid in any coordinate system. The physical phenomena we want to investigate are independent of the coordinate system, and we can choose the coordinate systems that are best suited to solve the problem depending on the problem. We will see that the coordinates chosen in this section will make significant progress in understanding the peculiarities at the event horizon $r = 2m$.

We first consider a light beam that is radially approaching the black hole. We again start from the line element of the Schwarzschild metric

$$ds^2 = -\left(1 - \frac{2m}{r}\right) dt^2 + \frac{1}{\left(1 - \frac{2m}{r}\right)} dr^2 + r^2 \left(d\vartheta^2 + \sin^2 \vartheta \, d\varphi^2\right).$$

Because of the radial movement, $d\vartheta = d\varphi = 0$, and for light $ds^2 = 0$, i.e., we get

$$0 = -\left(1 - \frac{2m}{r}\right) dt^2 + \frac{1}{\left(1 - \frac{2m}{r}\right)} dr^2$$

and from this

$$dt^2 = \frac{1}{\left(1 - \frac{2m}{r}\right)^2} dr^2,$$

so

$$\frac{dt}{dr} = \pm \left(1 - \frac{2m}{r}\right)^{-1}. \tag{27.4}$$

If the distance r is large and far from the horizon $2m$, the term on the right side becomes ± 1, and when we integrate the above equation, we get

$$t = \pm r + const.,$$

which is exactly what we expect for a light beam in Minkowski space. On the other hand, dt/dr becomes larger (for $+$) or smaller (for $-$) as $r \to 2m$. Since dt/dr determines the slope of the light cone, this means that the light cones become narrower, as one approaches the Schwarzschild radius. This is illustrated in Fig. 27.2.

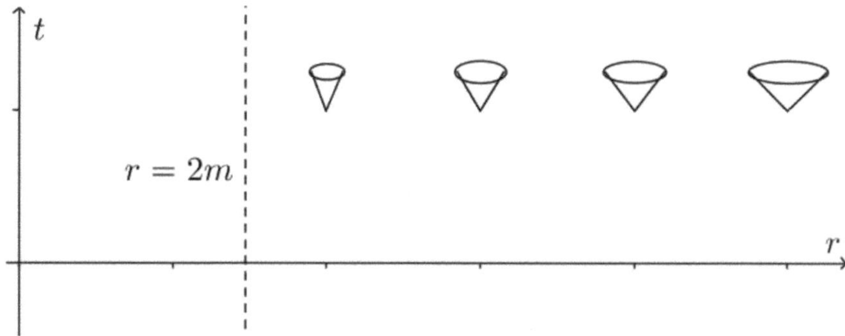

Figure 27.2.: Future light cone in the Scharzschild spacetime

If we integrate the above Eq. (27.4) for small r, we get

$$\int dt = \pm \int \frac{r}{r - 2m} \, dr.$$

To calculate the integral on the right side, we apply partial integration and note that the function $1/(r - 2m)$ has the primitive function $\ln |r - 2m|$:

$$\int \frac{r}{r - 2m} \, dr = r \ln |r - 2m| - \int \ln |r - 2m| \, dr \tag{27.5}$$

Applying partial integration again to the right integral yields with the primitive function $r - 2m$ of 1

$$\begin{aligned}
\int \ln |r - 2m| \, dr &= \int 1 \cdot \ln |r - 2m| \, dr \\
&= (r - 2m) \ln |r - 2m| - \int (r - 2m) \frac{1}{r - 2m} \, dr \\
&= (r - 2m) \ln |r - 2m| - r.
\end{aligned}$$

Substituting this into Eq. (27.5) leads to

$$\int \frac{r}{r - 2m} \, dr = r \ln |r - 2m| - ((r - 2m) \ln |r - 2m| - r) = r + 2m \ln |r - 2m|.$$

So overall, with a suitable integration constant, we get

$$t = \pm (r + 2m \ln |r - 2m|) + const. \tag{27.6}$$

The plus sign stands for light rays moving away from the black hole, the minus sign for light rays falling into the black hole. Fig. 27.3 demonstrates the course of the light rays, where $2m = 2.95$ was set.

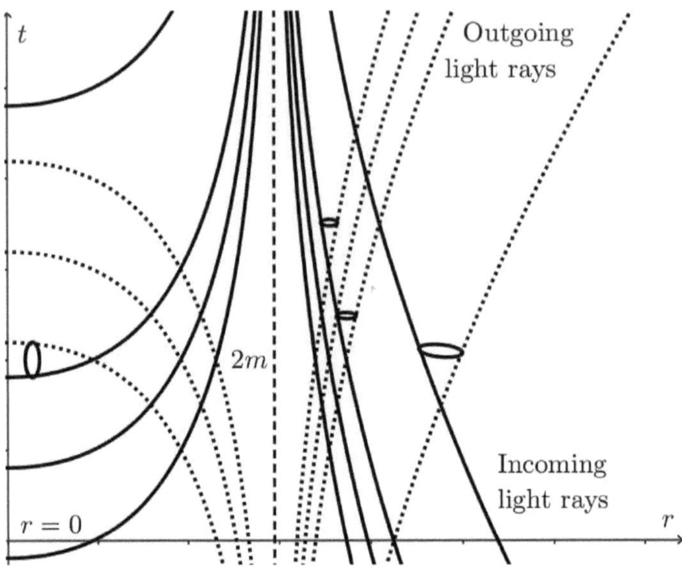

Figure 27.3.: Light rays in the Schwarzschild spacetime

Fig. 27.3 is complicated and requires some explanations. First, you can see the singularity at $r = 2m$ (dashed vertical line). All four incoming light rays (solid lines on the right side) from the region $r > 2m$ can never reach this boundary. If one assumes that the incoming light rays have overcome the singularity (solid lines on the left side), it seems that time and space reverse or time runs backwards, because the light rays land in backward-running, but finite time in the black hole (axis $r = 0$). For the outgoing light rays (four dotted lines), the right side is clear again, they move away from the Schwarzschild radius with increasing time. Also note the (decreasing) curvatures of the light rays! If one assumes on the left side, that the dotted light rays are moving away from the black hole, something similar happens as with the others: They approach - also in backward-running time - the Schwarzschild radius, but cannot overcome it. The light cannot escape the black hole. In Fig. 27.3 on the right side, three future light cones are also drawn, which are formed by one incoming and outgoing light ray each. You can see once again that the light cones narrow, the closer they come to the Schwarzschild radius. On the left side a future light cone is also drawn. You can see that it now lies spacelike and a particle or a light ray leaves no other choice but to fall into the black hole.

The phenomenon that space and time seem to reverse when transitioning from $r > 2m$ to $r < 2m$ can also be directly read from the Schwarzschild metric

$$ds^2 = -\left(1 - \frac{2m}{r}\right) dt^2 + \frac{1}{\left(1 - \dfrac{2m}{r}\right)} dr^2 + r^2 \left(d\vartheta^2 + \sin^2\vartheta \, d\varphi^2\right).$$

Because within the Schwarzschild horizon $(r < 2m)$ it is

$$g_{tt} = -\left(1 - \frac{2m}{r}\right) > 0$$

$$g_{rr} = \left(1 - \frac{2m}{r}\right)^{-1} < 0.$$

However, outside $(r > 2m)$ applies

$$g_{tt} = -\left(1 - \frac{2m}{r}\right) < 0$$

$$g_{rr} = \left(1 - \frac{2m}{r}\right)^{-1} > 0.$$

Looking at the geometry within the Schwarzschild radius, but close to $r = 2m$, and defining the size $\varepsilon = 2m - r$, the line element follows

$$ds^2 = -\frac{r - 2m}{2m - \varepsilon} dt^2 - \frac{2m - \varepsilon}{\varepsilon} dr^2 + (2m - \varepsilon)^2 \left(d\vartheta^2 + \sin^2\vartheta \, d\varphi^2\right).$$

Since $\varepsilon > 0$ for $0 < r < 2m$, it follows for a straight line with $t, \theta, \varphi = const.$

$$ds^2 < 0,$$

it is timelike, i.e., ε and thus also r have become timelike coordinates, while t has become spacelike. A massive particle falling onto the black hole moves on a timelike worldline, it must constantly change the distance r, namely reduce it. That means, once the particle has arrived within the Schwarzschild radius, it will inevitably fall into the black hole $(r = 0)$. But what happens if the particle within the Schwarzschild radius tries to send a signal outward? This photon must, no matter in which direction it was sent, move forward in "time" from the particle's point of view, but within the Schwarzschild radius the coordinate r indicates the time and r is always getting smaller. The photon cannot escape. Nothing can escape once it has arrived within the Schwarzschild radius. Even more, everything will fall onto the real singularity at $r = 0$, as $r = 0$ is the future of every time- or light-like worldline within $r = 2m$. Once a particle has crossed the horizon, it can no longer be observed from the outside. The Schwarzschild coordinates are therefore useless for describing the physics of the motion itself - they only describe the observed motion!

We now want to define coordinates that fix the coordinate singularity of the Schwarzschild metric at $r = 2m$. The new coordinates should be chosen so that the incoming light rays become straight lines. Since the incoming light rays according to Eq. (27.6) are represented by

$$t = -r - 2m \ln |r - 2m| + const.$$

the new coordinates are defined by

$$\bar{t} = t \pm 2m \ln |r - 2m|, \bar{r} = r, \bar{\vartheta} = \vartheta, \bar{\varphi} = \varphi. \tag{27.7}$$

We first choose the plus sign.

In a $(\bar{t}, r, \vartheta, \varphi)$ coordinate system, the equation with (27.6) becomes

$$\bar{t} = t + 2m \ln |r - 2m| = -r + const.$$

i.e., the incoming light ray is a straight line and forms an angle of $-45°$ with the r-axis. From the definition of the new time coordinate follows

$$\frac{d\bar{t}}{dt} = 1 + \frac{2m}{r - 2m} \frac{dr}{dt}$$

and from this

$$d\bar{t} = dt + \frac{2m}{r - 2m} dr.$$

Squaring and rearranging gives

$$dt^2 = d\bar{t}^2 - \frac{4m}{r - 2m} d\bar{t}\, dr + \frac{4m^2}{(r - 2m)^2} dr^2.$$

This is inserted into the line element of the Schwarzschild metric and obtained (terms with ϑ and φ omitted)

$$\begin{aligned}
ds^2 &= -\left(1 - \frac{2m}{r}\right) dt^2 + \frac{1}{\left(1 - \dfrac{2m}{r}\right)} dr^2 \\
&= -\left(1 - \frac{2m}{r}\right)\left(d\bar{t}^2 - \frac{4m}{r - 2m} d\bar{t}\, dr + \frac{4m^2}{(r - 2m)^2} dr^2\right)
\end{aligned}$$

$$+ \frac{1}{\left(1 - \dfrac{2m}{r}\right)} dr^2$$

$$= -\left(\frac{r - 2m}{r}\right) d\bar{t}^2 + \frac{4m}{r} d\bar{t}\, dr - \frac{4m^2}{r\,(r - 2m)} dr^2 + \frac{r}{(r - 2m)} dr^2$$

$$= -\left(\frac{r - 2m}{r}\right) d\bar{t}^2 + \frac{4m}{r} d\bar{t}\, dr + \frac{r^2 - 4m^2}{r\,(r - 2m)} dr^2$$

$$= -\left(\frac{r - 2m}{r}\right) d\bar{t}^2 + \frac{4m}{r} d\bar{t}\, dr + \frac{r + 2m}{r} dr^2.$$

So now with all terms:

$$ds^2 = -\left(1 - \frac{2m}{r}\right) d\bar{t}^2 + \frac{4m}{r} d\bar{t}\, dr + \left(1 + \frac{2m}{r}\right) dr^2 + r^2 \left(d\vartheta^2 + \sin^2 \vartheta\, d\varphi^2\right)$$

$$(27.8)$$

This is the **Eddington-Finkelstein form** of the metric. It is to be noted that the coefficient before dr^2 now at $r = 2m$ no longer becomes singular, i.e., the change to the new time coordinate $\bar{t}$ has extended the validity range of the radial coordinate r from $2m < r < \infty$ to $0 < r < \infty$. We now consider the worldline of a radial light beam in these coordinates and get

$$0 = -\left(\frac{r - 2m}{r}\right) d\bar{t}^2 + \frac{4m}{r} d\bar{t}\, dr + \frac{r + 2m}{r} dr^2.$$

Division by $d\bar{t}^2$ and $(r + 2m)/r$ results in

$$0 = \left(\frac{dr}{d\bar{t}}\right)^2 + \frac{4m}{r + 2m} \left(\frac{dr}{d\bar{t}}\right) - \frac{r - 2m}{r + 2m}.$$

This is a quadratic equation for dr/dt and of the form

$$x^2 + px + q = 0,$$

for which there is the (p, q) solution formula

$$x_{1/2} = -\frac{p}{2} \pm \sqrt{\left(\frac{p}{2}\right)^2 - q}$$

(see Chap. 29). We thus obtain the two solutions

$$\left(\frac{dr}{d\bar{t}}\right)_{1/2} = -\frac{2m}{r + 2m} \pm \sqrt{\left(\frac{2m}{r + 2m}\right)^2 + \frac{r - 2m}{r + 2m}}$$

$$
\begin{aligned}
&= -\frac{2m}{r+2m} \pm \sqrt{\frac{4m^2}{(r+2m)^2} + \frac{(r-2m)\,(r+2m)}{(r+2m)^2}} \\[2mm]
&= -\frac{2m}{r+2m} \pm \sqrt{\frac{4m^2}{(r+2m)^2} + \frac{r^2 - 4m^2}{(r+2m)^2}} \\[2mm]
&= -\frac{2m}{r+2m} \pm \frac{r}{r+2m} \\[2mm]
&= \begin{cases} -1 \\[1mm] \dfrac{r-2m}{r+2m} \end{cases}
\end{aligned}
\qquad (27.9)
$$

The solution $dr/d\bar{t} = -1$ once again states that incoming light rays fall onto the black hole on lines with a slope of -1. For outgoing light rays, we must take the plus sign in Eq. (27.6), so

$$
t = r + 2m \ln |r - 2m| + const.
$$

If we insert the Eddington-Finkelstein coordinates (27.7), we get

$$
\bar{t} = t + 2m \ln |r - 2m| = r + 4m \ln |r - 2m| + const.
$$

From this it follows

$$
\frac{d\bar{t}}{dr} = 1 + \frac{4m}{r - 2m} = \frac{r + 2m}{r - 2m},
$$

which shows that the second solution (27.9) belongs to the outgoing light rays. If we integrate the last equation, we get the function $\bar{t}(r)$ for outgoing light rays

$$
\bar{t} = \int d\bar{t} = \int \left(1 + \frac{4m}{r - 2m} \right) dr = r + 4m \ln |r - 2m| + const.
$$

Overall, the following picture emerges.

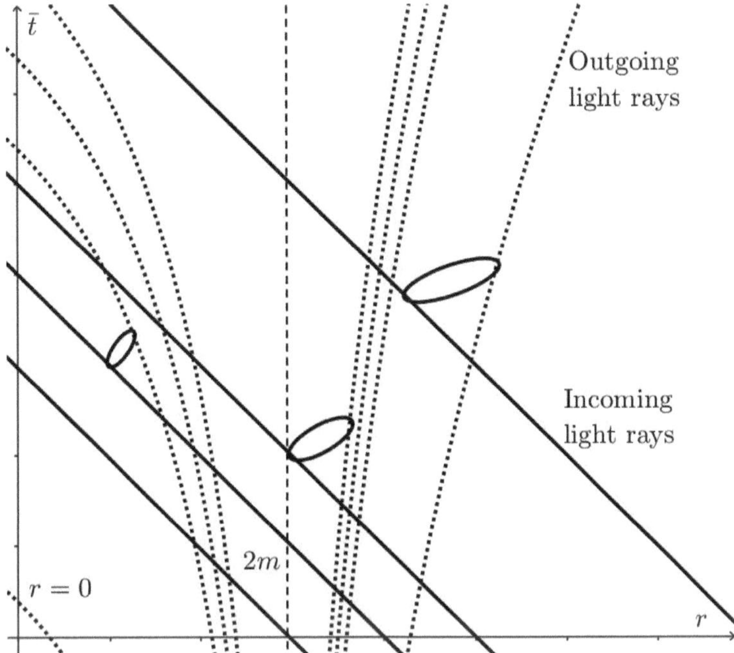

Figure 27.4.: Light rays in Eddington-Finkelstein Coordinates

Fig. 27.4 is again two-dimensional (ϑ and φ are not shown), and the coordinates are r and $\bar{t}$. The Schwarzschild radius at $r = 2m$ (dashed line) is again visible. Incoming (solid) light rays are straight lines. Outgoing (dotted) light rays in the area $r > 2m$ move away from the Schwarzschild radius and curve more and more with $r \to \infty$ until they finally (at infinity) assume the slope $+1$ and thus form the Minkowski light cones with the incoming light rays. Light rays that originate in the area $r < 2m$ cannot leave this region, they fall into the black hole. In Fig. 27.4, three future light cones are drawn. These become narrower as they fall onto the black hole and tilt by 45° to the left. The area $r = 2m$ acts like a one-way membrane: light from outside can pass through it, light from inside cannot penetrate it.

If we choose the minus sign in Eq. (27.7), i.e., define the time variable t^* by

$$t^* = t - 2m \ln |r - 2m|$$

and perform a similar calculation, we get for the line element in Eddington-Finkelstein coordinates

$$ds^2 = -\left(1 - \frac{2m}{r}\right) dt^{*2} - \frac{4m}{r} dt^*\, dr + \left(1 + \frac{2m}{r}\right) dr^2 + r^2 \left(d\vartheta^2 + \sin^2 \vartheta\, d\varphi^2\right),$$

i.e., a minus sign in front of the term with $dt^*\,dr$. For the light rays this means

$$\frac{dr}{dt^*} = 1 \quad oder \quad \frac{dr}{dt^*} = -\left(\frac{r-2m}{r+2m}\right).$$

The situation has thus reversed: light rays inside the Schwarzschild radius can pass through the surface at $r = 2m$ to the outside, but not vice versa. Also, particles cannot enter the interior of the surface. Such an object is called a **White Hole**, it is, so to speak, a temporally reversed Black Hole. It is quite likely that Black Holes exist, but unlikely that there are White Holes. The Einstein equations are indeed invariant under time reversal, but nature seems not to make use of this.

27.5. Kruskal Coordinates

By introducing the Eddington-Finkelstein coordinates, we have gained a better understanding of the Schwarzschild solution. However, there is a further development of this approach, by which the positive (Black Hole) as well as the negative (White Hole) coordinates are unified. We define the **advanced time parameter** v by

$$v = t + r + 2m \ln\left|\frac{r-2m}{2m}\right|. \tag{27.10}$$

Then follows

$$\frac{dv}{dt} = 1 + \frac{dr}{dt} + \frac{2m}{r-2m}\frac{dr}{dt} = 1 + \left(1 + \frac{2m}{r-2m}\right)\frac{dr}{dt},$$

thus

$$dv = dt + \left(\frac{r}{r-2m}\right) dr.$$

We insert this into the Schwarzschild metric and obtain with

$$d\Omega^2 = d\vartheta^2 + \sin^2\vartheta\, d\varphi^2$$

the relationship

$$\begin{aligned}
ds^2 &= -\left(1 - \frac{2m}{r}\right)dt^2 + \frac{1}{\left(1 - \dfrac{2m}{r}\right)}dr^2 + r^2 d\Omega^2 \\[2ex]
&= -\left(1 - \frac{2m}{r}\right)\left(dv - \left(\frac{r}{r-2m}\right)dr\right)^2 + \frac{1}{\left(1 - \dfrac{2m}{r}\right)}dr^2 + r^2 d\Omega^2
\end{aligned}$$

$$
\begin{aligned}
= \ & -\left(\frac{r-2m}{r}\right)\left(dv^2 - \left(\frac{2r}{r-2m}\right)dv\,dr + \left(\frac{r}{r-2m}\right)^2 dr^2\right) \\
& + \frac{r}{r-2m}\,dr^2 + r^2 d\Omega^2 \\
= \ & -\left(1 - \frac{2m}{r}\right)dv^2 + 2\,dv\,dr + r^2 d\Omega^2.
\end{aligned}
$$

This line element is not invariant due to the mixed term $dv\,dr$ when the advanced time is reversed ($v \to -v$). The advanced time reversal means "normal" time reversal ($t \to -t$) and

$$
r + 2m\ln\left|\frac{r-2m}{2m}\right| \to -r - 2m\ln\left|\frac{r-2m}{2m}\right|,
$$

which according to Eq. (27.6) means that incoming light rays are replaced by outgoing ones. Similarly, the **retarded time coordinate** is defined by

$$
w = t - r - 2m\ln\left|\frac{r-2m}{2m}\right|. \tag{27.11}
$$

This results in

$$
v - w = 2r + 4m\ln\left|\frac{r-2m}{2m}\right| \tag{27.12}
$$

and analogous to above

$$
ds^2 = -\left(1 - \frac{2m}{r}\right)dw^2 - 2\,dw\,dr + r^2 d\Omega^2.
$$

If we define the coordinate r^* by

$$
r^* = r + 2m\ln\left|\frac{r-2m}{2m}\right|,
$$

then we get

$$
v = t + r^*, \quad w = t - r^*,
$$

i.e., in a (r^*, t) coordinate system, incoming and outgoing light rays satisfy the equations $v = const.$ and $w = const.$; w and v are also called **null coordinates**.

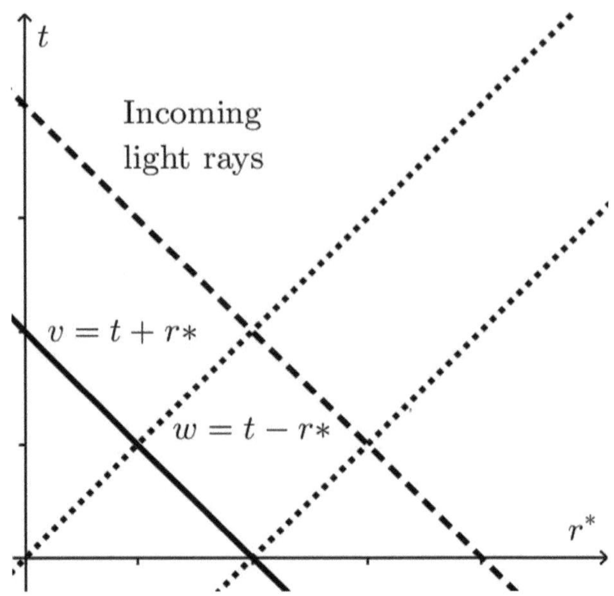

Figure 27.5.: Light rays in (r^*, t) Coordinates

In 27.5, the light cones in the (r^*, t)-plane remain constant, i.e., they no longer narrow. Now applies

$$
\begin{aligned}
dv\,dw &= \left(dt + \left(\frac{r}{r - 2m}\right) dr\right)\left(dt - \left(\frac{r}{r - 2m}\right) dr\right) \\
&= dt^2 - \left(\frac{r}{r - 2m}\right)^2 dr^2 = dt^2 - \left(1 - \frac{2m}{r}\right)^{-2} dr^2,
\end{aligned}
$$

from with

$$
\left(1 - \frac{2m}{r}\right) dv\,dw = \left(1 - \frac{2m}{r}\right) dt^2 - \left(1 - \frac{2m}{r}\right)^{-1} dr^2
$$

follows, i.e., the Schwarzschild metric can be expressed in null coordinates:

$$
\begin{aligned}
ds^2 &= -\left(1 - \frac{2m}{r}\right) dt^2 + \left(1 - \frac{2m}{r}\right)^{-1} dr^2 + r^2 d\Omega^2 \\
&= -\left(1 - \frac{2m}{r}\right) dv\,dw + r^2 d\Omega^2. \tag{27.13}
\end{aligned}
$$

This line element still has the disadvantage that the coefficient before $dv\,dw$ becomes zero at $r = 2m$. This can be avoided by using the so-called **Kruskal**

coordinates, which were found by M. D. Kruskal in 1960. They are defined for $r > 2m$ by

$$v' = \frac{1}{2}\left(e^{v/4m} + e^{-w/4m}\right)$$

$$w' = \frac{1}{2}\left(e^{v/4m} - e^{-w/4m}\right).$$

In Schwarzschild coordinates t, r, with (27.10) and (27.11) by replacing v and w

$$v' = \frac{1}{2}\left(e^{\left(t+r+2m\ln\left(\frac{r-2m}{2m}\right)\right)/4m} + e^{-\left(t-r-2m\ln\left(\frac{r-2m}{2m}\right)\right)/4m}\right)$$

$$= e^{r/4m}\left(\frac{r-2m}{2m}\right)^{1/2}\frac{1}{2}\left(e^{t/4m} + e^{-t/4m}\right)$$

$$w' = \frac{1}{2}\left(e^{\left(t+r+2m\ln\left(\frac{r-2m}{2m}\right)\right)/4m} - e^{-\left(t-r-2m\ln\left(\frac{r-2m}{2m}\right)\right)/4m}\right)$$

$$= e^{r/4m}\left(\frac{r-2m}{2m}\right)^{1/2}\frac{1}{2}\left(e^{t/4m} - e^{-t/4m}\right).$$

The radial coordinate r is then a function of v' and w', for which one finds the following implicit representation

$$v'^2 - w'^2 = e^{r/2m}\left(\frac{r-2m}{2m}\right)\frac{1}{4}\left[\left(e^{t/4m} + e^{-t/4m}\right)^2 - \left(e^{t/4m} - e^{-t/4m}\right)^2\right]$$

$$= e^{r/2m}\left(\frac{r-2m}{2m}\right). \tag{27.14}$$

For the time coordinates t, the relationship follows

$$\frac{v'}{w'} = \frac{e^{t/4m} + e^{-t/4m}}{e^{t/4m} - e^{-t/4m}} = \frac{1 + e^{-t/2m}}{1 - e^{-t/2m}}. \tag{27.15}$$

We want to express the line element of the Schwarzschild metric by the Kruskal coordinates and first calculate dv' and dw'. We calculate the differentials with Eq. (4.16):

$$dv' = \frac{\partial v'}{\partial r}\,dr + \frac{\partial v'}{\partial t}\,dt$$

and obtain

$$\frac{\partial v'}{\partial r} = \frac{1}{2}\left(e^{t/4m} + e^{-t/4m}\right)e^{r/4m}\left(\frac{1}{4m}\left(\frac{r-2m}{2m}\right)^{1/2} + \frac{1}{4m}\left(\frac{r-2m}{2m}\right)^{-1/2}\right)$$

$$
\begin{aligned}
&= \frac{1}{4m}\left(\frac{1}{2}\left(e^{t/4m}+e^{-t/4m}\right)\right)e^{r/4m}\left(\sqrt{\frac{r-2m}{2m}}+\frac{1}{\sqrt{\frac{r-2m}{2m}}}\right)\\
&= \frac{1}{4m}\left(\frac{1}{2}\left(e^{t/4m}+e^{-t/4m}\right)\right)e^{r/4m}\frac{1}{\sqrt{\frac{r-2m}{2m}}}\left(\frac{r-2m}{2m}+1\right)\\
&= \frac{1}{4m}\left(\frac{1}{2}\left(e^{t/4m}+e^{-t/4m}\right)\right)e^{r/4m}\frac{\sqrt{2m}}{\sqrt{r-2m}}\frac{r}{2m}\\
&= \frac{\sqrt{2m}}{8m^2}e^{r/4m}\frac{r}{\sqrt{r-2m}}\left(\frac{1}{2}\left(e^{t/4m}+e^{-t/4m}\right)\right)
\end{aligned}
$$

as well as

$$
\frac{\partial v'}{\partial t} = \frac{1}{4m}e^{r/4m}\left(\frac{r-2m}{2m}\right)^{1/2}\left(\frac{1}{2}\left(e^{t/4m}-e^{-t/4m}\right)\right).
$$

If we shorten the terms

$$
\frac{\sqrt{2m}}{8m^2}e^{r/4m}\frac{r}{\sqrt{r-2m}} = a,\quad \frac{1}{4m}e^{r/4m}\left(\frac{r-2m}{2m}\right)^{1/2} = b,
$$

we get

$$
dv' = a\left(\frac{1}{2}\left(e^{t/4m}+e^{-t/4m}\right)\right)dr + b\left(\frac{1}{2}\left(e^{t/4m}-e^{-t/4m}\right)\right)dt
$$

and similarly

$$
dw' = a\left(\frac{1}{2}\left(e^{t/4m}-e^{-t/4m}\right)\right)dr + b\left(\frac{1}{2}\left(e^{t/4m}+e^{-t/4m}\right)\right)dt.
$$

From this follows

$$
\begin{aligned}
dv'^2 - dw'^2 &= a^2 dr^2 - b^2 dt^2\\
&= \frac{1}{32m^3}e^{r/2m}\frac{r^2}{r-2m}dr^2 - \frac{1}{32m^3}e^{r/2m}(r-2m)\,dt^2\\
&= \frac{r}{32m^3}e^{r/2m}\left(\frac{r}{r-2m}dr^2 - \frac{r-2m}{r}dt^2\right).
\end{aligned}
$$

So for the Schwarzschild line element in Kruskal coordinates

$$
ds^2 = \frac{32m^3}{r}e^{-r/2m}\left(dv'^2 - dw'^2\right) + r^2 d\Omega^2. \tag{27.16}
$$

In deriving this formula, we assumed that $r > 2m$ is. The case $r < 2m$ is treated analogously by replacing the Kruskal coordinates by

$$
\begin{aligned}
v' &= e^{r/4m}\left(\frac{2m-r}{2m}\right)^{1/2}\frac{1}{2}\left(e^{t/4m}-e^{-t/4m}\right)\\
w' &= e^{r/4m}\left(\frac{2m-r}{2m}\right)^{1/2}\frac{1}{2}\left(e^{t/4m}+e^{-t/4m}\right)
\end{aligned}
$$

and by a similar calculation, the *same* line element (27.16) is obtained. The Kruskal line element now contains no term that points to a peculiarity at $r = 2m$. It is also called the maximal extension of the Schwarzschild solution. Of course, the real singularity at $r = 0$ remains. If we again restrict ourselves to radial light rays (ϑ, φ are constant), we get

$$
0 = \frac{32m^3}{r}e^{-r/2m}\left(dv'^2 - dw'^2\right),
$$

i.e.

$$
dv' = \pm dw',
$$

so here too, as with the null coordinates, the light rays are represented by straight lines in the (v', w')-plane. If we look again at the relationships between the Kruskal and the Schwarzschild coordinates (27.14) and (27.15), we get on the one hand

$$
v'^2 - w'^2 = e^{r/2m}\left(\frac{r-2m}{2m}\right)
$$

and on the other hand

$$
\frac{v'}{w'} = \frac{1+e^{-t/2m}}{1-e^{-t/2m}}.
$$

With these relations, we can design the Kruskal diagram 27.6. The diagram has four sectors, which are marked with I, II, III, and IV.

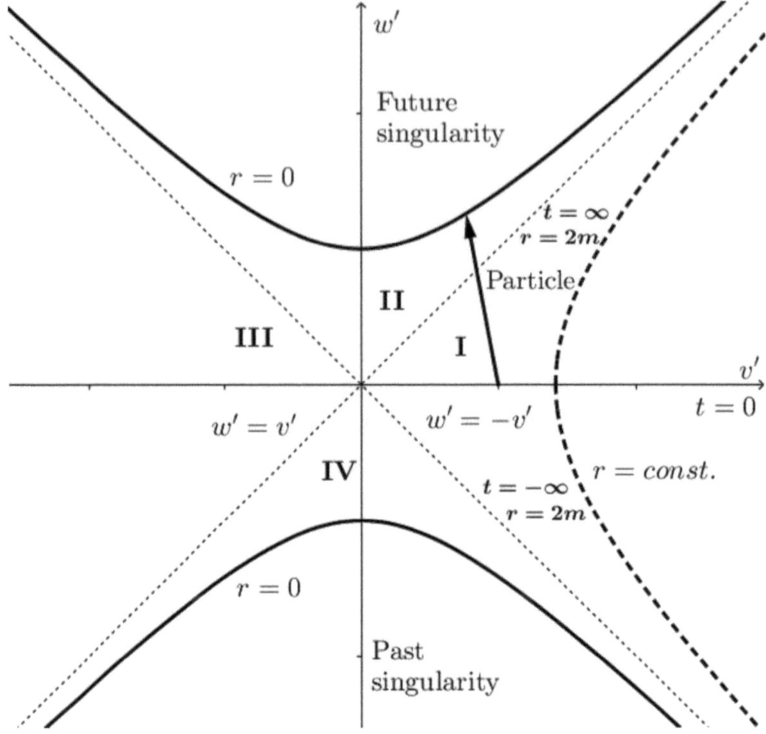

Figure 27.6.: Kruskal diagram of a black hole

Region I represents the space outside the Schwarzschild radius ($r > 2m$), it is bounded by the two right (dotted) bisectors ($r = 2m, t = \pm\infty$) or ($w' = \pm v'$), ($v' > 0$). The time coordinate t is for $t = 0$ the v' axis, other time values lie on straight lines through the origin, which lie between the v' axis and the bisectors (not shown in the figure). The bisectors themselves represent the (maximum) time values $\pm\infty$.

In Fig. 27.6 the timelike worldline of a particle (for simplicity a straight line) is drawn, which starts at $t = 0$, crosses the Schwarzschild radius $r = 2m$, thus reaches region II and ends in the real singularity (black hyperbola) at $r = 0$. If we imagine a light ray that starts from region I (i.e., parallel to and above the line $w' = -v'$), it can only reach region II, but not regions III and IV.

Once a particle (or a light ray) has arrived in region II ($r < 2m$), it eventually reaches the future singularity $r = 0$, i.e., the coordinate r shrinks uncontrollably and thus plays a similar role as in region I the time t, which however continues to grow unabated. Region II is the black hole, everything can fall in, nothing can come out.

Our world I cannot influence region IV, but it could be influenced by it, if outgoing light rays or timelike objects from IV reach I. Region IV also has a singularity, but nothing can fall into it, as time would have to run backwards, so region IV is a white hole. Sector III is completely separated from us. Nothing that happens there can ever reach us. And nothing that happens with us has an influence on what happens there.

The Kruskal diagram clearly shows the symmetry of the time reversal of the Schwarzschild solution. In the white hole, time runs backwards and gravity is not attractive, but repulsive. But, as mentioned above, nature apparently does not make use of this possibility.

Outlook

With these remarks on black holes, our expedition into Einstein's General Theory of Relativity comes to an end. For the treatment of further phenomena related to the GR, such as the study of rotating black holes or applications of GR in cosmology, we must refer to further texts, some of which are briefly introduced in Chapter 28.

28. Literature References and Further Reading on Part V

For Part V, I also recommend a popular science book titled "Black Holes and Time Warps" by Kip Thorne [34]. It describes in detail over 700 pages the subareas of gravitational waves, star collapse, stellar interiors, and especially black holes, but also the classical topics from Part IV are discussed in detail. The book, like Hawking's "A Brief History of Time [16]", comes without formulas and also deals (quite seriously) with phenomena known from science fiction literature such as wormholes and time travel.

For further deepening, especially of the topics of the last part, there are a number of advanced textbooks, which, however, require mathematical and physical prior knowledge to the extent that one learns in a bachelor's degree in physics.

- First of all the book by A. Zee, Einstein Gravity in a Nutshell [40], should be mentioned. The title of the book is somewhat misleading, as the book comprises 866 pages. It is amusing to read, as the author often gives anecdotes and little jokes, but also demanding, as the presentation of the facts often requires knowledge that a non-expert does not have. This book is comprehensive in content and also touches on current research areas such as quantum gravity and string theory at the end.

- The textbook by Wald, General Relativity [36] covers the entire spectrum of General Relativity and is written predominantly in a modern (abstract) mathematical language. It is therefore not easy to read and one must also work hard, as many details are omitted or formulated as exercises, for which unfortunately there are no solutions either in the book or elsewhere.

- A literature list on General Relativity should not be without the two classics, Gravitation and Cosmology [37] (just under 700 pages) by Weinberg, and *the* standard book of General Relativity, Gravitation [20] (just under 1300 pages!) by Misner, Thorne and Wheeler, both of which were created in the 1970s and have remained important sources and reference works to this day.

Of course, my second *German* book on General Relativity, Allgemeine Relativitätstheorie Schritt für Schritt [25], which was the main basis for the last part of this book, should not go unmentioned. It is written in as much detail as the present book, but delves deeper into mathematical (e.g., manifolds, differential forms) and physical (e.g., electrodynamics, gravitational waves, black holes, cosmology) topics and is therefore in many aspects a continuation or deepening of the "Einstein equations".

Part VI.

Appendix: Formulas and Tables

29. Functions, Formulas and Physical Laws

This chapter lists the formulas and physical laws used in the text but not explained or derived.

29.1. Functions

Trigonometric Functions

The trigonometric functions are defined on the unit circle $x^2 + y^2 = 1$, see Fig. 29.1.

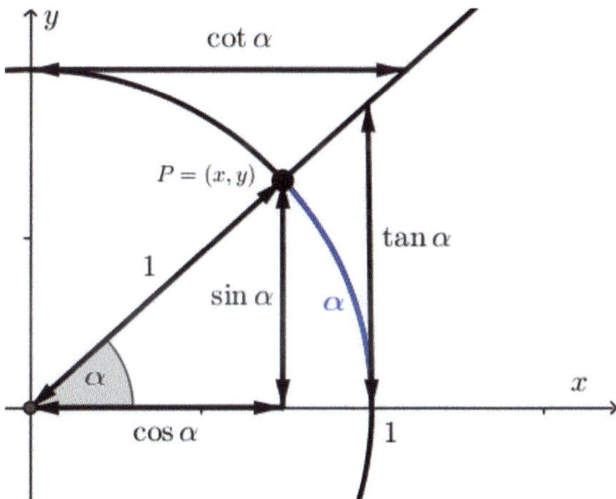

Figure 29.1.: The angular functions on the unit circle

If $P(x, y)$ is a point on the unit circle with the argument angle α, it is defined as follows:

- $\sin \alpha = y$

- $\cos \alpha = x$

- $\tan(\alpha) = \dfrac{\sin \alpha}{\cos \alpha}$

M. Ruhrländer, *Ascent to the Einstein Equations*, https://doi.org/10.1007/978-3-662-72672-3_29

- $\cot \alpha = \dfrac{\cos \alpha}{\sin \alpha}$

- If the radius of the circle is equal to r, it follows accordingly

  - $\sin \alpha = y/r$

  - $\cos \alpha = x/r$

  - $\tan(\alpha) = \dfrac{\sin \alpha}{\cos \alpha}$

  - $\cot \alpha = \dfrac{\cos \alpha}{\sin \alpha}$

Properties and addition theorems of trigonometric functions

1. $\sin^2 x + \cos^2 x = 1$

2. $\sin (x \pm y) = \cos x \sin y \pm \sin x \cos y$

3. For $x = y$ it follows: $\sin 2x = 2 \cos x \sin x$

4. $\cos (x \pm y) = \cos x \cos y \mp \sin x \sin y$

5. For $x = y$ it follows: $\cos 2x = \cos^2 x - \sin^2 x$

6. $\sin^2 x = \dfrac{1 - \cos(2x)}{2} = \dfrac{1}{1 + \tan^2 x}$

7. $\cos^2 x = \dfrac{1 + \cos(2x)}{2} = \dfrac{\tan^2 x}{1 + \tan^2 x}$

The Inverse Functions of the Trigonometric Functions, Arc Functions

The **arc functions** are the inverse functions of the trigonometric functions. These cannot be inverted in all of $\mathbb{R}$. Rather, the domain of the trigonometric functions must be restricted to an interval where the functions are strictly monotonic. If, for example, the sine function is restricted to the interval $\left[-\dfrac{\pi}{2}; \dfrac{\pi}{2}\right]$, it is strictly monotonically increasing there and thus the inverse function **arcsine** exists

$$\arcsin : [-1; 1] \to \left[-\dfrac{\pi}{2}; \dfrac{\pi}{2}\right]$$

with

$$\arcsin(\sin x) = x$$

and
$$\sin(\arcsin y) = y.$$
Similarly, the inverse functions **arccosine**
$$\arccos : [-1; 1] \to [0; \pi]$$
with
$$\arccos(\cos x) = x$$
and
$$\cos(\arccos y) = y$$
as well as **arctangent** $\arctan(x)$ and **arccotangent** $\operatorname{arccot}(x)$ are defined.

Exponential and Logarithm Functions

1. The function $\exp : \mathbb{R} \to \mathbb{R}$ with $\exp(x) = e^x$, where $e \approx 2.71828$ is the Euler's number, is called **exponential function**.

2. The exponential function is strictly monotonically increasing on its domain $\mathbb{R}$ and takes positive values. Therefore, its inverse function $\ln : \mathbb{R}_{>0} \to \mathbb{R}$, called the **natural logarithm**, exists on its range $\mathbb{R}_{>0}$.

29.2. Mathematical Formulas

Geometric Formulas

Pythagorean theorem : In a right-angled triangle with sides a, b and c, where the right angle is enclosed by the sides a and b, the following applies:
$$c^2 = a^2 + b^2 \tag{29.1}$$

Algebraic Formulas

1. **Binomial Formulas**: For any real numbers a and b the following applies:

    a) $(a + b)^2 = a^2 + 2ab + b^2$

    b) $(a - b)^2 = a^2 - 2ab + b^2$

    c) $(a - b)(a + b) = a^2 - b^2$

2. (p, q)-**Formula**: A quadratic equation of the form
$$x^2 + px + q = 0$$
   has the roots
$$x_{1/2} = -\frac{p}{2} \pm \sqrt{\left(\frac{p}{2}\right)^2 - q}.$$

3. **Vieta's Theorem**: The roots x_1 and x_2 of a quadratic equation of the form

$$x^2 + px + q = 0$$

satisfy

$$
\begin{aligned}
x_1 + x_2 &= -p \\
x_1 \cdot x_2 &= q.
\end{aligned}
$$

29.3. Physical Laws

This section lists the physical laws used in the text but not derived.

Einstein's Energy Law for Light Particles / Photons

$$E = h \cdot f,$$

where f denotes the frequency of the light wave and h the Planck constant.

Ideal Gas Law

In an ideal gas, the pressure p can be determined by

$$p = \frac{Nk_B T}{V}.$$

Here, N denotes the number of particles in volume V, T the temperature and k_B the Boltzmann constant.

30. Units and Constants

30.1. SI Units

Base Quantities

Base Quantity / Dimension	Quantity Symbol	Dimension Symbol	Unit	Unit Symbol
Time	t	T	Second	s
Length	l, s	L	Meter	m
Mass	m	M	Kilogram	kg
Current	I	I	Ampere	A
Temperature	T	$\ominus$	Kelvin	K
Amount of substance	n	N	Mole	mol
Luminosity	I_V	J	Candela	cd

Table 30.1.: Base quantities

Derived Quantities

All other quantities needed in physics are derived from these seven base quantities. Each physical quantity Q has a **dimension**, which can be represented as a product of powers of the dimensions of the base quantities:

$$\dim Q = \mathsf{T}^{\alpha} \cdot \mathsf{L}^{\beta} \cdot \mathsf{M}^{\gamma} \cdot \mathsf{I}^{\delta} \cdot \ominus^{\varepsilon} \cdot \mathsf{N}^{\zeta} \cdot \mathsf{J}^{\eta},$$

where the Greek exponents can be positive or negative integers (including zero). If all exponents are zero, then

$$\dim Q = 1$$

and Q is then called **dimensionless**. When applying more complicated functions (than e.g. addition or multiplication) to physical quantities, we always assume that the independent variables and the values of the functions are dimensionless. Thus, expressions like $e^{\alpha t}$, $\ln(\beta x)$ or $\sin(\gamma y)$ are dimensionless, as are the arguments αt, βx and γy.

The corresponding **derived SI unit** can be expressed analogously as a product of powers of the unit symbols:

$$[Q] = \mathsf{s}^{\alpha} \cdot \mathsf{m}^{\beta} \cdot \mathsf{kg}^{\gamma} \cdot \mathsf{A}^{\delta} \cdot \mathsf{K}^{\varepsilon} \cdot \mathsf{mol}^{\zeta} \cdot \mathsf{cd}^{\eta}$$

The derived quantities used in this book are summarized in Tab. 30.2.

Derived Quantity	Quantity Symbol	Unit	Unit Symbol	in SI Base Unit
Area	A	Square meter		m^2
Volume	V	Cubic meter		m^3
Frequency	f	Hertz	Hz	s^{-1}
Speed	v	Meter per second		$\mathsf{m\,s}^{-1}$
Acceleration	a	Meter pro square second		$\mathsf{m\,s}^{-2}$
Force	F	Newton	N	$\mathsf{kg\,m\,s}^{-2}$
Work, Energy	W	Joule	J	$\mathsf{kg\,m}^2\,\mathsf{s}^{-2}$
Power	P	Watt	W	$\mathsf{kg\,m}^2\,\mathsf{s}^{-3}$
Pressure	p	Pascal	Pa	$\mathsf{kg\,m}^{-1}\,\mathsf{s}^{-2}$
Density	ρ	Kilogram per cubic meter		$\mathsf{kg\,m}^{-3}$

Table 30.2.: Derived quantities

30.2. Natural Units

The definition of **natural units** uses the constancy of the speed of light $c = 3 \cdot 10^8\,\mathrm{m/s}$. Time is measured in meters: A **time meter** m is the time that light needs to traverse 1 meter. The value of the speed of light is then given by

$$c = \frac{1\,\mathrm{m}}{\text{time that light needs to cross 1 meter}} = \frac{1\,\mathrm{m}}{1\,\mathrm{m}} = 1.$$

So, the value of the speed of light is equal to 1 and moreover, c is dimensionless. The conversion from natural to SI units is obtained by

$$3 \cdot 10^8 \, \text{m/s} = 1 \Rightarrow 1\,\text{s} = 3 \cdot 10^8 \, \text{m} \Rightarrow 1\,\text{m} = \frac{1}{3 \cdot 10^8}\,\text{s}.$$

The SI system contains a lot of derived units, such as Joule and Newton. When using natural units, the dimensions of many SI units simplify considerably, e.g. for Joule in SI units

$$\text{J} = \text{N} \cdot \text{m} = \text{kg} \cdot \text{m}^2/\text{s}^2$$

and in natural units

$$\text{J} = \text{kg} \cdot \text{m}^2/\text{m}^2 = \text{kg}.$$

30.3. Physical Constants and Astronomical Sizes in SI Units

Name		Value
Speed of light in vacuum		$c = 299{,}792.458\,\text{km/s}$
Equatorial radius of the Earth		$R_E = 6.378 \cdot 10^6\,\text{m}$
Mass of the Earth		$M_E = 5.974 \cdot 10^{24}\,\text{kg}$
Radius of the Sun		$R_s = R_\odot = 6.96 \cdot 10^8\,\text{m}$
Mass of the Sun		$M_S = M_\odot = 1.99 \cdot 10^{30}\,\text{kg}$
Average temperature of the Sun		$T = 1.6 \cdot 10^7\,\text{K}$
Schwarzschild radius of the Sun		$r_S = 2 M_S\, G/c^2 = 2.96 \cdot 10^3\,\text{m}$
Boltzmann constant		$k_B = 1.38 \cdot 10^{-23}\,\text{JK}^{-1}$
Newton's gravitational constant		$G = 6.67 \cdot 10^{-11}\,\text{m}^3\text{kg}^{-1}\text{s}^{-2}$
Acceleration due to gravity		$g = 9.81\,\text{ms}^{-2}$
Planck's quantum of action		$h = 6.626 \cdot 10^{-34}\,\text{Js}$

Table 30.3.: Physical Constants and Astronomical Sizes

30.4. Mathematical Constants

Name	Abbreviation	Value
Number Pi	π	$3.14159\ldots$
Euler's constant	e	$2.71828\ldots$
Square root of 2	$\sqrt{2}$	$1.41421\ldots$
Square root of 3	$\sqrt{3}$	$1.73205\ldots$

30.5. Greek Alphabet

Letter	Name	Letter	Name
A, α	Alpha	N, ν	Nu
B, β	Beta	Ξ, ξ	Xi
Γ, γ	Gamma	O, o	Omicron
Δ, δ	Delta	Π, π	Pi
E, ϵ, ε	Epsilon	P, ρ, ϱ	Rho
Z, ζ	Zeta	Σ, σ	Sigma
H, η	Eta	T, τ	Tau
$\Theta, \theta, \vartheta$	Theta	Υ, υ	Upsilon
I, ι	Iota	Φ, φ, ϕ	Phi
K, κ	Kappa	X, χ	Chi
Λ, λ	Lambda	Ψ, ψ	Psi
M, μ	Mu	Ω, ω	Omega

Bibliography

[1] Abbott, B. P. et al. (2016) Observation of gravitational waves from a binary black hole merger, Phyical Review Letters **116**, 061102

[2] Blankenheim, T. (2018) Unterwegs in gekrümmter Raumzeit, printed by Amazon Fullfillment

[3] Born, M. (2012) Einstein's Theory of Relativity, Dover Publications

[4] Borthwick, D. (2016) Introduction to Partial Differential Equations, Springer

[5] Chiossi, S. G. (2021) Essential Mathematics for Undergrades, Springer Nature

[6] Collier, P. (2017) A Most Incomprehensible Thing, printed by Amazon Fullfillment

[7] Einstein, A. (2023) The Special and the General Theory, Prakash Books

[8] Epstein, L. C. (1983) Relativity visualized, Insight Press, San Francisco

[9] Feynman R. et al. (2011) The Feynman Lectures On Physics, Volume I, Basic Books

[10] Feynman R. et al. (2011) The Feynman Lectures On Physics, Volume II, Basic Books

[11] Fleisch, D. (2013) Vectors and Tensors, Cambridge University Press Cambridge

[12] Freund, J. (2008) Special Relativity for Beginners, World Scientific

[13] Fließbach, T. (2024) Mechanics For Physicists: An Introduction, Including Special Relativity, WSPC

[14] Gamov, G. (2002) Gravity, Dover Publications Mineola New York

[15] Halliday, D., Resnick, R. & Walker, J. (2013) Fundamentals of Physics, Wiley

[16] Hawking, S. W. (2001) A Brief History of Time, Transworld Publ. Ltd UK

[17] Jänich, K. (2001) Vector Analysis, Springer Berlin Heidelberg New York

[18] media.ligo.northwestern.edu/gallery/mass-plot, download: January 5, 2025

[19] Maggiore, M (2019) Gravitational Waves, Volume 1, Oxford University Press

[20] Misner, C. W., Thorne, K.S., Wheeler, J. A. (2017) Gravitation, Princeton University Press

[21] Moore, T. A. (2013) A general relativity workbook, University Science Book Mill Valley, California

[22] Observing.docs.ligo.org, download: January 5, 2025

[23] Ruhrländer, M. (2017) Lineare Algebra, Pearson Deutschland Hallbergmoos

[24] Ruhrländer, M. (2024) Brückenkurs Mathematik, 3. Auflage, Pearson Deutschland Hallbergmoos

[25] Ruhrländer, M. (2021) Allgemeine Relativitätstheorie Schritt für Schritt, Springer Berlin

[26] Ryder, L. (2009) Introduction to General Relativity, Cambridge University Press Cambridge

[27] Scheck, F. (2018) Mechanics: From Newton's Laws to Deterministic Chaos, Springer

[28] Schutz, B. (2007) Gravity from the ground up, Cambridge University Press Cambridge

[29] Schutz, B. (2003) A first course in general relativity, Cambridge University Press Cambridge

[30] Schutz, B. (1999) Geometrical methods of mathematical physics, Cambridge University Press Cambridge

[31] Sterling Education, (2024) High School Physics, Sterling Education

[32] Strang, G. (1998) Introduction to Linear Algebra, Wellesley Cambridge Press

[33] Thomas, G.B. et al. (2014) Thomas' Caluculus, Pearson Education Inc

[34] Thorne, K. S. (1995) Black Holes and Time Warps, W. W. Norton & Co

[35] Tipler, P. A. (2007) Physics for Scientists and Engineers, W. H. Freeman & Co Ltd

[36] Wald, R. M. (1984) General Relativity, The University of Chicago Press

[37] Weinberg, S. (Reprint 2017) Gravitation and Cosmology, Wiley India New Delhi

[38] Weisberg, J. M. & Huang, J. H. (2016) Relativistic Measurement from Timing the Binary Pulsar PSR B1916+16, The Astrophysical Journal **829**, 55

[39] Will, C.M. (1993) Was Einstein Right?, Basic Books; 2nd Edition

[40] Zee, A. (2013) Einstein Gravity in a Nutshell, Princeton University Press Princeton New Jersey

Index